Introduction to

Environmental Geotechnology

Introduction to

Environmental Geotechnology

Hsai-Yang Fang

CRC Press

Boca Raton   New York

Library of Congress Cataloging-in-Publication Data

Fang, Hsai–Yang
    Introduction to environmental geotechnology / author, Hsai–Yang
Fang.
        p. cm. — (New directions in civil engineering)
    Includes bibliographical references and index.
    ISBN 0-8493-8288-2 (alk. paper)
    1. Environmental geotechnology   I. Title. II. Series.
TD795.7.F36 1997
624.1′.51—dc20

96-41740
CIP

© 1997 by CRC Press LLC

No claim to original U.S. Government works
International Standard Book Number 0-8493-8288-2
Library of Congress Card Number 96-41740
Printed in the United States of America 1 2 3 4 5 6 7 8 9 0
Printed on acid-free paper

Dedicated
to my family
Julia, Andrew, and Janice
for their encouragement and support

Preface

There are two major reasons for the development of environmental geotechnology aspects as an engineering discipline. In the first place, the population growth, and in the second place, the rising living standards. When population increases, more land is needed and many soil deposits previously unfit for residential housing and other construction projects are now being used. In a progressive society, the living standard gives rise to industrial growth. As a consequence, hazardous pollution of air, water, land, and urban refuse production becomes inevitable, thereby endangering the global environment. To cope with the problematic soil deposits and adverse environmental conditions, present conventional construction technology has to, by necessity, take a new direction.

Problematic soil deposits on one hand and ground pollution problems on the other have challenged current soil mechanics concepts and methods of analyzing soil behavior under varied environmental conditions. For this reason, the environmental aspects of geotechnology have been expanded to include the study of soil and rock interaction with the atmosphere, biosphere, hydrosphere, lithosphere, as well as geo-microbiosphere, and the subsequent response as engineering behavior has paved the way for the emergence of environmental geotechnology.

Environmental geotechnology has grown rapidly since the first international symposium was organized in 1986 at Lehigh University. Subsequently, during this short interval of time, several international, national, and regional conferences and symposia on this and allied topics have been sponsored. Both theoretical and practical concepts have been developed. Recognizing the impact of this development, many universities took the lead and have instituted environmental geotechnology as a standard course within their curricular framework.

Since 1925, when Karl Terzaghi introduced the concept of soil mechanics, it has made rapid strides becoming a major discipline in the civil engineering field. Unfortunately, however, the major teaching effort has taken into account only the basis of physico-mechanical behavior of soil. In fact, in some institutions the geotechnical engineering courses have become a part of the engineering mechanics discipline, which implies that the fundamental aspects of soil behavior have been ignored totally disregarding the environment's role in soil behavior.

Environmental geotechnology is a highly interdisciplinary science. Since soil is very sensitive to the environmental conditions, its study should encompass areas of soil science, physical chemistry, mineralogy, geology, microbiology, etc. Curriculum development on this subject should, therefore, take serious consideration of this fact.

Air, water, and soil are in eternal interaction with each other. As such, any phase of the pollution process should not be viewed in isolation but should be integrated with the other phases. It is regrettable that at present, undue emphasis is placed on air and water pollution with little concern for the soil, which is also an inevitable life supporting system. It is, therefore, time for us to recognize this fact and give environmental geotechnology the place and importance it deserves.

In the text that follows, the author has attempted to generalize the soil/rock properties under diverse environmental conditions using a particle-energy-field theory based on mechanical, thermal, electrical, magnetic, radiation effects, etc. For a comprehensive analysis of the environmental effects on the soil and rock behavior and practical applications to soil-foundation-structure systems, a systematic correlation of experimental data is considered a prerequisite. These aspects form the special features of this book.

This book is intended to serve students, teachers, and practitioners of environmental geotechnology. As a result, the author feels it is necessary to make it a complete text of applied soil engineering broadly covering transportation, ocean engineering, water resource systems, irrigation, reclamation, and other areas of interest and also provide practical reference data and methodologies for analysis and design. It is hoped that this book will help to cater to these needs as well as to provide new direction for the future.

The author gratefully recalls the influence of the late Professor Hans F. Winterkorn, who as teacher, friend, and colleague sparked his interest in soil science and technology. The author also expresses his gratitude to all those who were generous enough to advise and encourage him in the preparation of the text, particularly late Professor Y.B. Acar, Professor C.H. Benson, Professor R.C. Chaney, Mr. J.L. Daniels, Professor J.C. Evans, Professor H.I. Inyang, Dr. A. Kaya, Professor R.M. Koerner, Professor G.A. Leonards, Dr. G.K. Mikroudis, Dr. Z.C. Moh, Professor R.N. Yong, and Professor L. Zeevaert. His thanks are also due to Professor G.Y. Luo of Nanjing University, Professor T.M. Yu, and Professor Z.X. Hu of Tongji University, and Professor Z.D. Zhu, Institute of Desert Research of the Chinese Academy of Sciences who provided technical publications and discussions. Special thanks to Ms. Eleanor Nothelfer, for her assistance during various phases of preparation of the manuscript and galley proof reading.

Hsai-Yang Fang

Table of Contents

Chapter 14
Wetlands, Coastal Margins, and Soil Erosion Problems 439

CHAPTER 1

Introduction

1.1 INTRODUCTION

Environmental geotechnology can be defined as an interdisciplinary science which covers soil and rock and their interaction with various environmental cycles, including the atmosphere, biosphere, hydrosphere, and lithosphere, as well as the geo-microbiosphere (Fang, 1986), which includes characteristics of tree and vegetation roots and bacterial activities in the ground soil and subsequent response to the engineering behavior of the soil-water system, as illustrated in Figure 1.1.

1.2 DEVELOPMENT OF ENVIRONMENTAL GEOTECHNOLOGY

1.2.1 General Discussion

There are two major reasons for the development of environmental geotechnology. First, is population growth, and second, rising living standards. When population increases more land is needed; many soil deposits previously claimed to be unfit for residential housing and other construction projects are now being used. In a progressive society, rising living standards indicate an increase in industrial growth. As a consequence, hazardous pollution of air, water, and land and urban refuse production become inevitable, thereby endangering the global environment. To cope with these problematic soil deposits and adverse environmental conditions, the present conventional construction technology has to take, by necessity, a new direction.

Problematic soil deposits, on one hand, and ground pollution problems on the other have challenged the current soil mechanics concepts and methods of analyzing soil behavior under varied environmental conditions. For this reason, the environmental aspects of geotechnology have been expanded, and their subsequent response to engineering behavior has paved the way for the emergence of environmental geotechnology. Further explanation of population growth and industrial progress is presented as follows.

1.2.2 Population Growth and Rising Living Standards

At the present time (1997), the estimated worldwide population is in excess of 5.9 billion. According to the United Nations's prediction, by the year 2050, conservative estimates give a population ranging from 15 to 20 billion. Approximately 80% of this growth will be in the developing countries.

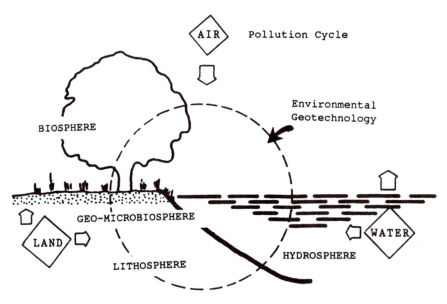

Figure 1.1 Environmental cycles and their interactions.

Depletion of productivity of agricultural land is directly related to the population growth. Uncontrolled urban expansion, conversion of agricultural lands for other purposes, desertification, and loss of productivity all combine to reduce agricultural productivity. Consequently, more land is needed and many soil deposits previously claimed to be unfit for residential housing or other construction projects are now being used. Such areas include: landfills, wetlands, collapsible soil regions, and mining subsidence areas. To overcome these natural or man-made problematic soil conditions, additional techniques are required to use conventional construction applications.

When society progresses, the living standard rises and, along with industrial progress, much of the air, water, and land becomes polluted. Open dumps and chemical and industrial wastes cause these problems as well as others listed in Table 1.1. The pressures of uncontrolled urbanization, industrial growth, and living standard have contributed to minimal environmental waste management practices, increasing the overall threat to a sustainable environmental resource base.

Environmental geotechnology has grown quickly since the first international symposium was organized in 1986 at Lehigh University. Subsequently, during this short interval of time, several international, national, and regional conferences and symposia on this and other related subjects have been promoted. Both theoretical and practical concepts have been developed. Recognizing the impact of this development, many universities took the lead and have instituted environmental geotechnology as a standard course within their curricular framework.

1.3 AIMS OF ENVIRONMENTAL GEOTECHNOLOGY

Environmental geotechnology is an emerging discipline which consists of a balance of the following subjects:

**Table 1.1 Some Causes of Land
and Groundwater Pollution
in the U.S**

Industrial wastes
   Hazardous chemical wastes
   Nuclear wastes
   Oil field brine
   Acid mine drainage
   Acid rain
   Highway deicing salts
   Saltwater intrusion
   Abandoned oil wells
Agricultural wastes
   Animal wastes
   Irrigation return flow
   Dry farming land
   Disposal well
   Evapotranspiration from vegetation
Urban solid wastes
   Landfills
   Septic tanks
   Sewage treatment plant discharges
   Waste lagoons
   Surface impoundments
Others
   Water from fault zones and volcanic origin
   Radon gas (Rn-222)
   Petroleum exploration and development

- Geotechnical Engineering
- Environmental Science and Engineering
- Soil Science (Agriculture)

As shown in Figure 1.2 each of these currently established disciplines has had its beginnings from easily traceable roots in the 1950s, e.g., Sanitary Engineering, Soil Mechanics, and Foundation Engineering. As seen in Figure 1.2, a veritable revolution took place to forge these disciplines which came from a number of topics all solidly encamped in established areas. Indeed, the revolution from the 1930s to 1950s was a tumultuous one. The reasons for the emergence of environmental geotechnology are very practical, logical, and noble, which are the following:

- Better productivity
- Healthier economy
- Safe and secure environment

They are, indeed, compatible goals and when properly approached via a unified discipline are mutually inclusive ones. Inputs into this new discipline and the reasons for its emergence have to do with natural resources such as ores, fuels, water, etc. and man-made resources such as metals, synthetics, polymers, etc., how they are processed and transported, and finally how they and their associated wastes are treated, stored, and disposed. These latter aspects, of course, are at the heart of the air-water-ground soil pollution and their interactions.

The state of the art of both educational and research programs in environmental geotechnology is given by Meegoda (1996). This program will train engineers to develop environmentally sound solutions to geotechnical problems and to solve environmental engineering problems unique to soil and subsurface conditions. The program is being developed by the

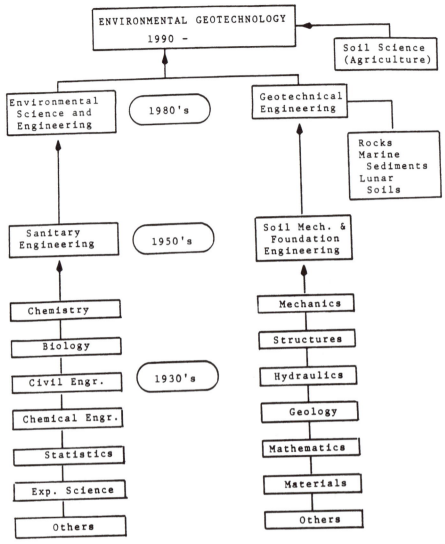

Figure 1.2 Environmental geotechnology — an emerging science. (Based on Koerner, R.M. [1987], Proc. 1st Int. Symp. on Environmental Geotechnology, v. 2, pp. 1–3.)

Department of Civil and Environmental Engineering at New Jersey Institute of Technology with a grant from the National Science Foundation.

1.4 ENVIRONMENTAL CYCLES AND THEIR INTERACTION WITH GEOTECHNOLOGY

1.4.1 General Discussion

The natural and man-made environments include a large variety, as illustrated in Table 1.1. Their interactions relating to the geotechnical engineering problems are presented in Figure 1.3. In examining Figure 1.3 indications are that many environmental geotechnical problems require knowledge from other disciplines in order to understand soil response to environments for both short- and long-term performance. In order to link these unrelated groups into a related system, the concept of particle energy field theory is introduced in the

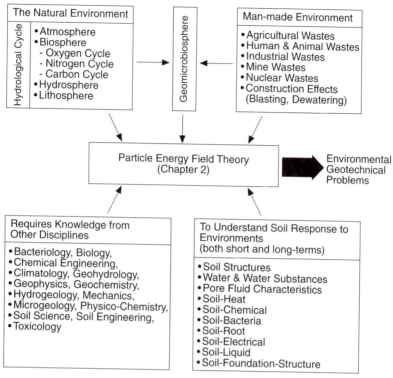

Figure 1.3 Soil-water-environment interaction relating to geotechnical problems — using particle energy field theory as a bridge to link these unrelated groups into a related system.

text, as indicated in Figure 1.3. Details of the concept of the particle energy field will be discussed in Chapter 2.

1.4.2 The Natural Environment

As indicated in Figures 1.1 and 1.3, the natural environment includes atmosphere, biosphere, hydrosphere, and lithosphere. A brief description of each element is presented as follows.

(1) The Atmosphere

The *atmosphere* (Greek: atmos = vapor) is the layer of air surrounding the earth. This layer of air also is divided into three sublayers as:

1. Troposphere: the portion of the atmosphere extending out from ground surface to about 12 km
2. Stratosphere: the portion of the atmosphere extending out from 12 km to about 50 km
3. Ionosphere: the portion of the atmosphere above 50 km

The atmosphere is most dense near the ground surface and decreases in density as it extends into space. This means the density of air in the troposphere is denser than in the stratosphere, and the air density in the stratosphere is denser than in the ionosphere. Air is a homogeneous mixture of gases. The major components of dry air, in percent by volume, are

- Nitrogen (N_2)   78.1%
- Oxygen (O_2)   21.0%
- Carbon dioxide (CO_2)   0.03%
- Noble gases   0.86% (Section 10.5)

The atmosphere also contains variable amounts of water vapor and dust (Section 3.8). The amount of water vapor varies from small amounts in desert regions (Section 18.6) to large amounts in tropical regions.

(2) The Biosphere

The *biosphere* (Greek: bio = life) includes all plant and animal life. As shown in Figure 1.1, it centers near the interface between the lithosphere and hydrosphere. The major elements which structure living matter are oxygen, hydrogen, and carbon. The biosphere can be further divided into three subcycles — oxygen, nitrogen, and carbon cycles — and will be further discussed in Section 1.4.3.

(3) The Hydrosphere

The *hydrosphere* (Greek: hydro = water) includes the vast amount of water in oceans, rivers, lakes, and underground deposits, also, the water contained in the ice and snow of the earth; and the water which makes up the clouds and vapor in the atmosphere. The hydrosphere contains dissolved gases and particles of the lithosphere. Waters of the hydrosphere cover and permeate portions of the lithosphere.

(4) The Lithosphere

The *lithosphere* (Greek: litho = stone) refers mainly to rock and soil and contains combined silicon, oxygen, aluminum, and many other elements. The chemical composition of the earth's crust and **A** and **B** horizons will be presented in Tables 3.7 and 3.8. Further discussions of this aspect will be presented in Chapter 3.

(5) The Geo-microbiosphere

The *geo-microbiosphere* is a new term and is not part of the main ecosphere. However, such a system is needed in the environmental geotechnology area to study the characteristics of tree-vegetation roots and bacterial activities in the ground soil and water and subsequent response to the engineering behavior of the soil-water system. Further discussions will be presented in Sections 5.9 and 5.10.

Since these systems are interrelated and form the environment, they are referred to as a whole as the *ecosphere* (Greek: oikos = house). The percentages by mass of the elements in the ecosphere are presented in Table 3.8.

1.4.3 The Cycles of Nature

Many elements in the ecosphere are exchanged between two or more of the subspheres. These exchanges result from physical or physicochemical interactions among these atoms or molecules. Such cyclic exchanges are called natural cycles. There are three cycles — namely, oxygen, nitrogen, and carbon cycles — which are briefly discussed as follows.

(1) Oxygen Cycle

In the earth, oxygen is found in combination with silicon (Si) and aluminum (Al) in aluminum silicates (rocks). Some are found in combination with metals such as metallic oxides, carbonates, sulfates, nitrates, and phosphates. The oxygen processes which occur in the biosphere stimulate the exchange of oxygen between the various spheres of the environments. Also, it utilizes molecular oxygen in biological oxidation in which food molecules are converted to carbon dioxide and water. The three substances, water (H_2O), carbon dioxide (CO_2), and molecular oxygen, along with the molecules of living systems, are the major substances in the *oxygen cycle*.

(2) Nitrogen Cycle

Nitrogen cycle is the fourth most abundant element next to carbon, hydrogen, and oxygen in the biosphere. The nitrogen cycle involves the transfer of nitrogen between the atmosphere, biosphere, lithosphere, and hydrosphere in various chemical forms. In the nitrogen cycle, the nitrogen fixation process is essential to microbial utilization. Fixation processes change inert N_2 into a usable form such as nitrate ion (NO_3), processed by plants into proteins and amino acids. Nitrogen fixation is reported in the natural life cycles of bacteria (Section 5.8) which convert N_2 into ammonia (NH_3) and other usable forms. Other bacteria convert ammonia into nitrite ion (NO_2). Fixed nitrogen manufacture involves the reaction of N_2 with oxygen from air or with hydrogen from water or methane (CH_4) to make ammonia.

(3) Carbon Cycle

The *carbon cycle* is the essential element of the biosphere. The main cycle is the oxidation of carbon-containing materials into atmospheric CO_2, with photosynthesis returning it to the biosphere. The most important features of the carbon cycle are

1. The amount of carbon 'fixed' annually by plants is oxidized back to CO_2.
2. There is a large reservoir of carbon as a carbonate in sedimentary rocks such as limestone and dolomite.
3. Carbon is being modified by fossil fuel combustion, resulting in increased atmospheric CO_2 concentrations.

Recent development on the global carbon cycle is given by Post et al. (1990). Natural cycles are all interrelated. It is difficult to predict what changes occur in one cycle from changes in another cycle.

1.4.4 Water Stability in Natural Environmental Systems

Water stability is directly related to the natural environmental systems. The upper and lower limits of water stability are characterized by E_h and pH values as reported by Garrels and Christ (1965) and as illustrated in Figure 1.4. The significance of E_h and pH values will be explained in Chapter 5. As shown in Figure 1.4, the natural environment is divided into two limits, the upper and lower limits of water stability. Two environmental conditions are presented: one is the environment in contact with the atmosphere and the other is isolated from the atmosphere. The characteristics of water contacts with the atmosphere include rainwater, streams, mine waters, and ocean water. Waters isolated from the atmosphere cover water-logged soils and organic-rich saline waters. Range of acid rain as characterized by E_h and pH value is indicated also in the figure. Both E_h and pH values are important parameters for characterization of approximate position of natural environments. Some of the relatively important soil types and phenomena listed in the figure will be further discussed in various chapters in the text.

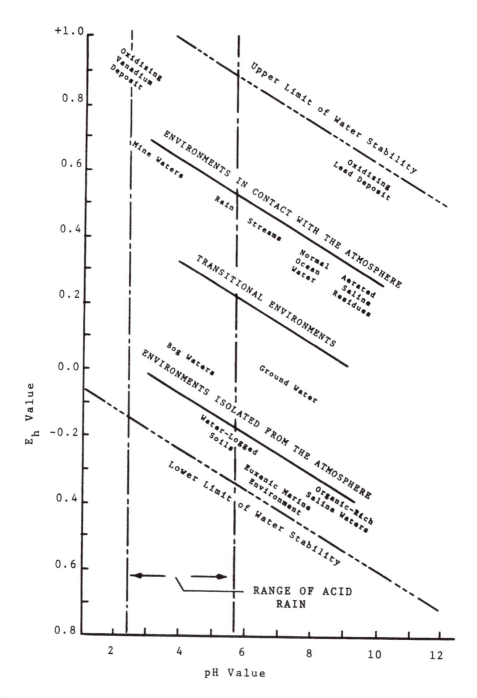

Figure 1.4 Approximate position of some natural environments as characterized by E_h and pH values. (Based on Garrels, R.M. and Christ, C.L. [1965], *Solutions, Minerals and Equilibria,* Harper & Row, New York.)

1.5 MAN-MADE ENVIRONMENT

1.5.1 General Discussion

The man-made environment covers many varieties, as indicated in Table 1.1. Figure 1.5 illustrates the current research effort placed on landfill and hazardous and toxic wastes. Of

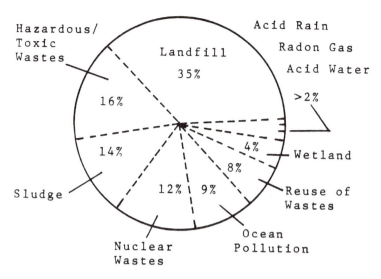

Figure 1.5 Current research areas within environmental geotechnology. (After Fang, H.Y. [1992a], Proc. Mediterranean Conf. on Environmental Geotechnology, Usmen and Acar, eds. pp. 11–19.)

course, these are two problems in which research is urgently needed. In addition to the landfill and hazardous wastes, acid rain, acid drainage, petroleum contaminated soils, and scrap waste rubber tires are emphasized herein because these problems are given very little attention by environmental geotechnical groups.

Other environmental geotechnical problems such as arid lands, desert regions, wetlands as well as problems on erosion, subsidence, and landslides caused by both man-made and natural causes also will be discussed in the text. Some of the special features of these problems will be pointed out in the following sections.

1.5.2 Solid and Liquid Wastes

Wastes can be grouped into solid and liquid forms. The majority of the solid waste is urban refuse. This refuse will decompose with time and generate gases and liquid waste, generally referred to as leachate. According to a U.S. Environmental Protection Agency (EPA) report, about 90% of wastes which includes industrial hazardous wastes is in liquid form. Therefore, control of solid and liquid wastes requires special emphasis on the control of liquid wastes.

The understanding of the interaction between pore fluids and clay behavior requires the knowledge of environmental aspects of geotechnical engineering. It is essential to the engineering utilization of naturally occurring materials for the containment of hazardous and toxic wastes. Without this knowledge of clay behavior in response to hazardous wastes, engineering system design, such as remedial action programs, can have no sound basis on which to project the long-term behavior of that system.

1.5.3 Solid Wastes (Urban Refuse)

Waste disposal material consists of anything that cannot be further used or recycled economically; thus, its composition varies from community to community, from country to country, as well as from season to season. The density varies from 50 to 400 pcf depending on the amount of metal and debris (Sowers, 1968 and Sarsby, 1995). Table 1.2 gives the average composition of solid waste based on U.S. national survey results and the city of New York, as well as Osaka, Japan (Yamamura, 1983). Usually, waste disposal material is used

Table 1.2   Average Composition of Solid Waste (Garbage)

	New York City (1968)	U.S. National (1973)	Osaka, Japan (1980)
Paper	58.8	51.6	37.1
Food wastes	9.2	19.3	13.2
Metal	7.6	10.2	5.5
Glass	8.6	9.9	12.3
Wood	2.5	3.0	2.5
Textiles	0.8	2.7	4.0
Rubber, leather	0.8	1.9	0.3
Plastic	0.8	1.4	15.2
Garden wastes	10.1	—	2.5

Data from Sowers, G.F. (1973), Proc. 9th ICSMFE, v. 4, pp. 297–310; and Yamamura, K. (1983), Waste Management and Research, v. 1, pp. 1–15.

for landfill and fuel, and in some countries it has been used for building blocks. However, due to human psychological effects regarding the use of this matter, the utilization has not been widespread. The most economical way to dispose of waste at the present time is for use as landfill.

1.5.4 Waste Rubber Tires

Waste rubber tires are a major part of solid wastes. The U.S. EPA indicates that production of rubber tires for automobiles increased 42% from 1960 to 1988. Currently, an estimated 280 million waste tires are discarded annually in the U.S. alone. From this amount, only a small number of tires are recovered for recycling purposes (5.6% recovered in 1988). Most waste tires are dumped into landfill space. The scrap tire piles are growing each year and pose two major threats to the public: fire hazard and health hazard. A state of the practice and evaluation of the selected waste products in highway construction is reported by Inyang and Bergeson (1992) and Collins and Ciesrelski (1994). Utilization of waste scrap rubber tire for roadway materials and other applications is reported, also. Further discussions on engineering applications of waste scrap rubber tires will be presented in Section 17.10.

1.5.5 Petroleum-Contaminated Soil

An oil spill in most cases is accidental — during transportation both on land and sea, as leakage from storage tanks, or during the oil drilling process. Also, there are some cases where oil was spilled purposely as in the Gulf War in 1991. The worst tanker disaster in northern America was in 1989, when the supertanker Exxon Valdez sank in Alaska's Prince William Sound, spilling 11 million gallons of oil. The worst oil spill in Europe occurred March 1978 off the northwest coast of France when the tanker Amoco Cadiz lost 68 million gallons of crude oil. A serious oil spill in Britain involved the tanker Torrey Canyon which ran aground in 1967, spreading 40 million gallons of oil into the Atlantic, fouling beaches and coastlines in Cornwall and southwestern England. More recently, in 1993, a tanker spilled 26 million gallons of crude oil among the Shetland Islands in the North Sea, near Scotland, after losing power and being battered by hurricane force winds and seas. The largest oil tanker spill in history took place in 1979 off Trinidad and Tobago, when two ships collided spilling 97 million gallons of oil.

According to an EPA report in 1979, every year more than 10 million gallons of oil escapes into U.S. waters as a result of more than 10,000 spills. In the U.S., there are several million underground storage tanks of various sizes containing petroleum products. It is estimated that up to 25% of these tanks are leaking petroleum liquids into the ground. Amendments to CERCLA (Comprehensive Environmental Response, Compensation and Liability Act) in 1986 show increased recognition of this problem. Leakage of petroleum hydrocarbon in the ground poses a significant potential hazard to the public and environment. It can be a source of contamination to runoff water and groundwater supplies, as well as a cause of possible adverse effects on plants, fish, and wildlife.

When an oil spill or a leakage occurs, soils around the source of leakage are contaminated. Some major tasks need to be performed for remediation and reclamation of the contaminated area. Removal of the contaminated soil is a significant part of the overall treatment. However, limitations in the disposal facilities and receiving landfills complicate this problem. There are emerging technologies for *in situ* cleanup, such as electrokinetic and electromagnetic (Sections 9.10 and 9.13) removal or biodegradation (Norris et al., 1994). However, they are either in an experimental stage or may require a long time to produce results. An economical and beneficial solution seems to stabilize the excavated contaminated soil and render it a useful material for construction (Section 17.8).

1.6 ACID RAIN AND ACID DRAINAGE

1.6.1 Acid Rain (Acid Deposition)

Among all the types of pollutants listed in Table 1.1, acid rain is most common and covers a larger area than any other, but very little attention is placed on this subject; therefore, more discussion is presented herein.

Acid rain is one of the major multimedia man-made environmental problems. It can cause all phases of air-water-ground pollution. *Acid rain* is generated by the emission of various pollutants into the air with the principle sources being fossil fuel power plants, automotive exhausts, and industrial facilities. These emissions and exhausts contain sulfur dioxide (SO_2), nitrogen (NO), and other gases. The gases combine with oxygen (O), and water vapor (HOH) in the air to form sulfuric acid (H_2SO_4), nitric acid (HNO_3), hydrochloric acid (HCl), and other acids and elements.

There are two basic forms of deposition as these acids fall to the earth, and in addition to the acid content, acid rain also has two unique characteristics which cause more complications:

1. Acid rain has no borders; because the acids are suspended high above the earth, they can be carried hundreds of miles by the wind before they drop or wash out from the air.
2. The compositions of acid rain are not uniformly distributed.

Since the main source of acid rain is from fossil fuel power plants, 44% of all SO and NO produced in the U.S. comes from the five states in the coal-burning Ohio River Valley combined with Missouri and Tennessee. The wind carries the gases in a northeasterly direction toward the New England states, mid-Atlantic states, and Canada (DER, 1990). Figure 1.6 shows the acid rain distribution in eastern U.S. There are significant increases from 1955 to the present day. Studies of acid rain in Europe also have shown an increasing trend toward higher acidity with pH values varying between 2 and 5. Reports from China in 1983 indicate that in their southwestern region, the pH of rainwater had dropped to 2.7.

Acid rain kills aquatic life by acidifying water and poisoning plants and animals with the minerals it has leached from the soil. Also, it damages trees and crops by burning their foliage and diverting their energy from growth to repair. Nutrient loss from soils due to rainfall acidity is reported by Reuss (1978). In 1991, a nationwide study of thousands of lakes and streams indicated that 75% of the acidic lakes are caused by acid rain as well as 47% of the

(a)   (b)

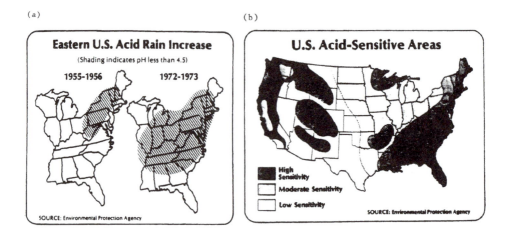

Figure 1.6 Acid rain and acid-sensitive area distribution in the U.S. (a) Acid rain (1955–1973); (b) acid-sensitive areas. (Data from U.S. Environmental Protection Agency.)

streams. This work used data from a survey by the EPA of 1180 lakes and 4670 streams in what the EPA determined to be acid-sensitive areas of the U.S.

Acid rain causes problems in almost every aspect of the environment (Gunnerson and Willard, 1979) and it is related to an interdisciplinary study between atmosphere, biosphere, hydrosphere, and lithosphere. However, from a geotechnical engineering viewpoint, current study on this aspect focuses on how acids affect volume change, cracking behavior, stress-strain relationship, and hydraulic conductivity of subgrade soil in relation to the performance of pavement components, and the foundations of roadway structures which influence the infrastructure of the life cycle of soil-pavement systems.

1.6.2 Acid Mine Drainage

Acid mine drainage also covers large areas (Piwowar and Work, 1997). This problem is in the water draining from old abandoned coal mines. The iron sulfides (FeS_2) or pyrites in the mines are oxidized to sulfuric acid (H_2SO_4). Then sulfuric acid dissolves in water and drains through the mines. When these acidic waters enter the streams, lakes, or ground soil-water, pollution results. In 1969, acid mine drainage pollution of an estimated 20,000 km of waterways in the U.S., mostly in Pennsylvania, West Virginia, and Kentucky, was reported.

Acid mine water is a chemically complex liquid, having no definite quantitative relationship between its constituents. It transports and deposits mainly inorganic materials as solids, in solution or in suspension. Millions of gallons of acid mine drainage flow into Pennsylvania's streams and rivers each year. These acid mine waters range from almost alkaline to strongly acid. Small particles of insoluble solids in the mine water are transported by water through the mine and emerge with the effluent stream. The larger particles settle out with a decrease in stream velocity; the smaller particles remain in suspension for a considerable period of time or they may seep through into river embankments, earth dams or wells.

1.7 ENVIRONMENTAL GEOTECHNICAL PROBLEMS

As discussed in Section 1.2 and Figure 1.3, environmental geotechnology requires knowledge from other disciplines in order to understand soil responses to environments. Likewise, other disciplines need our knowledge to assist in solving their problems. Some areas have been evaluated and assisted by geotechnical groups to some extent and some are of less concern. The following sections are brief discussions of such areas.

1.7.1 Arid Lands and Desert

Based on a United Nations report in 1983 and others (Zhu, 1987), the deterioration of productive land into desert-like conditions continues without letup at the rate of 14.8 million acres a year. Desertification currently affects approximately 100 nations. The spread of deserts threatens 35% of the earth's land surface — or about 45 million square miles — and 20% of its population (about 850 million people); 75% of this area and 60% of the population are already affected. Worldwide food and agriculture organizations estimate that the annual loss of food production caused by desertification amounts to $26 billion. Because of the present desertification, the United Nations estimates that one billion of the world's people already are underfed, and many are dying from starvation in many developing countries.

1.7.2 Marginal Lands (Wetlands)

Utilization of marginal lands for urban development has the most significant impact on social-economic-environmental programs as indicated by many case studies. These case studies include lower Manhattan Island and the New York Museum of Modern Art; in addition, 20,000 acres of New Jersey inland swamp areas which have been constructed into large shopping centers with multistory stores and parking garages.

Utilization of wetlands for various engineering uses can create additional useful lands; however, there are many detrimental effects for such uses. For example, highway earth fills on wetlands may produce various ecological effects. Research findings reported by Shuldiner et al. (1978) indicated the ecological effects of earth fills on wetlands include:

1. Inhibition of storm water and tidal distribution
2. Increased water turbidity
3. Alteration of water circulation patterns
4. Removal of natural filtration systems
5. Introduction of exotics
6. Inhibition of movement of animals
7. Alteration of biological productivity
8. Alteration of nutrient flux

Determination of the impact of highway earth fill or other structure on the environmental field of a specific wetland or in the problematic soil deposits is a complex problem. However, these considerations can be used as a guidance in making highway location and design decisions when wetlands and flood plains are involved. Some of these problems will be further discussed in Chapter 14.

1.7.3 Coastal Margins

In 1992, the LMER program (Section 14.2) indicated that human activities have brought many changes to coastal environments, such as:

1. Loss of wildlife habitat: Among many examples are the loss of 50% of Connecticut's and 90% of California's coastal wetlands.
2. Interception of water and sediment: Waterworks for irrigation, storm protection, and power have reduced the area of wetlands on many coasts and shifted channel locations in the Mississippi deltas.
3. Increased pollution of nearshore environments: Industrial and agricultural chemicals.
4. Increasing rate of sea level rise: Greenhouse effect and sea level rise (Section 14.7).

1.7.4 Soil Erosion

Soil erosion is caused by the drag action of wind, rainfall, or wave action (beach erosion) on the surface of bare or unprotected soil surface. It involves a progress of both soil particle detachment and transport. The sediment that may be produced by erosion restricts the efficiency of drainage and irrigation ditches, fills, and reservoirs; pollutes surface water; damages agriculture land, homes, roadways, and bridges; and upsets the natural ecology and environment. Based on U.S. Department of Agriculture estimates, the total cost of sediment damage and dredging resulting from soil erosion was approximately $500 million yearly in 1980. It costs approximately $16 million annually just to remove sediment from irrigation ditches. These figures do not take into account the damage to agricultural land, homes, roadways, bridges, and recreational areas which results from the loss of soil.

1.8 POLLUTION PROCESSES AND SOIL-POLLUTION INTERACTION

1.8.1 General Discussion

All types of pollution have direct or indirect effects on ground soil properties. For example, rain falling on a garbage dump will pollute both surface water and groundwater systems. The polluted water will attack foundation structures such as footings, caissons, piles, and sheet pilings. If the polluted water is used for mixing concrete, it will affect the workability and durability of the concrete. In embankment construction, the moisture-density relationship of soil will also be affected. On the other hand, the evaporation from surface water and evapotranspiration from ground surface both enter into the biosphere and atmosphere. The percolation from rainwater into ground soil due to sorbed ions and the sorption processes will cause some soil solution to leach into drainage systems or will evaporate into the air. Figure 1.7 presents the effects of air-water-ground (land) pollutions on the engineering behavior of construction materials.

The pollution cycles are not isolated but rather closely related. The pollution problems cannot be solved without considering all interrelated aspects. Unfortunately, present emphasis is placed in water and air pollution with very little concern for ground soil pollution. Figure 1.8 shows the extent of current activities on three major pollution research areas. These three activities must take place simultaneously and must be equally emphasized.

1.8.2 Ground Pollution Sources or Routes

Ground pollution sources can be divided into three major routes as: (1) point, (2) line and (3) surface sources. Each source can be further divided into fixed and movable points, the fixed points being such as landfill area, storage area, and industrial sites, as well as mine strips, and movable points of pollution sources including leaking of hazardous materials during transportation (TRB, 1986) such as oil spills during transportation both on land and at sea. Figure 1.9 presents the ground pollution sources' classification. It must be pointed out that some pollution is unexpected, such as the oil spills during the Gulf War in 1991.

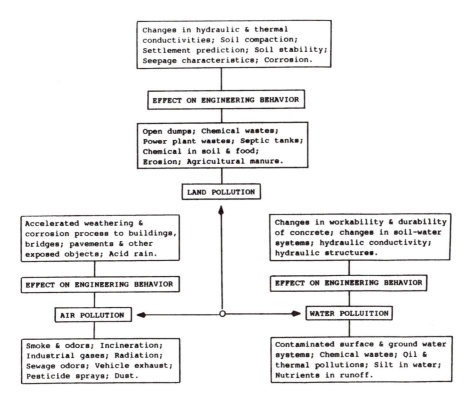

Figure 1.7 Effect of air-water-ground pollutions on the engineering behavior of construction materials.

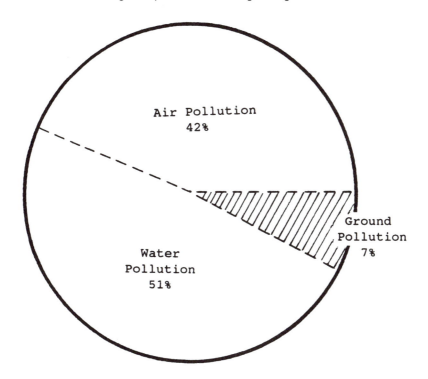

Figure 1.8 1986 survey indicates efforts on various pollution research activities. (After Fang, H.Y. [1989a], Proc. 2nd Int. Symp. on Environmental Geotechnology, v. 1, pp. 167–194, and Fang, H.Y. [1995b], *Encyclopedia of Environmental Control Technology,* v. 9, Cheremisinoff, ed. Gulf Publishing, Houston, TX, pp. 13–117.)

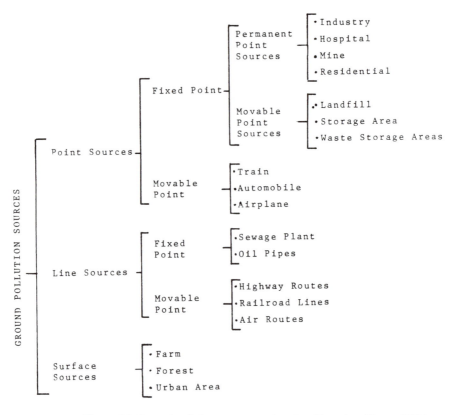

Figure 1.9 Ground pollution sources and routes. (Based on Wang, 1985.)

1.8.3 Pollution Process Mechanisms

Regardless of the sources, there are three basic mechanisms by which the ground soil can be contaminated:

1. Contamination may occur from rainfall such as acid rain or rain falling into a sanitary landfill, or oil or chemical wastes spilled onto the ground.
2. Pollutants are introduced as leakage from well disposal or construction of waste disposal facilities such as landfills, septic tanks, laterals, and lagoons.
3. Hydraulic or chemical alterations allow polluting substances to move within or between soil layers. In this category, the phenomena cover chemical, physicochemical, and microbiological aspects.

Categories 1 and 2 are direct processes. Category 3 covers the physicochemical decomposition process, chemical alterations and bacterial attack, leaching, and ion-exchange reaction. These processes will be discussed in various chapters within the text.

1.9 METHODS FOR ANALYSIS OF ENVIRONMENTAL GEOTECHNICAL PROBLEMS

1.9.1 General Discussion

Based on previous discussion, it is clearly understood that a geotechnical engineer has responsibility to tackle this highly interdisciplinary subject, and, on the other hand, it is a challenge dealing with complex problems. Environmental geotechnology requires knowledge from other disciplines in order to understand soil response to environments for both short- and long-term problems. To do this, the following facts are what a geotechnical engineer must realize:

1. Most geotechnical projects occur in nature and, therefore, must be considered as long-term installations and constructed outdoors where they will be open to various environmental effects.
2. Soil is more sensitive to environmental effects than any other construction material. Loading may greatly affect soil properties but it is not the only parameter to be considered. Other variables caused by environmental factors must also be considered such as heat, wet and dry, pollution intrusion, etc.

It has been demonstrated in the text and Figure 1.3 that soil-environment interaction is a complex phenomena, and current approaches based on loading alone cannot evaluate effectively the phenomena involved. Therefore, additional modifications or supplemental information are urgently needed. In order to overcome these shortcomings in current concepts, information such as thermal-electric-magnetic effects, physicochemical, chemical, etc. is imperative. If we do not consider these factors, there will be more "geotechnical surprises".

1.9.2 Review of Current Soil Mechanics Concepts

Since 1925, when Karl Terzaghi introduced the concept of soil mechanics into the civil engineering field, thereby becoming a major subject in the instructional curricula, basic concepts and theories have been established which greatly improve modern design and construction technology in civil engineering. However, the majority of effort is based on the physical-mechanical behavior of soil. In some institutions, the geotechnical course becomes a part of engineering mechanics courses, thereby ignoring the fundamental behavior of soil itself.

Since soil is very sensitive to local environmental conditions, it cannot be evaluated by a mechanical approach alone. Figure 1.10(a and b) (Fang, 1995b) illustrates the current research and instruction efforts in geotechnology and on soil behavior. In examining these two figures, the majority of effort is placed on mechanical (loading) and short-term studies. Very little effort is made to study the environmental effects such as physical-chemical-biological on the properties of soil.

Also, at present, design and construction of most geotechnical projects are based on test results following ASTM and AASHTO standards. These standards are based on controlled conditions at room temperature with distilled water as the pore fluid. In addition, most geotechnical analyses concentrate on loading conditions tested under short-term duration but projected into long-term performance. Since field situations and standard control conditions are significantly different, premature or progressive failures frequently occur. These concepts and methods currently used are outlined as follows:

1. Soil parameters such as specific gravity (ASTM D854-72) and Atterberg Limits (ASTM D423-66; D424-71) for a given soil under any environmental conditions are assumed to be constant.
2. Parameters commonly used are the void ratio or porosity as indicators of soil deformation under load.
3. The water content in the soil mass is mainly based on gravity water only, and other types of water in the pore space, such as environmental water (Sections 4.9 and 5.2), are not included.
4. Flow through a soil mass considers the hydrostatic potential only. Other causes, such as thermal, electrical, magnetic and chemical generally are not considered in analysis and design.

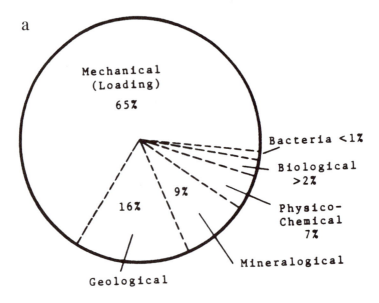

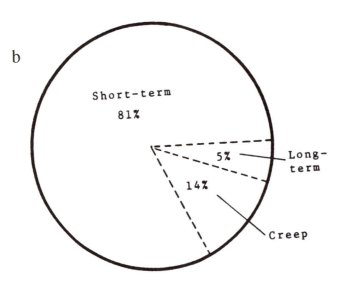

Figure 1.10 Current instruction and research efforts on geotechnology. (a) Instruction effort; (b) research activities on soil behavior.

5. Constitutional models based on soil's stress-strain relationship are arbitrarily assumed. In some cases, the assumption may hinge on an individual's preference not based on the soil behavior.

1.9.3 Proposed Approaches or Methods for Evaluation of Soil Behavior

(1) General Discussion

At the present time, the subject of geotechnology stands at a crossroad. One road still dogmatically follows the classical concept developed by Terzaghi, as mentioned in previous sections, and the other, the environmental situations that have challenged these concepts and

Table 1.3 Comparison of Controlling Parameters for Soil Classifications

Present controlling parameters[a] (environmental factors not considered)	Proposed controlling parameters (environmental factors considered)
Grain-size distribution (ASTM D422)	% passing #200 sieve (ASTM D11450)
Soil consistency (moisture content)	Specific surface
Liquid limit (ASTM D423)	pH in pore fluid
Plastic limit (ASTM D424)	Ion-exchange capacity
Plasticity Index (ASTM D424)	Absorption and adsorption
	Dielectric constant

[a] Unified Soil Classification System (ASTM D2487-83).

methods for analyzing soil behavior under varied environmental conditions, because they cannot effectively explain all the soil-water-environment phenomena and soil-structure interaction the modern world presents. The situation cannot be changed or corrected dramatically overnight, but it must change eventually without a large gap between the two directions. To cope with this issue, a combined approach is presented which is why the particle energy field theory is introduced in this textbook. In other words, this new textbook is presented within an old framework with new information blended into it as necessary.

(2) Instruction

Methods for evaluation of soil behavior under various environmental conditions have been discussed in previous sections and illustrated in Figure 1.10. To further understand the soil-water-environmental interaction, courses such as soil science, physicochemistry, mineralogy, geology, microbiology, etc. are recommended for students to take.

(3) Laboratory Testing

Table 1.3 presents the comparison of controlling parameters for soil classifications. At present, soil classification such as unified (ASTM D2487-83) and AASHTO (M145-88) are all based on particle size and soil consistency. Since soil is sensitive to local environments, some additional parameters such as specific surface, pH in pore fluid, adsorption, dielectric constant, etc. must be evaluated. The polluted soil can be identified by soil surface cracking patterns, colors, odor, and volume changes. These characteristics must also be examined and given consideration. Identification and characterization of contaminated soil will be presented in Section 3.8.

(4) Data Interpretation and Analysis

For interpretation of various soil-water-environmental interaction, the particle energy field theory as pointed out in Figure 1.3 are suggested. Detailed concepts will be presented in Chapter 2 and their applications will be discussed in various chapters throughout the text.

1.10 SCOPE AND ORGANIZATION OF THE TEXT

The main purpose of this text is the generalizing of basic concepts, test methods, soil properties, and analysis related to environmental aspects of geotechnology. The systematic correlation of experimental data with theoretical explanation is considered a prerequisite for comprehensive analysis of the environmental factor's effect on soil or rock for practical application of soil-foundation-structure systems in the environment. The text contains 18 chapters and can be categorized into three groups:

Table 1.4   A Flowchart for Characterization and Identification of Contaminated Ground Soil for Engineering Purposes

```
┌─────────────────────────────┐              ┌──────────────────────────────────┐
│ Possible Factors Affecting  │              │ Major Controlling Parameters     │
│ on Soil Characteristics     │              ├──────────────────────────────────┤
│                             │              │ . Silica/sesquioxide Ratio       │
│ . Grain Size                │              │   reflects following follow-     │
│   Specific Surface          │              │   ing properties:                │
│   % passing #200 sieve      │              │   Absortion & adsorption         │
│ . pH in Pore Fluid          │    ┌──────────┐   Dielectric constant        │
│ . Ion Exchange Capacity     │ ►  │SOIL-WATER│◄  Ion exchange                 │
│ . Absorption & Adsorption   │    │  System  │   pH solution                  │
│ . Dielectric Constant       │    └──────────┘   Mineral types or            │
│ . Mineral Characteristics   │              │      characteristics             │
│   Isomorphous substitution  │              ├──────────────────────────────────┤
│   Linkage between sheet     │              │ Particle Size and Water Content  │
│   Charge density            │              │ . Particle Size & Water Content  │
│   Particle shape            │              │   reflects following properties: │
└─────────────────────────────┘              │   Activity                       │
                                             │   Liquid limit                   │
                                             │   Plastic limit                  │
                                             │   Plasticity index               │
                                             └──────────────────────────────────┘

              ┌─────────────────────────────────┐
              │ ENGINEERING BEHAVIOR            │
              │ . Conductivity (Hydraulic,      │
              │   Thermal, Electric, )          │
              │ . Volume Change (Shrinking,     │
              │   and Swelling)                 │
              │ . Compaction, Consolidation     │
              │ . Stress-Strain & Strength      │
              └─────────────────────────────────┘
```

1. Basic concepts of environmental geotechnology which include the fundamental theoretical considerations
2. Distinguishes the importance of environmental effects on soil/rock relating to various basic aspects of soil mechanics such as compaction, consolidation (settlement), bearing capacity, and earth pressure
3. Illustrates by using these environmental aspects the various environmental geotechnical problems such as landslides, ground subsidence, erosion, waste control facilities, and ground improvement techniques

Numerical examples and problems are also provided in each chapter. The primary purpose of this textbook is to serve both for teaching the engineering student as well as for reference. Table 1.4 summarizes the various tests for obtaining the environmental geotechnical properties of soils and potential applications.

1.11 SUMMARY

1. Ground pollution problems challenge current soil mechanics concepts and methods for effectively analyzing soil behavior under various environmental conditions; therefore, the environmental aspects of geotechnology have been expanded to cover soil and rock and their interaction with various environmental cycles including the atmosphere, biosphere, hydrosphere, and lithosphere, as well as geo-microbiosphere.
2. The importance of environmental cycles, relating to various geotechnical problems as well as required knowledge from other disciplines, is presented in Table 1.1.
3. A new concept named the 'particle energy field' is proposed in the text for the purpose of explaining various soil behavior under different environmental conditions. The details of the concept will be presented in Chapter 2. All explanations will be made by the same concept throughout the text.

4. Both natural and man-made pollutions are outlined and discussed.
5. Pollution process mechanisms and soil-pollution interactions are discussed, including ground pollution sources and routes.
6. A flowchart for characterization and identification of contaminated ground soil for engineering purposes is presented.
7. The scope and organization of the text are outlined and discussed.

PROBLEMS

1.1 Comment on the current concepts and approaches of soil mechanics related to ground pollution problems.

1.2 Why is mechanical energy considered a short-term process and why are the chemical, physicochemical, and geo-microbiological processes long term?

1.3 Define the term 'ecosystem' and how the ecosystem relates to environmental geotechnology.

1.4 Why are some soils more sensitive to environment than others?

1.5 Distinguish between toxic wastes and hazardous wastes.

1.6 Explain why and how oil spills cause extensive environmental damage along the seashore environments.

1.7 Explain the pollution process mechanism in the soil-water system.

1.8 What are the major sources of ground pollution?

1.9 Explain why the air-water-ground soil pollution is interrelated?

1.10 Why does air-water pollution affect the geotechnical behavior of ground soil?

1.11 What are acid rains? How do acid rains affect steel and concrete structures?

1.12 How does acid water affect the concrete mixtures?

CHAPTER 2

Particle-Energy-Field Theory and Its Applications

2.1 INTRODUCTION

As pointed out in Chapter 1 and illustrated in Figure 1.3, environmental geotechnology is an interdisciplinary science that requires knowledge from other fields. The present knowledge of geotechnical engineering concepts and methods is not effectively analyzing soil behavior under various environmental conditions; therefore, a theory called the "particle energy field" has been introduced as a unified approach for analyzing soil behavior under various environmental conditions (Fang, 1989a and 1991a). A brief review of the fundamental concept and its applications to environmental geotechnology are presented as follows.

2.2 THE CONCEPT OF PARTICLE-ENERGY-FIELD THEORY

2.2.1 Fundamentals

The theory of particle energy field consists of three major components:

- Elementary particles
- Particle systems
- Energy fields

The combination of these three components into one system is called the particle-energy field. Basically, the theory combines the concepts of solid-state physics and chemistry on one side; physicochemical, organic chemistry, and biology on the other side. Interacting between these two groups is the common denominator known as the particle. Particles are the fundamental building units of all types of materials, including soil, water, gas (air), and pollutants. In addition, some environmental phenomena such as ion-exchange reaction, adsorption, redox reaction, soil-bacteria interaction, etc. which pose difficulties in the current geotechnical approach are incorporated in this theory. The theory is based on the following assumptions.

2.2.2 Assumptions

1. The physical world is constructed of particles such as atoms, ions, and molecules.
2. These particles may attract or repel each other.

3. Bonding energies such as ionic, covalent, and chemical and linkage such as cation, water dipole, and dipole-cation control the stress-strain-strength and durability between particles.
4. The relative movement of particles is caused by these energies.
5. Particle systems can be
   - Solid state if attraction   (A) >> repulsion (R)
   - Liquid state if attraction   (A) ≈ repulsion (R)
   - Gaseous state if attraction   (A) << repulsion (R)

2.2.3 Elementary Particles

Elements are composed of tiny, fundamental particles of matter called *atoms*. Ordinary atoms are neutral, i.e., they do not carry an electrical charge. However, under certain circumstances, atoms can become electrically charged. Such charged atoms are called *ions*. Some elements form positive ions, called *cations*, and some form negative ions, called *anions*.

The atom as a basic particle of matter is composed of still smaller particles called subatomic particles. The *neutron, electron,* and *proton* are classified as subatomic particles. The positive subatomic particle present in the atom is called a *proton*. Units of negative charges are known as *electrons*. A third subatomic particle found as a constituent of atoms which carries no electrical charge (neutral) is known as the *neutron*.

The sharing of a pair of electrons binds the atoms together to form a new kind of particle called a *molecule*. Molecules are stable particles and are characteristically chemical particles of many compounds. A summary of basic types of particles which serve as building units of matter is presented in Table 2.1.

In addition to the basic particles as indicated in Table 2.1, it must be noted that new particles have been discovered named *quarks*. Reports by the Fermi National Accelerator Laboratory in Illinois indicated that particles such as protons and neutrons are composed of triplets of quarks. The finding confirms a prediction based on a theory known as the Standard Model, that nature has provided the universe with six types of quarks; the other five — the up, down, strange, charm, and bottom quarks — had all been known or discovered by 1977.

Since the infancy of the universe shortly after the Big Bang only the up and down quarks have survived in nature, and the protons and neutrons that make up the nuclei of all atoms are built from combinations of these two quarks. The other quarks disappeared from the observed universe, but have been recreated by modern particle accelerators. Physicists hope that their newly acquired knowledge about the quark will help them to understand why all matter has mass, and why the masses of the different atoms that make up the universe have the specific values they do. The quark theory of matter has successfully explained many aspects of physics that previously defied understanding. Maybe someday this theory can further explain soil-water-gas pollution problems.

2.2.4 Particle Systems

Since the physical world consists of three states of matter — solid, liquid, and gas (air) — as shown in Figure 2.1, any other elements existing are these in combinations. When soil is dry, it is in a solid state; when it is saturated, it becomes a two-phase system; if soil is partially saturated, it is in a three-phase system, involving solid, liquid, and gaseous states. Regardless of the state of matter, the microstructure is composed of particles. Stress-strain relationships of soil hinge on the bonding behavior of two or more particles. Water content of the soil and flow of water through soil are dependent on the energies between particles. Therefore, the particle system is influenced by bonding energies between particles.

Table 2.1   Basic Types of Particles Which Serve as a Building Unit of Matter

Subatomic Particles

Name of particles	Electric charge	Mass
Electron	Negative charge (−)	1/1837 (of H atom)
Proton	Positive charge (+)	1 (of H aton)
Neutron	Neutral (0)	1 (of H atom)

Atoms, Molecules, and Ions

Name of particles	Electric charge	Examples
Atoms	Neutral	Sodium (Na), magnesium (Mg), carbon (C), nitrogen (N), oxygen (O), hydrogen (H)
Molecules A group of covalently bonded atoms	Neutral	Methane (CH_4), ammonia (NH_3), hydrogen chloride (HCl)
Ions (simple) Charged atoms	Charged particles	
Cation Anion	Positive charge (+) Negative charge (−)	Sodium ion (Na^+), magnesium ion (Mg^{2+}), oxide ion (O^{2-}), chloride ion (Cl^-)

Polyatom ions
  Groups of covalently bonded atoms
  which carry charges
	Charged particles	Carbonate ion (CO_3^{2-}), nitrate ion (NO_3^-), hydroxide ion (OH⁻)

Data from Dickson, T.R. (1974), *Understanding Chemistry: From Atoms to Attitudes,* John Wiley & Sons, New York; and others.

2.3 ENERGY, ENERGY FIELD, AND PARTICLE ENERGY FIELD

2.3.1 Energies between Particles

Basic physics and chemistry indicate energy gradients are the main causes for particle movement from one place to another. Particle motion, whether it is monotonic or dynamic, originates from particle behavior under energies such as potential, thermal, electrical, magnetic, etc.

(1) Bonding Behavior between Particles

There are two major types of bonds existing within atoms and molecules comprising soil particles: the primary bond and the secondary bond. The primary bond is the bond between atoms forming molecules. The secondary bond is the atoms in one molecule or ion bonding to another. Further discussion on bonding strength will be presented in Sections 2.5 and 4.5.

(2) Attractive and Repulsive Forces

All microparticle soil carries an electric charge. When two particles are close to each other, net attractive or repulsive forces exist between particles. Several formulas for estimating these forces are presented in Section 4.6.

2.3.2 Energy and Energy Field

In the previous section, we discussed the fundamental concept of particles. However, particles react differently within various energy fields; therefore, further modifications are made to include various energy fields as follows.

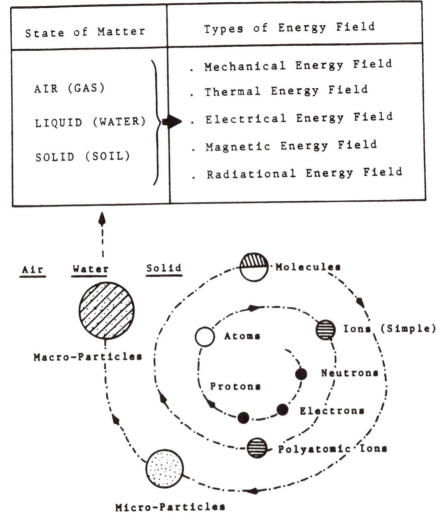

Figure 2.1 The particle systems, states of matter, and types of energy fields.

Energy is the quality possessed by an object that enables it to do work. The sources of energy can be derived from surface force and body force. The surface force creates energy sources including potential, kinetic, thermal, electrical, magnetic, and radiation, as well as the body force such as gravity force.

Energy field is defined as a space in which each energy charge reacts to another charge (Figure 2.2). In other words, the energy field is the area of influence in the vicinity of the energy charge and the interaction among the other energy charges.

2.3.3 Particle Energy Field

Particle energy field is the collection or assemblage of individual particles in space which interact and exhibit surface and/or body forces. For practical purposes, let the energy fields be divided into five basic groups, namely:

- Mechanical energy field
  - Potential energy field (energy of position)
  - Kinetic energy field (energy of motion)

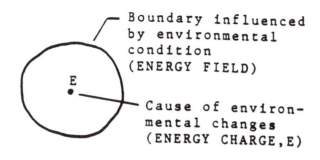

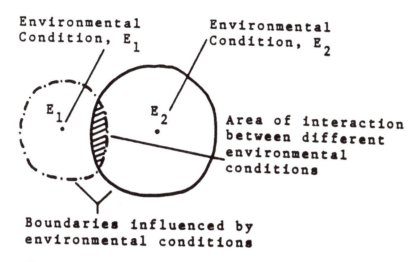

Figure 2.2 Schematic diagram illustrates the energy charge and energy field.

- Thermal energy field
- Electrical energy field
- Magnetic energy field
- Radiation energy field

2.3.4 Energy Fields and Environments

Particles are the basic structural units for all materials; however, each particle reacts differently at various energy fields. In other words, particles respond to various environments differently. As indicated in Figure 2.1, there is a similarity between environment phenomena and energy fields. Combining both environment phenomena and energy fields into one system is shown in Table 2.2. Table 2.2 covers five different energy fields and represents five environmental phenomena. Some important parameters in each energy field to be considered together with typical geotechnical problems are listed. In many cases, some environmental phenomena or reactions cross into two or more energy fields such as those that occur in hazardous and toxic waste and landslide problems.

For proper evaluation of environmental characteristics in each environmental zone or in each energy field, proper laws or principles are required as indicated in Table 2.2. For example, the flow movement due to hydraulic gradient (potential energy) will follow Darcy's law; however, if flow movement is caused by thermal gradient, then it should follow Fourier's law, and if it is due to electric potential, then it should follow Ohm's law. Because environmental condition changes, the soil behavior changes also, consequently, the method of interpretation will also change.

Table 2.2   Identification and Classification of Environmental Zones Based on Particle Energy Field Theory

Environmental zones and parameters to be considered	Environmental geotechnical problems		Laws/principles required for evaluation
	Single phase	Multiphase	
	Environment (Natural and Man-Made)		
Mechanical energy Load, deformation, velocity, weight, mass, wave, sound, etc.	Simple foundation design Flow-through porous medium	Acid rain; hazardous/toxic wastes Radon gas emanation and mobilization Nuclear wastes control and storage Landslides; earthquake	Darcy's law Hooke's law Newton's law Laws of motion
Thermal Energy Hydration Heat of wetting Kinetic dispersive forces Thermoosmosis	Freezing-thawing Wet-dry cycle Thermal pollution Underground fire Soil cracking		Laws of thermodynamics Fouriers's law General gas law
Electrical energy Polarization Proton migration Electromotive force Electric conductivity	Creep, viscosity Stress hardening and softening Aging effect Migration Decontamination		Coulomb's law Joule's law Ohm's law Ampere's law
Magnetic energy Electromagnet Ferromagnetism Electromagnetic induction Electromagnetic waves			Faraday's law Lenz's law Biot-Savart's law Gauss's law
Radiation energy Decay process Radioactivity Nuclear reactions	Construction operations		Atomic physics Nuclear physics

After Fang, 1989a and 1991b.

There are five basic energy fields as discussed in a previous section. Even though each energy field has its own identity with individual characteristics, they are interconnected, as shown in Figure 2.3. Interaction between various energy fields, as indicated in the figure, such as electric-viscous effect and thermo-electric effect will be discussed in detail in Section 9.13.

2.4 THE STATES OF AGGREGATION OF IDEALIZED PARTICLES

2.4.1 General Discussion

In Section 2.2, we have introduced fundamentals and assumptions of particle energy field theory which include elementary particles, particle systems, and energy field. In this section, further discussion on the states of aggregation of idealized particles, for the purpose of comparing particle characteristics between idealized and practical conditions, is presented. The states of aggregation of idealized particles also include macrometric solids and liquids, and the structure of crystals, the true solid.

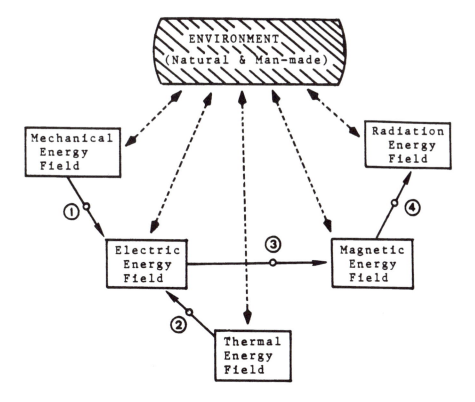

①   **Electro-viscous Effect**

②   **Thermo-electric Effect**

③   **Electro-magnetic Effect**

④   **Electro-magnetic Radiation**

Figure 2.3 A schematic diagram illustrates particle energy fields and their interactions with environment.

2.4.2 Types of Bonds and Energy Ranges

Energy ranges for different types of bonds are shown in Tables 2.3 and 2.4. Table 2.3 summarizes stability conditions for various interacting pairs that are of interest for our theme. For example, the bond energy of 4.84 kcal/mol of water dipole-dipole interaction (Table 2.3) is about one half of 9.7 kcal required to evaporate 1 g-mol of water at its boiling point under normal atmospheric pressure. Table 2.4 shows the ranges for different types of bonds expressed in electron volts (eV). If the particles composing a system fall within the same limit size or mass range, e.g., atomic, colloidal, or macro, then the strength of the bond will primarily decide the solid or liquid character of the system. Further discussion on bond strength of soil particles together with numerical example will be presented in Section 4.5.

2.4.3 Volume Relationships for Idealized Particles

Types of packing and consistency relationships between materials of the same general character are important for evaluation of states of aggregation of particles. Table 2.5 presents

Table 2.3   Type of Attraction Relating to the Energy of Interaction

Interaction pair	Type of attraction	Equilibrium separation (Å)	Energy of interaction (kcal/mol)
Na^+ F^-	Ion–ion	1.88	157
Na^+ OH_2	Ion–dipole	2.14	21.6
OH_2 OH_2	Dipole–dipole	2.37	4.84
Ne Ne	Coupled electronic oscillation	3.30	0.0613

After Moelwyn-Hughes; cited by Winterkorn, H.F. (1963), The Princeton University Conf. on Engineering and World Water Resources, Princeton University, Princeton, NJ, pp. 15–34.

Table 2.4   Energy Ranges for Different Types of Bonds

Type of bond	Energy (eV[a])	Typical properties	Examples
Ionic	7–11	Low electric conductivity at low temperature; strong ansorption in the infrared	NaCl; KBr; LiF; NaF
Covalent	7–12	Great hardness	Diamond (C) Carborundum (SiC)
Metallic	1–4	High electric conductivity	Na; Fe; Al; Cu
Ion-dipole	0.93	Hydration or strongly restrained water	Na^+ (OH_2)
Hydrogen bond	0.3–0.5	Tendency to polymerization	Ice, HF
Dipole–dipole	0.208		H_2O–H_2O, silt, loess
Molecular	0.08–0.3	Low melting and boiling points; great compressibility in solid state	CH_4
Van der Waals	0.0026		Noble gases; clay minerals (Ne–Ne)

[a] 1 eV = 1.602×10^{-12} erg = 0.386×10^{-19} cal.

After Lüscher (1967), (in German); see Winterkorn, H.F. (1974), Notes on Construction Materials, Department of Civil and Geological Engineering, Princeton University, Princeton, NJ.

Table 2.5   General volume relationships for Different States of Aggregation of Simple Compounds

	Volume ratios for corresponding states			
	Critical	Boiling	Melting	Densest
V_o (critical)[a]	1	2.66	3.12	3.78
V_b (boiling)[b]	0.38	1	1.17	1.42
V_m (melting)[b]	0.32	0.86	1	1.21
V_o (densest state)[c]	0.27	0.71	0.83	1

[a] At critical temperature and pressure.
[b] At 1 atmosphere pressure.
[c] Extrapolated to absolute zero temperature.

After Trautz (1922), (in German); see Winterkorn, H.F. (1974), Notes on Construction Materials, Department of Civil and Geological Engineering, Princeton University, Princeton, NJ.

the general volume relationships for different states of aggregation of simple compounds. In examining Table 2.5, volume ratios are indicated for corresponding states between the volumes of a particular substance at its densest state and its melting, boiling, and critical states (Trautz, 1922). For simple chemical compounds we have the general relationships as shown in Table 2.5 which are also related to different types of packing as shown in Table 2.6. The existence of these relationships, combined with those founded by Batschinski (1913) on the connection between the internal friction of liquids at different temperatures and their respec-

Table 2.6 Volume Relationships for Different Types of Packing and Their Corresponding States

| | Phase volume (%) | | | | | |
| | Densest | | Melting | | Boiling | |
Packing	Solids	Voids	Solids	Voids	Solids	Voids
Hexagonal-rhombohedric	74.1	25.9	61.2	38.8	52.2	47.8
Cubic	52.4	47.6	43.3	56.7	36.9	63.1

Based on Batschinski (1913), *Z. Phys. Chem.* v. 84, pp. 643–706; and Winterkorn, H.F. (1970), *Chemical Dynamics,* Wiley-Interscience, New York.

tive mole volumes, leads to the fertile concept of the solid and liquid states of macromeritic systems and the consistency properties within these states as functions of volume relationships (Winterkorn, 1970).

2.5 CRYSTAL — A TRUE SOLID

2.5.1 General Discussion

A normal condition of *true solids* is the crystalline state characterized by a well-defined internal structure and by typical external features. The science of the natural forms and structure of true solid matter is traditionally called crystallography or crystal physics. More recently the term solid-state science has come into existence. The particular term employed indicates the background of the user and his area of special interest in this science.

A *crystal* is a solid substance bound by plane faces with definite geometrical shapes, such as triangles, parallelograms, or trapezoids. However, this outside form must have been obtained by growth from a nucleus and must reflect the internal arrangement of the atoms, ions, or molecules of which the crystal is composed. The first law of crystallography states that: 'the angles between corresponding faces of all crystals of the same chemical substance are always the same and characteristics of the substance.'

2.5.2 Crystal Symmetry

The form elements of crystals — the faces, edges, and corners — repeat themselves in a regular manner. In order to recognize the symmetry of a crystal, it is necessary to rotate it and look at it from all sides. In this way, one easily recognizes that certain faces, edges, and corners are identical with others of the crystal and can be exchanged for them by simple operations such as rotation around an axis, or rotation combined with reflection which gives a mirror image after definite angles of rotation. While all possible crystal symmetry can be expressed by these two operations, one commonly differentiates between the following four types of symmetry:

1. Center of symmetry, which acts like the focus of an optical lens creating to every face a parallel but inversed double on the opposite side of the crystal.
2. Symmetry axes which pass through the crystal and around which one can turn the crystal so that by a rotation through an angle of 60°, 90°, 120°, or 180°, the crystal is in the same relative position as at the start. Since in a complete rotation of 360°, the same relative position will have been obtained 6, 4, 3, and 2 times, respectively, the corresponding rotation axes are called 6-, 4-, 3-, and 2-valent. A 1-valent axis would indicate that each face of a crystal is unique and has no corresponding face on any other part of the crystal. Symmetry axes may or may not be combined with a symmetry center.

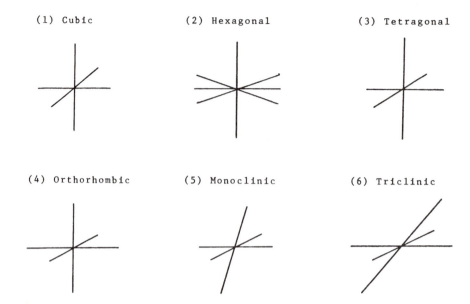

Figure 2.4 Crystal systems. All crystals belong to one of the six crystal systems.

3. Symmetry planes which divide crystals into two parts that are related to each other as an object and its mirror image are also called bilateral. Symmetry planes may exist in combination with symmetry axes and centers.
4. Rotoreflection axes in which a rotation of a face by a certain angle is followed by reflection on a plane perpendicular to the rotation axis to create an equivalent face.

The highest symmetry is possessed by a crystal that has the greatest number of symmetry elements. A systematic study of the various symmetry elements and their combinations has shown that there are 32 different symmetry classes. Representative of the class of highest symmetry is the hexakis-octahedron, the 48 faces of which can be derived from one single face by the operation of the symmetry elements involved. The lowest class is represented by a form in which every face is different from every other. This form of crystallization is typical for potassium bichromate.

2.5.3 The Crystal Systems

These 32 classes can be arranged into the 6 symmetry systems (Figure 2.4). The ordering principle for these is the type of coordinate system which lends itself best to expression of the respective crystal faces by simple rational indices.

1. Cubic system: In the cubic system which embraces crystal types of the highest degree of symmetry, the reference coordinates are the same as introduced into space geometry. They are three axes of common origin, each at an angle of 90° with respect to the others. All directions are equivalent, which means that the space or body contained in the space has the same properties in all directions.
2. Hexagonal system: The hexagonal system is characterized by a 6- or 3-valent principal axis perpendicular to which are three axes forming angles of 60° between each other.
3. Tetragonal system: The tetragonal system has one 4-valent principal axis and two axes at right angles to each other and to the principal axis. The minor axes are equivalent to each other, but not to the principal axis.
4. Orthorhombic system: The orthorhombic system has three axes at right angles to each other; these axes, however, are not equivalent.

Table 2.7   Symbols for Typical Symmetry Combinations of Crystal Classes

	Symbol	Symmetry combination
(1)	O	Rotation group of the octahedron
(2)	T	Rotation group of the tetrahedron composed of four 3-valent and three 2-valent axes
(3)	Cn	The cyclic groups: C = rotation axis, n = its valency
(4)	Dn	The dihedron group: an n-valent axis with n perpendicular bivalent axes
(5)	V	The group of four: three nonequivalent axes perpendicular to each other

Note: • Superscripts: h, d, v, i designate, respectively, horizontal, diagonal, and vertical symmetry planes and inversion of symmetry center.
• Numerical subscripts designate valency of symmetry axes.
• C_2 is a center of symmetry and S_2 represents a 2-valent and S_4 a 4-valent rotation plus reflection axis.

5. Monoclinic system: The monoclinic system differs from the above in that it has the principal axis and one minor axis at a right angle to each other, while the second minor axis forms with the principal axis an angle that is different from 90°. No axis is equivalent to any other.
6. Triclinic system: In a triclinic system, all three axes are nonequivalent and are at angles that differ from each other and from 90°.

Axes are called equivalent when the physical properties of the crystal are the same in the direction of the axes. They are called nonequivalent if the physical properties are different in the different axis directions. These physical properties may be mechanical strength, elasticity, thermal and optical characteristics, and others. The optical characteristics of crystals are especially important. Their study forms a branch of crystallography which is called crystal optics.

2.5.4 Crystal Classes

Before giving a simple schematic presentation of the relationship between the 32 crystal classes and the 6 symmetry systems, it is desirable to introduce a few symbols for typical symmetry combinations (Table 2.7). The distribution of the 32 crystal classes among the 6 symmetry systems is shown in Table 2.8.

2.5.5 Crystal Lattices

The preceding treatment of symmetry elements, classes, and systems did not require any assumption concerning the internal structure of actual crystals, although a relationship between internal structure and exterior geometric features of a crystal had been recognized since the time of Steno in 1669. Employing the hypothesis that matter was composed of very small discrete particles called atoms, Bravais in 1848 showed that all crystals can be built up by orderly arrangement of atoms in space in so-called space lattices. Specifically, he showed that only 14 types of unit lattices or lattice cells were necessary to build up all possible types of space lattices. These basic lattice types or cells have been called Bravais lattice. They are

1. The simple cubic lattice with an atom or ion at each corner of the cube
2. The body-centered cubic lattice, the same as (1), but with an additional atom in the center of the cube
3. The face-centered cubic lattice, the same as (1), but with an additional atom in the center of each of the six faces
4. The base-centered hexagonal lattice
5. The rhombohedral lattice
6. The simple tetragonal lattice
7. The body-centered tetragonal lattice
8. The simple rhombic lattice

Table 2.8   The Distribution of the 32 Crystal Classes Among the Six Symmetry Systems

Class	Symbol	Class	Symbol
I. Regular or Isometric System			
(1)	O^h	(4)	T^d
(2)	O	(5)	T
(3)	T^h		
II. Hexagonal system			
(6)	D_6^h	(12)	C_3^i
(7)	C_8^v	(13)	D_3^h
(8)	C_8^h	(14)	C_3^v
(9)	C_6	(15)	C_3^h
(10)	D_6	(16)	C_3
(11)	D_3^d	(17)	D_3
III. Tetragonal System			
(18)	D_4^h	(22)	D_4
(19)	C_4^v	(23)	V^d
(20)	C_4^h	(24)	D_4
(21)	C_4		
IV. Rhombic System			
(25)	V^n	(27)	V
(26)	C_2^v		
V. Monoclinic System			
(28)	C_2^h	(30)	C_2
(29)	C_2		
VI. Triclinic System			
(31)	S_2	(32)	C_1

9. The single-face-centered rhombic
10. The body-centered rhombic
11. The face-centered rhombic
12. The simple monoclinic
13. The side-centered monoclinic
14. The triclinic

He had anticipated physical knowledge which could be proven only after the discovery of X-rays, and with his 14 space lattice types given an ordering scheme for the 230 possible space groups whose existence was shown only in 1891 by Schonfliess after extensive geometrical research. For further information see standard textbooks on theoretical inorganic chemistry such as Day and Selbin (1969).

2.6 GRANULAR PARTICLES AND THEIR PACKING CHARACTERISTICS

2.6.1 General Discussion

In general, soil structure can be grouped into two categories, the macro- and microstructures. The macrostructure, dealing with large sizes of soil particles, ranges from 0.074 mm to 0.3 m. The microstructure has soil particles smaller than 0.074 mm. For practical application, the macrostructured materials are referred to as granular soil and are also called noncohesive soil. The structure of the noncohesive granular soil is an important part of the discipline of granulometry. The structures are different and, in general, they are less affected by local environments in comparison with cohesive soil.

The primary structure of noncohesive soil refers to the natural arrangement of the constituent particles of a soil in what may be regarded as a continuous system, such as a body of sand or gravel or a natural secondary unit in a cohesive soil such as the 'peds'. It embraces the study and measurement of the size, shape, and surface features of individual particles as well as the influence of these properties and of the gradation of the particles on the packing characteristics, mechanical resistance properties, and permeability of multiparticle systems. While granulometry deals mainly with the type of materials listed above, many of its laws are of a geometric nature and are equally true for particles of atomic size as they are for gravels and boulders. Thus knowledge originally obtained on assemblies of atoms and molecules can be utilized for sand and gravel systems, and vice versa.

2.6.2 Granular Particle Measurement

Granular particles are measured by particle size, shape, and packing characteristics. Often particle shape is given in general terms such as: spherical, rounded, angular, and irregular. Particle size is simply given by differentiation between boulders, cobbles, gravel, and sand and will be discussed in Section 3.4.

(1) Size Measurement

Granular particles are measured by particle size and shape. Size of particles can be determined by their volumetric relationships or by direct measurement. The volumetric relationships can be expressed by phase volume, absolute volume, void ratio, or porosity. The measurement methods include slide calipers for large pieces. Smaller particles can be determined by sieve analysis, sedimentation, and elutriation methods, or indirect estimate from material's surface area.

(2) Shape Measurement

Shapes may be characterized by the closest geometric form, such as: cylindrical, elliptic, cubic, prismatic, plate, and needle shape. Some commonly used parameters for the description of granular materials include effective size and uniformity coefficient. The term *effective size*, D_{10}, as determined from sieve analysis (ASTM D422-72), is the size of the screen opening that permits 10% of the granular material to pass and retains 90%. The *uniformity coefficient*, C_u, is the ratio of the size of screen openings passing 60% to that passing 10%. Further discussion with numerical example will be presented in Section 4.4.

The particle shape can also be expressed numerically, such as: sphericity, volumetric coefficient, elongation and flatness ratios, and shape factor. Size and shape measurement for granular soils are summarized in Table 4.2. The computation procedures are given by Winterkorn and Fang (1991).

2.6.3 Packing Characteristics of Granular Materials

Packing characteristics of granular materials depend mainly on the bulk density of the packing systems. Upon collection of granular particles placed in a container, the total volume of the system can be measured when divided into the weight of the sample, resulting in the bulk density. Confusion may result from using the term density, when unit weight is the actual measured quantity. However, bulk density has attained wide engineering usage, especially in the geotechnical engineering field. The measured weight includes the solids in the system plus any moisture or gas within the particle or adhering to its surface. The measured volume includes both the solids and the voids between the particles. Thus, bulk density can vary from a maximum value with the particles packed as closely as possible, to a minimum value when there are large, void spaces.

Factors affecting packing state include mechanical factors, moisture content, container characteristics (Gray, 1968), particle shape, surface condition, and size distribution as have been reviewed and discussed by Fowkes and Fritz (1974). The effect of particle characteristics on soil strength has been studied by Koerner (1970). The soils with more angular particles and lower sphericities had significantly higher angles of shearing resistance. The friction angle increases with decreasing effective size, D_{10}. This increase is significant with particle sizes less than 0.06 mm. Further discussions on mechanical properties and strength characteristics on granular systems are given by Farouki and Winterkorn (1964) and Siddiqi et al. (1987).

2.6.4 Interparticle Characteristics and Their Primary Structure

(1) The Laws of Granulometry

The *laws of granulometry* (Winterkorn, 1970) are frequently used for evaluation of particle packing characteristics; therefore, a brief presentation of these laws are stated as follows: These laws were derived from observations and experimentation on actual physical systems. However, in their most encompassing form they reflect abstractions necessary to obtain the simplest possible mathematical expression. Therefore, in the application of these laws to actual physical systems, the abstractions that have been made in their derivation must be properly corrected for. Also, these laws may be expected to best express the behavior of actual physical systems if their constituent particles approach closely the ideal, i.e., spherical shape.

(2) Interparticle Forces

The interparticle forces for various soil types range from practically nonexistant to strong electrostatic bonds as indicated in Tables 2.5 and 2.6. However, for granular soil, the packing of particles and their primary structure are important because their interparticle forces are influenced by packing systems (Pauling, 1960).

2.6.5 Particle Packing Systems

(1) Uniform Packings

The theoretically possible types of 'continuous, incompressible, uniform packings' of identical spheres are given by Kézdi (1964). Their porosities range from 26 to 72%. With sand-size spheres and rounded sand particles, it is very difficult in practice to get uniform packing with porosities higher than 50% or less than 36%, nor is the packing between these limits of a uniform character. Rather, photographs by Mogami (1967) show that in a sand mass of a certain bulk porosity, there are domains of regularly packed particles (cubic, e = 0.91; orthorhombic, e = 0.65; etc. where e = void ratio) in various orientations, separated by interphases of more loosely and irregularly packed particles. The lower the void ratio, the less the volume proportion of the interphases and the greater that of the more orderly and densely packed domains. In contrast to their geometric abstractions, material spheres in terrestrial environment possess packing characteristics that are influenced by sphere size surface/volume ratio, adsorbed gas and water layers, and interparticle friction. In general, packing density decreases with decreasing particle size.

(2) Polygonal Shapes

For simple polygonal shapes that can be fitted together without intervening cavities, the minimum theoretical porosity approaches zero. However, normal handling and compaction methods (Section 11.2) usually result in higher maximum and minimum porosities than obtained for spheres of equivalent size, which is especially true for crushed stone particles having sharp corners and edges and rough surfaces. If the corners and edges have been rounded off by natural transportation or by milling, such particles pack easily to greater densities than spherical ones.

(3) Binary Systems

For the binary systems, the densest system obtainable with spheres of two different sizes is one on which the larger spheres are packed to lowest porosity (26%) and the interstices filled in densest packing by spheres so small that the extent and curvature of the pore walls do not affect their packing, the minimum theoretical porosity would then be 6.75%. A practical though very imperfect approach to this type of packing is the macadam pavement, in which first layer of uniformly sized broken stone is laid with as much interlocking as possible and the interstices filled with successively smaller-sized stone and finally sand, and the process is repeated for the next layer. When mixtures of larger and smaller particles are used, the smaller always interfere with the packing of the larger.

The normal porosity of the beds of the single-sized components is 50%, which indicates their nature as angular, broken solids. The contribution of the smaller-sized particles to the total volume ranges only from 21.6 to 25.6%, while that of the large particles ranges from 28.9 to 50% as reported by Furnas (1931).

(4) Ternary Systems

Packing interference is also evident in ternary systems. The greatest density is achieved by proportions of about one part of the smallest and two parts of the larger-size materials. Systems composed of about equal parts of all three components exhibit the smallest difference between practical obtainable maximum and minimum porosities. Other characteristics of ternary systems have been described by Kézdi (1964) and Winterkorn (1970).

(5) Continuous Gradings

Continuous gradings yield low porosity mixtures requiring little compactive effort and are therefore of great practical importance in soil stabilization and in the making of concretes (Holl, 1969). The greater the range from the maximum to the minimum particle size, the less the porosity of the system. For most natural materials of relatively narrow gradation, the range of easily obtainable and reproducible porosity lies between 36 and 46%. The porosity of the practically densest state is usually 9.5% lower than that of the loosest.

2.7 CHEMICAL FUNDAMENTALS

In the previous sections, we have shown that the different states of aggregation of matter were consequences of the relative magnitudes of the respective dispersive and attractive forces acting on the particulate material components. For true solids, liquids, and gases, the ultimate particulate components for our purposes are the atoms and molecules. The order of magnitude of the attraction or bond energies between atoms or molecule pairs is presented in Tables 2.3 and 2.4. The main dispersive force per atom or molecule is the kinetic portion of their thermal energy content which amounts to:

$$KE = 3/2 \ N \ T \ per \ mole \qquad (2.1)$$

or

$$KE = 3/2 \ k \ T \ per \ particle \qquad (2.2)$$

where KE = kinetic energy per mole or per particle;
     N = Avogadro constant;
     k = Boltzmann constant; and
     T = absolute temperature.

Equations (2.1) and (2.2) are well-established facts in chemistry. Using these concepts applied to geotechnical engineering, some additional justifications must be made and readers must keep in mind that actual field and theoretical conditions are different. It is important to also look at some of the other clues, not to add or repeat superannuated arguments, but to point out those physical and chemical properties of matter that can be most simply presented and understood on the basis of the atomic concept. On the chemical side, these clues come mainly from the proportions in which different substances react with each other, while on the physical site they come from the properties of gases and solids. With our present tabulated knowledge on size, shape, structure, and bond and thermal energies of atoms and molecules, we can consider chemistry as molecular architecture, using the atoms and molecules as the smallest building stones.

2.8 PHENOMENA OF SOLID-LIQUID-GAS AT INTERFACE

Various phenomena at the solid-liquid-gas interface are important for evaluation of soil-water interaction in the environment as well as mass transport phemonena of unsaturated soil. Some relatively important terms closely related to environmental geotechnology are presented as follows. Further discussions together with specific problems will be presented in Chapter 10.

2.8.1 Single-Phase Interface

Single-phase interface covers liquid-liquid, solid-solid, and gas-gas. Among these three interfaces, the liquid-liquid interface occurs in environmental geotechnology most commonly as clean water interacting with polluted water, salt water intrusion, and oil-water mixtures. In solid-solid interfaces such as dry sand-gravel mixtures, coal, and crushed stones, if moisture is presented between them, then the single-phase interface becomes the double-phase or even the multiphase interface. Gas-gas (air) interfaces can be evaluated by the kinetic molecular theory (KMT). However, in many cases, gas particles will be absorbed by dust (Sections 3.10 and 10.7); then the behavior of gas-gas becomes a gas-solid interface. Oil-water interface is more complicated than any other single-phase interface because oil itself lies between the liquid-solid-gas form. The degree of consistency of oil itself will affect oil-water interface mechanisms.

Salt water intrusion is a dynamic equilibrium of groundwater movement along the coastal region. It is a shoreward movement of salt water from the ocean into coastal aquifers due to

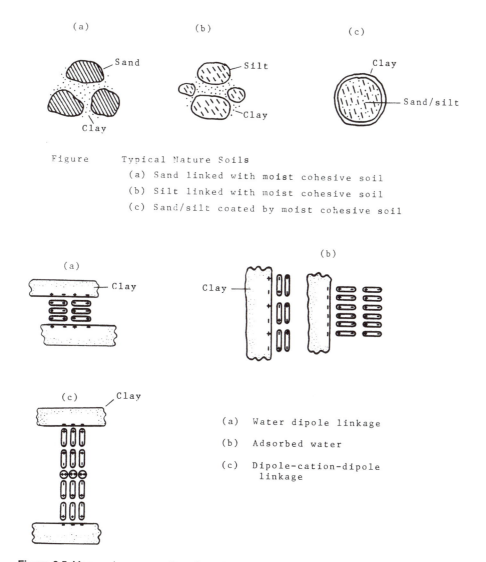

Figure   Typical Nature Soils
        (a) Sand linked with moist cohesive soil
        (b) Silt linked with moist cohesive soil
        (c) Sand/silt coated by moist cohesive soil

        (a) Water dipole linkage
        (b) Adsorbed water
        (c) Dipole-cation-dipole
           linkage

Figure 2.5 Macro-micro composite soil particles. A typical weak linkage between soil particles.

the overpumping of groundwater. Estuaries are also a saltwater-fresh water interaction problem. This phenomenon occurs where rivers (fresh water) meet the sea (salt water). It is a fragile system and easily destroyed, yet an important part of the earth's ecosystem (Section 14.4).

2.8.2 Two-Phase Interface

In the two-phase interface, the characteristics of adsorption play an important role. Adsorption is defined as the concentration of a substance on a surface. It is the uniform penetration of molecules of one phase in-between the molecules of a second phase. All solids or mineral particles tend to adsorb gases and solutes with which they are in contact. In order for the adsorption to be appreciable, it is necessary that the adsorbent have a large surface area. Therefore, fine-grained soil has more adsorption capacity than larger soil particles. In general, the amount of adsorption depends on the nature of the adsorbent, the nature of the substance being adsorbed, the surface area of the adsorbent, the temperature, and the pressure

Table 2.9 Types of Interface Relating to Energy Fields with Examples

Type of interface	Typical environmental geotechnical engineering problems	Energy field
Single-phase		
Solid-solid	Sand-gravel mixtures (dry)	
Liquid-liquid	Gasoline leak into groundwater aquifer	
	Saltwater intrusion	
	Estuaries	
	Petroleum spills into river, ocean	
	Leachate seeping into wells	
	Acid rain deposition into rivers, lakes	
	Acid drainage seeping into rivers, streams, lakes	
Gas-gas (air)	Various air pollution problems	Mechanical energy field
Two-phase		
Solid-liquid	Erosion; scouring	
	Silt in river; sedimentation	
	Various water pollution problems	
Solid-gas	Radon gas adsorbed on dust or suspension	
	Various air pollution problems	
Gas (air)-liquid	Gas (air) in water	Thermal energy field
Multiphase		
Solid-liquid-gas (air)	Saturated or unsaturated soil (vadose zone)	
	Mass transport phenomena	
	Thermal-electric-magnetic effects on soil-water system	
	Radon gas in groundwater	Multimedia energy field

in the soil/water system. A decrease in the temperature or increase in the pressure increases the amount of adsorption.

Some natural soils such as sand-silt link together with moist cohesive soil or sand-silt coated by moist cohesive soil to form a composite particle (Figure 2.5). The linkage between two particles is through adsorbed water, water dipole, or dipole-cation-dipole. In many cases, they are only temporary and once the soil becomes dry, the linkage force between two particles can be dismissed.

2.8.3 Multiphase Interface

The soil-water interaction is commonly treated as a two-phase interface; however, in the natural case, this interaction is a multiphase interface because whether or not soil is saturated or dry, it always contains some gases. Other cases include water-repellent soils (Section 4.11), where water movement is in a water-repellent soil and the wetting phenomenon is a vapor-liquid-solid interaction. All types of pollution transport in the soil-water layers belong to the multiphase interface group. Table 2.9 summarizes all three interfaces together with typical environmental geotechnical engineering problems.

Multiphase phenomena also occur in natural environments. Water vapor exists in the soil-water system due to the relative humidity of the air in soils. The pressure of the water vapor in the soil voids increases with temperature. In general, water vapor moves from the warmer zone and condenses in the cooler soil. In the case of the summer season of hot weather which warms the soil to considerable depth, followed by a cool spell which cools the surface soil rapidly, appreciable amounts of water vapor move up from the warm soil below and condense in the upper soil layer. Such movement may also occur in the autumn season when the lower soil horizons have not yet cooled to the temperature of the surface soil. Likewise, some moisture may condense onto the soil surface from warm atmosphere with high humidity.

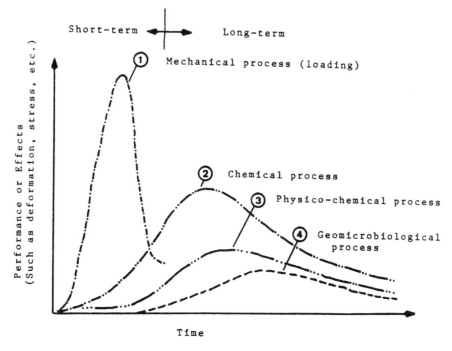

Figure 2.6 Schematic comparison of relative effects on soil performance by various geomorphic processes.

2.9 PARTICLE BEHAVIOR UNDER LOAD — MECHANICAL ENERGY FIELD

2.9.1 General Discussion

Mechanical energy includes both potential and kinetic energies which dominate today's geotechnical engineering concepts and approaches. Potential energy mainly is loading which includes compaction, consolidation, distortion, bending, crushing, kneading, shearing, and other construction processes. Kinetic energy includes the movement of water through porous media, vibrations, etc.

It must be noted that all mechanical energy field-related problems are considered as short-term performances. Figure 2.6 illustrates schematically comparisons of characteristics of various energy fields. Of course, mechanical energy plays the most important role; however, such problems do not seriously consider the local environments. Since most geotechnical projects occur in nature, problems must be considered as long-term installations and where they will be open to various environmental effects. Also, soil is more sensitive to environmental effects than any other construction material; therefore, many premature or progressive failures frequently occur due to unexpected soil environment interaction.

2.9.2 Load-Deformation Process

When a soil mass is subjected to a loading, whether static or dynamic, the increase in density or volume change will change the total internal energy of the soil. It is a process involving potential energy (energy of position), kinetic energy (energy of motion), and heat energy (such as change in moisture content of soil). These energies can cause the reorientation of soil particles which possess forces of repulsion or attraction due to their adsorbed ions and adsorbed water molecules as discussed by Lambe (1958 and 1959), Yong and Warkentin (1966), Winterkorn and Fang (1991), and Mitchell (1993). The physicochemical changes during the geomorphic process may be the compression of gases in the voids, thereby increasing the amount of dissolved gases in pore water, the elastic strain of solid particles, and the characteristics of double-layer thickness.

During the load-deformation process, there must be a relative motion of soil particles. Force is required to overcome the frictional resistance developed between particles during the motion. Energy spent to overcome the frictional resistance is heat energy. There are three basic types of frictional resistance which characterize the relative motion between soil particles during loading process:

1. When the particle surface is dry
2. Partially saturated, known as hydrodynamic or thick-film lubrication, surface friction
3. Saturated surface-boundary or thin-film lubricated surface friction (Eshbach, 1952)

On a molecular scale, a compression wave is started whenever a solid particle is struck. For example, when a drop hammer strikes the pile cap, the molecules of pile material (say, a steel pile) at the top surface are subjected to a net force caused by the hammer. According to Newton's Second Law, this force causes an acceleration, and the molecules start to move downward. At this point, they push on neighboring molecules and a pulse is transmitted to the tip of the pile. When a shock wave or impact load travels along the pile, a pressure is momentarily built up wherever the molecules are closer together than is normal (before the pile was driven). The behavior for particles (molecules) around the pile and soil will depend upon the type of pile, soil types, and local environmental conditions. Distortion phenomena during the pile-driving process are mainly due to that which gives rise to an elastic force that pushes the next molecule along.

The Law of Conservation of Energy has been used for interpreting the consolidation theory (Geuze and Bruyn, 1948) and the theory of compaction (Li, 1956). In these systems it is indicated that the work done on the system is equal to the change in total energy plus the heat transferred from the system.

2.9.3 Particle Dynamics

In the previous section we discussed the pile-driving process. Other dynamic problems relating to geotechnical engineering projects such as machine vibration, quarry blasting, dynamic consolidation, wave action, liquefaction, acoustics (sound wave), echo technique, and many others are also important. Some of these problems will be further discussed in Section 11.9.

The basic parameters of particle dynamics are velocity, acceleration, mass, force, work, energy, wave, vibration, etc. as listed in Table 2.2. In a liquid or gas, compression waves are called sound waves. The characteristics of sound waves include wave pulse, transverse and longitudinal waves, and the frequency of sound waves. When Newtonian mechanics is applied to the motion of a system, it is found that motion can be regarded as wave motion, called normal modes of vibration. The frequency of oscillation in a normal mode is termed as the natural frequency of the system. The lowest natural frequency is called the fundamental frequency. When the driving frequency is near a natural frequency of the vibrating body, the amplitude of these forces oscillating becomes exceptionally large. This large response at a certain driving frequency is called resonance. Resonances of great variety occur in the interactions of molecules.

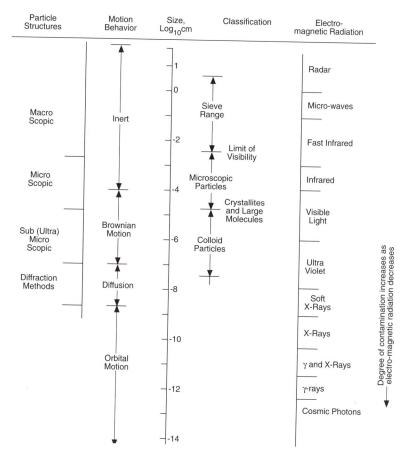

Figure 2.7 Soil particle sizes, their motion behavior and measuring techniques (After Freudenthal [1950], [in German]; see Winterkorn, H.F. [1974], Notes on Construction Materials, Department of Civil and Geological Engineering, Princeton University, Princeton, NJ.)

2.10 MULTIMEDIA ENERGY FIELD

2.10.1 General Discussion

As time goes on, the soil behavior is no longer controlled by mechanical energy alone. Figure 2.6 illustrates the mechanical and multimedia energy fields as reflected on deformation vs. time. The mechanical alteration may significantly cause soil deformation; however, it is considered a short-term process, while the others are considered long-term processes with the chemical process playing the most important role among these processes. Soil not only is a complex material, but also is very sensitive to local environments more than any other construction material. As indicated in Figure 2.7, if soil particle size changes, it also can change particle motion behavior; consequently, it will change bonding and energy characteristics between soil particles.

As time goes on, local environments such as temperature changes, freezing-thawing, wetting-drying, pollution intrusion, etc. will change soil particle characteristics. Then, soil behavior changes under the influences of thermo-electric-magnetic energy fields because they are very closely related in the natural environment; however, each energy field holds its own distinct character and identity. The energy sources indicate that some are naturally inherited

Table 2.10 Sources of Energy in the Thermal-Electric-Magnetic Energy Fields

Thermal Sources
  Internal heat sources
    When solid particles are wetted
    When ion or molecules are dissolved
  External heat sources
    Nature
      Solar energy
      Geothermal
    Man-made
      Heat pump
      Steam pipes
      Electrical cable lines
      Chemical or nuclear wastes
      Carbon dioxide produced in landfill sites
      Underground fires
Electrical sources
  Internal sources
    Soil-water interaction
    Thermal-electric effect
  External sources
    Applied electrical current directly to soil mass
    Applied electrical current to soil-water system
Magnetic sources
  Natural (permanent) — iron ores
  Artificial (temporary) — electromagnets

and some are man-made as shown in Table 2.10. Some of the relatively important sources affecting the soil-water behavior will be presented in the following sections and some more detailed information is again presented in various chapters throughout the text.

2.10.2 Thermal Energy Field

There are three basic characteristics of the thermal energy field on the soil-water system:

1. Additional forces produced when water is added to dry or partially saturated soil. Such forces include heat of wetting force and the kinetic dispersive force. These forces are referred to as environmental forces or stresses (Sections 5.7 and 5.8).
2. Heat transfer process due to the thermal energy field will affect soil-water-air behavior.
3. The physical state of matter existing in the thermal energy field will change as a function of the change in temperature. These changes include heat of fusion, heat of vaporization, and heat of sublimation as shown in Figure 2.8.

2.10.3 Electric and Magnetic Energy Fields

Electric energy plays an important role relating to the basic soil-water behavior as illustrated in Figure 2.3. Some fundamental characteristics are outlined as follows:

- Polarization and proton migration
- Electrokinetic process
- Electro-viscous effect

Further discussions of these effects will be presented in Chapters 8 and 9. As discussed in a previous section, indications are that thermo-electric energies are closely related in the natural soil-water system. Their distribution in the system is in a random pattern due to

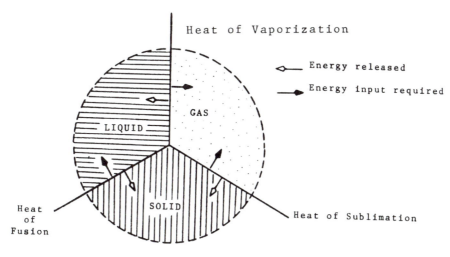

Figure 2.8 The changes of state of matter in the thermal energy field.

complex soil-water interactions in the environment. The ion movement in the soil-water system is also in a random motion. This is due to the bombardment of the dispersed particles by molecules of the medium and is called Brownian movement. When additional electric current is applied to the soil-water system, the particles remain in random motion, but the energy field boundary will change. Because of these changes, when two or more moving electric charges interact in the system, then the thermo-electric energies are changed into thermo-electric-magnetic energies. Further discussion on electro-magnetic energy will be presented in Section 9.13.

2.10.4 Radiation Energy Field

Geotechnical problems interacting with the radiation energy field can be grouped into four general areas:

1. Disposal or management of nuclear wastes
2. Control of radioactive and toxic radon gas (Rn-222)
3. Utilization of nuclear energy for large construction operations
4. Utilization of gamma-ray as a nondestructive testing method

Further discussions and applications to environmental geotechnology will be discussed in Chapter 10.

2.11 JUSTIFICATION FOR APPLICATION TO GEOTECHNOLOGY

Particle energy field theory as proposed in this text mainly applies to geotechnical engineering. In nature, soil is normally composed of solid, liquid, and gaseous phases consisting of soil particles with various sizes, ranging from small boulders (0.3 m) to colloidally dispersed mineral and organic particles (>2 μm), with mineral character ranging from that of practically unchanged fragments of igneous, sedimentary, and metamorphic rock through a wide range of weathering products to clay minerals and hydrous oxides of natural soil formations in the biosphere.

Whatever the size of the soil particles, laws of behavior should be determined for this size or size range. However, since the properties at one size or structural level are usually determined, or at least greatly influenced, by those of lower size levels, the influence of structural units at lower size levels must be studied and taken into account.

Explanation of the solid, liquid, and gaseous stages of matter by particle energy field theory represents the relationship of the volume of the solid particles to the volumes of the material as a whole in the solid, liquid, and gaseous states. It is a general concept that unites all solid engineering materials which are systems of particles that are glued together. The behavior of particles that are glued together depend on:

1. Sizes, shapes, and mutual arrangement of component particles of a system
2. Cementing agents or forces acting to hold the particulate components together

If the particles concerned are atoms, ions, or molecules, then the study under consideration falls into the realm of crystal physics or particle physics (Born, 1946). If the particles fall in the visible range, the respective discipline is called granulometry (Winterkorn, 1967). However, the basic geometric, symmetry, and size distribution-packing interrelationships are independent of particle size. The glue in the broad sense may range from electrons to hydrated particles of Portland cement. Since basic geometrical and symmetrical relationships are independent of particle size, pertinent geometrical relationships found in assemblies of atom-sized components are applicable to assemblies of sand- and gravel-size particulate components and vice versa. Liquid portions of soil are usually an aqueous electrolyte solution and are an intrinsic component of soil as a destructive agent by itself or a vehicle of other destructive agents acting on soil or any other construction materials.

Since all matter — whether solid, liquid, or gas — is constructed from various types of particles, it is logical and proper to use a common denominator as a base for the evaluation of the engineering behavior of soil. In addition, soil is very sensitive to local environments such as pollution, more so than any other construction material. These pollutant substances are also formed from various types of particles.

2.12 SUMMARY

1. Currently, the subject of soil mechanics itself has no unified theory or concepts to analyze all soils under various conditions. Therefore, a new theory called 'particle-energy-field theory' is proposed in the text for the purpose of a unified approach for analyzing soil behavior under various environmental conditions.
2. The theory of particle energy field consists of three major components: elementary particles, particle systems, and energy fields. Combination of these three components into one system is called 'particle energy field'. The energy fields are divided into five basic groups, namely: mechanical, thermal, electrical, magnetic, and radiation energy fields. Table 2.2 summarizes the characteristics of an energy field, environmental phenomena, together with illustration of typical geotechnical engineering problems and laws and/or principles required for evaluation.
3. All energy fields are interrelated to each other as illustrated in Figure 2.3. Phenomena at solid-liquid-gas interface relating to various environmental geotechnical problems are discussed.
4. Since the term 'environment' is generally used very loosely, based on the particle energy field concept which redefines and conforms this abstract term, a relatively concise term is presented in this chapter.
5. It is intended that all explanations of soil-water-environment interaction be based on this proposed particle energy field theory throughout the text. Comparisons between existing and the new concept are also made when it is possible. It must be noted that the existing mechanical energy concept is part of the proposed particle energy field theory and one of the five energy fields.

PROBLEMS

2.1 Why do ground pollution problems challenge current soil mechanics concepts and methods for effectively analyzing soil behavior under various environmental conditions?

2.2 Comment on particle-energy-field theory as proposed in the text.

2.3 What is the fundamental basis for development of the particle-energy-field theory?

2.4 Why is mechanical energy considered a short-term process and why are thermal, electric, and magnetic energies considered long-term processes?

2.5 Define the energy, energy field, particle energy field, and particle systems.

2.6 Explain surface and body forces.

2.7 Scientists reveal secrets of top quark — what are the secrets?

2.8 How would you define the term 'environment'?

2.9 Discuss the similarity between environmental phenomena and energy fields.

2.10 Discuss the energy flow in the changes of state matter of acid rain.

CHAPTER 3

Nature of Soil and Environment

3.1 SOIL AS A NATURAL GENETIC SYSTEM

Soils are formed from rock by physical, chemical, and biological forces that are usually grouped under the headings of climate, parent material, topography, and organisms. The degree of change from a parent material to a soil system is a function of time and of the rate of reaction of the dominant processes (Chapter 8). The definition of natural soil can be generalized into three basic categories such as:

1. Engineering: From an engineering viewpoint, soil is any earthy material that can be removed with spade and shovel and is the product of natural weathering. This soil includes gravel and sand deposits, agricultural-pasture-forest and desert soils, and finely grained colluvial and alluvial valley and delta deposits (Winterkorn and Fang, 1991).
2. Geology: From a geological viewpoint, the soil may be considered as the superficial uncon-solidated mantle of disintegrated and decomposed rock material, which, when acted upon by organic agencies and mixed with varying amounts of organic matter, may furnish condi-tions necessary for the growth of plants. In its broadest sense the term 'soil' has been used to include all the mantle of rock decay (Keller, 1984; Kiersch, 1991).
3. Pedology: From a pedological viewpoint, soil is the climatically conditioned petrographic and biogenic transformation product of the outermost layer of the solid crust (Byers et al. 1938); it is a natural body, differentiated into horizons varying in type and amounts of mineral and organic constituents, usually unconsolidated and of various depths (Joffe, 1949); soil is a unique creation that differs from the parent material below in morphology, physical properties, and biologic characteristics; and the soil mantle of the earth may be termed the 'pedosphere' alongside the atmosphere, the lithosphere, and the hydrosphere, as shown in Figure 1.1.

The pedologic soil is a dynamic system subject to temperature, moisture, and biologic cycles and developing in a certain genetic direction under the influence of climate. The rate of this development is influenced by parent material, vegetation, and human activity (Thom and Smith, 1938; Winterkorn, 1944).

3.2 SOIL AS A POLYDISPERSE SYSTEM

Soil is a polydisperse system composed of: (1) solid inorganic and organic particles; (2) an aqueous phase carrying matter in solution, and sometimes in dispersion; and (3) a gaseous phase of varying composition and is functionally related to biological activity. The aqueous

**Table 3.1 Soil Particle Ranges —
International Terms
and Definitions**

Designation	Diameter (mm)
Stone	>20
Gravel	20–2
Coarse sand	2–0.2
Fine sand	0.2–0.02
Silt	0.02–0.002
Clay	<0.002

and the gaseous phases are usually considered together as pore space or porosity. The porosity varies in amount and in dimensional characteristics from soil to soil, from layer to layer, and in the surface layers from season to season.

3.2.1 Characteristics of the Solid Phase

Soils as polydisperse systems may contain particles ranging from ionic size (10^{-8} cm) to gravel and stone size. Since the soil information of interest is found mostly in the international pedologic literature, it is indicated to use the accepted international terms and definitions in presenting granulometric composition (Table 3.1). Materials larger than 0.02 mm are usually called granular; those smaller than 0.02 mm are called silty-clay materials. Soil containing more than 65% of coarse material are called granular soils; those containing more than 35% of silt and clay are called silty-clay materials. In a compacted state, granular soils contain a granular skeleton giving them good volume stability and friction properties, while the silty-clay materials show volume changes with changing moisture content and possess low angles of internal friction.

Soils are given textural names in accordance with the following identification chart of the Bureau of Public Roads (BPR), which was taken from U.S. Department of Agriculture (Davis and Bennett, 1947) and also represents international pedologic usage as shown in Table 3.2. Combining the silt and clay fractions as silty-clay materials, we may establish three major physical groups of mineral soils (Table 3.3).

3.2.2 Characteristics of the Liquid Phase

The chemical formula for water is H_2O, which represents a composition of 88.8% by weight of oxygen (O) and 11.2% of hydrogen (H). The H_2O molecules, however, are associated as well as disassociated as a function of temperature (Rao, 1933). If liquid phase is contaminated, then additional information on environmental chemistry must be considered. This information includes soil chemistry, the role of microorganisms in aquatic chemical phenomena, atmospheric chemistry, the geosphere and hazardous substances, etc. (Palmer et al. 1991) and some of these properties will be discussed in Section 5.3.

3.2.3 Characteristics of Soil-Air Interface

The portion of the soil porosity not filled with water represents the soil-air. Soil-air is in constant exchange with atmosphere and its composition reflects the atmosphere, except for the concentration of those components that are used up or produced by microbiological activities in the soil. Such substances are oxygen (O), which is used up, and carbon dioxide (CO_2), which is produced. Atmospheric air contains 20,939 + 0.004 vol% of oxygen and 0.031 + 0.0016 vol% of carbon dioxide. According to Ebermayer (1890) cited by Winterkorn (1955), the proportion of CO_2 in the soil atmosphere is a function of all the physical and chemical factors that condition soil activity and productivity. Typical data found for the CO_2

Table 3.2 Mechanical Composition of the Principal Soil Classes Adopted by Bureau of Public Roads (BPR)

Soil class	Limits in the proportions of the soil separates		
	Sand (%)	Silt (%)	Clay (%)
Sand	80–100	0–20	0–20
Sandy loam	50–80	0–50	0–20
Loam	30–50	30–50	0–20
Silty loam	0–50	50–100	0–20
Sandy clay loam	50–80	0–30	20–30
Clay loam	20–50	20–50	20–30
Silty clay loam	0–30	50–80	20–30
Sandy clay	55–70	0–15	30–45
Silty clay	0–15	55–70	30–45
Clay	0–55	0–55	30–100

Based on U.S. Geological Survey data.

Table 3.3 Percentage of Silt-Clay Content of Three Groups of Mineral Soils

Designation	Silt-clay content (%)
Granular soils	0–20
Cohesive-granular soils	20–35
Cohesive-nongranular soils	35–100

Table 3.4 Soil Type, Pore Space, and Volume of Solid Particles

Soil type	Pore space (%)	Volume of solid particles (%)
Moor soil	84.0	16.0
Sand soil	39.4	60.6
Loess loam	45.1	54.9
Clay soil	52.7	47.3

Data from Blanck, E., ed. (1930, 1939), *Handbuch der Bodenlehre*, v. 6, Julius Springer, Berlin.

content at atmosphere and some soils are discussed in Section 13.6. Further discussions on soil-air related to unsaturated soil-water systems are presented in Section 5.13.

The oxygen content of soil-air decreases as a function of the increase of the carbon dioxide content, since the carbon dioxide is derived from the oxygen. Table 3.4 presents the soil type, pore space, and volume of solid particles of natural soil. It must be noted that natural soils always possess air spaces even if allowed to take in all the water they can. Of course, after long-time flooding this air space may be rather small.

3.2.4 Characteristics of Soil at *In Situ* Condition

Soils are creations of climatic forces and may be considered attuned to them. These forces derive from daily and seasonal temperature variations, from fluctuations in moisture content, from the annual swell and sink of the biologic potential, and from any other periodic phenomenon that affects the surface layer of the earth. As a result of these factors, soils *in situ* are not mixtures of their components but are natural organized systems, as shown in the section on soil genesis. These systems continue to be exposed to the forces that formed them and their properties are in a continuous state of flux. As a result of this condition, soils *in situ*

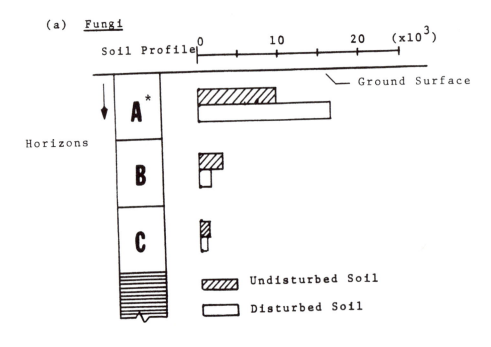

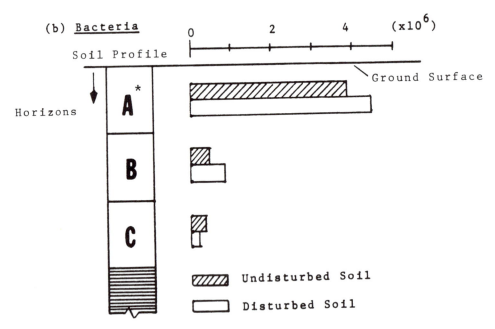

Figure 3.1 Fungi and bacteria distribution within the soil profiles and horizons. (a) Fungi; (b) bacteria. (Data from Iowa Research Bulletin 132; cited by Millar, C.E. and Turk, L.M., [1943], *Fundamentals of Soil Science*, John Wiley & Sons, New York.)

share many essential properties with 'living' systems and may almost be considered as living even if we disregard the large microflora and microfauna dispersed in soils that render soils actual living systems.

Figure 3.1 shows fungi and bacteria distributions in the soil profiles and horizons. Significant differences of fungi and bacteria contents for undisturbed and disturbed conditions within the soil profiles and horizons are indicated. These phenomena also indicate the reason undisturbed and remolded soil samples have significant structural differences.

3.3 SOIL TEXTURE, STRATA, HORIZONS, AND PROFILES

3.3.1 Soil Strata and Texture

Soils are three-dimensional systems; they have a two-dimensional areal extent and a third depth dimension. Whether they are geological depositions or formed on site by the interaction of geologic parent material, climatic factors, topography, and living organisms, soils show areal limitations and changes with depth. Horizontal as well as vertical transition into another soil type may be gradual or abrupt depending on the geologic and soil-forming factors. When vertical changes are due to differing geologic processes, the resulting layers are called strata.

A soil may be composed of only one size fraction of narrow range such as beach sand or loess or any number of size fractions in continuous or irregular grading. The size composition of a soil is called its texture. Stones or gravel retained on U.S. #4 sieve (4.76 mm) are called *coarse aggregate* (C.A.). Materials passing #4 sieve are called *fine aggregate* (F.A.). The fractions that pass the U.S. #200 sieve (0.074 mm) are called soil *fines*.

3.3.2 Horizons and Profile

When vertical changes are due to differing geologic processes, the resulting layers are called *strata*. When vertical differentiation is caused by soil-forming factors, the resulting layers are called *horizons*. The set of horizons from the soil surface to the original or physically altered parent rock is called the *profile*. The horizon containing the parent material or substrata is commonly called the *C-horizon*. The top layer which spans from the surface deposit of decaying plant litter to a depth at which the organic matter is completely humified is called the *A-horizon*. Between the A- and C-horizons lies the *B-horizon* which is usually a locus of accumulation of material in suspension or colloidal solution washed down from the A-horizon by percolating precipitation water. Both the A- and B-horizons develop at the expense of the C-horizon or parent material. If distinct differentiation has taken place in the three primary horizons, they are subdivided into subhorizons and are denoted, respectively, as A_{00}, A_0, A_1, A_2, A_3, B_1, B_2, B_3, and C_1, C_2. The theoretical soil profile showing the principal horizons is presented in Figure 3.2.

3.3.3 Podzol Soil Profiles

The degree of change from a parent material to a soil system is a function of time and of the rate of reaction of the aging processes (Section 5.8). The relative maturity of a soil is judged from the development of its characteristic profile or assembly of horizons. Considering only the gross temperature and moisture components of climate, we find the following five primary correlations between climate and soil type:

- Wet and cold
- Wet and hot
- Cold-dry and hot-dry
- Balanced climate
- Salinization, alkalization, and dealkalization

In wet and cold climates acid-decomposition products even of sparse vegetation dissolve iron and aluminum silicates of parent material in the surface mineral layer and deposit them

Plant debris on the soil usually absent on soils developed under grasses	A_{00}	Loose undecomposed plant debris
	A_0	Matted plant debris partially decomposed
A. Horizons of dominantly inorganic materials characterized by • Maximum humus accumulation • Maximum eluviation (removal of clay) • Development of granular, crumb, or platy structures	A_1	Dark colored horizons of maximum organic matter content best development of crumb or granular structures • May or may not be eluviated • May be absent from some soils altered by cultivation, etc. • Commonly thick in chernozems and prairie soils, very thin or absent from podzols
	A_2	Light colored horizon within the A group having maximum eluviation (minimum clay), minimum accumulation of organic matter, and minimum development of granular structure • Commonly has weakly developed structural units, platy and crumb are most common prominent in pozdols, planosols, and solodized-solonetz soils
	A_3	Transition horizon, more like the A than the B, sometimes absent • Commonly has coarse granular to poorly developed prismatic or block structure
B. Horizons of illuviation characterized by • Accumulation of silicate clay and oxides of iron or aluminum • Or blocky, prismatic or columnar structures • Or development of stronger red or yellow colors • Or some combination of these features	B_1	Transition horizon, more like the B than the A, sometimes absent • Commonly has weakly deveoped nutty to prismatic or blocky structure
	B_2	Horizon of • Maximum accumulation of silicate clay or oxides of iron and aluminum, or • Maximum deveopment of blocky or prismatic or columnar structure, or • Deveopment of stronger red or yellow colors, or • Some combination of these factors
	B_3	Transition horizon, more like B than C
C. Relatively unaltered, unconsolidated parent material	C_1	Slightly altered parent material
	C_2	Parent material of variable length
D. Any stratum underlying the C or the B unlike the material from which the soil has formed	D	D_r is used for consolidated rock like that from which the C has developed

Figure 3.2 Theoretical soil profile showing the principal horizons. (After HRB [1957], HRB Special Report 25, 32p.)

in lower layers. As a result, under a dark, organic layer of vegetable mold there is a leached whitish silicic layer overlying a brownish or reddish layer containing an accumulation of iron and aluminum oxides often cementing other grains together. This type of soil is named podzol and the process of its formation is called podzolization (USDA, 1938; Fanning and Fanning, 1989). The general characteristics of podzol soil profiles are shown in Table 3.5.

3.3.4 PCA Simplified Soil Profile and Horizons System

(1) General Discussion

Over the years, the system of letter destinations of the different horizons has been changed and extended several times. The designations shown in Figure 3.3 are termed Master Horizons (PCA, 1992). There are 22 further subdivisions within the Master Horizons that are termed Subordinate Distinctions. A complete description of these horizons and their subordinates is given by the U.S. Department of Agriculture (USDA, 1981 and 1986). Since the Master Horizons is too extensive to describe here, only the general characteristics of the O-, A-, B-,

Table 3.5   General Characteristics of Podzol Soil Profiles

Horizon	Thickness	Description
A_{00}	1 in. or more	Loose leaf litter
A_0	1/2 in. or more	Fermenting leaves, twigs, and wood; humified acid organic matter
A_1	1/4 in. or more	Dark-gray mixture of acid humus and mineral soil; usually very thin and entirely lacking in many places
A_2	1/2–30 in. or more	Whitish-gray or pale pinkish-gray, highly leached, acid, phylliform (finely laminated) soil of light or medium texture
B_2	2–10 in.	Dark coffee-brown silty or loamy soil containing much organic matter and iron oxides; sometimes cemented
B_3	4–10 in.	Yellowish-brown or brownish-yellow loam or clay loam with little organic matter
B_{31}		Transition and parent material usually more or less acid and sandy, but may be somewhat calcareous in places

After USDA (1938), Soils and Men, Yearbook of Agriculture, U.S. Department of Agriculture, Washington, D.C.

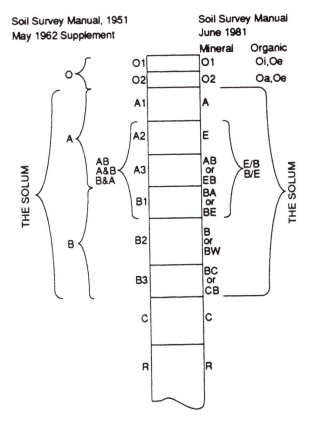

Figure 3.3 A simplified pedological soil profile showing the major horizons. (After PCA [1992], *PCA Soil Primer*, Portland Cement Association, Skokie, IL, 40p.)

C-, and R-horizons are summarized. The O-, A-, and B-horizons are layers that have been modified by weathering, while the C-horizon is unaltered by soil-forming processes. The R-horizon, below the other soil layers, is the underlying material in its original condition of formation.

(2) Horizons

1. O-horizon: O-horizon is composed primarily of organic litter, such as leaves, twigs, moss, and lichens, that has been deposited on the surface.
2. A-horizon: The original top layer of soil has the same color and texture throughout its depth. It is usually 3 to 4 m thick but may range from 5 to 75 cm. It is also referred to as the topsoil or surface soil.
3. B-horizon: B-horizon is the soil layer just below the A-horizon. It is usually 25 to 35 cm thick but may range from 10 cm to 3 m. The B-horizon is also referred to as the subsoil.
4. C-horizon: This soil layer is just below the B-horizon. It is quite different from the B-horizon. It may be of indefinite thickness and extend below any elevation of interest to the geotechnical engineer. The C-horizon may be clay, silt, sand, gravel, combinations of these soils, or stone. The C-horizon is also referred to as parent material or soil material.
5. R-horizon: The R-horizon is the layer of solid bedrock underlying the C-horizon. It is of indeterminate depth and is in its original condition of formation.

3.4 IDENTIFICATION AND CLASSIFICATION SYSTEMS OF SOILS

3.4.1 General Discussion

The classification system provides a language which communicates information in a brief manner without the necessity of a lengthy description. In the geotechnical engineering field, numerous soil identification and classification systems exist such as: Unified (ASTM D2487-93), American Association of State Highway and Transportation Officials (AASHTO M145-88), Federal Aviation Administration (FAA), and U.S. Department of Agriculture (USDA) soil classification systems. In comparison among these methods, the USDA soil classification system is relatively closely related to environmental geotechnology. Especially, the USDA system is useful for dealing with shallow foundations, landfills, hazardous waste controlling facilities, wetlands, surface and subsurface drainage systems, and erosion investigations. Therefore, the USDA method will be presented in great depth in this chapter in comparison with other methods.

3.4.2 Unified Soil Classification System (ASTM D2487-83)

The Unified Soil Classification System grew out of the soil classification developed by Casagrande in 1948. The system was significantly revised in 1983 (Howard, 1984). The Unified System divides soils into coarse-grained and fine-grained. The coarse-grained soil contains more than 50% retained on the #200 sieve, and the fine-grained soil having more than 50% passing through this sieve. For the fine-grained soil, further division includes inorganic and organic soils. Highly organic soils contain primarily organic matter; those having organic odor and dark color are classified as peat. Lengthy, detailed classification systems and charts can easily be obtained in any standard geotechnical textbooks and will not be reproduced herein.

In the area of environmental geotechnology, the fine-grained soil size is more important in comparison with granular soil, because the fine-grained soil is more sensitive to environments (Section 5.12). Therefore, the part of soil classification, based on the plasticity chart as illustrated in Figure 3.4, is needed. The plasticity chart is used for classification of fine-grained soils and fine-grained fraction of coarse-grained soils.

In examining Figure 3.4, the A-line separates the inorganic clays from the inorganic silts. The inorganic clays lie above the A-line, and the inorganic silts lie below the A-line. Organic silts plot in the same region below the A-line and with liquid limit (LL) ranging from 30 to 50 as the inorganic silts of medium compressibility. Organic clays plot in the same region as inorganic silts of high compressibility below the A-line and LL greater than 50. Equation

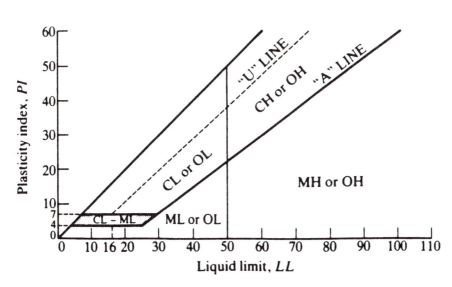

Figure 3.4 The plasticity chart for classification of fine-grained soil. Equation of "A" line: horizontal at *PI* = 4 to *LL* = 25.5, then *PI* = 0.73 (*LL* − 20). Equation of "U" line: vertical at *LL* = 16 to *PI* = 7, the *PI* = 0.9 (*LL* − 8).

of A-line: horizonal at PI = 4 to LL = 25.5, PI = 0.73(LL-20). Equation of U-line: vertical at LL = 16 to PI = 7, then PI = 0.9(LL-8). This U-line is the upper limit of the relationship of the plasticity index to the liquid limit for commonly used fine-grained soils.

3.4.3 AASHTO Soil Classification System (AASHTO M145-88)

The American Association of State Highway and Transportation Officials (AASHTO) classification system is derived from the U.S. Bureau of Public Road (BPR) system of classification. They have classified soils in accordance with their performance as subgrade beneath highway pavements. There are seven basic groups, A-1 to A-7. The members of each group have similar load-bearing values and engineering characteristics under normal traffic conditions. The best soils for road subgrades are classified as A-1, the next best A-2, etc., with the poorest soils classified as A-7. Group A-2 is subdivided into A-2-4 to A-2-7 subgroups — the last number identifying the type of #200 sieve fraction present (ASTM D422-72). Differentiation between the quality within a certain group is made by the group index (GI), which is calculated as follows:

$$GI = (F − 35) [0.2 + 0.005(LL − 40)] + 0.01(F − 15)(PI − 10) \qquad (3.1)$$

where F is the percent passing the #200 sieve (ASTM D1140-71), LL is the liquid limit, in percent (ASTM D423-66), and PI is the plasticity index, also in percent (ASTM D424-71). The group index is given in parentheses after soil groups and should be rounded to the nearest whole number. The values of the group index range from 0 to 20. The smaller the value, the better quality of the soil for highway construction use within that subgroup. The AASHTO subgrade soil classification is shown in Table 3.6.

3.4.4 Other Soil Classification Systems

The Federal Aviation Agency (FAA) soil classification is based on the soil gradation, soil consistency, soil expansive characteristics, and California bearing ratio (CBR). This system mainly is used for airfield pavement design. Classification based on tensile strength will be presented in Section 6.7. This classification is useful for selecting the stabilizing materials

Table 3.6 AASHTO Soil Classification System

General classification	Granular materials (35% or less passing No. 200)							Silt-clay materials (more than 35% passing No. 200)				
	A-1		A-3	A-2				A-4	A-5	A-6	A-7	
Group classification	A-1-a	A-1-b		A-2-4	A-2-5	A-2-6	A-2-7				A-7-5	A-7-6
Sieve analysis, percent passing												
No.10	50 max	—	—									
No. 40	30 max	50 max	51 min									
No. 200	15 max	25 max	10 max	35 max	35 max	35 max	35 max	36 min	36 min	36 min	36 min	36 min
Characteristics of fraction passing No. 40												
Liquid limit	—		—	40 max	41 min	40 max	41 min	40 max	41 min	40 max	41 min	41 min
Plasticity index	6 max		NP	10 max	10 max	11 min	11 min	10 max	10 max	11 min	11 min	11 min
Usual types of significant constituent materials	Stone fragments, gravel, and sand		Fine sand	Silty or clay gravel and sand				Silty soils		Clay soils		
General rating as subgrade	Excellent to good							Fair to poor				

[a] Plasticity index of A-7-5 subgroup is equal to or less than LL minus 30. Plasticity index of A-7-6 subgroup is greater than LL minus 30.

From AASHTO (1986, 1988).

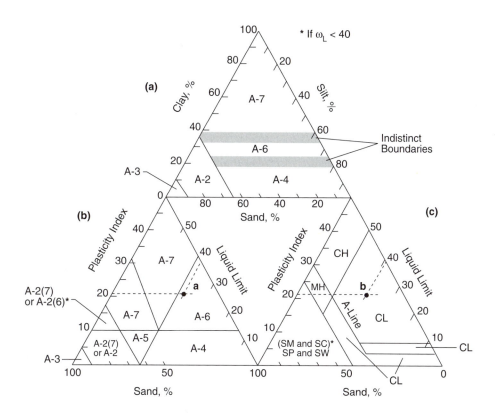

Figure 3.5 Interrelationships of (a) USDA triangle chart, (b) AASHTO soil classification, and (c) unified soil classification. (After Handy, R.L. and Fenton, T.E. [1977], TRR 642, pp. 13–19.)

for ground improvement engineering. In addition, the tensile test is also useful for estimation of degree of sensitivity of soil toward the environment (Section 5.12).

For comparison purposes, Figure 3.5 plots both the Unified and AASHTO soil classification systems on a USDA textural classification triangle chart. This chart shows the interrelationships of these three classification methods, where Figure 3.5(a) represents the USDA classification, Figure 3.5(b) represents AASHTO, and Figure 3.5(c) represents the Unified classification system. Example 3.1 is used to illustrate the procedures of these three methods.

Example 3.1

A soil has liquid limit (LL) = 38, plasticity index (PI) = 21, and 82% passing #200 sieve. Use the Unified and AASHTO classification systems to classify this soil.

Solution

(a) AASHTO classification
Using Figure 3.5(b), when liquid limit = 38 and plasticity index = 21, the AASHTO classification is A-6 (A-6: clay soils), as shown at point **a**. The group index (GI) then is computed from Equation 3.1 as follows.
(b) Group Index (GI)
Substitute LL, PI, and % passing #200 sieve data into Equation 3.1:

GI = (82 − 35) [0.2 + 0.005 (38 − 40)] + 0.01 (82 − 15) (21 − 10)

    = (47) [0.2 + 0.005(−2)] + 0.01 (67)(11)

    = 8.93 + 7.37

    = 16.3. The group index is 16.

The complete AASHTO classification of this soil is A-6[16].

(c) Unified classification

For the Unified classification, when PI = 21 and LL = 38. Using Figure 3.5(c), the soil is classified as CL as shown at point **b**. CL means the soil is a clay with low plasticity, such as inorganic clays, gravelly clays, sandy clays, silty clays, or lean clays.

The USDA soil classification system and identification and classification of contaminated soils are closely related to environmental geotechnology; therefore, further discussions of these two methods are presented in Sections 3.5 and 3.7.

3.5 USDA SOIL CLASSIFICATION SYSTEM

3.5.1 General Discussion

The U.S. Department of Agriculture (USDA) soil classification system was devised by Russian agricultural engineers in about 1870 to permit close study of soils with the same agricultural characteristics. Around 1900 this system was adopted by the U.S. Department of Agriculture. Highway engineers found that this system and the resulting valuable soil information could be used in identifying soils (HRB, 1957). However, this system can be used only as a preliminary step in soil investigation since the engineering properties of soil must be determined after identification.

3.5.2 Orders — Zonal, Intrazonal, and Azonal

In the USDA system, soils are divided into three main orders — zonal, intrazonal, and azonal — depending on the amount of profile developed.

1. Zonal soils are mature soils characterized by well-differentiated horizons and profiles. They are found where the land is well drained but not too steep.
2. Intrazonal soils are those with well-developed characteristics resulting from some influential local environmental factors. Bog soils, peats (Section 13.5), and saline-alkali soils (Section 13.11) are typical examples.
3. Azonal soils are relatively young and reflect to a minimum degree the effects of environment. They do not have profile development and structure developed from the soil-forming processes. Alluvial soils of flood plains and dry sands along large lakes are typical examples.

3.5.3 Great Soil Groups and Soil Series

(1) Great Soil Group

The USDA system is subdivided into suborders as discussed in the previous section, and then further subdivided into great soil groups on the basis of the combined effect of climate, biologic factors, and topography. Thorp and Smith (1949), based on the Marbut system (USDA, 1938), developed a soil classification in the higher categories as shown in Table 3.7. The essential features for the definition of a soil unit are (a) number, (b) color, (c) texture, (d) structure, (e) thickness, (f) chemical and mineral composition, (g) relative arrangement of the various horizons, and (h) the geology of the parent material.

Table 3.7 Soil Classifications in the Higher Categories

Order	Suborder	Great soil groups
Zonal soils	1. Soils of the cold zone	Tundra soils
	2. Light-colored soils of arid regions	Desert soils
		Red desert soils
		Sierozem
		Brown soils
		Reddish-brown soils
	3. Dark colored soils of semiarid, subhumid, and humid grasslands	Chestnut soils
		Reddish chestnut soils
		Chernozem soils
		Prairie soils
		Reddish prairie soils
	4. Soils of the forest-grassland transition	Degraded chernozem
		Noncalcic brown or Shantung brown soils
	5. Light-colored podzolized soils of the timbered regions	Podzol soils
		Gray wooded or gray podzolic soils[a]
		Brown podzolic soils
		Gray-brown podzolic soils
		Red-yellow podzolic soils[a]
	6. Lateritic soils of forested warm-temperature and tropical regions	Reddish-brown lateritic soils[a]
		Yellowish-brown lateritic soils
		Laterite soils[a]
Intrazonal soils	1. Halomorphic (saline and alkali) soils of imperfectly drained arid regions and littoral deposits	Solonchak or saline soils
		Solonetz soils
		Soloth soils
	2. Hydromorphic soils of marshes, swamps, seep areas, and flats	Humic-glei soils[a] (includes wiesenboden)
		Alpine meadow soils
		Bog soils
		Half-bog soils
		Low-humic-glei[a] soils
		Planosois
		Groundwater podzol soils
		Groundwater laterite soils
	3. Calcimorphic soils	Brown forest soils (braunerde)
		Rendzina soils
Azonal soils		Lithosols
		Regosols (includes dry sands)
		Alluvial soils

[a] New or recently modified great soil groups.

After Thorp, J. and Smith, G.D. (1949), *Soil Sci.,* v. 67, pp. 117–126.

(2) Soil Series

Soils within each great soil group are divided into soil series. A soil series comprises all soils that have the same:

1. Parent material
   (a) Solid rock (igneous, sedimentary, metamorphic)
   (b) Loose rock (gravels, sands, clays, other sediments)
2. Special features of parent material: residual or transported by wind, water, ice, or combinations
3. Topographic position: rugged to depressed
4. Natural drainage: excessive to poor
5. Profile characteristics

The different series usually have geographic names indicative of the location where they were first recognized and described (e.g., Sassafras, Putnam, Cecil, Hagerstown, etc.). For example, Hagerstown is a name of a town in the state of New Jersey where the Hagerstown soil was first recognized and described. Further discussions of soil morphology, genesis, and classification are presented by Fanning and Fanning (1989).

3.5.4 Soil Taxonomy

In 1951 when soil taxonomy was initiated, there were approximately 5500 soil series recognized in the U.S. Soil taxonomy is intended to be a logical, well-defined classification system. It is a comprehensive soil classification system and continually updated to the present. In 1965, USDA adopted what was then known as the 7th Approximation as its soil classification system. To understand more about soil taxonomy for engineering applications see Johnson and McClelland (1977), SMSS (1990), and PCA (1992).

3.6 CHEMICAL COMPOSITION OF NATURAL SOIL AND CLAY FRACTION

3.6.1 General Discussion

Considering the great variability of the chemical composition of the parent materials from which soils are formed, an equally great variability in soil composition may be expected. The most general chemical soil classification was made by Marbut (1920) by dividing soils into two classes, namely:

1. Pedalfers: Soils of humid climates (precipitation > evaporation), therefore, water percolation and elutriation resulting in a relative concentration of iron (Fe) and aluminum (Al) compounds in one of the profile or/and horizons
2. Pedocals: Soils of semiarid and arid climates (precipitation < evaporation) resulting in a $CaCO_3$ concentration within or on the surface of the soil profile

Significant differences in color exist not only between different soils, but also between different horizons of the same soil. The color may be inherited from the parent material, it may represent chemical weathering products whose differential transmission to different horizons produces their differences in color, it may be due to organic matter in various amounts and degrees of humification (Stevenson, 1986), or it may be due to surface or subsurface drainage conditions.

3.6.2 General Chemistry of the Soil

The chemistry of the soil involves many and very different chemical compounds. The most important group is composed of the compounds of the element silicon, which make up more than 75% of the various soil layers. The silicon compounds of soil are derived from the decomposition of every kind of silicate rock on the earth's surface. Table 3.8 presents the approximate chemical composition of the earth's crust and A- and B-horizons of the soil profile as reported by the U.S. Department of Agriculture (Byers et al. 1938b). Similar results are also reported from other sources (NISS, 1978). However, they indicate that larger varieties of chemical compositions exist in various soil profiles and horizons. Average compositions of the earth's crust and of the A- and B-horizons of soils, expressed in percentages by weight, moles, and ionic volumes of the main constituent elements, are presented in Table 3.9. It is indicated that the minerals in the earth's crust, and especially those in the soil layers, are essentially oxygen ion structures held together by smaller cations. Further discussion on ionic and electrical structure of soil minerals is presented in Section 4.2.

Table 3.8   The Approximate Chemical Composition of the Earth's
Crust and A- and B-Horizons

Composition	Earth's crust (%)	A-horizon (%)	B-horizon (%)
Aluminum oxide (Al_2O_3)	15.34	9.97	11.77
Iron oxide (Fe_2O_3)	6.26	3.59	5.74
Calcium oxide (CaO)	4.96	1.04	1.01
Magnesium oxide (MgO)	3.90	0.60	0.85
Potassium oxide (KO_2)	3.06	1.73	2.06
Sodium oxide (Na_2O_3)	3.44	0.82	0.83
Manganese oxide (MnO)	0.10	0.14	0.14
Titanium oxide (TiO_2)	0.78	0.90	0.90
Phosphorous oxide (PO)	0.29	0.14	0.14
Sulfur oxide (SO_2)	0.25	0.09	0.09
Silicon oxide (SiO_2)	61.28	76.46	76.46

After Byers, H.G., Anderson, M.S., and Bradfield, R. (1938b), "Soils and
Men, Yearbook of Agriculture", U.S. Department of Agriculture, pp.
911–928.

Table 3.9   Average Elementary Composition of the Earth's
Crust and of the A- and B-Horizons of Soils

Element	Earth's crust			Soil horizons (% by weight)	
	Weight	(mol%)	Ionic volume	A	B
0^a	48.18	63.40	94.00	52.78	49.73
Si	28.60	21.50	0.83	35.70	35.70
Al	7.13	5.62	0.69	5.30	6.24
Fe	4.38	1.64	0.44	2.35	4.05
Ca	3.52	1.85	1.42	0.74	0.71
Mg	2.35	2.04	0.62	0.36	0.51
K	2.54	1.37	2.07	1.43	1.71
Na	2.52	2.32	1.40	0.61	0.61
Mn	0.06	0.023	0.01	0.09	0.09
Ti	0.47	0.206	0.03	0.54	0.54
P	0.13	0.088	0.001	0.06	0.06
S	0.13	0.086	0.002	0.05	0.05

[a] Oxygen percentage does not include contribution from free water
content.

Data from USGS with modifications; cited by Winterkorn, H.F. and
Fang, H.Y. (1991), *Foundation Engineering Handbook,* 2nd ed. (1st
ed. 1975) Van Nostrand Reinhold, New York, pp. 88–143.

3.6.3 Soil Organic Matter

Soil organic matter is a mixture of many different compounds, the more important of
which are carbohydrates, proteins, fats, and resins. The absolute amounts and relative proportions of these compounds vary with the climate and macro- and microbiologic activities
in the soil. The organic matter is concentrated in the surface layers and decreases with
increasing depth in the soil. It ranges from less than 1% in inorganic solids and sands to
almost 100% of the solid matter in peat bogs.

In the semiarid chernozen soils, the organic matter may run up to 10% of the weight of the
dry soil. The average carbon content of humus is 56%. The average carbon-nitrogen ratio (C/N
ratio) is 10:1 (Lyon and Buckman, 1939). A value of 10% of organic matter, therefore, corresponds to 5.6% of carbon and 0.56% of nitrogen. Lateritic soils of the wet tropics, despite their
often brilliant red and yellow inorganic appearance, have been found to contain up to 2% and

Table 3.10   Percentage of Elements in Common Clay Minerals

Mineral element	Kaolinite	Halloysite	Montmorillonite	Beidelite	Montronite	Mica-like minerals
Si	20.6–25.3	18.9–21.7	22.6–24.2	21.4–22.3	14.7–22.5	23.7–24.5
Al	15.9–21.3	17.9–20.8	10.6–14.4	6.4–14.7	0.21–12	11.5–17.4
Fe	0.22–1.46	0–0.29	0.15–1.02	0.58–13.5	11.0–29.8	0–4.6
Mg	0–0.6	0–18	1.26–3.96	0.12–1.8	0.06–2.4	1.2–2.7

Data from Blanck, E., ed. (1939), *Handbuck der Bodenlehre*, v. 6, Julius Springer, Berlin.

Table 3.11   Relationship of Ion-Exchange Capacity, Silica-Sesquioxide Ratio, and Si/Al(Fe) Ratio of Some Natural Soils

Clay type	Ion-Exchange[a] capacity (meq/100 g)	Source[b]	Silica-sesquioxide ratio	Si/Al(Fe) ratio
Cecil	13	Alabama clay loam	1.3	0.65
Susquehanna	47	Alabama well-oxidized soil	2.3	1.15
Putnam	65	Heavy Missouri silt-loam	3.2	1.60
Wabash	78	Missouri alluvial clay	3.3	1.60
Lufkin	82	Alabama black belt soil	3.8	1.90
Montmorillonite	95	Wyoming bentonite	5.0	2.50

[a] The base exchange capacity represented the milliequivalents of cations adsorbed on the surface of 100 g of colloid and exchangeable for other cations that may be introduced into the system.
[b] Data from Winterkorn, H.F. and Baver, L.D. (1934), *Soil Sci.,* v. 38, no. 4, pp. 291–298; and Winterkorn, H.F. (1955a).

more of organic matter. The total amount of carbon in a soil is the sum of that contributed by the organic matter and that present in the profile as inorganic carbonates, especially those of Ca and Mg. Horizons enriched in Mg are typical of pedocal soils (Figure 3.3).

Clay contains various types of natural minerals. In most clays these minerals are either montmorillonite, illinite, kaolinite, halloysite, or combinations thereof. The strength of a clay system is a function of the number and strength of interparticle bonds of the minerals. Table 3.10 presents the percentage of elements in some common clay minerals as reported by Blanck (1939). The typical geotechnical properties of some basic clay minerals are presented in Table 4.7. Further discussions on the ionic and electric structures of clays will be presented in Section 4.3.

3.6.4 Silica/Sesquioxide Ratio (SSR) of Clay Fraction

As discussed in Section 3.4 and Tables 3.10 to 3.11, it is indicated that most clay minerals contain various amounts of SiO_2, Al_2O_3, and Fe_2O_3. Many of the properties of clays are related with the *silica/sesquioxide ratio* (SSR) (Mattson, 1932; Winterkorn and Baver, 1934) as:

$$SSR = SiO_2/Al_2O_3 + Fe_2O_3 \qquad (3.2)$$

The SSR defined in Equation 3.4 is a useful parameter for identification and characterization of clay minerals. It is indicated in Figure 3.6 that when SSR value increases, clay content increases as reflected by the activity and plasticity index. The higher the SSR value of clay, the more the sensitivity toward the environment. Correlation between SSR value and ion-exchange capacity for various natural soil deposits is presented in Table 3.11, and Figures 3.6 and 3.7.

Equation 3.2 also can be written as:

$$SSR = \frac{\dfrac{x}{A}}{\dfrac{y}{B} + \dfrac{z}{C}} \qquad (3.3)$$

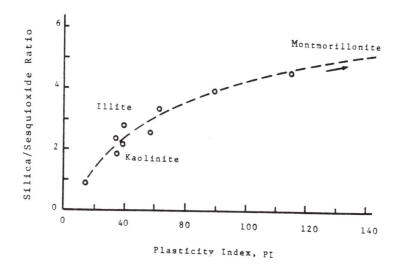

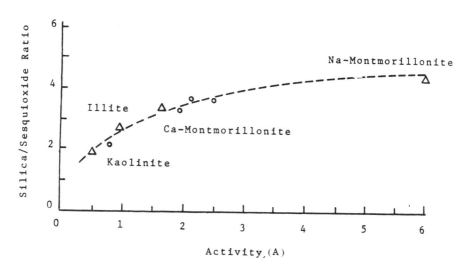

Figure 3.6 Silica/sesquioxide ratio (SSR) vs. plasticity index (PI) and activity (A) for various types of soils and clay minerals.

where x = % of SiO_2
      y = % of Al_2O_3
      z = % of Fe_2O_3
      A = molecular weight of SiO_2
      B = molecular weight of Al_2O_3
      C = molecular weight of Fe_2O_3

From Appendix 2D, the atomic weights of these elements are Si = 28.1; O = 16.0; Al = 27.0; and Fe = 55.8.

Then the molecular weight of these molecules can be calculated:

$A = SiO_2 = (28.1) + (16.0) (2) = 60.1$
$B = Al_2O_3 = (27.0) (2) + (16.0) (3) = 102.0$
$C = Fe_2O_3 = (55.8) (2) + (16.0) (3) = 159.6$

Substituting these numerical values into Equation 3.3,

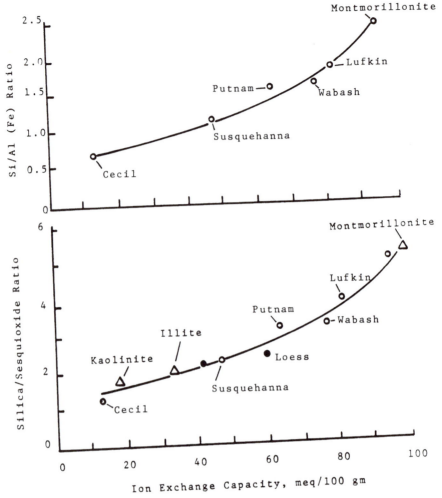

Figure 3.7 Silica/sesquioxide properties vs. ion-exchange capacity for various types of soils and clay minerals. (a) Si/Al ratio vs. ion-exchange capacity; (b) SSR vs. ion-exchange capacity.

$$SSR = \frac{\dfrac{x}{60.1}}{\dfrac{y}{102.0} + \dfrac{z}{159.6}} \tag{3.4}$$

3.7 GRANULAR SOILS (COBBLE, BOULDER, GRAVEL, SAND, AND SILT)

3.7.1 General Discussion

Types of soil minerals represented in the various size fractions with exception of the clays will be discussed in this section. The bearing of the mineralogical character of soil particles on the performance of many engineering systems indicates the desirability of a brief exposition of the types of soil minerals that are normally represented in the various size fractions. This category includes cobbles, boulders, gravels, sands, and silts. From an environmental-geotechnical viewpoint, the smaller the soil particles, the more sensitive they are to the environment; however, all these smaller particles are derived from larger particle materials. Therefore, the origin and sources of these larger particle soils are also important to the environment.

Table 3.12   Engineering Nomenclature for Size Categories

Name	Size range mineral, or U.S. sieve no.	In mm
Boulders	>12 in.	>305
Cobbles[a]	3–12 in.	76.2–305
Gravel[b]	No. 4–3 in.	4.76–76.2
Coarse	3/4–3 in.	19.1–76.2
Fine	No. 4–3/4 in.	4.76–19.1
Sand	No. 200–No. 4	0.074–4.76
Coarse	No. 10–No. 4	2.0–4.76
Medium	No. 40–No. 10	0.42–2.0
Fine	No. 200–No. 40	0.074–0.42
Soil fines or silt-clay materials[c]	<No. 200	<0.074

[a] If rounded, otherwise field stones or rock fragments.
[b] If used as components of concretes and mortars, gravels are called "coarse aggregate" and sands "fine aggregate".
[c] Previous differentiation between silt and clay on basis of particle size (clay <5 or 2 μm has been abandoned because clay or silt behavior for fine materials depends essentially on mineralogical character.

For the purpose of mutual understanding, the engineering nomenclature of the various sized categories of soil mineral particles is presented in Table 3.12. An engineering soil may consist of only one size fraction or of any number of them in every possible proportion. The mineral nature of cobbles, boulders, and gravel is essentially that of their parent rock with a film or layer of weathering products on their surface, the composition and thickness of which depend on the environmental factors and the time of exposure.

3.7.2 Aggregates

In geotechnical engineering, the term aggregates is frequently used because it covers various sizes of soil mineral particles such as crushed stone, sand, and gravels. The aggregate industry means the crushed stone, sand, and gravel industries as a whole. Furthermore, the term aggregate is divided into two subterms: fine aggregate and coarse aggregate.

1. Fine aggregate (F.A.) is material that passes through a 3/8-in. U.S. sieve, and is retained on the #200 sieve (0.074 mm).
2. Coarse aggregate (C.A.) refers to crushed stone or gravel, all of which is retained on a #4 sieve (0.187 in.).
3. Construction aggregate is defined as any combination of sand, gravel, and/or crushed stone sold to or used by the construction industry (Huhta, 1991).

Crushed stone can be composed of limestone, granite, traprock, or any other hard, sound rock that is produced by blasting and then crushing. Crushed stone is screened to a specific size for specified purposes. According to the U.S. Bureau of Mines (Tepordei, 1990) 1.2 billion tons of aggregate produced in the U.S. in 1989 is crushed stone. Of this, approximately 68% is limestone, 14.5% granite, 8.3% traprock, 3% dolomite, 2.3% sandstone and quartzite, and the remaining is made up of various rocks such as marl, shell, and slate.

3.7.3 Sand and Gravel

Sand and gravel are found in natural deposits. Most sand and gravel deposits are formed by deposition in water and found along stream channels. They may be just mechanically comminuted parent rock materials or may represent the mechanically and chemically most resistant mineral constituents of the parent rock. The extent to which the mechanical and chemical breakdown takes place depends on the environmental conditions and the length of exposure to them.

Fluvial gravels and sands become more quartzitic the longer the path of transportation. In humid climates, gravels and sands tend to be siliceous and quartzitic, but they may be any type of mineral in dry climates. The white sands of New Mexico are gypsum, coral, and shell beach sands and may consist almost exclusively of $CaCO_3$. The beach sands along the Atlantic coast change from Maine to the Keys of Florida from essentially quartzitic to predominantly calcitic, and the black sand of Yellowstone Park and of some of the blue and purple beaches of the Pacific Islands consist of obsidianite or similar glasses, as do some of the Alaskan and Aleutian beach sands. However, these sands become whiter and more quartzitic the less the distance from the low water line, i.e., the more intense the leaching process. The 'sinking' beach at Los Angeles contains more unweathered feldspars than old stable beaches of the same latitude. Dolomite sands are found in the Bavarian Jura and in other locations. Sand- and gravel-sized particles used by the engineers may also be synthetic pedogenic products such as the pisoliths in laterite soils (Section 13.9). Table 3.13 presents a percentage of sand size in sand, loamy sand, and sandy loam.

3.7.4 Silt (0.02–0.002 mm)

The silt fractions in soils may be essentially unaltered, mechanically comminuted minerals, as in glacier-ground fines, or may fall within a wide range of weathering products, depending on environmental conditions and time of exposure. They may also be products of synthesis or of chemical flocculation of more finely subdivided original weathering products. In temperate climates, silts ordinarily bear close relationship to the minerals of the parent rock such as feldspars, mica, muscovite, and quartz.

The chemical compositions of silt are given in Tables 3.14 and 3.15. Table 3.14 shows the elementary composition of major clay minerals resembles closely the composition of the silt. Table 3.15 shows comparisons of the percentage of dominant elements among the clay, silt, and sand. It is indicated that oxygen (0), aluminum (Al), and iron (Fe) contents in both sand and silt are identical, except the silicon (Si) content. However, it must be noted that for special sands (desert, coral, shell, and other beach sands) the chemical composition must be individually ascertained.

3.8 IDENTIFICATION AND CHARACTERIZATION OF CONTAMINATED SOILS

3.8.1 General Discussion

For design, construction, and maintenance of hazardous and toxic controlling facilities, or utilization of abandoned landfill site(s) used as building sites, roadways, recreation parks, etc., the condition of the site(s) must be known. The identification and characterization of ground soil at the site(s) is one of the prerequisites necessary to a more complete understanding of the soil-pollutant interaction in the environment. The following methods including visual identification, characterization, and classification of contaminated soil for engineering uses are proposed by Fang (1993).

3.8.2 Visual Identification of Contaminated Soils

During reconnaissance and field investigations at the hazardous/toxic waste site(s), certain special items must be observed such as:

Table 3.13 Percentage of Sand Sizes in Subclasses of Sand, Loamy Sand, and Sandy Loam

Basic soil class	Subclass	Soil separates				
		Very coarse sand, 2.0–1.0 mm	Coarse sand, 1.0–0.5 mm	Medium sand, 0.5–0.25 mm	Fine sand, 0.25–0.1 mm	Very fine sand, 0.1–0.05 mm
Sands	Coarse sand	25% or more		Less than 50%	Less than 50%	Less than 50%
	Sand		25% or more		Less than 50%	Less than 50%
	Fine sand				50% or more	
				— or —		
			Less than 25%			Less than 50%
	Very fine sand					50% or more
Loamy sands	Loamy coarse sand	25% or more		Less than 50%	Less than 50%	Less than 50%
	Loamy sand		25% or more		Less than 50%	Less than 50%
	Loamy fine sand				50% or more	
				— or —		
			Less than 25%			Less than 50%
	Loamy very fine sand					50% or more
Sandy loams	Coarse sandy loam	25% or more		Less than 50%	Less than 50%	Less than 50%
	Sandy loam	Less than 25%	30% or more		Less than 30%	Less than 30%
				— and —		
	Fine sandy loam				30% or more	
				— or —		
		Between 15 and 30%			More than 40%[a]	
	Very fine sandy loam	Less than 15%				30% or more
				— or —		
					30% or more	More than 40%[a]

[a] Half of fine sand and very fine sand must be very fine sand.

After USDA (1938); and PCA (1992), PCA Soil Primer, Portland Cement Association, Skokie, IL, 40p.

Table 3.14 Elementary Composition of Some Clay Minerals Resembles Closely the Composition of the Silt

Minerals	Formula	Percentage elementary composition						
		O	Si	Al	K	Na	H	F
Feldspars								
Orthoclase	$KalSi_3O_8$	41.5	33.0	10.4	15.0	0	0	0
Albite	$NaAlSi_3O_8$	44.3	35.0	11.1	0	9.5	0	0
Muscovite	$(OH,F)_2 KAl_2(Si_3AlO_{10})$	44.0	21.0	20.2	9.8	0	0.24	4.7
Quartz	SiO_2	53.2	46.8	0	0	0	0	0

Table 3.15 Average Elementary Composition of the Different Size Fractions of Soils

Dominant elements	Percentage of dominant elements		
	In clay	In silt	In sand[a]
O	50–57	41–53	41–53
Si	15–25	21–47	41–53
Al	02–21	0–20	0–20
Fe	0–30	0–5	0–5

[a] For special sands (desert, coral, shell, and other beach sands) the chemical composition must be individually ascertained.
Data from Winterkorn (1955).

1. Characteristics of site(s)
   (a) Number and condition of abandoned buildings
   (b) Conditions of roads in surrounding areas
2. Ground soil/water characteristics
   (a) Soil/water color and odors
   (b) Soil erosion features (Section 14.5)
   (c) Ground soil cracking patterns (Section 6.5)
3. River/stream conditions
   (a) Flow velocities of surface water
   (b) Color/odor of water
   (c) Water bubbles and temperature of water

3.8.3 Color of Ground Soil and Water

(1) Color of Ground Soil

Significant differences in color exist not only between different soil horizons of the same soil. The color may be inherited from the parent material or represent chemical weathering products, or it may be due to organic matter in various amounts and degrees of humidification. In many cases, once soil-water is polluted it also produces various colors. One differentiates between uniform, spotted, streaked, and mottled colors, all of which have physical or chemical significance. For color description, the Munsel notations should be used whenever possible. They take into account:

1. Hue-dominant spectral (rainbow) color
2. Value-relative lightness of color (approximately square root of total amount of light)
3. Chroma-relative purity of spectral color

Table 3.16 summarizes characteristics of soil related to its color under normal environmental conditions. However, if ground soil is heavily contaminated, then these colors will

Table 3.16 Characteristics of Soil Related to its Color

Soil colors	Soil characteristics
Black to dark gray	Most organic soils[a]
Red color	Due to the presence of nonhydrated hematite or bloodstone
Yellow and brown colors	Indicate good drainage and aeration
Gray colors	Reduced or bivalent iron compounds indicate bad drainage and anaerobic conditions
Mottled iron colors	Show restricted permeability and aeration
Black and dark brown colors	Show characteristics of organic matter, except in rare cases when they are due to manganese (Mn), titanium (Ti), and other dark-colored minerals
White colors	May be due to preponderance of silica, lime, gypsum, and relatively pure clay deposits such as kaolinite

[a] In tropical soils, the dark color may or may not be due to organic matter; when not due to organic matter, it is associated with poor drainage. Dry inorganic clays develop an earthy odor upon moistening, which is distinctive from that of decomposed organic matter.
Data from Winterkorn, H.F. and Fang, H.Y. (1991), *Foundation Engineering Handbook,* 2nd ed. (1st ed. 1975) Van Nostrand Reinhold, New York, pp. 88–143; and others.

change again. In most cases, these colors are shiny with multicolors. Further discussion on this aspect is presented in Section 6.6.

(2) Color of Water and Characteristics of Water Bubbles

Water bubbles are formed by certain polluted waters because of the differences of dielectric constants and their surface tension. Dull-like yellowish colored bubbles in slow-moving flow areas indicate where the water is polluted. Therefore, number, color, and size of water bubbles floating on the surface of water also can be used to identify the degree of contamination of surface water and also can be indirectly used to predict the degree of pollution of ground soil.

3.8.4 Odors in the Environments

(1) General Discussion

Odor of the natural gases also can assist to identify or characterize soil-water pollution. On the other hand, some highly poisonous gases exist in nature or as a by-product of manufacturing which are colorless and odorless, such as carbon monoxide (CO), methane (CH_4), and radon gas (Rn-222). Therefore, odor alone cannot be used for identification or classification of polluted air-water-ground soil systems. However, if additional information on the origin or sources of odor is available, then the odor information is valuable for identification purposes. Some natural or by-product hazardous and toxic gases are briefly presented as follows and used as additional information for identification of polluted soil-water systems.

(2) Odor of Natural or By-Product Gases

1. Ammonia (NH_3) is a colorless gas having a strong odor. It is soluble in water and forms water solutions. Ammonia is used in the manufacture of fertilizers, medicines, explosives, synthetic fibers, and plastics. It is also used in dyeing, tanning, paper-making, and electro-plating processes.
2. Carbon monoxide (CO) is a colorless, odorless, highly poisonous gas. It is often formed when incomplete combustion occurs.
3. Chlorine (Cl_2) is a highly poisonous gas. It is a green-yellow color with an irritating odor. Chlorine is used in many industrial projects such as the manufacture of paper, textiles, dyes, paints, plastics, insecticides, and medicines, as well as to kill bacteria in water to make it safe for drinking.

4. Hydrogen chloride (HCl) is a colorless gas having a sharp odor. It is mixed with water to form hydrochloric acid, and is used in the manufacture of metal chloride, glues, dyes, and various other chemicals. It is also used in electroplating processes and steel manufacturing.
5. Hydrogen sulfide (H_2S) is a colorless, highly poisonous gas having a terrible rotten egg-like odor. It is produced in large amounts by anaerobic decay of organic matter in coastal mudflats and boggy areas (Sections 13.5 and 14.1).
6. Methane (CH_4) is a colorless, odorless gas formed in the decomposition of organic matter and is found in large quantities in landfill sites.
7. Nitrogen dioxide (NO_2) is a highly poisonous red-brown gas with an irritating odor. It is formed when nitrogen oxide (NO) gas is combined with oxgen gas (O_2).
8. Sulfur dioxide (SO_2) is a colorless, poisonous gas with a choking odor. It is used in the manufacture of paper, and bleach for paper, textiles, and flour. The important use is to produce sulfur trioxide (SO_3), which is used in the manufacture of sulfuric acid (H_2SO_4).

(3) Odor of Soil and Water

Fresh, wet organic soils usually have a distinctive odor of decomposing organic matter. This odor can be made more noticeable by heating the wet sample. Odor also can be used to identify many naturally existing gases or polluted substances trapped in the soil voids. If soil is contaminated by petroleum oil it can be easily identified by its odor. Tests for odor in water have been standardized by ASTM (D1292-80) as shown in Table 3.17.

3.8.5 Cracking Patterns of Ground Soil

Cracking patterns and soil color are closely related. They both will be affected by contaminated soil-water, especially by organic acid contamination such as acetic acid ($C_2H_4O_2$), aniline ($C_6H_5NH_2$), and carbon tetrachloride (CCl_4). Significant differences of cracking patterns of ground soil between nonpolluted and polluted soil samples are observed. Therefore, ground surface cracking patterns are a useful tool for evaluation of contaminated soil. For further discussion on this aspect see Section 6.6.

3.8.6 Properties Employed in Characterization of Contaminated Soil

Almost all geotechnical engineering classification such as AASHTO, Unified, FAA, USGS, etc. are based on the soil particle size composition of the solid constituents on their interaction with the water substance. Since soil-water interaction is dominated by total amount of surface present in a sample, the ratio of surface to volume (specific surface) increases with decreasing particle size. These interaction and concomitant consistency changes are normally determined by the grain size distribution curve.

Since soil is sensitive to local environment, some additional parameters such as specific surface, pH in pore fluid, silica/sesquioxide ratio (SSR), sorption, dielectric constant, etc. must be considered. Some of these parameters are briefly discussed as follows:

1. Silica/sesquioxides ratio (Section 3.6.4)
2. Sorption (absorption + adsorption) characteristics of soil (Section 6.4).
3. Specific surface of soil — The specific surface of soil is the surface area per unit volume of solid. It is a function of the particle size. Accordingly, the smaller the soil particles, the larger is the amount of surface per unit volume at which soil-water-environment interaction can take place.
4. Dielectric constant (Section 9.7)

Table 3.17   Odor Classified by Chemical Types

Odor characteristics[a]				Odor class	Chemical types	Examples
Sweetness	Pungency	Smokiness	Rottenness			
100	50	0–50	50	Estery	Esters Ethers Lower ketones	Lacquer, solvents, most fruits, many flowers
100	50–100	0–100	50	Alcoholic	Phenols and cresols Alcohols Hydrocarbons	Creosote, tars, smokes, alcohol, liquor, rose and spicy flowers, spices, and herbs
50	50	0–50	50	Carbonyl	Aldehydes Higher ketones	Rancid fats, butter, stone fruits and nuts, violets, grasses, and vegetables
50	100	0–50	50	Acidic	Acid anhydrides Organic acids Sulfur dioxide	Vinegar, perspiration, rancid oils, resins, body odor, garbage
100	50–100	50–100	0–100	Halide	Quinones Oxides and ozone Halides Nitrogen compounds	Insecticides, weed killers, musty and moldy odors, husks, medicinal odors, earth, peat
50	50	100	100	Sulfury	Selenium compounds Arsenicals Mercaptans Sulfides	Skunks, bears, foxes, rotting fish and meat, cabbage, onion, sewage
100	50	50	100	Unsaturated	Acetylene derivatives Butadiene Isoprene	Paint thinners, varnish, kerosine, turpentine, essential oils, cucumber
100	50	0–50	100	Basic	Vinyl monomers Amines Alkaloids Ammonia	Fecal odors, manure, fish and shellfish, stale flowers such as lilac, lily, jasmine, and honeysuckle

[a] The degree of odor characteristic perceived is designated as follows:
100 indicates a high level of perception.
50 indicates a medium level of perception.
0 indicates a low level of perception.

3.9 CLASSIFICATION OF CONTAMINATED SOILS

3.9.1 Basic Concept

The basic concept of classification for contaminated soil hinges on soil particle size. The particle size of spherical and cubic shape is $3/r$ and $6/d$, respectively, in which r = radius of sphere and d = edge length of cube. Values for the amount of surface in centimeter square per unit solid volume in cubic centimeter are given in Table 3.18 for the typical soil sizes.

If clay and colloids are present in a soil even in small percentages, they may contribute the overwhelming portion of the total solid surface. The type and intensity of interaction

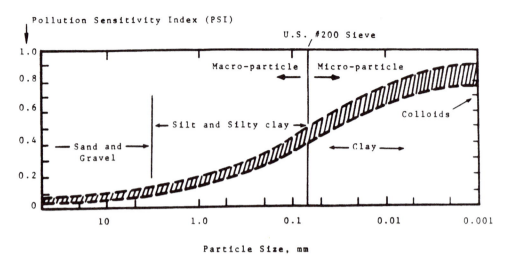

Figure 3.8 Pollution sensitivity index (PSI) relating to soil particle size. (After Fang [1987].)

between soil and pore fluid depend on the chemical composition and physicochemical properties of the particle surfaces.

3.9.2 Sensitivity of Soil to Environment

The sensitivity of soil to environment not only depends on the local environment, but also is influenced by a naturally inherited mineral structure, such as bonding characteristics between particles (Section 4.5), ion-exchange capacity, etc. The weaker the bonding energy between the particles or higher the ion-exchange capacity, the higher the sensitivity of the particles to the environment. For example, montmorillonite is potentially more sensitive to the environment than illite and kaolinite, because the montmorillonite group has larger surface areas, weaker bonding energy, and higher ion-exchange capacity.

3.9.3 Pollution Sensitivity Index (PSI)

Figure 3.8 shows the relationship between pollution sensitivity index (PSI) vs. particle size as proposed by Fang (1987). As particle size decreases, pollution sensitivity increases. Table 3.18 proposes a simple classification system for contaminated soil. This table is prepared based on soil particle sizes and surface areas.

In examining Table 3.18, it is indicated that when clay content increases, the pollution sensitivity potential increases significantly. However, the information contained therein is to be used with caution because of the possible modification or counteracting effects due to the previously mentioned naturally inherited mineral structure characteristics.

3.10 DUST: MICROSCOPIC DEBRIS

3.10.1 General Discussion

A flake floating in the air or on the water surface we generally refer to as dust; we also call it microscopic debris. Dust is one of most important elements relating to our day-to-day environment next to the air and water. They are not as simple a matter as we thought. Dust acts as a 'middleman', storing bacteria, pollens, ashes, and fine soil particles. Using this

Table 3.18   Classification of Contaminated Soil as Reflected by Pollution Sensitivity Index (PSI) and Soil Particle Surface Area

PSI	Soil type	Size (cm)	Surface area (cm^2/cm^3)	Sensitivity
0–2	Gravel	0.2–8.0	15–0.125	Very low
2–4	Sand	0.005–0.2	600–15	Low
4–6	Silt	0.0005–0.005	6,000–600	Medium
6–8	Clay	0.0001–0.0005	30,000–6,000	High
8–10	Colloids	<0.0001	>30,000	Very high

After Fang, H.Y. (1987, 1993).

middleman as a vehicle, it then transports these unwanted pollutant substances from one place to another.

The first scientific approach for analyzing dust was by French microscopist Edmond Locard in 1930. Then Rhodes (1931) analyzed dust as a clue for criminal acts. Later, dust used by archaeologists (Deetz, 1967, Shackley, 1981); when fragments in the dust are analyzed they provide valuable clues to the existence and behavior of persons that have gone before them. The identification and characterization of dust particles, whether relating to a forensic investigation or to the discovery of a prehistoric culture, have many similarities, not the least of which is the use of microscopy. In fact, the identification of dust particles through microscopy is not restricted to criminal environment, as pointed out by Bisbing (1989). Dust is a bridge linking various pollutants into the environment, and further discussion will be presented in Section 10.7.

3.10.2 Characteristics of Dust

Dust has distinguishing characteristics: it is produced from disintegration of various materials and the composition of a dust particle is quite variable and complicated. Some take the form of true liquid aerosols and appear as mists or fogs. Some take the form of solid smokes, dusts, and fly ash. On the other hand, particulates may be a conglomeration of solids and liquids. Particles are found to contain a variety of substances such as Al^{3+}, Ca^{2+}, Fe^{2+}, Pb^{2+}, Mg^{2+}, and Na^+ in various states of combination with ions such as nitrate ion (NO_3^-), sulfate ion (S^{2-}) and chloride ion (Cl^-). Some are found to contain organic compounds, such as sulfuric acid (H_2SO_4) and nitric acid (HNO_3). Most of the lead (Pb) in particulate pollution comes from lead alkyl fuel additives (Dickson, 1974). Some particulates absorb gaseous pollutions from polluted air such as radon gas (Rn-222).

3.10.3 Dust in the Environment

A solid particle that blows into air is called dust and the solid that falls into water is called suspension. Figure 3.9 presents the comparison of particle size of dust-suspension and other clay minerals and elements commonly used in geotechnical engineering. Table 3.19 presents the classification of dust based on state of matter and environment.

3.11 SUSPENSION: DUST IN THE WATER

3.11.1 Characteristics of Suspension

When dust blows into water, it is referred to as a suspension, as stated in the previous section. Suspension is a loose term; it is defined as a substance dispersed in another substance

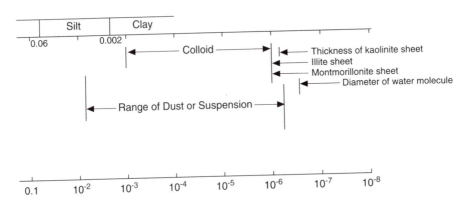

Figure 3.9 Comparison of particle size of dust/suspension and other clay minerals and elements.

Table 3.19 Classification of Dust Based on State of Matter and Environment

State of matter	Outdoor dust	Indoor dust
Liquid type	Mist, fog, chemical fumes	Vapor from cooking, showers
Solid type	Pollen	Human or animal hairs
	Solid smokes	Ashes from cigar/cigarette
	Very fine silt (such as loess)	Ash and soot from fireplace and wood stoves
	Fly ash	Lint (fabric fibers)
	Construction debris	Dander (dry skin flakes)
		Pollen (hay fever)
		Cosmetic powder
		Flaking paint
		Dry rot
		Spider webs
		Dead insects
Solid/liquid composites	Conglomeration of solids and liquids such as smog	Damp dust in nonheated basement corner
		Dust around the bathroom or in the kitchen
		Dust in a high humidity room
		Air freshener

in a form in which the dispersed particles are larger than colloidal particles (>1000 Å). This dispersion is called a suspension. A suspension system consists of two phases, one of which is dispersed in the other. The dispersed phase is a discontinous phase, and the dispersion medium is a continuous phase. The usual range of size of the dispersed particles is between 1 and 1000 μm. If the particles are larger than 1000 μm, the system is a coarse suspension. If the particles are smaller than 1 μm, the system is a true solution. The interaction among suspension, solution, and colloid state will be further discussed in Section 4.9.

3.11.2 Suspension in the Environment

Suspension particles consist of organic and inorganic particles. Some pollutants are decomposed by chemical and biological processes occurring in water. These are called degradable or biodegradable pollutants. Degradation refers to breaking down into simpler substances. Most organic pollutants are degradable. However, some organics such as certain pesticides and detergents are nondegradable, or they decay very slowly in water. These types of suspension are called hard or refractory organics. Hard pesticides include chlorinated pesticides such as DDT, chlordane, and endrin. The classification of dust-suspension based on origin, particle size, and degree of decomposition is presented in Table 3.20.

Table 3.20 Classification of Dust-Suspension Based on Origin, Particle Size, and Degree of Decomposition

Based on Origin	
Organic	**Inorganic**
Oil, food, chemicals	Tree leaves, straws, fine silt

Based on Particle Size

Larger than 1000 µm:	coarse suspension
Less than 1 µm:	solution
Less than 1×10^{-3} µm:	colloid

Based on Degree of Decomposition

Degradable material:	leaves, roots, food, and other biological matter
Nondegradable material:	DDT, Chloridance, Endrin, and fine silt

The mechanism of dust-suspension interacts with the environment through sorption reaction (Section 6.4). The sorption process covers both absorption and adsorption. The absorption is a mechanical process and the adsorption has both physical and physicochemical processes, but the adsorption is a bridge linking dust and the environment. For example, dust-suspension relating to radioactive and toxic radon gas in the air and in the water will be discussed in Section 10.5.

3.12 SUMMARY

1. Definition of natural soil is generalized into three basic categories such as engineering, geology and pedology.
2. Soil is a polydisperse system composed of solid, liquid, and gaseous phases. Characteristics of each phase are discussed.
3. Mechanical composition of the principal soil classes adopted by Federal Highway Administration and U.S. Geological Survey are tabulated.
4. Soil texture, strata, horizons, and profiles are discussed. PCA-simplified soil profile and horizon systems are introduced as shown in Figure 3.3.
5. The USDA soil classification system is discussed in detail, because this system is relatively closely related to environmental geotechnology.
6. Chemical composition of natural soil includes organic matter, sand, silt and clay fraction and is discussed and summarized.
7. The term silica/sesquioxide ratio (SSR) relating to natural soil properties is discussed, and this ratio correlating to standard soil parameters such as plasticity index and activity is presented.
8. Visual identification procedures of contaminated soils for engineering purposes are proposed. The procedures include characteristics of site(s), ground soil and water. These characteristics include soil-water color, odor, and ground soil cracking patterns.
9. A simple classification system for contaminated soil based on pollution sensitivity index (PSI) and soil particle sizes is proposed. This system may be used to predict the pollution potential of soil(s) for preliminary investigations for design, construction, and maintenance of hazardous- and toxic-controlling facilities or other related applications.

PROBLEMS

3.1 Distinguish clearly between soil identification and soil classification systems. To which of these does the Unified Soil System belong and why?

3.2 Without the aid of laboratory facilities, how would you:
(a) Identify an organic soil.
(b) Identify an inorganic silt.
(c) Distinguish between a silty clay and a highly plastic clay.
(d) Distinguish between a very fine sand and an inorganic silt.
(e) Determine whether a damp fine sand was clean or dirty.

3.3 Soil is a polydisperse system composed of solid, liquid, and gaseous phases. Discuss briefly the characteristics of each phase. Also discuss briefly the general characteristics of podzol soil profiles.

3.4 "It has been recognized that the Agricultural Soil Survey Map is more than a productivity classification map and it cannot be depended upon as a reliable map of soil conditions." Please comment on this statement.

3.5 A profile being examined in central Pennsylvania has the following characteristics: very little organic debris on the surface, an A-horizon consisting of yellow clay and a B-horizon of an iron oxide layer underlain by bauxite. The profile has developed to a depth of 15 ft.
(a) Is this profile indigenous to the area?
(b) What order is the profile in? and what suborder?
(c) What great soil group?
(d) What explanation can you give for the profile developing in this area?

3.6 Specifications for a fill material have been set up. Essentially the material specification consists of a maximum and minimum grain size requirement and a plasticity index (PI) requirement. Would you recommend material specifications of this type? Why? Be as explicit as possible.

3.7 Define silica/sesquioxide ratio (SSR) and discuss the significance of this ratio with respect to clay mineral identifications. Determine the SSR values for kaolinite and illite soils.

3.8 A soil has a liquid limit (LL) of 55 and a plastic limit (PL) of 25. The moisture content of the soil as it is excavated for use in a fill is 30%:
(a) Is the soil likely to be stiff or soft when compacted at its existing moisture content in a fill?
(b) What would a light rain do to the consistency of this soil?

3.9 Why are pollution sensitivity of clays related to the clay particle sizes? Why is montmorillonite clay more sensitive to environment than illite or kaolinite?

3.10 Discuss what types of parameters are needed and why for characterization of contaminated soil. Why are color and odor of soil-water related to the characteristics of the contaminated soil?

CHAPTER **4**

Soil Technology

4.1 INTRODUCTION

To understand the environmental aspects of soil-water systems or interaction within the contaminated soil and water, a knowledge of soil technology as understood by the geologists and pedologists is required. There are three basic soil particle systems, namely micro-soil, macro-soil, and macro-micro-composite soil particle systems as discussed in Chapter 3. Since smaller particles are more sensitive to environment, emphasis is placed on the micro-soil particle system in this chapter, which includes clay mineralogy, structural units of soils and their interactions, the colloidal state, clay-water-electrolyte systems, hydrophobic and hydrophilic soils, as well as homoionic and man-made soils.

Also, to understand the interrelationship between the test results and the fundamental clay behavior on a microscopic scale, a knowledge of chemistry, physicochemistry, and biology is needed. This working knowledge of chemistry is essential to the actual understanding of a given test result, which quantitatively measures the effect of a certain pore fluid on a certain clay or a certain clay property. It is essential to the engineering utilization of naturally occurring materials for the containment of hazardous and toxic wastes. Without this knowledge of clay behavior in response to hazardous wastes, any engineering systems design, such as remedial action programs, can have no sound basis on which to project the long-term behavior of that system.

4.2 CLAY MINERALOGY

4.2.1 General Discussion

The concepts of clay mineralogy are necessary to study the behavior of the soil and soil-water system; work began in the field about 1887 by Le Chatelier and Lowenstein in 1909. These concepts were well established by Hadding of Sweden in 1923 and Rinne of Germany in 1924. U.S. Geological Survey began studies sometime in 1924 by Ross and Shannon. Engineering applications of clay mineralogy concepts relating to foundation engineering were given by Casagrande (1932) and to the soil stabilization and highway construction materials by Winterkorn and Baver (l934) and Winterkorn (1937). Further applications in various geotechnical problems are presented by Grim (1953) and Lambe (1953). Recent developments on physicochemical properties of soil minerals are given by Yong and Warkentin (1966), Ingles (1968), Winterkorn and Fang (1975 and 1991), Mitchell (1976 and 1993), Bennett et al. (1991), and Velde (1992).

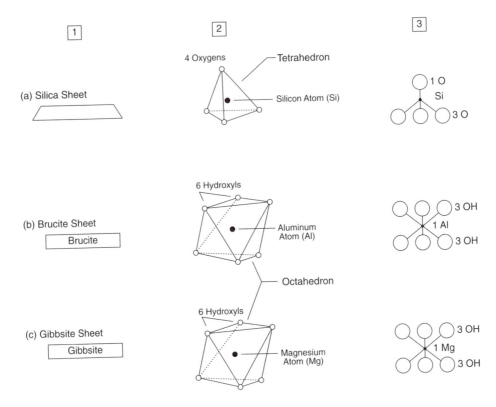

Figure 4.1 Structural units of clay minerals. (a) Silica sheet; (b) brucite sheet; (c) gibbsite sheet.

4.2.2 Mineralogical Data Requirements

Soil contains various natural minerals. The mineralogical composition determines particle size distribution and many physicochemical properties of clay systems. The following major factors are needed for soil engineering investigations:

- Structures of layer silicates
- Particle size distribution
- Interparticle forces
  - Attractive forces
  - Repulsive forces
  - Interaction energy
- Water retention and hydraulic conductivity (Chapter 7)
- Heat capacity and heat conductivity (Chapter 8)

4.3 STRUCTURAL UNITS OF SOILS

4.3.1 General Discussion

Structural units of soils are composed primarily of silicon oxide (SiO_2), and aluminum oxide (Al_2O_3) or iron oxide (Fe_2O_3) with varying amounts of other elements. The silicate layer minerals are built up of two fundamental structural units, the *silica* sheets and the *brucite* (B) or *gibbsite* (G) sheets as shown in Figure 4.1. They are essentially oxygen structures that are glued together by Si, Al, Mg, and also other cations.

In examining Figure 4.1, three types of symbols which represent silica, brucite, and gibbsite sheets are listed. These symbols frequently appear in various literatures. In general,

Table 4.1 Structural Classification of Layer Silicates

Group	Interlayer constituents	Minerals	
		Dioctahedral	**Trioctahedral**
2-layer 1:1 7Å	None	Kaolinite group Kaolinite Dickite Nacrite	Serpentine group Chrysotile Lizardite Antigorite Chamosite etc.
	Variable H_2O	Halloysite	
3-layer 2:1 10Å	None Na, Ca, Mg, H_2O (expandable 2:1 minerals) Cations + H_2O mixtures: mica + expandables K, Na, Ca (micas)	Pyrophyllite Montmorillonite group Mixed-layer minerals Muscovite Illite Glauconite Margarite	Talc Montmorillonite group Vermiculite Lepidolite Phlogopite Biotite
4-layer 14Å	None		Chlorites
3-layer plus 4-layer	Mg + H_2O		Corrensite "Swelling Chlorite"

After Warshaw, C.M. (1962), *McGraw-Hill Yearbook of Science and Technology, Annual Review of Scientific Developments,* Mc-Graw-Hill, New York, p. 462.

type 1 is commonly used in geotechnical engineering; types 2 and 3 are used in geological and agricultural sciences.

4.3.2 Clay Mineral Structures

The clay minerals are formed by various combinations of the *tetrahedral* and *octahedral* sheets as illustrated in Figure 4.1. If the mineral consists of one octahedral sheet and one tetrahedral sheet, it is termed a *one-to-one* (1:1) mineral. In like manner, clay minerals which consist of a combination of two tetrahedral sheets and one octahedral sheet to form a single layer are known as *two-to-one* (2:1) minerals. The third major mineral type is known as a *two-to-one-to-one* (2:1:1) mineral and consists of alternately a tetrahedral sheet and an octahedral sheet, a tetrahedral sheet and an octahedral sheet. Hence, the three main mineral types of 1:1, 2:1, and 2:1:1 are then subdivided into groups or subgroups depending upon the octahedral sheet configuration. The structural classification of layer silicates proposed by Warshaw (1962) is presented in Table 4.1.

The structural arrangement of the oxygen ions (O^{2-}) is the predominant feature into which electric neutralization must fit as well as possible. If cationic glue does not provide enough positive charges within the mineral structure, then the excess negative charges of the oxygen structures must be satisfied on the mineral surfaces by adsorption of positive ions including H from the environment. These ions are exchangeable for others. This property is called the *base exchange capacity* or *ion-exchange capacity* of the clay mineral.

4.3.3 Identification and Determination of Clay Minerals

Basic building blocks, the silica tetrahedral sheet and octahedral sheet of clay minerals can be represented schematically as shown in Figures 4.1 and 4.14. Various schematical diagrams frequently found in various literatures representing tetrahedral and octahedral units are also presented for comparison. Utilizing the symbols for the basic building blocks, symbolically, a review of the structure of the primary mineral groups can be made. For identification of mineral types, various laboratory techniques such as ultrapac techniques, scanning electron, and others can be used. Also, interpreting the physical properties and microstructure of clay particles can be done by microanalysis techniques. A brief presentation of each method commonly used is presented as follows.

There are numerous techniques for determination of composition of clay minerals, including X-ray diffraction analysis, differential thermal analysis, optical microscope, and electron microscope methods. Among these four methods, the X-ray diffraction analysis and differential thermal analysis are the most common methods used in geotechnology.

(1) X-Ray Diffraction (XRD) Analysis

XRD analysis is used to study the form and structure of crystal aggregates. Average interplanar distance and the average composition of the atomic planes can be derived from XRD studies. A set of typical XRD analysis of four common clay minerals — chlorite, illite, kaolinite, and montmorillonite is shown in Figure 4.2.

(2) Differential Thermal Analysis (DTA)

DTA study is concerned with the water molecules in clay. It consists of simultaneously heating a test sample and a thermally inert substance at constant rate about 10°C/min to over 1000°C and continuously measuring differences in temperature between the sample and the inert material. Differences in temperature between the sample and inert substance reflect reactions in the sample brought about by the heating. Typical test results of DTA are presented as the difference in temperature between sample and inert substance vs. temperature. Analysis of test results consists of comparing the known materials. Detailed discussions are provided by Tan et al. (1986).

(3) Scanning Electron Microscope (SEM)

The SEM has a ×20 to ×150,000 magnification range and a depth of field some 3000 times greater than of the light microscope. The clay particles themselves and cracking surfaces through soil masses may be viewed directly, and it is an effective method for measuring clay particles. The morphology of aggregates is also determined by SEM. Principles, measuring techniques, and typical examples of SEM are given by Sudo et al. (1981).

There are numerous methods available commercially which include thermogravimetric analysis (TGA) and transmission electron microscopes (TEM). TGA study is also concerned with the water molecules in clays. The TEM investigates the shapes of individual crystals and the compositions of these individual crystals.

4.4 SIZE AND SHAPE MEASUREMENT OF SOIL

4.4.1 Effective Size and Uniformity Coefficient

In Section 2.5, the basic characteristics of granular particles and their measurements have been discussed. The differentiation between boulders, cobbles, gravels, and sand also have been discussed and classified in Section 3.4. Further discussion on determination of size and shape by effective size (D_{10}) and uniformity coefficient (C_u) and surface area of soil particles per unit volume for both macro- and micro soil particles will be presented as follows.

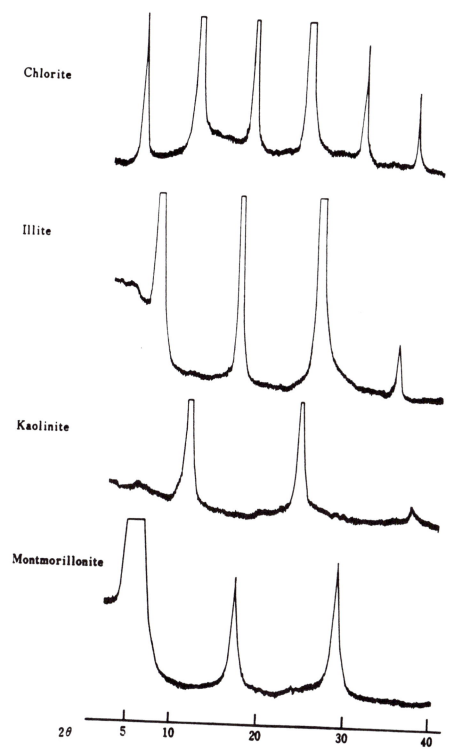

Figure 4.2 Typical X-ray diffraction analysis of some clay minerals. (After Shibuya, T. [1973], Geological Study of Landslide Clay, KICT Report no. 10, Kajima Institute of Construction Technology, Tokyo, 37p.)

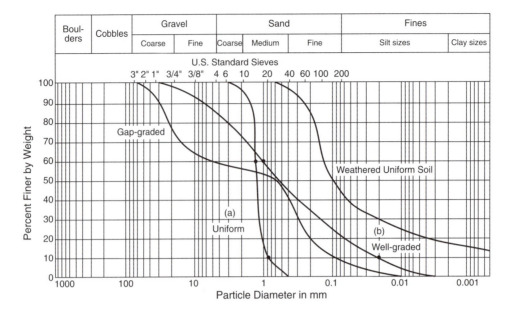

Figure 4.3 ASTM-ASCE grain size distribution chart.

The terms of effective size and uniformity coefficient commonly used in geotechnology are useful for evaluation of soil particle size and shape relating to environment for both macro- and micro-soil particles. The term of effective size, D_{10}, as determined from sieve analysis is the size of the screen opening that permits 10% of the granular material to pass and retains 90% as indicated in Figure 4.3.

The uniformity coefficient (C_u) is the ratio of the size of screen openings passing 60% to that passing 10% as shown in Figure 4.3, as: $C_u = D_{60}/D_{10}$.

Example 4.1

Determine the effective size, D_{10}, and the uniformity coefficient, C_u, for the uniformly graded soil (Curve **a**) and well-graded soil (Curve **b**) as indicated in Figure 4.3.

Solution

From uniform-graded soil of Curve **a** in Figure 4.3, the $D_{10} = 0.8$ mm, and $D_{60} = 1.5$ mm, thus

$$C_u = 1.5/0.8 = 1.875$$

From the well-graded soil of Curve **b** in Figure 4.3, the $D_{10} = 0.025$ mm, and $D_{60} = 1.0$ mm, thus

$$C_u = 1.0/0.025 = 40.0$$

From the above data in Example 4.1, it is shown that for uniform-graded soil, the uniformity coefficient is 1.875, and the well-graded soil is 40.0. Therefore, the smaller the ratio, the more uniform the material. The lowest theoretical value for this ratio is 1.0. For beach sand, C_u values ranging from 1.2 to 3.4, and the D_{10} values varying from 0.15 to 1.43 mm are reported (Section 14.4). C_u values for gap-graded or weathered soils can reach as high as 400.

Table 4.2 **Summary of Size and Shape Measurement Methods**

Direct methods
Slide caliper (large pieces)
Light microscope
Scanning electron microscope
X-ray diffraction (Figure 4.2)
Sieving and screening
Particle size ranging from 10 to 4×10^{-3} cm
Filtration (as a form of sieving)
Such as: filter papers, nitrocellulous films, molecular filters
Sedimentation and elutriation methods
Standardized sedimentation methods and elutriation methods
Computing from surface area (indirect methods)

4.4.2 Size and Shape Measurement for Granular Soils

Surface area of soil particle is an important parameter relating to environment. For granular soil, the surface area per unit volume is small in comparison with fine-grained soil. Size of granular materials can be determined by their volumetric relationships or by direct measurement. The volumetric relationships in granular systems can be expressed by phase volumes, absolute volumes, void ratio, or porosity. The phase volume is the portion of the total volume contributed by the various solid, liquid, and gaseous components of the system, expressed as a fraction or percent of the bulk volume.

The absolute volume is the actual volume occupied by the various phases, usually expressed in volumetric units employed in engineering such as cubic feet, cubic yard, or cubic meter. The sum of the absolute phase volumes must give the total or bulk volume. Size and shape measuring techniques for both macro- and micro-soil particles are summarized in Table 4.2.

4.4.3 Surface Area for Fine-Grained Soil

Soil surface area for the fine-grained soil is much more important than granular soil for studying soil-water interaction in the environment. Because the surface area-mass ratio of most clay particles is high for the electric forces at the particle surfaces and influences the engineering behavior of these particles and their aggregations, numerous methods have been proposed for determination of surface area for fine-grained soils. Most of the methods require elaborate and delicate equipment and sophisticated techniques to measure the surface area. A simple procedure with reasonable accuracy called the desiccator method is proposed by Sridharan and Rao (1972) as follows:

$$S = \frac{\omega}{M} \frac{N}{10^4} A. \, 10^{-16} \tag{4.1}$$

where S = specific surface area, m^2/g,
     ω = equilibrium moisture content in g water adsorbed per g of soil;
     N = Avogadro's number (6.025×10^{23});
     M = molecular weight of water (18.016 g); and
     A = area in cm^2 Angstroms per water molecule (10.8 Å).

Hence:

$$S = 3612 \, \omega \,\, (m^2/g) \tag{4.2}$$

4.5 PARTICLE BONDS, BOND ENERGIES, AND LINKAGES

4.5.1 Particle Bonds

There are two major bonds existing within the atoms and molecules comprising clay particles: the primary bond and secondary bond. The primary bond, also called interatomic bond, is the bond between atoms forming molecules. These bonds are strong and not easily broken during conventional engineering works.

The secondary bond is those atoms in one molecule bonding to atoms in another molecule. There are two types of secondary bonds, the van der Waals forces and hydrogen bonds.

1. The van der Waals force is weaker than the hydrogen bonds. This force is generally attributed to electrical or electromagnetic attraction between systems of molecules.
2. The hydrogen bonds are a specific type of dipole force. If hydrogen is the positive end of a permanent dipole, then the resultant attraction of a negatively charged dipole is termed a hydrogen bond. Hydrogen bonds are stronger than ordinary dipole forces, as they are formed with atoms which are strongly electromagnetive to approach the hydrogen atom more closely. Hydrogen bonds significantly affect the physicochemical properties of clay minerals as will be subsequently discussed.

If the particles composing a system fall within the same size or mass, the strength of the bond will primarily decide the solid or liquid character of the system, as discussed in Section 2.4. Energy ranges for different types of bonds are shown in Tables 2.3 and 2.4. For most inorganic compounds, bond strength values which are sufficiently accurate for engineering purposes can be calculated under the assumption of simple ionic or dipole attraction, and in most soil minerals as well as in water, the bulk of the solid atomic volume is contributed to by the oxygen ions.

4.5.2 Bond Energies

For the sake of convenience, bond energies are normally expressed in kilocalories per mole of bonds. By division through the Avogadro number (6.02×10^{23}) one obtains the energy per single bond which can be converted into ergs or other appropriate energy units. Finally, by dividing through the length of bond or an appropriate multiple thereof, one can obtain the bond force which divided by the pertinent molecular or ionic cross section gives the bond strength in force per unit cross section. The type of attraction relating to the energy of interaction has been discussed in Section 2.3 and tabulation of stability conditions for various interacting pairs that are of interest for our theme are summarized in Table 2.3. The bond energy ranges are also summarized in Table 2.4.

To illustrate the computation procedure, the bond strength of water as reflected by tensile strength is used as shown in Example 4.2. Figure 4.4 presents the relationship between bond energy and degree of contamination. The degree of contamination of soils increases as bond energy between particles decreases. The bond energy data used in Figure 4.4 are obtained from Pauling (1960) and Table 2.4.

Example 4.2

Calculating tensile strength of water

Solution (After Winterkorn, 1974)

Method [A] Calculating Tensile Strength of Water — A Rough Approximation

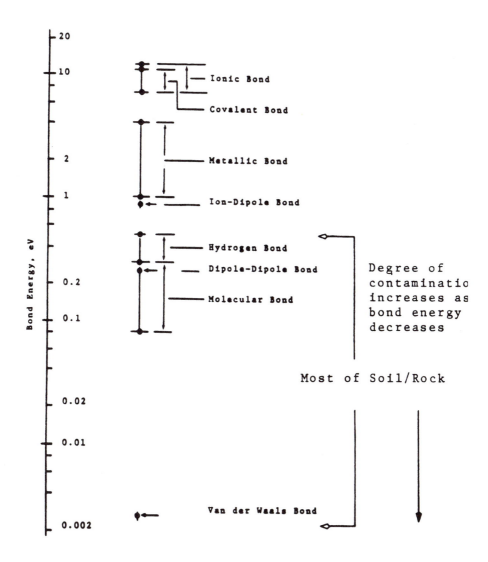

$$* \quad 1 \; eV \; = \; 1.602 \; x \; 10^{-12} \; erg \; = \; 0.386 \; x \; 10^{-19} \; cal$$

Figure 4.4 Energy ranges for different types of bonds relating to the degree of contamination of soil.

[1] Determination of Size of Water Molecule
    Assume: 1 mol of H_2O (18 cm³) is subdivided into 6.02×10^{23} cubes of equal size each containing 1 molecule H_2O. Then:
    Volume of each cube $= 18/6.02 \times 10^{23} = 29.9 \times 10^{-24}$ cm³
    Length of cube side $\quad = 3\sqrt{29.9 \times 10^{-24}} = 3.1 \times 10^{-8}$
    Area of cube side $\quad\quad = 9.6 \times 10^{-6}$ cm²

[2] Determination of Bonding Energies
    Assume: Total bonding energy = heat of evaporation (H_v). Then:
    H_v at normal boiling point (100°C)
    $\quad\quad\quad\quad = 539$ cal/g $= 9.7$ kcal/mol
    Heat required to bring 18 g of water from 20° to 100°C
    $\quad\quad\quad\quad = 80 \times 18 = 1,440$ cal
    $\quad\quad\quad\quad = 1.44$ kcal
    Total heat required $= 9.7 + 1.44 = 11.14$ kcal

Work against atmosphere for 1 mol H_2O gas:

W = PV + RT, where R = gas constant

$\qquad\qquad\qquad\qquad$ = 1.9 cal (T = 373 K)

W = 710 cal $\qquad\qquad$ = 0.71 kcal

[3] Actual Energy of Molecular Separation

$\quad$ E = 11.14 - 0.71 = 10.43 kcal/mol

$\qquad$ = $10.43 \times 4.185 \times 10^{10}$ = 44.8×10^{10} erg/mol

$\qquad$ = $44.8 \times 10^{10}/6.02 \times 10^{23}$ = 7.44×10^{-3} erg/cube

[4] Bonding Energy Distribution

$\quad$ Assume: Bonding energy is uniformly distributed over six sides of cube:

$\quad$ 7.44/6 = 1.24×10^{-13} erg/cube face

$\quad$ Cube face = 9.6×10^{-6} cm^2

$\quad$ Hence: 1.24×10^{-13}: 9.6×10^{-16} = 1.3×10^2 erg/cm^2

[5] Tensile Strength of Water

$\quad$ If for complete separation cubes have to be moved by one side length, i.e., 3.1×10^{-8}, cm, then 1.3×10^2 : 3.1×10^{-8}

$\qquad\qquad\qquad\qquad$ = 42×10^8 dyn/cm^2,

$\quad$ Since 981 dyn $\qquad$ = 1 g force

$\quad$ 9.81×10^5 $\qquad\quad$ = 1 kg

$\quad$ 42×10^8: 9.81×10^5 = 4270 kg/cm^2

$\qquad\qquad\qquad\qquad$ = 60,600 psi

Method [B] Computed from Surface Tension

[1] Approach

$\quad$ Assume: We create 2 surfaces each of 1-cm^2 size by pulling apart a water column of 1 cm^2 cross section. Then:

$\quad$ The surfaces represent an energy of 2×72.8 = 145.6 ergs.

[2] Tensile Strength of Water

$\quad$ Assume: Further that we had to separate the new surfaces by two bond lengths during which the tensile stress (TS) decreased uniformly from TS to 0 with mean value of TS/2. Then:

$\quad$ 145.6 erg = (TS/2) $(2 \times 3.2 \times 10^{-8}$ cm)

$\quad$ TS $\qquad$ = $145.6/3.1 \times 10^{-8}$ dyn/cm^2

$\qquad\qquad$ = 47×10^9 dyn/cm^2

$\qquad\qquad$ = 4.78×10^3 kg/cm^2

In addition to Example 4.2, there are other various approaches for computing the tensile strength of water. Such approaches include modulus of volume elasticity method, and a centrifuge approach is proposed.

4.5.3 Linkage Between Particles

The linkage between particles mainly are adsorbed water, water-dipole, or dipole-cation-dipole as illustrated in Figure 2.5. These linkages are much weaker than bonding strengths as discussed in a previous·section. Many natural problematic soil deposits such as residual soils, clay shales, dispersive clays, collapsible silts, and soluble carbonate rocks belong to these types of linkage. In addition, man-made composite materials such as gravel-lime and fly ash-stabilized soil also belong to these types of linkage.

4.6 INTERPARTICLE ENERGIES OR FORCES

4.6.1 Dipole Moment and Orientation Effect

All microparticle soils carry an electric charge. When two particles are close to each other in face-to-face arrangement, an attractive force exists between the charged surfaces and the intervening exchangeable cations. If the atoms in the adjacent surface approach each other

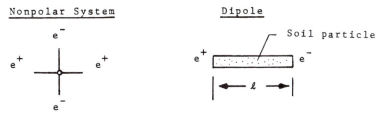

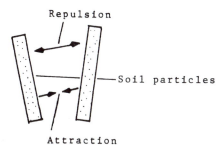

Figure 4.5 Basic characteristics of interparticle structures. (a) Nonpolar system and dipole movement; (b) repulsion and attraction forces.

so closely that their outer electron shells overlap, a net repulsive force results. When the various attractive and repulsive energies are summed algebraically, the net energy of interaction is obtained. Some basic characteristics of interparticle structures of soil illustrated in Figure 4.5 are presented.

If the center of action of the positive charges coincides with the center of action of the negative charges, the system has no dipole moment and is termed a *nonpolar system*; otherwise, it is called *polar* (Figure 4.5(a)). The effect consisting of an attraction between the oppositely charged ends of the permanent dipoles with a distance (ℓ) is called *orientation effect*. This effect is highly dependent on temperature. The unit of the dipole moment is a debye. *Debye* is defined as a molecule in the unit that electrical charges separate by 0.21 Å and has a dipole moment of 1 debye unit. Typical data for various materials are listed as:

Substance	Dipole moment (debye)
Water	1.89
Acetic acid	1.74
Acetone	2.90
Aniline	1.55
Benzene	0
Carbon tetrachloride	0

4.6.2 Attractive Energy or Force

When two particles are close to each other in face-to-face arrangement, an attractive force exists between the negatively charged surfaces and the intervening exchangeable cations.

Figure 4.5(b) illustrates both attractive and repulsive forces between two clay particles. Several formulas proposed for estimating these forces are presented below.

(1) Norrish Equation (1954)

According to Norrish (1954), if the relationship of the cation to the surface is similar to that of a point charge to a thick plane conductor, the attractive energy, E_A, is given by

$$E_A = \sigma ve/2d\varepsilon \tag{4.3}$$

where E_A = attractive energy, erg cm^{-2};
  σ = surface charge density;
  v = ionic valence;
  e = electronic charge;
  d = distance between the ions and surface; and
  ε = dielectric constant.

If the interparticle cations and charged surface are regarded as the plates of a parallel-plate condenser, the E_A, becomes

$$E_A = 2\pi\sigma^2 d/\varepsilon \tag{4.4}$$

When the cations are far apart, Equation (4.3) would apply; when they are close together, Equation (4.4) would apply. Actually, the attractive energy will likely be between Equations (4.3) and (4.4).

(2) MacEwan Equation (1954)

MacEwan postulated that at surface separations less than 10 Å, the intermediate cations are arranged in a single layer which attracts the opposing surfaces electrically, presumably as indicated above. But at greater distances, when more space is available, segregation occurs. The result is that two layers of cations are formed, one being attached to each mineral surface.

$$E_A = \frac{2\sqrt{\sigma^2 ev}}{\varepsilon}\ \exp\left[-2\pi r\ \frac{\sqrt{\sigma}}{ev}\right] \tag{4.5}$$

where r = separation between cationic layers. Others symbols are the same as Equation (4.3).

(3) Jordine et al. (1962) and Jordine Equation (1963)

Jordine et al. (1962) and Jordine (1963) believe that the attractive energy arising from interaction between the charged surfaces and intervening cations is described by:

$$E_A = \frac{(\varepsilon_\omega - \varepsilon_m)\sigma ve}{\varepsilon_\omega(\varepsilon_\omega + \varepsilon_m)4d} - \frac{z\pi\sigma^2 d}{(\varepsilon_\omega + \varepsilon_m)} \tag{4.6}$$

where ε_ω = dielectric constant for water and ε_m = dielectric constant for montmorillonite.
Equation (4.6) is valid only if (**d**) is less than the thickness of the clay layer; the first term represents the interaction energy between the intervening cations and their induced electrical images in the clay; the second term represents the interaction energy between these cations and the permanent charges on the clay.

4.6.3 Repulsive Energy or Force

If the atoms in adjacent surfaces approach each other so closely that their outer electron shells overlap, repulsion occurs. This repulsion is the so-called Born repulsion. Several equations are proposed as follows.

(1) Pauling Equation (1945, 1960)

$$E_R = \frac{B e^2}{R^n}$$

(4.7)

where E_R = repulsive energy;
$\quad$ B = constant;
$\quad$ e = electronic charge;
$\quad$ R = distance between atoms in adjacent surfaces; and
$\quad$ n = 9 (but which depends on the kinds of atoms involved).

Equation (4.7) indicates that repulsive forces are caused by their outer electron shells overlapping in the so-called Born repulsion. A second type of repulsion, suggested by Van Olphen (1954), is that caused by dipole-dipole interaction. The repulsive energy is given by the following.

(2) Van Olphen Equation (1954)

$$E_R = \frac{8\mu^2}{\varepsilon}\left[h(s, \ell)\right]$$

(4.8)

where μ = dipole moment per unit area;
$\quad$ s = length of the dipole (Figure 4.5(a));
$\quad$ ℓ = the closest distance between planes of like poles; and
$\quad$ ε = dielectric constant.

ℓ is the closest distance between planes of like poles and $h(s,\ell)$ is a complicated function of s and ℓ which, for simplicity, will not be given here. Since μ and s depend on the number and location of the charge sites in the crystal lattice, the mineralogical composition determines the magnitude of the relevant repulsive energy.

4.7 PARTICLE ARRANGEMENT AND CLAY STRUCTURES

4.7.1 Particle Arrangement and Structures

Two basic types of particle arrangement are frequently used, the flocculated orientation, which is edge-to-face arrangement, and the dispersive orientation, which is face-to-face arrangement (Figure 4.6). Major factors which cause soil to tend toward flocculation include increasing the concentration of the electrolyte, valence of the ion, and temperature, or decreasing the dielectric constant of the pore fluid, size of the hydrated ion, pH value, and anion adsorption as shown in Table 4.3. Table 4.3 is based on the Gouy-Chapman theory of the electric double-layer and explanations will be presented in Section 9.8.

For the idealized clay structures, various researchers have proposed several structural systems. Barden and Sides (1971) proposed that clay structures include card-house, book-

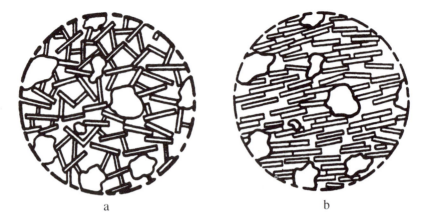

Figure 4.6 Flocculated and dispersed structures. (a) Flocculated structure; (b) dispersed structure.

Table 4.3   Environmental Factors as Reflected on Physicochemical Parameters on Soil Structure and Behavior

Parameter	Double-layer thickness		Soil structure	
	Increase	Decrease	Deflocculated	Flocculated
Electrolyte concentration	Decrease	Increase	Decrease	Increase
Ion valence	Decrease	Increase	Decrease	Increase
Dielectric constant	Increase	Decrease	Increase	Decrease
Temperature	High	Low	Increase	Decrease
Size of hydrated ion	Large	Small	Increase	Decrease
pH value	High	Low	Increase	Decrease
Anion adsorption	Increase	Decrease	Increase	Decrease

Based on Evans et al. (1983) and Evans, J.C. (1991), *Foundation Engineering Handbook*, 2nd ed. Van Nostrand Reinhold, New York, pp. 750–777.

house, honeycomb, dispersed, turbostratic, and stack. Yong and Sheeran (1973), Collins and McGown (1974), and others have proposed micro-fabric and macro-fabric systems for idealized clay structures including marine clays. In addition, Bennett et al. (1991) present microstructure of fine-grained sediment which covers from mud to shale.

4.7.2 Isomorphous Substitution

Isomorphous substitution is frequently used in the characterization of the clay mineral structure. It simply is the replacement or substitution of cations within the unit structure. Isomorphous substitution or replacement of one cation for another frequently results in charge deficiencies. That is, if, for example, one cation of a +3 charge is substituted with a cation of a +2 charge, the charge deficiency of +1 exists in the area of that substitution. The cation exchange capacity of a clay mineral is a quantified measure of that clay mineral's capacity for cation substitution. Some typical values are presented in Table 4.4.

A clay adsorbs cations with a fixed total charge under a given condition. These cations can be replaced with a group of different ions having the same total charge, by means of isomorphous substitution, broken bonds, as discussed by Kelley (1948) and Grim (1967). These replacements depend on the valence, ion types, ion size, etc. A typical replaceability series is as follows:

$$Na^+ < Li^+ < K^+ < Mg^{2+} < Ca^{2+} < Ba^{2+} < Al^{3+} < Fe^{3+} < Th^{4+} \qquad (4.9)$$

Table 4.4 Typical Values of Isomorphous Substitution

Mineral types	Isomorphous substitution (nature and amount)
Chlorite	Al for Si, Fe; Al for Mg
Halloysite (2H$_2$O)	Al for Si (1 in 100)
Halloysite (4H$_2$O)	Al for Si (1 in 100)
Illite	Al for Si (1 in 7); Mg, Fe for Al: Fe, Al for Mg
Kaolinite	Al for Si (1 in 400)
Montmorillonite	Ng for Al (1 in 6)
Muscovite	Al for Si (1 in 4)
Talc	None
Vermiculite	Al, Fe for Mg; Al for Si

Data from Grim, R.E. (1968), *Clay Mineralogy,* 2nd ed. (1st ed. 1953), McGraw-Hill, New York; Lambe, T.W. and Whitman, R.V. (1979), *Soil Mechanics,* SI Version, John Wiley & Sons, New York.

In some cases, the above statement may not be true when local environmental conditions change. Further discussion relating to environmental geotechnology will be discussed in each chapter as pertinent.

4.8 ION-EXCHANGE REACTION AND ION EXCHANGE CAPACITY

4.8.1 General Discussion

The importance of the phenomenon of ion exchange capacity has been recognized in the areas of soil science, biochemistry, microbiology, food technology, and medicine; however, very little attention has been given to geotechnical engineering. As reviewed by Kelley (1948), Grim (1968), Koryta (1991), and Wachinski and Etzel (1996), the phenomenon of ion exchange was discovered by Thompson in 1850 and Way in 1850–1852 illustrating that clay minerals have the property to absorb certain cations and anions and retain these ions in an exchangeable state. The quantitative studies made by Gedroiz between 1912 and 1913 and Hissink between 1922 and 1932 called this exchange phenomenon a base exchange or ion exchange. First introducing the ion-exchange technique into highway construction, soil stabilization, and soil engineering was Professor H. F. Winterkorn in early 1930 (Winterkorn, 1937).

There are two types of ion-exchange mechanisms: cation exchange and anion exchange. A great deal of information is available regarding cation exchange; however, very little data are found on anion exchange due to the complex reaction during the exchange process, because it is difficult to isolate these reactions for measurement. From a geotechnical engineering point of view, the ion-exchange reaction produces two important phenomena:

1. This reaction can cause changes in the soil-water structure from dispersive to flocculative structures or vice versa (Figure 4.6).
2. It can change the water behavior, such as change hard water into soft water by the removal of calcium (Ca^{2+}) and (Mg^{2+}) ions.

4.8.2 Characteristics and Mechanisms of Ion-Exchange Reaction

(1) General Discussion

As discussed in Section 2.2 and Table 2.1, ordinary atoms are neutral; however, under certain circumstances atoms can become electrically charged. Such charged atoms are called ions. Some atoms form positive (+) ions, called cations, and some forms negative (−) ions, called anions. Charged particles formed by the loss or gain of electrons by atoms are called *simple* ions. Groups of covalently bonded atoms which carry charges are called *poly-atomic* ions. In clay minerals the most common exchangeable cations and anions are Ca^{2+}, Mg^{2+}, H^+, K^+, NH_4^+, Na^+, SO_4^{2-}, Cl^-, PO_4^{3-}, and NO_3^-.

The term exchangeable ions is loosely defined (Bohn et al. 1985; Sposito, 1989) as those released from the soil by solutions of neutral salts in the soil-water system. Soluble salt can be removed by water alone and can also remove some anions from soils, but because most soil colloids are negatively charged, the major ion-exchange reaction is the exchange of soil cations for the cations of the extracting solution.

(2) Ion-Exchange Reaction Mechanisms

The principle of ion exchange may be illustrated in the use of synthetic exchange silicates (zeolites), or permutites, in softening hard waters (Section 5.5). This is a typical illustration to show how the ion exchange works. Usually sodium (Na) is the exchangeable element of an active permutite (Z). As hard water, whose hardness is usually due to soluble calcium (Ca) and magnesium (Mg), passes through the softening material, the calcium and magnesium are absorbed, while sodium is set free and goes into solution, as shown in the following reaction:

$$Na_2Z + CaSO_4 = CaZ + Na_2SO_4 \tag{4.10}$$

where Na_2Z = sodium zeolite and $CaSO_4$ = calcium sulfate.

It is the calcium sulfate in the reaction which causes the hardness of water. CaZ is the calcium zeolite or calcium-saturated permutite, and Na_2SO_4 is sodium sulfate. Calcium is a very active displacing agent, so that in dilute solution it displaces the sodium of the permutite. Numerous typical examples are available in standard chemical textbooks shown as the ion-exchange reactions (Bohn et al. 1985; Sposito, 1989).

(3) Ion-Exchange Capacity (IEC)

IEC of soil as defined by the U.S. Department of Agriculture (USDA, 1938) and Millar and Turk (1943) states that the IEC is the maximum quantity of base that the soil is capable of adsorbing from a neutral solution of a nonvalent cation, and expressed in terms of milliequivalents per 100 g of soil (meq/100 g). ASTM (D1129-78) states that the ion exchange is a reversible process by which ions are interchanged between an insoluble material and a liquid with no substantial structural change of the material. There are two forms of IEC as:

1. Volume basis: The volume basis is the number of milliequivalents of exchangeable ions per milliliter of backwashed and settled bed of ion-exchange material in its standard form.
2. Weight basis: The weight basis is the number of milliequivalents of exchangeable ions per dry gram of ion-exchange material in its standard form.

4.8.3 Factors Affecting Ion-Exchange Capacity

Many factors affecting IEC include particle size, temperature, concentration of pore fluid, and environmental conditions. These factors are discussed as follows.

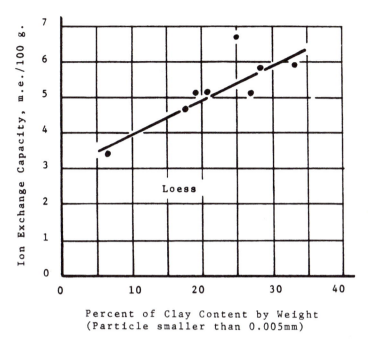

Figure 4.7 Effects of particle size on ion-exchange capacity. (After CAS [1961], Research on Loess, Civil Engineering Research Report no. 13, Chinese Academy of Science, Beijing, 157p.)

(1) Particle Size

The cation-exchange capacity of kaolinite and illite increases as the particle size decreases. It has generally been considered that the cation exchange capacity of montmorillonite does not change substantially with particle size. Experiments shown in Figure 4.7 indicate that the percent of particles less than 0.005 mm of loess increases as IEC increases.

(2) Specific Surface Area

The smaller the particle, the larger the specific surface and the larger the IEC as shown in Figure 4.8. In most clay minerals, the specific surface is greater than 50 m/g, and the IEC is larger than 18 meq/100 g.

The grinding causes a variation in particle size as reported by Kelley and Jenny (1928) cited by Kelley (1948). Decrease in particle size causes an increase in surface and an increase in the number of broken bonds. There is little further increase in the cation-exchange capacity of kaolinite after 96 hr of grinding, indicating that the kaolinite structure has been destroyed at this point and a new permutite-like structure has developed.

(3) Temperature or Heat

Kelley (1948) reported the temperature effect of cation exchange is generally small. Hofmann and Klemen (1950) reported the change in cation-exchange capacity of montmorillonite saturated with the calcium ion (Ca^{2+}), the sodium ion (Na^+), and the lithium ion (Li^+) on heating to various temperatures. The exchange capacity is reduced on heating but the reduction is not uniform and varies with the cation present (Figure 4.9). It also shows that the exchange capacity of montmorillonite is reduced to a considerable extent by heating before the swelling property

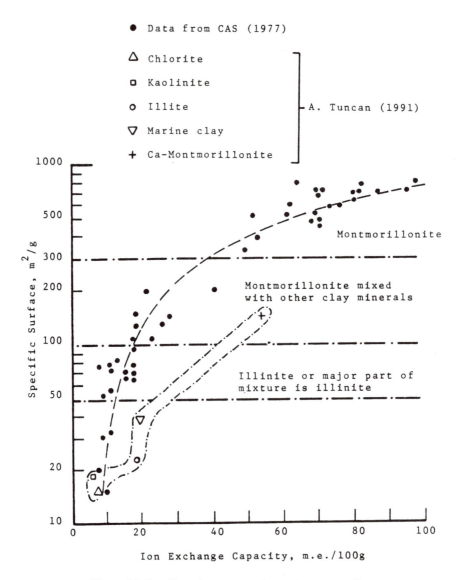

Figure 4.8 Specific surface area vs. ion-exchange capacity.

is lost. In the case of nonexpanding clay minerals indications are that a gradual reduction in cation exchange capacity occurs with increasing temperature of heating.

(4) Effect of Size of Ions

Table 4.5 shows the relative sizes of major ions. It has been found that if ions do not differ in size by more than 15% and in valency by no more than one unit (Paton, 1978), they can substitute for one another in the silicate minerals. It can be seen in Table 4.5 that sodium (Na^+) and calcium (Ca^{2+}) or magnesium (Mg^{2+}) and ferrous iron (Fe^{2+}) can replace one another completely.

(5) Correlation with Other Parameters

Section 3.7 indicated that silica/sesquioxide ratio (SSR) is a useful parameter for identi-fication and characterization of clay minerals. This SSR is related with IEC for some natural

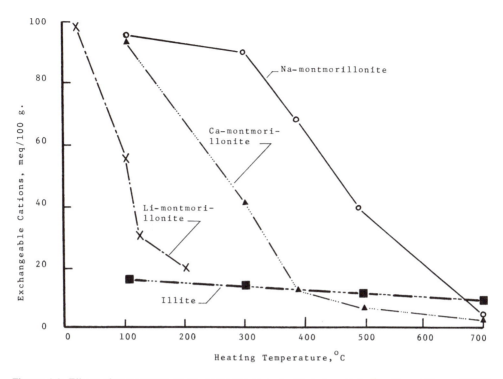

Figure 4.9 Effects of temperature on ion-exchange capacity. (Data from Hofmann, U. and Klemen [1950]; cited by R.E. Grim [1968].)

Table 4.5 Relative Size of Major Ions

Element	Ions	Relative size	
Oxygen	O^{2-}	1.40	
Potassium	K^{1+}	1.33	
Calcium	Ca^{2+}	0.99	
Sodium	Na^{1+}	0.97	
Ferrous ion	Fe^{2+}	0.74	
Magnesium	Mg^{2+}	0.66	
Ferric ion	Fe^{3+}	0.64	
Aluminum	Al^{3+}	0.51	
Silicon	Si^{4+}	0.42	

Data from Paton, T.R. (1978), *The Formation of Soil Material,* George Allen & Unwin Ltd., London.

soils shown in Figures 3.6 and 3.7, which indicate that SSR increases as ion-exchange capacity increases for all types of soils. It also indicates that the cation-exchange capacity is related to the soil types as reflected by the plasticity index (PI), as shown in Figure 4.10. A linear relationship between these two parameters is obtained for most soils except for montmorillonite, because montmorillonite has unique properties as discussed in Sections 4.10 and 13.12.

4.8.4 Anion Exchange

A great deal of information is available regarding cation exchange; however, very little data are found on anion exchange due to the complex reaction during the exchange. Some

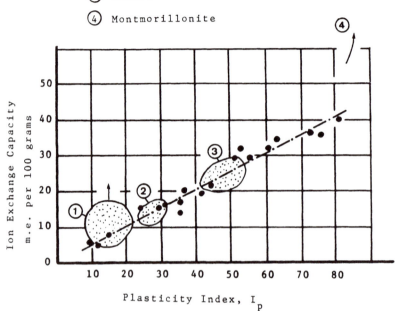

Figure 4.10 Ion-exchange capacity vs. plasticity index.

comments from various researchers (Kelley, 1948; Rich, 1968; Yu and Chiang, 1984) on this aspect are presented and are briefly stated herein.

1. Both cations and anions are involved in reactions of soils with electrolytes.
2. As the pH is increased, a negative charge is released. Thus, anion retention is diminished as the pH is increased.
3. The maximum negative charge of soils is developed after they are phosphated and adjusted to pH = 8.2. In the determination of net negative charge in soil containing abundant hydroxy groups, one needs to determine anion retention as well as cation-exchange capacity.
4. In soils the material most reactive with anions are amorphous (Al) and (Fe) hydrous oxides or hydroxides.

4.8.5 Ion-Exchange Reaction in Contaminated Soil-Water Systems

(1) General Discussion

Ion exchange usually takes place in an aqueous environment, and the ions generally have considerable solubility. However, it has been shown that clays may take ions from water suspensions of very insoluble substances and resistant minerals by means of ionic sorption reactions, and it is probable that the reaction can take place in suspension of high concentration in the unsaturated soil layers. Table 4.6 lists most common ions found in soil-water-pollutant systems.

Table 4.6   Most Common Ions Found in Soil-Water-Pollutant Systems

Most Common Exchangeable Ions in Soil/Water System	
Water	**Soil**
Cl^-, SO_4^{2-}, Ca^{2+}, H^+, K^+, CO_3^{2-}, Na^+, HCO_3^-, Mg^{2+}	Ca^{2+}, Mg^{2+}, H^+, K^+, NH_4^+, Na^+, SO^{2-}, Cl^-, PO_4^{3-}, NO_3^-
Common Ions Found in Polluted Pore Fluids	
As, F, Pb, Cr, Hg, Cl, I, Br, NH_3, H_2S, HNO_3, H_3PO_4, $KBrO_3$, KBr	

(2) Ion-Exchange Reaction between Solids

In general, cation exchange was shown under a given set of conditions, but various cations were not equally replaceable. Kelley (1948) showed that ion exchange can take place directly between plant roots and clays without the intermediate solution of the ions. The cation moves directly from the clay to the plant in return for another ion, which moves directly from the plant to the clay, and found that sodium saturated clays (Na-clays) were able to take enough barium sulfate ($BaSO_4$) to fill about one fifth of the exchange positions of the clay.

(3) Ion-Exchange Reaction between Phases

It has been shown that the cation-exchange reaction can take place in alcohol and that H-montmorillonite will react with dry NH_3 gas to form NH_4 montmorillonite. pH value effects on the IEC of loess is illustrated in Figure 4.11. In examining Figure 4.11, when pH increases, the IEC also increases for all three cases.

4.9 THE CLAY-WATER-ELECTROLYTE SYSTEM

4.9.1 Characteristics of Diffuse Double-Layer

It has been shown in Section 4.2 that the clay layer made up of tetrahedral and octahedral sheets often carries a net negative charge as a result of substitutions of certain of the cations within the sheet structure. This net negative charge is often compensated by cations located on the layer surfaces. When the clay is in the presence of water, these compensating cations have a tendency to diffuse away from the layer surface. The phenomenon for cations to diffuse away from the layer surface in solution is termed diffuse double-layer, also called electric double-layer (Figure 4.12). The diffuse double-layer is the major character in the clay-water-electrolyte system. The characteristics of the zeta potential and other terminology will be discussed in Section 9.8.

Double-layer has a constant charge which is solely determined by the type and degree of isomorphous substitutions and the resulting net negative charge in the mineral layers. The understanding of the ion distribution relative to the layer surface of the clay mineral is predicated upon work in colloidal chemistry. Although several theories have been proposed to describe characteristics of double-layer and their interactions (Van Olphen, 1977; Mitchell, 1993), the most commonly used theory in geotechnical engineering is the Gouy-Chapman theory. Major factors influencing the thickness of a double-layer is summarized as shown in Table 4.3. Further discussion on this aspect will be presented in Section 9.8.

4.9.2 The Clay Micelle and its Structure

The clay micelle is an important structural unit in aqueous clay suspension and comprises the solid clay particle itself as well as its sphere of influence in the surrounding water or

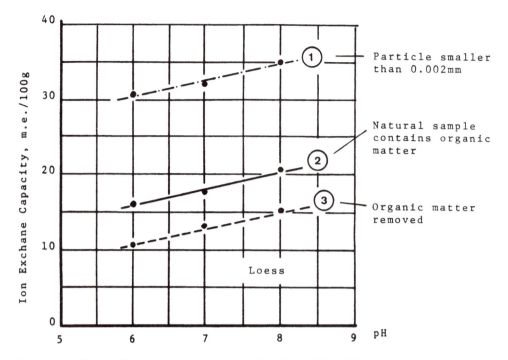

Figure 4.11 pH value effects on ion-exchange capacity of loess. (After CAS [1961], Research on Loess, Civil Engineering Research Report no. 13, Chinese Academy of Science, Beijing, 157p.)

aqueous solution (Figure 4.13). The sphere of influence corresponds to the distance in which exchange ions originating from the particle surface may be found in a dynamic equilibrium between the electric attraction forces from the charged particles and the inherent kinetic energy of the ions which themselves may be surrounded by strongly held water molecules.

4.9.3 Free Water and Environmental Water

Water existing in the soil/water system can be divided into two groups: free water and environmental water (Fang, 1987). *Free water* also called *gravity water,* can be removed by gravity force. *Environmental waters* consist of double-layer water, hydrated water, hydratation water, osmotic water, and oriented water (Figure 4.13). Environmental waters are strongly influenced by local environmental conditions, especially in polluted soil/water environments. Brief discussions of each type of environmental water are presented as follows.

(l) Adsorbed Water

Clay particles in moist soils are always hydrated. Because the particle is surrounded by layers of water molecules, it is called *adsorbed water.* Exchangeable cations are the positively charged ions from the clay mineralogical structure which can be replaced or substituted.

(2) Osmotic Water

Normal water located between the hydrated ions themselves and between the hydrated mineral surfaces from which they issue and whose attraction sphere is held by electrostatic forces has been called *osmotic water.* Osmotic water is due to the osmotic tendency of the

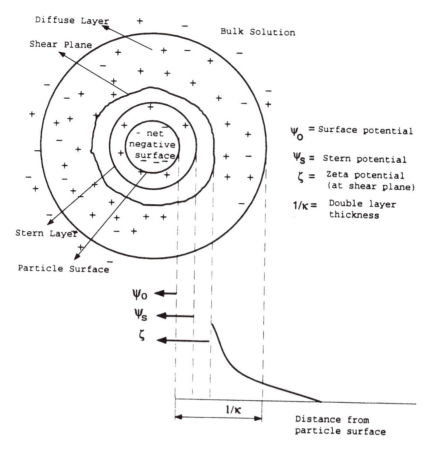

Figure 4.12 Schematic diagram illustrating electric-double-layer and its zeta potential. (After Kaya and Fang [1995].)

cations to disperse as much as possible in the available solvent space but is restrained by the electrostatic attraction to the charged clay particles.

(3) Hydration Water and Hydratation Water

Clay particles are always surrounded by a layer of water molecules, and by bonding of water to ions or electrically charged surface. When the water molecules are no longer able to hold all the thermal energy previously stored in the system and the difference is given off as heat of solution, or heat of wetting in the case of dispersion of solid particles, such water is called *hydratation water* (Winterkorn, 1942). The *hydration water* is held only by the dissociated exchange ions (Eucken, 1940).

Strong binding of water by dissolved ions may manifest itself in volume contraction, decreased specific heat, and decreased dielectric constant of the solvent water. While the cause for these phenomena are mainly of an electric nature, the observed results are greatly influenced by the size and shape of the interacting ions and molecules and by the effect of certain anions which decrease the viscosity of their solvent water.

(4) Oriented Water

Oriented water is located at the air-water interface as indicated in Figure 4.13, and it is several hundred molecules thick at normal temperature.

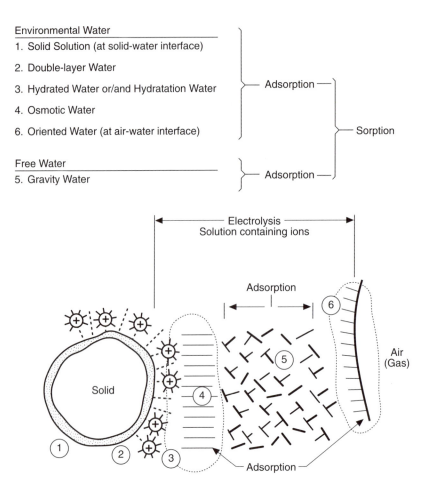

Figure 4.13 Soil-water-air interaction in the environment. (Based on Fang, H.Y. [1986], Proc. 1st Int. Symp. on Environmental Geotechnology, v. 1, pp. 1–14.)

4.10 CHARACTERISTICS AND STRUCTURES OF SOME TYPICAL CLAY MINERALS

4.10.1 General Discussion

The basic structural units of soils have been discussed in Section 4.2 and Figure 4.1. The structural classification of layer silicates is presented in Table 4.1. Basically, there are three main clay mineral types: 1:1, 2:1, and 2:1:1; however, it can be further subdivided into groups or subgroups depending primarily upon the character of their octahedral sheet (Figure 4.1). From an engineering viewpoint, three groups of clay minerals are generally recognized: kaolin, illite, and montmorillonite together with the chlorite group as shown in Figure 4.14. Further discussions on characteristics and structures of some common minerals relating to environmental geotechnology are presented as follows:

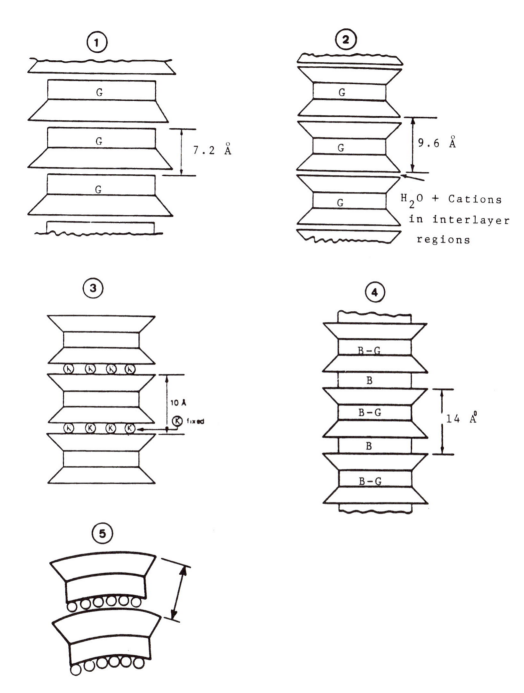

Figure 4.14 Diagrammatic sketch of the structures of some common clay minerals. (1) Kaolinite; (2) montmorillonite; (3) illite; (4) chloride; (5) halloysite.

4.10.2 Kaolin, Montmorillonite, and Illite Groups

(1) Kaolinite Group

The *kaolin* group consists of clay whose crystal lattice is rigid and contains one sheet of silica and one sheet of alumina, known as the 1:1 minerals. The idealized chemical formula of this mineral, which is a two-layer clay consisting of one gibbsite sheet (Figure 4.14(1)) joined with a silica tetrahedral sheet, is $(OH)_8Si_4Al_4O_{10}$. Aluminum occupies the cation sites within the octahedral layer; the tetrahedral layer cations are silicon. Because the interlayer bonding is primarily a hydrogen bond, the interlayer attraction forces are substantial relative to other types of secondary bonding. Isomorphous substitution, the replacement of a cation in the ideal structure with a cation similar in size and charge density, is generally very low in kaolinite. Minerological analyses measuring the basal spacing of the kaolinite structure, the distance from the base of one layer to the base of the next, or the total thickness of a layer, is 7.2 Å. The cation-exchange capacity of the kaolinite clay mineral is generally considered to be 3 to l5 meq/l00 g (Table 4.4).

(2) Montmorillonite Group

Two common mineral groups are found within this group of the 2:1 family of clay minerals, the smectite group and the illite group. The smectite group of clay minerals was formerly known as the *montmorillonites*, but recent studies have identified montmorillonite as just one of the clay minerals found within the smectite group. Montmorillonite consists of two sheets of silica tetrahedra on either side of an octahedral gibbsite sheet. The idealized chemical formula of this mineral is $(OH)_4Si_8Al_4O_{20}\cdot nH_2O$, where nH_2O is the interlayer (n layers) of adsorbed water. In reality, montmorillonite virtually always exhibits some ionic substitution, commonly with sodium (Na) or calcium (Ca). Na-montmorillonite is a common form of bentonite clay (Section 13.12).

Montmorillonite consists of three sheets stacked (Figure 4.14(2)) such that the bonding between successive layers consists of van der Waals forces. Cations may be present between the layers to balance change deficiencies created by ionic substitutions primarily in the octahedral layer. These interlayer bonds are, therefore, relatively weak and easily separated by imposed stresses such as the adsorption of water or other polar liquids. Because the lattice substitutions within the basic sheets and the cations within the interlayer need to balance charge deficiencies, the cation-exchange capacity is relatively high. It is generally considered that the cation-exchange capacity of montmorillonite is between 80 and l50 meq/100 g. The basal spacing, the distance between layers, can vary from a minimum of 9.6 Å to complete separation or infinity. Montmorillonite clay is considered highly sensitive to environments more than the kaolinite or illite groups.

(3) Illite Group

The *illite* group is intermediate in its properties and has a much more complicated chemical formula. They are also known as the mica-like clay minerals. Illites, like montmorillonites, are 2:1 minerals. They consist of a gibbsite sheet between two silica tetrahedral sheets, and the successive layers are ionically bound by potassium cations which populate the interlayer region. The idealized chemical equation of illite is, therefore, $(K,H_2O)_2(Si)_8(Al,Mg,Fe)_{4.6}O_{20}(OH)_4$. The octahedral cations are aluminum, magnesium, or iron, and the tetradral cations can be aluminum or silicon. Cations-exchange capacity of illinite is generally from 10 to 40 as compared to 80 to 150 meq/100 g for montmorillonite and 3 to 15 for kaolinite (Figure 4.14(3)).

4.10.3 Chlorites and Halloysites

(1) Chlorites

Chlorites occur extensively in soils and are examples of 2:1:1 layer silicates (Figure 4.14(4)). The elemental composition of chlorites varies widely. They occur commonly in

Table 4.7 Summary of Geotechnical Properties of Basic Clay Minerals

Properties	Kaolinite	Illite	Montmorillonite
Specific gravity, G_s	2.61	2.60	2.51
Liquid limit, LL	50–62	95–120	150–700
Plastic limit, PL	33	45–60	55
Plasticity index, PI	20–29	32–67	100–650
Activity, A	0.2	0.6	1–6
Compressibility index, c_c	0.2	0.6–1.0	1–3
Friction angle drained, degree	20–30	20–25	12–20
Particle shape		Platy	
Specific surface, m²/g	10–20	65–100	50–800
Cation-exchange capacity, meq/100 g	3–15	10–40	80–150
Isomorphous substitution, (nature/amount)	Al for Si 1 in 400	Al for Si 1 in 7 Mg, Fe for Al Fe, Al for Mg	Mg for Al 1 in 6
Linkage between sheets	H bonding + secondary valence	Secondary valence + K linkage	Secondary valence + exchangeable ion linkage
SiO_2/R_2O_3 ratio R = Al + Fe	2	2–3	4

sedimentary rocks or marine sediments. Chlorites are basically similar to the illites except that an organized octrahedral sheet of chlorites replaces the area populated by cations in the case of illite. The idealized chemical formula is $(OH)_4(Si,Al)_8(Mg,Fe)_6O_{20}$ (for 2:1 layer) and $(Mg,Al)_6(OH)_{12}$ (for interlayer). Basal spacing of this mineral, which has four sheets to a layer, is 14 Å. Cation-exchange capacity is found to vary from 10 to 40 meq/100 g. The basic chlorite including its several varieties is quite common.

(2) Halloysites

As indicated in Table 4.3, *halloysite* is a form of kaolinite in which water is held between structural units in the basal plane, and consists of crystals of hollow cyclinders typically 1 m in length and of a diameter of about 1/5 to 1/10 of the length. Halloysite can be divided into hydrate and dehydrate halloysites as follows.

Hydrate halloysite ($4H_2O$) contains a mono-molecular layer of water within its mineral layers; when this water is completely removed by oven drying the result is *dehydrated* halloysite ($2H_2O$). The Atterberg limits of the two halloysites are different, but both plot below the A-line of Casagrande's plasticity chart (Figure 3.1). Hydrated halloysite, which has also been termed halloysite, can dehydrate irreversibly to halloysite, sometimes known as metahalloysite. Mineral structures of halloysites for both hydrate and dehydrate are shown in Figure 4.14(5).

4.10.4 Geotechnical Properties of Clay Minerals

Since all natural soils are derived from these basic clay minerals, the geotechnical properties of each individual mineral must be evaluated. Table 4.7 summarizes the geotechnical properties of basic clay minerals.

A clay in which the principal adsorbed ion is hydrogen (H) is called a *hydrogen clay* or *H-clay*. If the adsorbed ion is sodium (Na^+) it is a *sodium clay* or *Na-clay*, etc., for example, Na-montmorillonite and Ca-montmorillonite. These symbols mean that we have montmorillonite clay particles with sodium (Na^+) and calcium (Ca^{2+}) cations, respectively, adsorbed on the particle surfaces. It must be noted there is no guarantee, of course, that all the cations adsorbed on the internal surface of a clay will be of the same sort — all sodium or all calcium, etc.

If a clay having an adsorbed ion of one kind is brought into contact with ions of another kind, it may release some or all of the ions of the first kind and hold those of the second kind in their places. This process is called base exchange as discussed in Section 4.8. Other environmental geotechnical properties of these clay minerals such as water adsorption, hydraulic conductivity, compaction, and strength will be discussed in various chapters in the text.

4.11 HYDROPHILIC AND HYDROPHOBIC SOILS

4.11.1 General Discussion

Generally speaking, all soil can be grouped into two categories, namely, *hydrophilic* (likes water) and *hydrophobic* (dislikes water) soils. In most cases, soils belong to the hydrophilic type. However, in this section, discussions focus on how these two types of soil can be interrelated and how they can be used in environmental geotechnical projects.

Some hydrophobic soils are natural and some are man-made. Most of the desert soils belong to the hydrophobic group. This type of soil will repel water faster than other soil. Desert soils do not always occur in nature, some are man-made due to poor land management. In the following section, two distinct approaches are presented as:

1. Converting from hydrophobic soil into hydrophilic conditions for the purpose of changing desert soil (Chapter 18) into agricultural farm land
2. Using reversible processes changing hydrophilic soil into hydrophobic conditions for the purpose of use as a clay liner (Chaper 17)

4.11.2 The Changing Mechanisms between Hydrophobic-Hydrophilic Conditions

Hydrophobic soil type is useless from an agricultural point of view, because these soils cannot absorb the water. Figure 4.15 presents a diagram to explain the water droplet at mechanical equilibrium on a solid surface. In soils, the liquid-solid contact angle (θ) is the angle formed by a water meniscus in contact with the solid pore walls. According to DeBano (1969), if the angle (θ) is greater than 90°, water will not penetrate or move through the soil mineral fabric. On the other hand, we can manipulate the change in angle (θ), as illustrated in Figure 4.15.

Gauffreau (1989) proposed a heating method based on the concept illustrated in Figure 4.15(5) for changing a material from hydrophilic into hydrophobic conditions for the purpose of using hydrophobic material as a clay lining material for hazardous/toxic waste control system (Section 17.7). The heating technique is used to change the angle (θ). The concept of water repellence (hydrophobia) by the soil particles themselves is uniquely applied to the problems associated with waste containment. This water repellence is achieved through the use of an organic coating which, to maintain the low-cost aspect of this method, is chosen to be activated sludge. Several heating techniques are used to intensify the hydrophobic effect and various tests are taken to observe the degree of water repellence in each soil sample.

4.11.3 Characteristics of Hydrophobic and Hydrophilic Soils

The major differences of characteristics between hydrophobic and hydrophilic soils are listed as follows:

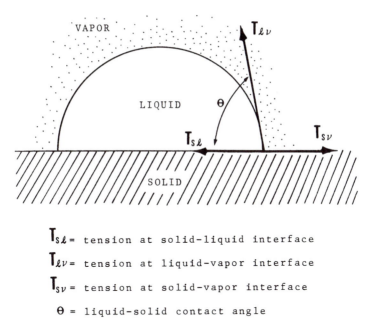

$\mathsf{T}_{s\ell}$ = tension at solid-liquid interface

$\mathsf{T}_{\ell\nu}$ = tension at liquid-vapor interface

$\mathsf{T}_{s\nu}$ = tension at solid-vapor interface

Θ = liquid-solid contact angle

Figure 4.15 Vapor-liquid-soil interaction. (After DeBano, L.F. and Letey, J., eds. [1969], Proc. Symp. on Water-Repellant Soils, University of California at Riverside, 354p.)

1. Hydrophobic soil
   - Very small amount of electrolyte is required for stability.
   - Particles have a given electrical charge.
   - Particle migration is under applied potential.
   - Concentration of dispersed phase is low.
2. Hydrophilic soil
   - No electrolyte is required for stability; no precipitation by small amounts of electrolyte.
   - Particles may be positive, negative, or neutral depending on pH value in the solution.
   - Particles may or may not migrate under an applied potential.
   - Concentration of dispersed phase can be high.

4.12 HOMOIONIC MODIFICATIONS (PURE SOIL)

4.12.1 General Discussion

For engineering applications, the relationship between clay fraction and the engineering behavior of soils depends not only on its quantity and the physicochemical properties of its micelles, but also on the relative amounts and characteristics of the other soil constituents, including the aqueous and gaseous phases with which the clay particles interact. In order to illustrate clay's effects on the physicochemical properties of a natural soil, a series of experiments were carried out at Princeton University. The purposes of the homoionic modifications are

1. To aid in the recognition, understanding, and separation of the component factors that produce the engineering properties of clays and clay soils as normally defined
2. To provide for certain physicochemical properties of pure clays and their homoionic variants which may be directly useful for certain specific engineering purpose
3. To define probable ranges of particular engineering properties in cases where specific data are not available

Table 4.8   Physical Properties of Some Typical Natural and Homoionic Soils

Soil type	Natural soil	Homoionic soil							
		H	Na	K	Mg	Ca	Ba	Al	Fe
Putnam subsoil									
Plasticity index	17	16	32	18	22	21	19	15	16
% clay (>0.005 mm)	33								
Activity	0.52	0.48	1.00	0.55	0.67	0.64	0.58	0.45	0.48
Loess pampaneo subsoil									
Plasticity index	45	27	37	20	34	29	34	32	26
% clay (>0.005 mm)	50								
Activity	0.90	0.54	0.74	0.40	0.68	0.58	0.68	0.64	0.52
Marshall subsoil									
Plasticity index	28	21	48	18	36	30	28	21	17
% clay (>0.005 mm)	37								
Activity	0.76	0.57	1.30	0.49	0.97	0.81	0.76	0.57	0.46
Loess pampaneo subsoil									
Plasticity index	13	6	12	—	12	12	9	7	6
% clay (>0.005 mm)	27								
Activity	0.48	0.22	0.44	—	0.44	0.44	0.33	0.26	0.22
Cecil subsoil									
Plasticity index	37	34	30	38	34	32	36	31	36
% clay (>0.005 mm)	76								
Activity	0.49	0.45	0.39	0.50	0.45	0.42	0.47	0.41	0.47

Data selected from Winterkorn (1942) and Winterkorn and Fang (1975).

4.12.2 Method for Preparation of Homoionic Soil Samples

Homoionic modifications from a natural soil were prepared in batches using the method described by Scheffer and Schachtschabel (1959) and modified by Vees and Winterkorn (1967). A brief description of the procedures is presented:

1. A 10-liter aqueous solution was made containing ten times the amount of cations required for ion exchange in the form of a soluble salt.
2. The solutions were placed in the mixer bowl along with a paddle previously coated with a chemically inert synthetic resin film.
3. The powdered clay was added slowly while the mixer was running, and the resulting dispersion was mixed for 2 days; then it was placed in a container and left at least for 7 days.
4. Subsequently, the supernatant clear solution was decanted and the dispersion or paste was concentrated and washed either in a filter press or by means of large Buchner funnels connected to a vacuum.
5. The filter was frequently tested for anion content, and the process was considered complete when no anion presence could be detected by a pertinent chemical method.

4.12.3 Typical Homoionic Modification Data

Extensive laboratory experiments on homoionic modifications were reported by Winterkorn (1942). Physical properties including heat of wetting of several natural and homoionic soil materials are summarized by Winterkorn and Fang (1975). A condensed form of these data is presented in Table 4.8. Figure 4.16 is a graphical presentation of the data from Table 4.8. It indicates that homoionic modifications could significantly change the engineering properties for a given natural soil. A high shrinkage limit combined with a large heat of wetting indicates soil that is especially valuable to slaking and loss of stability when placed in contact with water while in the dry state.

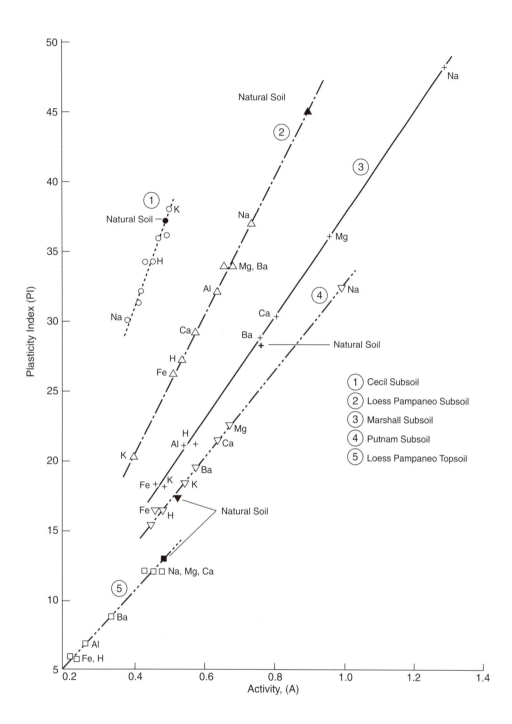

Figure 4.16 Interrelationships of activity and other physical properties of natural and homoionic soils.

4.13 THE MAN-MADE SOIL

4.13.1 General Discussion

Soil, a natural substance together with air and water, is a necessity for human survival. Soil is not only used for farming, it also is used as a major construction material. Soil is

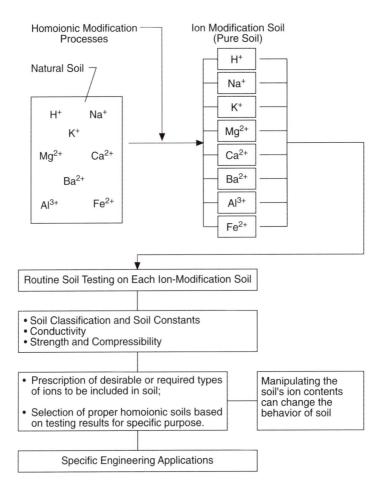

Figure 4.17 The concept and approaches for man-made soils. (After Fang, H.Y. [1992b], Topic 6, Geo-environmental Engineering, Recent Accomplishments and Future Trends in Geomechanics in the 21st Century, U.S.-Canada Workshop, University of Oklahoma and U.S. NSF, pp. 309–317.)

taken from nature whatever way nature produced it. In many cases, the soil taken from nature is not exactly suitable for certain construction purposes; therefore, premature or progressive failures frequently occur. In this study, a method is proposed that engineers can use to make the decision regarding what type of soil is needed for certain application(s). The principle of *man-made soil* is based upon soil's inherent genetic characteristics, which by manipulating the soil's ion content can change behavior of the soil.

4.13.2 The Concept and Approach of Man-Made Soil

The concept and approach of man-made soil as proposed by Fang (1992b) is a method of converting the natural soil into a man-made soil by manipulating its ion contents so as to

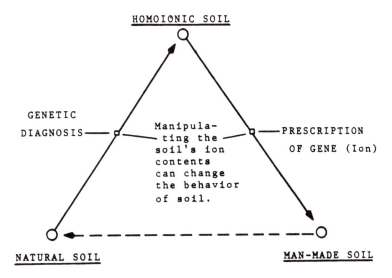

Figure 4.18 Process of man-made soil.

change its engineering behavior, as illustrated in Figure 4.17. This conversion process involves extraction of the individual or homogeneous ions (may be called *genes*) from the natural soil. This extracted soil is termed here as *homoionic* soil or *pure* soil, as discussed in Section 4.12.

The approach is hinged on controlling the ion substitution or manipulating (addition or removal) ions in the soil element to fit into engineering needs. Figure 4.18 is the simplified version of Figure 4.17 and assumes that natural soil contains various types of ion as shown in Tables 3.11 and 3.12.

Laboratory studies have been made using typical clay minerals and natural soils. Routine geotechnical tests are performed on each of the ion-modified soils (pure soils) and also natural soils and a comparison of their engineering properties is made to form guidelines for choosing proper soil types for specific projects. Specific projects such as hazardous/toxic control facilities, hydraulic barriers, and top and bottom seals can benefit from such modified soil. Also, it can be used for specific ground improvement projects including soil stabilization and grouting.

4.14 SUMMARY

1. To understand the environmental aspects of soil-water interaction in the environment, a knowledge of soil technology is necessary. To understand the interrelationship between the experimental results and the basic clay behavior on a microscopic scale, a knowledge of chemistry, physicochemistry, and biology is required. This information is summarized and discussed.

2. Major factors relating to clay minerals are outlined and discussed, including structures of layer silicates, interparticle forces, and particle bonds and linkages.

3. Ion-exchange reactions and their applications to soil-water systems are discussed together with various factors affecting ion-exchange capacity.

4. The behavior of clay-water-electrolyte systems is examined. The characteristics of diffuse-double-layer and its uses are also presented.

5. Engineering properties of some typical clay minerals are summarized and discussed.

6. Hydrophilic and hydrophobic soils and man-made soils are introduced. Specified applications relating to environmental geotechnology are proposed and discussed.

PROBLEMS

4.1 How can the dispersion of grain sizes be noted? Name at least two ways. Do these methods give you a good indication of the degree of dispersion? Why is the degree of dispersion worth knowing?

4.2 A 'more than' plot rather than a 'less than' plot is used on a cumulative grain size frequency diagram (Figure 4.3). What difference does it make? Do the modes of a frequency plot show on a cumulative frequency plot?

4.3 Using the symbols for gibbsite, brucite and the silicate tetrahedron, sketch the commonly accepted structure for kaolinite and montmorillonite.

4.4 For the following changes in the soil-water system, list the effect on the double-layer thickness.
(a) Decrease in pH
(b) Decrease size of attracted cation
(c) Decrease valence of attracted cation
(d) Increase in ion concentration

4.5 Which clay structure would you expect to have the highest value of the following soil properties? (a) Permeability, (b) Sensitivity, (3) Rebound upon load removal, (d) Shear Strength, and (e) Compressibility.

4.6 Two quartz particles are essentially the same size, but not the same shape. The first particle has a settling velocity of 0.95 cm/sec, the second has a settling velocity of 0.80 cm/sec. Which particle has the lowest sphericity? What is sphericity?

4.7 Explain why the surface area of fine-grained soil particle is so important in comparison with larger soil particles.

4.8 What is a homoionic modified soil and what is a pure soil?

4.9 Comment on man-made soil. Is there any merit to developing such soil to be used in relation to the hazardous/toxic waste controlling systems?

4.10 Explain the characteristics and mechanism of soil-water interaction between hydrophilic and hydrophobic soils.

CHAPTER 5

Soil-Water-Air Interaction in the Environment

5.1 INTRODUCTION

5.1.1 General Discussion

Water has a most important relationship with all construction materials. It may serve as an essential component as in hydraulic cements and concretes. Water and aqueous solutions may interact physically and chemically with the materials of structures designed to contain or conduct them, such as water supply and sanitary installations, including dams and conduits constructed of earth, concrete, wood, or metals. Because of its ubiquity, even in desert air, and its physical and chemical characteristics, water is one of the main deteriorating agents acting on practically all construction materials employed on this earth and greatly influences or even determines their durability under service conditions. Obviously, the better we understand the water substance the better we can gauge its effect on construction materials. If soil-water is polluted, then the complexity of each phase is multiplied. Since soil-water environment interaction is the major part of geotechnology, we must make an effort to understand as much as possible about the interaction and its response to local environments during soil's useful life period.

5.1.2 Soil Moisture Terminology

Researchers in various fields have come to use a variety of terms to designate important soil-water systems. Figure 5.1 shows the approximate relationship of commonly used soil moisture terms. This scale proposed by Olmstead and Smith (1938), with the moisture content increasing upward, has no top. There are five divided lines marked from **A** to **E**. A brief explanation of each point is presented as follows.

(1) Suspensions

Above point **A** are the soil suspensions (Section 3.10). The settling volume is marked point **A** which is space-occupied by a unit weight of soil after settling from a suspension.

(2) Gravitational Water

From points **A** to **B** is the region of gravitational water. The maximum water-holding capacity is the moisture content of a 1-in. layer of loose soil after gravitational water has

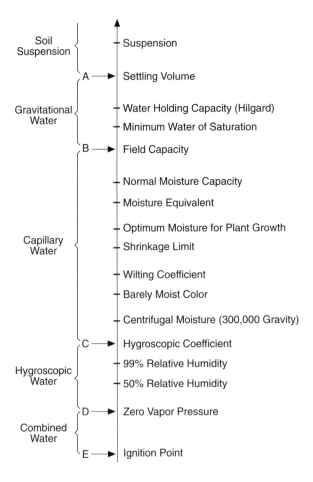

Figure 5.1 Soil moisture scale showing the approximate relationship of commonly used soil moisture terms. (After Olmstead, L.B. and Smith, W.O. [1938], Soils and Men, Yearbook of Agriculture, U.S. Department of Agriculture, pp. 897–910.)

drained from the largest pores. The minimum water of saturation is the lowest moisture content at which a soil may be saturated with water when the soil grains are worked into the position of closest packing.

(3) Capillary Range

From points **B** to **C** is the capillary range. Capillary tensions vary from zero at point **B** to several hundred pounds per square inch at point **C**. In this region, several important terms must be pointed out as:

1. Field capacity representing maximum amount of water that can be held against the force of gravity (Section 7.10)
2. Moisture equivalent (Section 6.5)
3. Optimum moisture content (Section 11.2)
4. Shrinkage limit (Section 6.2)
5. *Wilting coefficient* is the moisture content just low enough to cause plants to wilt. The points at which soils show a barely perceptible difference in color from the air-dry condition is indicated for reference.

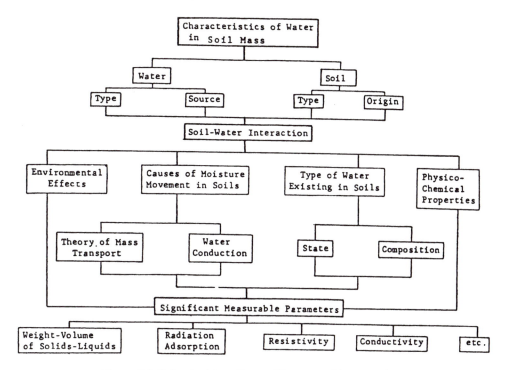

Figure 5.2 Soil-water interaction and its measurable parameters.

(4) Hygroscopic Range and Ignition Point

From points **C** to **D** is the hygroscopic range. The hygroscopic coefficient marks the upper limit of this range. The water remaining after all hygroscopic water has been removed is called *combined water*. It is held by chemical rather than physical forces and can be driven off only by heating. At the ignition point, a bright red heat, all soil water is finally lost.

Since plants cannot extract water from the soil at moisture contents below the wilting point, the amount of water in the soil at any time in excess of the wilting coefficient is referred to as available water. The maximum available water is the difference between the wilting coefficient and the field capacity, as indicated in Figure 5.1.

5.1.3 Soil-Water System and Its Measurable Parameters

Soil-water interaction is complex and its behavior is not only dependent on soil types and their origin, but also related with the environmental conditions and stress history. In geotechnical engineering, the mechanism of soil-water interaction related to the stability of the soil mass is essential in progressive failure and erosion, surface creep, landslide, etc. Figure 5.2 summarizes environmental factors affecting soil-water interaction and their measurable parameters. This chapter on soil-water interaction will cover soil-water interaction mechanisms in both thermal and electric energy fields.

5.2 NATURE OF WATER

5.2.1 Molecular Dimensions and Bulk Properties

Water is a peculiar substance; it should be a gas at normal pressure and temperature conditions in analogy with hydrogen sulfide (H_2S), hydrogen selenium (H_2Se), and hydrogen tellurium (H_2Te), but it is a liquid and even as a liquid it has certain 'solid' structural properties as revealed by X-ray patterns. These peculiarities which are observed in pure bulk water also make themselves known in interactions with other substances, such as dissolved ions, polar and nonpolar substances, and with contacting surfaces. These interactions result in effects that are observable on the various dimensional levels such as the molecular, the colloidal, the microscopic, and the macroscopic or normal visual level.

Pure water is a colorless, odorless, tasteless liquid which boils at 100°C and freezes at 0°C. The smallest structural unit of the water substance is the H_2O molecule. It is composed of two H-atoms and one O-atom and can be designated as hydrogen oxide, and oxygen hydride or a hydrogen hydroxide. Its molecular weight is the sum of that of its component atoms, i.e., 2(1.0081) + 16.00, which places it among the six lightest molecules known to chemistry.

The water molecule HOH can be considered as approximately spherical with a radius of about 1.38 Å while that of the O^{2-} ion is 1.32 Å. The H^+ ions are so located on the surface of the O^{2-} ion that they form an angle of about 105° with the center of the O^{2-}. This arrangement gives the water molecule a bilateral symmetry and a permanent dipole of 1.87 × 10 esu-cm (esu = electro static unit). This bilateral electric structure and dipole moment are the basic reasons for the peculiar nature of the water substance. They result in a strong and direct interaction between adjoining water molecules which, because of its directional characteristics, prevents a close packing of the water molecules.

5.2.2 Packing Characteristics

As indicated in a previous section, due to strong and direct interaction between water molecules, water molecules cannot be closely packed. The following example is used to further explain the packing characteristics of water molecules.

Example 5.1

Compute the absolute volume of water

Solution

Assuming that the water molecules are homogeneous spheres, then 1 g mol of water, i.e., 18 g or 6.02×10^{23} molecules, has an actual solid volume of:

$$6.02 \times 10^{23} \times (1.38)^3 \,(4/3)\, \pi = 6.62 \text{ cm}^3$$

In the various possible forms of space symmetrical packing of such spheres, we obtain the following values for the volumes occupied by 1 g mol of such a sphere:

Absolute volume of solids 6.62 cm³

Hexagonal rhombohedral packing	8.94 cm³
Tetragonal spheroidal packing	9.49 cm³
Orthorhombic packing	10.92 cm³
Cubic packing	12.64 cm³
Normal water	18.00 cm³
Ice	19.65 cm³

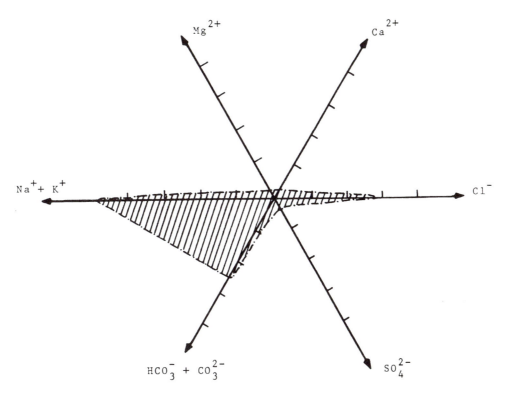

Figure 5.3 Typical characteristics of water chemistry. (Data selected from Petroleum Geology [1979] and Wang [1985].)

The data in Example 5.1 show the looseness of the water and ice structures which contain only 36.8 and 33.7% of actual water substance per unit volume. This looseness is caused by the low order of symmetry of the electric force field of the water molecule combined with the strong interaction between water molecules if arranged in accordance with their electric structures. What is the order of magnitude of the attraction between water dipoles and how does it compare with that of other molecular or ionic interactions or bonding are not clearly understood at the present time. However, the bond energy of the water molecular has been discussed in Section 4.2.

5.2.3 Chemical Composition of Water

A brief discussion of the characteristics of the liquid phase of soil has been presented in Section 3.2. Further discussions on chemical composition of water and its related phenomena are presented as follows.

Water molecules do not only associate into larger structural units, but they also dissociate into H^+, OH^-, and O^2 ions. Although the degree of this dissociation is of very small magnitude, it is of great theoretical and practical significance for pure water and also polluted water, which contains aqueous solutions of acids, bases, and salts. Higher H^+ concentrations are due to exchangeable H ions in the soil particles and/or the presence of mineral or organic acids. Figure 5.3 shows typical characteristics of water chemistry. In natural water they contain more than 60 known elements. Table 5.1 presents the major chemical contents between sea and river waters.

Table 5.1 Comparison of Major Chemical Contents Between Sea and River Waters

Elements	Seawater (ppb)	River water (ppb)
Al	1.0	400
B	4,450	10
Br	67,300	20
Ca	4.11×10^5	15,000
Cl	1.94×10^7	7,800
Fe	3.4	670
Hg	0.15	0.07
I	64	7.0
K	3.92×10^5	2,300
Li	170	3
Mg	1.29×10^6	4,100
Na	1.08×10^7	6,300
Pb	0.3	3.0
S	9.04×10^5	5,600
Sr	8,100	70
U	3.3	0.3
Zn	2.0	20

Data obtained from Wang, C.H. (1985), *Water Pollution,* Scientific Publishing, Beijing.

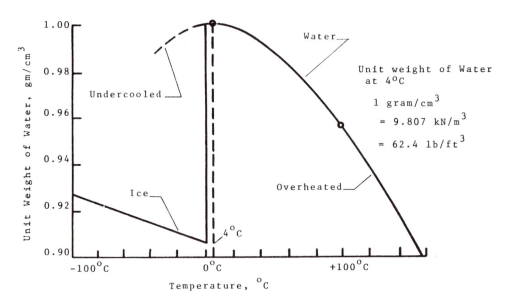

Figure 5.4 Effect of temperature on unit weight of pure water. (Data from Rao [1933].)

5.3 PROPERTIES OF WATER AND WATER SUBSTANCES

5.3.1 Density, Unit Weight, and Specific Gravity of Water

(1) Density and Unit Weight

The *density* is defined as mass per unit volume, kilograms per cubic meter (kg/m³), or grams per cubic centimeter (g/cm³). The *unit weight* is pound per cubic foot (pcf) and is

Table 5.2   Typical Surface Tension Data at Various Interphases

Interface	Temperature (°C)	Surface tension (dyn/cm)
Acetone-air	20	23.7
Benzene-water	20	35.0
Carbon tetrachloride-air	20	27.0
Ethyl alcohol-air	20	22.7
Glycerin-air	20	63.4
Mercury-air	15	487.
Water-air	0	75.6
Water-air	20	72.8
Water-air	100	58.9

Data from Sears, F.W. and Zemansky, M.W. (1955), *University Physics,* Addison-Wesley, Cambridge, MA; and Hough (1967).

defined as a body force unit or force per unit volume. In the SI system, the unit will be Newton per cubic meter (N/m³). The density of water at 4°C is 62.4 pcf or 1.94 slugs per ft³. In the metric system, it is equal to 1 g/cm, or for the SI system, it is equal to 9.807 kN/m. Figure 5.4 shows the relationship between unit weight vs. temperature for pure water.

(2) Specific Gravity

The *specific gravity* of a substance is the ratio of its density to the density of water at 4°C. Dimensionally, specific gravity is a pure number. The *specific weight* of a material is defined as the weight of the material per unit volume, and is, therefore, expressed in the units of force divided by volume, or force divided by length cubed in the foot-pound-second (fps) system, which is expressed in pound per cubic foot (pcf).

5.3.2 Surface Tension

The molecules on the surface of a liquid are attracted to each other; therefore, it is evident that a tensile force may be considered to be acting across any line in the surface of the liquid. The intensity of the molecular attraction per unit length along any line in the surface of the fluid is called the *surface tension*. There are several methods for measuring surface tension of water or liquid as suggested by ASTM (D1590-77). The numerical value of the surface tension of liquid is dependent upon the nature of the liquid itself, the type of the adjacent fluid, and the temperature. The value of surface tension decreases as the temperature increases. The unit of surface tension in the SI system is Nm and the commonly used cgs unit is dyn cm (1 Nm = 10 dyn cm). The typical value for water at 20°C is 75.8 dyn/cm. Table 5.2 presents typical surface tension data at various interphases. In addition to surface tension of liquid, a device for measuring tensions in water was developed by Gilbert (1960).

5.3.3 Viscosity of Water and Other Pore Fluids

(1) Absolute or Dynamic Viscosity

The viscosity of fluid is used as an index of the relative fluidity in comparison with the ideal fluid and ideal solid. The ideal fluid is zero and the ideal solid infinity. The *viscosity* of the liquid is defined as the ratio of shear stress to the rate of shear strain. The shear stress is the shear force divided by shear area. The rate of shear strain is the change in velocity, divided by the distance between shear surfaces. The viscosity determined under such conditions is called *absolute* or *dynamic viscosity* (μ). Figure 5.5 shows the effects of temperature on dynamic viscosity of various liquids commonly used.

Dynamic or absolute viscosity may be expressed in pound-second per square foot (lb-sec/ft^2) or slugs per foot-second. In the metric system, it may be expressed in dyn-second per square centimeter, gram-second per square centimeter, or in poise. The term poise is in honor of Poiseuille, a French scientist. The centi-poise, or 0.01 poise is a common unit. The commonly used units for dynamic or absolute viscosities are listed as follows:

1 dyn-sec per cm^2	=	1.0 P
1 g-sec per cm^2	=	980.7 P
1 lb-sec per ft^2	=	478.7 P
1 poise		= 1000.0 mP

(2) Kinematic Viscosity

The term *kinematic viscosity* (ν) is the ratio of the absolute viscosity divided by the density of fluid. For the kinematic viscosity, the units can be expressed in terms of feet square per second, or centimeters squared per second. In honor of Sir George Stokes, an English scientist, 1 c^2/sec is called a stoke. The conversions between the English and metric systems for dynamic and kinematic viscosities are shown in Example 5.2.

Example 5.2

(a) Convert the units of viscosity from poise to slugs per foot-second, and from pound-second per square foot to poise.
(b) Convert from 1 stoke into square foot per second or from square foot per second to stoke.

Solution

(a) 1 in. = 2.54 cm; 1 dyn = 2.248 × 10^{-6} lb, then:

$$1\,P = 1\frac{dyn - sec}{cm^2}$$

$$= \left(2.248 \times 10^{-6}\right)\left(2.54 \times 12\right)^2 lb - sec/\,ft^2$$

therefore,
1 P = 2.088 × 10^{-3} lb-sec/ft^2
       = 2.088 × 10^{-3} slugs/ft-sec, or
1 lb = 454 g; g = gravity = 980 cm/sec^2;
1 ft = 30.48 cm, then

$$1\frac{lb - sec}{ft^2} = \frac{(454)(980)}{(30.48)^2} \times 1\frac{dyn - sec}{cm^2} = 479\,P$$

(b) By definition, 1 stoke = 1 cm^2/sec
     1 cm^2 = 1.076 × 10^{-3} ft^2, then
     1 stoke = 1.076 × 10^{-3} ft^2/sec, or
     1 ft^2/sec = (30.48)^2 (1 cm^2/sec) = 929.03 stokes

5.3.4 Dielectric Constant of Water

The *dielectric constant* (ε) is a measure of a material's ability to perform as an insulator. That is, it is a measure of the capacity of a material to reduce the strength of an electric

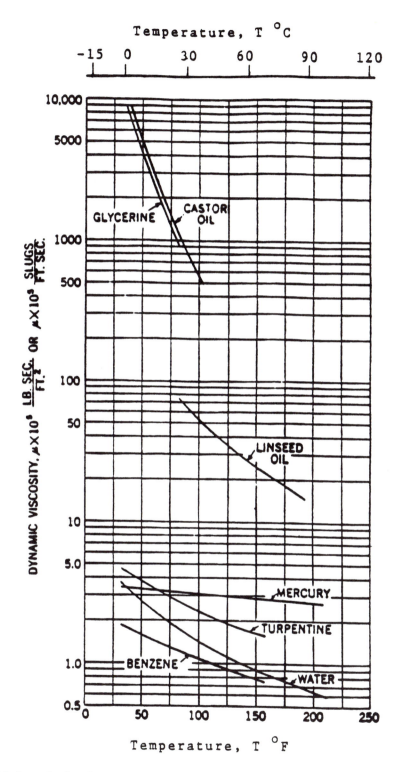

Figure 5.5 Dynamic viscosity vs. temperature of some commonly used liquids. (1 lb-sec/ft.^{2} = 478.7 poises.) (Data from Brady, G.S. [1947], *Material Handbook,* McGraw-Hill, New York; and others.)

energy field. The higher the dielectric constant of the material, the more the material behaves as an insulator. The dielectric constant of water is very high in comparison with other liquids. The high dielectric constant of water is the reason for its great solvent power for strong electrolytes. Further discussion on the dielectric constant includes measuring techniques, typical data, and correlation with other soil parameters, and will be presented in Section 9.4.

5.4 SOLUTIONS, COMPOUNDS, MIXTURES, AND ELECTROLYTES

5.4.1 Solutions, Solvents, and Solute

(1) Solutions

Solutions are mixtures of substances which dissolve in one another to become single-phase mixtures. Solutions can be unsaturated, saturated, and supersaturated depending on the amount of solute that is dissolved in the solvent. Solutions are also called dilute or concentrate, depending on the amount of solute dissolved in a given amount of solvent.

(2) Solvent and Solute

Solvent is a substance, usually a liquid, in which another substance called the *solute* is dissolved. Solubility is the amount of a solute that will dissolve in a given amount of solvent. The concentration of a solution expresses the quantity of solute relative to the quantity of the solvent or of the solution. A quantitative expression of the concentration of a solution is a ratio of two measurements as:

$$A = B/C \tag{5.1}$$

or

$$A = B/D \tag{5.2}$$

where A = concentration of solution; B = quantity of solute; C = quantity of solvent; and D = quantity of solution.

There are three units commonly used to indicate the concentration: (1) the percent by weight, (2) mole fraction, and (3) molarity. A brief explanation of each term together with common units frequently used to describe the characteristics of solutions, compounds, and mixtures follows.

1. Mole: The *mole* or chemical mole is amount of a substance in grams that contains as many chemical particles as there are carbon atoms in exactly 12 g of carbon-12. The chemical particles must be specified and may be atoms, molecules, or ions, or any specific groups of such entities.
2. Molarity (M): The *molarity* is defined as the number of moles of a solute per liter of solution. Solution concentrations can be expressed in terms of the number of solute particles (moles) that will be present in each liter of solution, or as Example 5.3.
3. Parts per million (ppm): Solutes found in very low concentrations in solutions are sometimes expressed in terms of the number of milligrams of solute per kilogram of solution, or the number of milligrams of solute per liter of solution.
4. Equivalent per million (epm): This is unit chemical equivalent weight of solute per million unit weights of solution (ASTM D1129-90).

Example 5.3

A solution contains 117 g of sodium chloride (NaCl) dissolved in water to make 2.0 liter of solution. Use molarity (*M*) to express such a solution.

Solution

(1) Determine the number of moles of NaCl.
From Appendix 2D, the atomic weight of Na = 23.0; Cl = 35.5; then:
117 g/(23.0 + 35.5) g/mol of NaCl = 2.0 mol of NaCl;

(2) Molarity $(M) = \dfrac{\text{Number of moles of solute}}{\text{Volume of solution in liters}}$

$$= \frac{2.0\,\text{mol of NaCl}}{2\,\text{liter}}$$

$$= 1.0\,\text{mol of NaCl/liter}$$

Notes: Chemists use molarities commonly dealing with solutions. A concentration is often expressed as 1.0 M of NaCl and it means that the solution contains a concentration of sodium chloride, so that each liter of solution will contain 1.0 mol of sodium chloride. In analysis for water pollution or contaminated pore fluid and chemical stabilization or grouting, these units frequently appear in literature.

5.4.2 Compounds and Mixtures

Compounds have definite properties and compositions and are homogeneous. Under certain conditions, they can be separated into two or more chemical elements. Water (H_2O) is composed of the chemical elements hydrogen (H) and oxygen (O). This compound always contains hydrogen and oxygen and any sample will contain 8 g of oxygen to every 1 g of hydrogen.

If we mix sugar or salt with water, it produces a compound. Mixtures of this kind do appear to be homogeneous, but do not have a definite composition. Sugar and water mixtures or salt and water mixtures can contain variable amounts of sugar in water or salt in water. When such variable compositions occur, they are called *mixtures*. *Solution* is classified as a mixture.

5.4.3 Electrolytes

Substances that form ions in water solution are called *electrolytes*. Some substances that ionize completely are called *strong* electrolytes. Others ionize only slightly and are called *weak* electrolytes. The following discussion will classify substances as strong or weak electrolytes:

1. Salts: All salts are strong electrolytes. For example, KBr, NaCl, NH_4NO_3, $CaCl_2$, and Na_2CO_3 are strong electrolytes. Some salts such as AgCl do not produce highly conductive solutions because they have a very low solubility.
2. Base: All soluble metal hydroxides (OH) are strong electrolytes. For example, NaOH, KOH, and $Mg(OH)_2$ are strongly electrolyted. But bases such as NH_3 and the amines tend to be weak. The strongest base that can exist in water is OH. This is because any stronger base will attract a proton from water to form a hydroxide ion.
3. Acids: Some acids are strong electrolytes, others are weak. Most acids are somewhere between the two extremes.

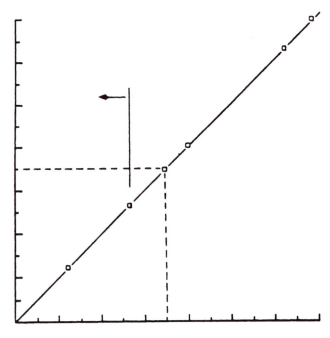

Figure 5.6 pH vs. H⁺ ion concentration.

Electrolysis is an important aspect of the electrochemical process. It is the relationship between the quantity of electricity passed through the cell and the amounts of substances produced by oxidation and reduction at the electrodes (Section 9.5). The quantity of electrical charge associated with 1 mol of electrons is called the *faraday* (F). The relationship between faraday and other units commonly used are

$$1 \text{ faraday} = 1 \text{ mole of electrons}$$
$$= 96,500 \text{ coulombs} \tag{5.3}$$

5.4.4 Acidity and pH Value

All soils contain H ions in their aqueous phase since water itself is dissociated into H^+ and OH^- ions. Soil acidity varies with the season. It is normally expressed as the negative logarithm of the H^+ ion concentration, which is called the *pH*. Accordingly, a pH of 7 indicates neutrality and a value less than 7 denotes acidity which increases with decreasing pH value. The relationship between pH and H^+ is shown in Figure 5.6. The pH of water and other liquids can be tested by ASTM standard procedure (ASTM D1293-95).

Soil-water acidity is important to the geotechnical engineer because of its corrosive effect on foundation structures and construction materials; on the other hand, acidity is sometimes desirable because of its catalytic effect on certain reactions employed in soil stabilization for ground improvement. At present, pH value is used for indicating the degree of contamination in most aqueous phases including acid rain, acid mine drainage, and leachate. Since the pH indicates only the concentration of the H ion as shown in Figure 5.6, other ions existing in the solution cannot be detected. Therefore, the validity of pH as the unified indicator for all contaminated pore fluid is questionable.

Example 5.4

What is the pH value of a 0.01 mol/l solution of sodium hydroxide (NaOH)?

Solution

(OH^-) $= 1 \times 10^{-2}$ mol/l
(H^+) $= 1 \times 10^{-12}$ mol/l
pH $= -\log_{10} (H^+) = -(-12) = 12$

5.5 TYPES AND SOURCES OF WATERS

5.5.1 Groundwater

Groundwater use in the U.S. is approximately 20% of the total national water demand. It accounts for more than 85% of the public water supply in several southwestern states and furnishes the total or partial water supply for most of the nation's l00 largest cities. It is estimated that more than 50% of the national population and more than 95% of the rural population receive their drinking water from groundwater resources. Also, indications are that some of the groundwater has been used for construction projects.

The implication of allowing groundwater resources to become polluted cannot be over-stated. A literature review on polluted groundwater is given by Todd and McNulty (1976) which includes causes, occurrence, procedures for control, and methods for monitoring. Some 30 specific topics are covered within the chapter headings of urban, industrial, agricultural pollutions, pollution from well, salt water, and surface water, pollutants, and effects and evaluating pollution.

In addition, McNabb et al. (1977) reported how the nutrient, bacterial, and virus control is related to groundwater contamination and their mechanisms. Groundwater contamination problems in the U.S. have been reviewed. More recently, problems concerning groundwater pollution have been expanded and cover more areas such as radon gas (Rn-222) (Section 10.4), petroleum products, metals, and other hazardous/toxic chemicals (Section 14.4). Both monitoring and *in situ* measurements on groundwater and vadose zone investigations are further discussed (Section 7.10).

5.5.2 River and Lake Waters

River and lake waters have various composition, often containing water which has come into contact with various geological formations or contains dissolving minerals and substances of decaying vegetations, dead animals, or mixed with polluted surface and subsurface waters. Some inland lakes are known as salt lakes, because they have accumulated vast amounts of dissolved mineral salts. Some additional causes such as acid rain and acid drainage will also affect the water components of lakes and rivers. Lakes containing high concentrations of salts in their waters are called *salt* lakes. If rich in sodium (Na) and potassium carbonate (K_2CO_3), they are called an *alkali* lake, or rich in sodium sulfate (Na_2SO_4) a *bitter* lake.

5.5.3 Hard, Soft, Acidic, and Alkali Waters

(1) Hard and Soft Waters

Hard water is water containing calcium ion (Ca_{2+}) and magnesium ion (Mg_{2+}) in concentrations greater than about 60 ppm. Hard water can produce scales in drainage pipes, water tanks, boilers, and various types of foundation structures. Such scales are a result of the formation of solid calcium and magnesium carbonates ($MgCO_3$). Water can be softened by removal of calcium and magnesium ions. The mechanism of how hard water changes into soft water has been discussed in Section 4.8.

Table 5.3   Typical Chemical Composition of Acid Rain

Acid	pK_a	Acid	pK_a	Note
HCl	0	$Al(H_2O)_6^{3+}$	5	
HNO_3	0	H_2CO_3	6.4	
H_2SO_4	0	H_2S	6.9	K_a = acid ionization constant
H_2SO_3	1.9	HSO_3^-	7.2	(A large ionization constant
HSO_4^-	2.0	$H_2PO_4^-$	7.2	value indicates an extensive
H_3PO_4	2.1	$B(OH)_3$	9.0	reaction with water)
HF	3.2	H_4SiO_4	9.2	pK_a = negative logarithms of K_a
$Fe(H_2O)_6^{3+}$	3.0	NH_4^+	9.3	(It is also used to indicate the
$Fe(H_2O)_5OH^{2+}$	4.0	HCH	9.4	strength of weak electrolyte)
Organic acids	3 5	HCO_3^-	10.3	
* H_2O (Water)	14			

After Gunnerson, G.G. and Willard, B.E., eds. (1979), *Acid Rain,* ASCE, New York, 166 p.

Water from which most of the calcium and magnesium ions have been removed is called *soft water.* Some waters are naturally soft, while others have varying degrees of hardness. One common method for softening water is by use of ion-exchange techniques (ASTM D1126-86).

(2) Acidic and Alkali (Basic) Waters

Any fluid can be acidic, basic (alkali), or neutral. Acidic solutions have concentrations of hydronium ion (H_3O^+) that are higher than pure water, and basic solutions have concentrations of hydronium ion that are less than pure water (Section 5.4.3 and Figure 5.6).

The majority of acid water is derived from acid rain and acid drainage. The importance of these acid waters related to the environmental problems has been briefly discussed in Section 1.6. Effect of climates on acid rain will be discussed in Section 8.3. Table 5.3 presents the typical chemical composition of acid rain.

Acid drainage in general is referred to as acid mine drainage. This problem is in the water draining from old abandoned coal mines. The iron sulfides or pyrites in the mines are oxidized into sulfuric acid. Then the sulfuric acid dissolves in water and drains through the mines. The acidic solution containing dissolved minerals is called acid mine drainage. When these acidic waters enter the streams, lakes, or ground they cause soil-water pollution. In 1969, there was acid mine drainage pollution of an estimated 16,000 km of waterways in the U.S., mostly in Pennsylvania, West Virginia, and Kentucky. Assessment of acid mine drainage pollution in southern Illinois has also been reported.

Acid mine water is a chemically complex liquid, having no definite quantitative relationship between its constituents. It transports and deposits mainly inorganic materials as solids, in solution, or in suspension. These acid mine waters range from alkaline to strongly acid (Stefanko et al. 1965). The principle contaminants in the acid mine water are sulfuric acid (H_2SO_4) and ferric hydroxide (Fe[OH]$_3$). Ferric hydroxide is a rust-colored gelatinous precipitate that coats stream bottoms and destroys aquatic life and is the major river pollution problem in Pennsylvania.

In addition to iron and sulfuric acid, other ions are present in acid mine water but generally in much smaller quantities. The more important of these impurities are silica, aluminum, manganese, calcium magnesium, and sulfate. The acid mine water is a suitable habitat for some types of microorganisms. A majority of these microorganisms are bacteria belonging to the bacillus group. These bacteria are autotrophic, they require no organic material for their existence, but derive energy by oxidizing inorganic substances. Acid drainage water will cause corrosion in various steel and concrete structures, highway pavements, sign posts,

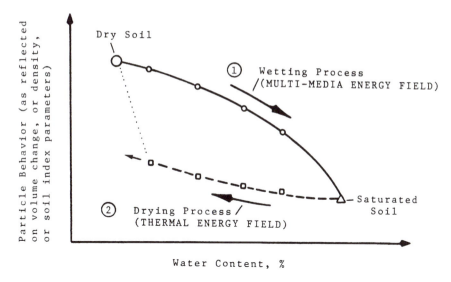

Figure 5.7 Effect of wetting or drying processes on soil behavior. (1) Wetting process; (2) drying process.

drainage pipes, culverts, and underground structures. For prevention of acid mine drainage, sealing the abandoned mines or diverting the water which flows through the mines is commonly suggested.

Other types of polluted waters are listed and discussed in Chapters 7 and 8. Seawater pollution is discussed in Section 14.4. Identifications and measurements of quality of these waters can be done through proper laboratory analysis provided by ASTM Standards Section 11 (1997).

5.6 ELECTROCHEMICAL CHARACTERISTICS OF SOIL-WATER SYSTEM

5.6.1 Surface Electrochemical Characteristics of Soil

Interaction between a liquid and a solid can occur only on the solid's surface. The surfaces of the soil solids may affect the water structure not only by electrical interaction with the H_2O dipoles and their directional bond-forming capacity, but also by lack of such interaction. The surface of a solid soil particle may serve as something to lean on and be protected by the large and loose cage structures that water molecules tend to build. The essential components of a soil-water system reveal themselves as:

1. Surfaces of the soil minerals present with their characteristic geometric and electric patterns.
2. Cations located on or within the mineral surface layer in the case of very dry soils, but dissociating to a greater or lesser extent into a surrounding water phase until equilibrium is established between the kinetic dispersive forces of the cations and their Coulombic attraction to the charges on the mineral surfaces.
3. The water molecules occur in different forms of association of different packing densities on the mineral surfaces, around the dissociated ions and the free water domains, with the distribution of the various forms being a function of the amounts and activities of the components and the temperature. Thus, in an aqueous clay suspension we may have water simultaneously in the conditions, as discussed in Section 4.9 and Figure 4.8. Soil-water interaction is a complex phenomenon which includes ion-exchange reaction, leaching, osmotic, hydration, adsorption, heat of wetting, oxidation-reduction reactions, etc. Further discussions on environmental aspects are presented in the following sections.

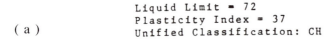

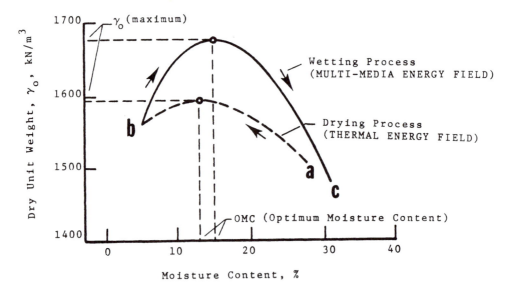

Figure 5.8 Effect of flow path direction on engineering properties of soils. (a) Compaction test result; (b) liquid limit test result. (Data from Tamez, E. [1957], Some factors affecting the dynamic compaction test, ASTM STP 232, pp. 54–66; and Krebs, R.D. [1964], Discussion of study of reproducibility of Atterberg limits, HRR no. 63, pp. 29–30.)

5.6.2 Flow Routes Relating to Soil-Water Interaction

Soil equilibrium conditions differ when in contact with liquid water and water vapor. Even at the same moisture content, the structure and physicochemical properties of a soil-water system may differ considerably, depending upon the flow path from which this moisture content has been attained. These paths may reflect absorption of water vapor by a dry soil, drying of a wet soil to the desired moisture content (Figure 5.7), or the squeezing of liquid water from a wet soil by external pressure.

(1) Flow Path-Wetting Process

From Figure 5.7(1), the entrance of water into a porous system is due to its affinity for the internal surface of the system; then the resulting phenomena are governed by the following factors:

1. The driving force or the affinity of the internal soil surface for water, the magnitude of which is indicated by the heat of wetting and which is the product of the amount and activity of the internal surface area
2. The total porosity and the pore size distribution, which determines the permeability to pore fluid (or water) and the ease or difficulty of escape of free and adsorbed gas (or air) in the original soil system
3. The rate of destruction of the bonds between the solid particles which changes the geometry of the pore spaces
4. The swelling capacity of the clay minerals which may result either in internal swelling that decreases the permeability or may be large enough to result in surface exfoliation of the systems

(b)

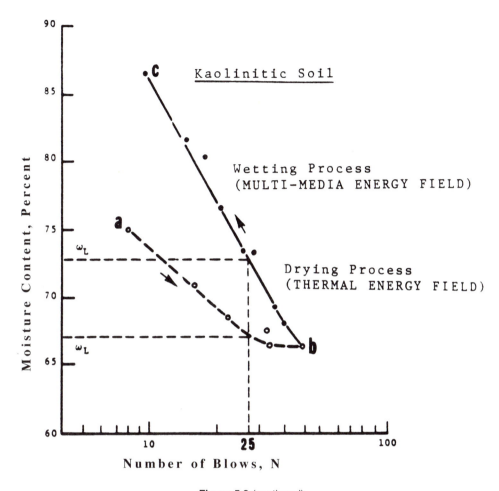

Figure 5.8 (continued)

(2) Flow Path-Drying Process

From Figure 5.7(2), as a wet or moist cohesive soil system loses water, the soil particles move closer and closer together. If the drying proceeds from the surface down as in a mud flat, the dehydrated surface layer shrinks while the water resistance between the upper and lower layers and the layers themselves prevent an adjustment to the volume decrease of the surface layer. As a result, tensile stresses are developed in the surface layer. This cracking pattern produces the greatest stress release with the least amount of work.

Flow path directions are an important factor and it is pointed out that the same water content of a soil is not necessary to give the same engineering properties. Figures 5.8(a) and (b) show the significant effects of flow path direction on compaction and liquid limit test results.

5.6.3 Soil-Water Interaction in Various Energy Fields

Soil-water interaction is complex. If polluted water is involved, the complication will significantly increase. For evaluation of the soil/water system, we must treat the soil element in three phases: solid, liquid, and gas. The liquid portion can be divided into two parts, gravity water and environmental waters. Gravity water is a great distance from dissolved ions and

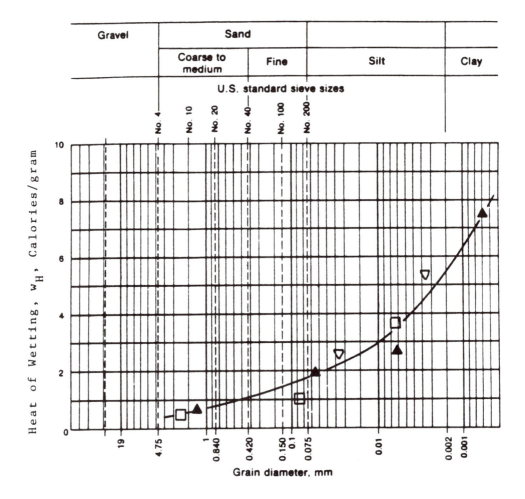

▽ Fang (1994c)

▲ Landolt-Bornstein (1952)
  cited by Winterkorn and Fang (1975)

Figure 5.9 Effect of soil particle size on heat of wetting.

molecules and existing soil-liquid and liquid-gas interfaces. Environmental waters are strongly influenced by local environmental conditions such as pore fluid character, temperature, and pressure. In this section comments are added with respect to the soil-water interaction in the thermal and electric energy fields.

5.7 SOIL-WATER INTERACTION IN THE THERMAL ENERGY FIELD

5.7.1 Heat of Wetting

The heat of wetting is an important reaction in the soil-water system. When water contacts dry or partially saturated soil, heat is produced. The *heat of wetting* is the evolution of heat from hydroscopic substances when they become wet. This phenomenon occurs in connection with fine grained soil, especially the colloids so that the quantity of heat evolved on wetting

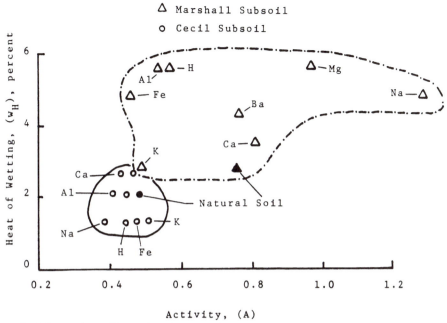

Figure 5.10 Comparison of heat of wetting vs. activity between natural and homoionic soils. (Data from Winterkorn (1942) and Table 4.8.)

the soil increases with fineness. Colloidal soil materials give comparatively high heats of wetting, though varying according to the type of material and pore fluid. The heat of wetting can be measured by a simple apparatus developed by Winterkorn and Baver (1934), which has been used widely for determination of heat of wetting in the laboratory (Vees and Winterkorn, 1967). Factors affecting heat of wetting are presented as follows.

The effects of particle size on heat of wetting are shown in Figure 5.9. The comparison between polluted and nonpolluted soil samples is also discussed. As reported by Winterkorn and Fang (1991), the wetting of 1 g of bentonite clay dried over phosphoric acid (P_2O_5) with 0.3 g of water results in the release of about 20 cal or about 1200 cal/mol of water thus interacting; this is about one fourth of the energy of the hydrogen linkage per mole of water.

The effect of exchangeable ions on heat of wetting is also important. The facts are important when the heat is released or absorbed as ions, molecules are dissolved, or solid particles are wetted. Figure 5.10 presents the comparison of heat of wetting vs. activity between natural and homoionic soils. Two soil types are used and significant differences between natural and homoionic soils are found.

Since many underground structures such as slurry walls and clay liners are made mainly of bentonite, the surface behavior of bentonite is sensitive to the type of pore fluid, and the thermal gradient is one of the major factors in the hazardous and toxic-controlling facilities, therefore, the characteristics of heat of wetting for types of soil and pore fluid must be evaluated. Figure 5.11 presents the relationship between soil types as reflected by silica/sesquioxide ratio and heat of wetting. Significant differences between Cecil soil and Ca-bentonite are found. The natural characteristics of these natural soil deposits used in Figure 5.11 can be found in Section 3.6 and Table 3.11.

5.7.2 Water Intake Ability

Soil absorbs water and the water intake ability of each soil type is important to an engineer. The mechanism of water intake ability will be discussed in this section. Further discussion on saturation and sorption (absorption + adsorption) of soil will be presented in Section 6.5.

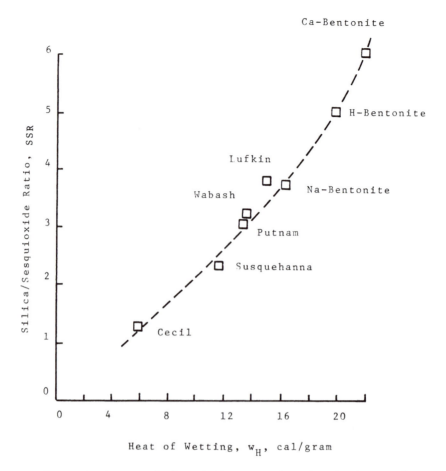

Figure 5.11 Relationship between silica/sesquioxide ratio vs. heat of wetting for several natural soil deposits.

Figure 5.12 presents water intake and Figure 5.13 shows the change in volume of water held per cubic centimeter of soil as a function of the consolidating loads and of the ions. Sodium (Na) soil suffers a large decrease in volume of water as loads are applied. The volume of water in the natural soil decreases considerably under the first load increment, but, even as the loads become larger, the volume change is relatively small. Calcium (Ca) soil has the least amount of total volume change of all the homoionic soils in both consolidation and expansion. This phenomenon indicates that the water films around the calcium soil particles are compact and are subject to only moderate reduction by increase in load. The small increases of moisture content during reduction of pressure indicate that the calcium has but small attraction for additional water.

In the case of potassium (K), the water film is also compact and slightly reduced by pressure; however, when the load is entirely removed, a large volume increase results which is due either to a strong attraction of the potassium ion for water or to an increased springiness of the soil system. Sodium soil, if unconfined, will attract a large amount of water. However, this attraction is not sufficiently forceful to produce a marked volume increase when consolidation loads are removed. As has been indicated in the foregoing discussion, the effects of the ions on the engineering properties of soils may be due not only to the change in water affinity by the ions, but also to changes of the elastic properties and of the aggregation of the solid soil constituents.

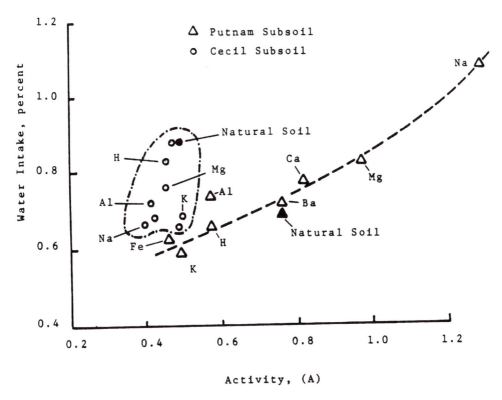

Figure 5.12 Water intake ability vs. activity between natural and homoionic soils. (Data from Table 4.8.)

5.7.3 The Kinetic Dispersive Force

In the thermal energy field, the kinetic dispersive forces around the soil particle are important. This force can be estimated from classical thermodynamic irreversible processes. The fundamental principles of thermodynamics are the law of the conservation of energy, and the second law of thermodynamics which is that a system alone either remains constant (reversible process) or increases (irreversible process). This function is called the entropy of the system. The theorem also permits the calculation of absolute entropies if the specific heats of the components of the system are known for the entire range from absolute zero to the working temperature. Therefore, the dispersive and disturbing forces of the thermal energy, E_T, can be estimated as:

$$E_r = \int_{T_0}^{T} C_v \, dt \tag{5.4}$$

where E_T = total thermal energy content per mole; C_V = mole heat at constant volume; T = temperature of interest; T_0 = absolute zero temperature; and dt = temperature differential.

In Equation (5.4), where the capacity factor is the mole-heat, C_V,

$$C_V = (\text{Specific Heat}) \times (\text{Mole Weight}) \tag{5.5}$$

The intensity factor is the absolute temperature measured preferably in °K. The total thermal energy, E_T, content per mole can be calculated from Equation (5.4). The mole heat, c_V, itself

Putnam Clay

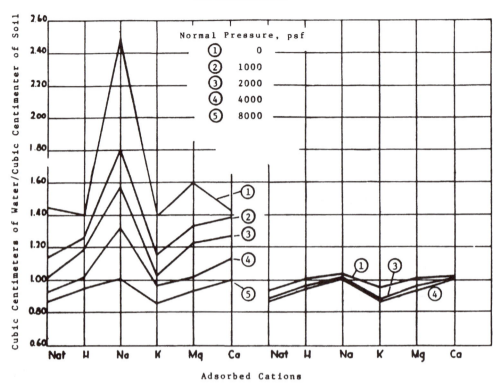

Figure 5.13 Water-holding capacity of Putnam clay as a function of the exchange ions and of the normal pressure. (After Winterkorn, H.F. and Moorman, R.B.B. [1941], Proc. 21st Annual Meeting, HRB, pp. 415–434.)

is a function of temperature and the state of matter (gas, liquid, or solid), since the latter determines the various manners in which thermal energy can be utilized.

In the solid state, we may have oscillation of the atoms in the three dimensions of Cartesian space around their equilibrium positions. In liquids we may have similar oscillations around temporary equilibrium positions; also, rotation and translation. In gases we have translations in monatomic gases to which rotations and oscillations are added in the cases of di-, tri-, and polyatomic molecules, and finally electron oscillations at elevated temperatures. For detailed accounts of the various phenomena involved and their theoretical treatment, reference is made to the standard works on the theory of specific heats. The disordering and dispersive forces increase with increasing temperature, and these forces are not the same for every molecule in a particular system held at a particular temperature, but are distributed around a mean value in accordance with the Boltzmann probability function.

5.8 SOIL-WATER INTERACTION IN THE ELECTRIC ENERGY FIELD

5.8.1 Polarization

All atoms and, hence, the molecules formed thereof contain positive and negative electric charges, the former being associated with the atomic nucleus, while the latter is embodied in the electrons that surround the nuclei. The size of the nuclei is of the order of 10^{-12} Å,

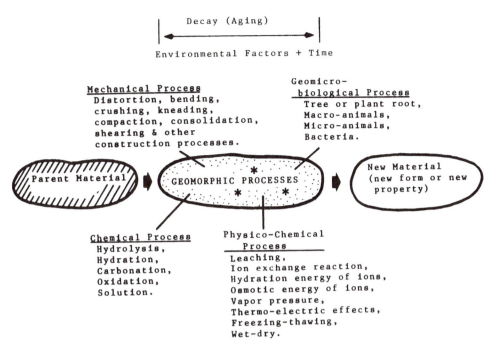

Figure 5.14 The stages of the geomorphic process. (After Fang, H.Y. [1986], Proc. 1st Int. Symp. on Environmental Geotechnology, v. 1, pp. 167–194.)

with the size of the atoms being of the order of 10^{-8} Å; the electrons occupy most of the atomic and molecular volumes.

In strong electric fields, the volume occupied by the electrons can be distorted and shifted with respect to the position of the associated nuclei. This phenomenon is called *polarization*. While positive and negative charges associated with matter try to arrange themselves in a spatial order which leads to mutual neutralizations, their volume and structural properties interfere with this tendency. For further discussion, see a standard textbook on physical-chemistry.

5.8.2 Proton Migration

In the highly electric or polar system composed of soil minerals, dissolved ions, and water dipoles, the various electric fields associated with these constituents tend to readjust themselves in the direction of lowest free energy or highest entropy of the system. A primary role in this readjustment is played by *proton migration* and *proton exchange*. According to Eyraud et al. (1965), this explains:

1. Slow hydratation of oxides by thermal proton diffusion
2. Manifestation of Lewis acidity by exchange of a surface water molecule with any other molecule from the gaseous or liquid phase
3. Manifestation of a Bronsted acidity by the exchange of a surface proton with a certain solution

This model also explains the significant role that internal hydratation plays in the formation of tropical soils or laterite soil (Section 13.9). Responsible for this is the increase in kinetic activity of the proton with increasing temperature, which expresses itself also in an increase in degree of dissociation of the water molecule with increase in temperature.

Table 5.4 Geomorphic Processes by Chemical and Physicochemical Processes

Carbonation

(1) $Ca(OH)_2 + 2CO_2 = Ca(HCO_3)_2$
(2) $2KOH + CO_2 = K_2CO_3 + H_2O$

Hydration

(3) $CO_2 + H_2O = CH_2O + O_2$
(4) $CaSO_4 + 2H_2O = CaSO_4 \cdot 2H_2O$
(5) $2Fe_2O_3 + 3H_2O = 2Fe_2O_3 \cdot 3H_2O$

Hydrolysis

(6) $CaSiO_3 + 2HOH = H_2SiO_3 + Ca(OH)_2$
(7) $KAlSi_3O_8 + HOH = HAlSi_3O_8 + KOH$

Carbonation and Hydrolysis (Solution)

(8) $2KalSi_3O_8 + 2H_2O + CO_2 = H_4Al_2Si_2O_9 + 4SiO_2 + K_2CO_3$
(9) $4K(AlSi_3O_8 + 4H_2O + 2CO_2 \rightarrow Al_4(Si_4O_{10}) \cdot (OH) + 8SiO_2 + 2K_2CO_3$
(10) $Ca_3(PO_4)_2 + 2H_2O + 2CO_2 = Ca_2H_2(PO_4)_2 + Ca(HCO_3)_2$
(11) $Ca_3(PO_4)_2 + 4H_2O + 4CO_2 = CaH_4(PO_4)_2 + 2Ca(HCO_3)_2$
(12) $CaCO_3 + H_2O + CO_2 \rightarrow Ca(HCO_3)_2$
(13) $Ca(soil)_2 + 2H_2O + 2CO_2 \rightarrow 2H (soil) + Ca(HCO_3)_2$

Oxidation

(14) $2FeO + O = Fe_2O_3$
(15) $2FeS_2 + 7O_2 + 2H_2O \rightarrow 2FeSO_4 + 2H_2SO_4$
(16) $FeS_2 + 7O + H_2O \rightarrow FeSO_4 + H_2SO_4$
(17) $4FeSO_4 + O_2 + 2H_2SO_4 \rightleftharpoons 2Fe_2(SO_4)_3 + 2H_2O$
(18) $6FeSO_4 + 3O + 3H_2O \rightleftharpoons 2Fe_2(SO_4)_3 + 2Fe(OH)_3$

Ion Exchange and Others

(19) $Na_2\% + CaSO_4 = Ca\% + Na_2SO_4$
(20) $Ca\% + 2NaCl = Na_2\% + CaCl_2$
(21) $2H(Soil) + Ca(HCO_3)_2 \text{ (or } CaCO_3) \rightarrow Ca(soil) + 2H_2CO_3$
(22) $\text{Organic nitrogen} \rightarrow NH_4^+ \rightarrow NO_2^- \rightarrow NO_3^-$
(23) Soil solids $\rightarrow$ soil solution $\rightarrow$ higher plants
                                    $\rightarrow$ drainage losses

5.9 GEOMORPHIC PROCESS (AGING PROCESS)

5.9.1 The Concept of Geomorphic Process

A parent material in a given location, after a certain time period and due to various environmental factors, will gradually change its form or properties. The stages or phenomena of such changes are called the '*geomorphic process*' as introduced by Fang (1986) and illustrated in Figure 5.14. The rate of change depends on how severe the environmental conditions are. There are five possible processes which could happen to soil and rock, namely: (1) mechanical (load); (2) chemical; (3) physicochemical; (4) geomicrobiological (flora and fauna) processes; and (5) soil-forming.

5.9.2 Geomorphic Process Mechanisms

(1) Mechanical Causes and Short-Term Process

As illustrated in Figure 5.14, mechanical processes include compaction, consolidation, shear, and much currently used construction equipment. Results of mechanical processes change the physical properties of soil as reflected on load-deflection, stress-strain, or void ratio vs. pressure relationships. In general, the mechanical alteration is considered a short-term process, while the others are long-term processes.

(2) Chemical and Physicochemical Processes

Chemical alteration can change, as indicated in Figure 5.14, including hydrolysis, hydration, carbonation, oxidation, solution, and others. Some of these processes are illustrated in Table 5.4. Of course, all the processes function according to the local environment. For example, in humid climates where moisture is present in the soil for long periods of time, the chemical and physical processes force rock decay. The combination of mineral constituents, especially iron with oxygen, produces the transformation of ferrous salts to ferric compounds as shown in Equation (14) of Table 5.4. Minerals containing strong basic elements are inclined to exchange a part of the base from hydrogen in water such as Equations (6) and (7) in Table 5.4 illustrate. The decomposition of organic matter in the soil produces large quantities of CO_2 in the soil/water system. This gas readily combines with bases producing carbonates or bicarbonates, and typical cases are shown in Equations (1) and (2) in Table 5.4.

As the bicarbonates of calcium and magnesium are fairly soluble, carbonation is the primary process in accumulation of residual soil material through the decomposition of limestone and in the decomposition of minerals with a high calcium content. The sinkhole (Section 13.6) is a typical case of the decomposition of limestone. Equation (8) in Table 5.4 is commonly associated with geotechnical engineering. Under the combined action of carbonation and hydrolysis, orthoclase ($2KAlSi_3O_8$) with CO_2 and $2H_2O$ will become kaolinite ($H_4Al_2Si_2O_9$), potassium carbonate (K_2CO_3) and amorphous silica ($4SiO_2$). As time passes and local environmental conditions change, the properties and forms of soils will continue to change. Orthoclase can change into kaolinite and potassium. Likewise, kaolinite and others will change into something else if local environmental conditions change. The rate of change depends on how severe the environmental conditions are.

(3) Geomicrobiological Process

The term called 'geomicrobiological' process is proposed by Fang (1986) as illustrated in Table 5.5. The geomicrobiological process is divided into two major categories, flora and fauna. The flora is mainly tree or plant roots which have some degree of influence on the soil structure and behavior, but not as significantly, in comparison with the bacterial effects (Table 5.6) on soil which may be directly or indirectly related to the chemical or physicochemical alteration of soil behavior. For example, decomposition in landfill causes the ground temperature to increase. Because of aerobic conditions normally prevailing in landfill or in polluted soil above the groundwater level, these bacteria oxidize the sulfide to sulfuric acid which attacks the underground structures. Also, the autotrophic type of bacteria requires energy for their survival and takes it from oxidation of mineral constituents. As a result it may change the clay mineral structures and alter soil behavior during the geomorphic processes as illustrated in Table 5.6.

Table 5.5   Engineering Classification of Geo-Microbiology and Its Effects on Geotechnical Problems

Types and descriptions	Influenced depth Horizons A	B	C	Effects on geotechnology
Flora				
Tree or plant roots				
Tree-root system	x	x		Ground settlement; broken
Soil-root system				drainage pipes; damaged
Root-water mechanics				highway pavement; damaged landfill liners
Fauna				
Animals				
Macro-animals				
Rodents, millipedes,	x	x		Damage top seals of clay liners;
centipedes, spiders, mites,				top portion of slurry wall;
ticks, slugs, snails, insects,				surface erosion
ants, and earthworms				
Micro-animals				
Protozoa	x	x		Bacteria corrosion; landfill liners
Amoeba, flagellates, ciliates, infusoria				
Nematoda				
Feed on decaying organic	x	x		Bacteria corrosion
matter; feed on earthworms,				
protozoa, and bacteria				
Bacteria				
Heterotrophic				
Energy and carbon directly	x	x	x	Bacteria corrosion; clay liners;
from the soil organic matter				slurry walls
Autotrophic				
Energy from oxidation of	x	x		Bacteria corrosion
mineral constituents				

After Fang, H.Y. (1986), Proc. 1st Int. Symp. on Environmental Geotechnology, v. 1, pp. 1–14.

Table 5.6   Effect of Bacteria on Pore Fluids

(a) $H_2S + 2O_2 \xrightarrow{\text{bacteria}} H_2SO_4$

(b) $CH + O_2 \xrightarrow{\text{bacteria}} CO_2 + H_2O$

(c) $CH_2O + O_2 \xrightarrow{\text{bacteria}} CO_2 + H_2O$

(d) $C_6H_{12}O_6 + 6O_2 \xrightarrow{\text{bacteria}} 6CO_2 + 6H_2O$

(e) $SO_4^{2-} + \text{organic matter} \xrightarrow[\text{bacteria}]{\text{anaerobic}} S^{2-} + H_2O + CO_2$

(f) Organic sulfur compounds $+ O_2 \xrightarrow{\text{bacteria}} CO_2 + H_2O + SO_4^{2-}$

(g) Organic nitrogen compounds $+ O_2 \xrightarrow{\text{bacteria}} CO_2 + H_2O + NO_3^-$

(h) Organic phosphorous compounds $+ O_2 \xrightarrow{\text{bacteria}} CO_2 + H_2O + PO_4^{3-}$

(i) Organic sulfur/nitrogen compounds $+ H_2O \xrightarrow[\text{decay}]{\text{anaerobic}} CO_2 + H_2S + CH_4 + NH_4^+$

(4) Soil-Forming Process

Soil-forming processes basically are the combination of physicochemical, chemical, and geomicrobiological processes. The term called '*soil-forming*' is commonly used in soil science and agricultural engineering. The factors concerned in the development of soils from accumulated materials that originate from rock weathering can be grouped into four categories, namely: climate, biotic, edaphic, and topographic as discussed in Chapter 3. Climatic factors include temperature, rainfall, and biotic factors include flora and fauna (Table 5.5). The edaphic factors refer to parent soil materials and topographic factors mainly concern surface relief. Soil-forming processes also can be grouped into the following subgroups:

1. Podzolization: Podzolization is principally a dissolving process which involves both true solutions and soil, and hence is a deteriorative process which takes place in acid mediums in all humid regions. It begins when sufficient basic elements have been leached out to affect acidity.
2. Lateritization: Lateritization is characterized by the removal of silica and residue of the sesquioxides to the extent that the alumina-silica ratio (Section 3.6) becomes 1.0 to less than 2.0, or falls below the siallitic stage. Sometimes it is called desilicification.
3. Carbonation: Carbonation is a soil-forming process of subhumid, semiarid, and also arid regions. It is the conversion of soil calcium, principally, into carbonates which accumulate in the subsoils at different depths as determined by quantity of annual precipitation.
4. Alkalization: Alkalization is the dominant soil-forming process of arid regions. It consists mainly of accumulation of soluble salts which form in the weathering of mineral soil particles, owing to insufficient rainfall or drainage to carry them away, and in the action of the salts, particularly those of sodium on the soil base-exchange compounds. In these types of soil, the ion-exchange compounds contain considerable sodium and potassium; in some of these compounds the percentage of these two alkali cations (exchangeable) may reach 100% (Kelley and Brown, 1924 cited by Kelley, 1948). Degradation of sodium alkali soils (solon-chaks) may take place when such soils, owing to lowering of the groundwater table, become subject to leaching, in which exchangeable alkali cations (Na, K) may ultimately become more or less displaced by hydrogen.

5.10 CORROSION PROCESS

5.10.1 General Discussion

The corrosion process is part of the geomorphic process or the 'aging process'. It is related to time and local environments as discussed in Section 5.9 and Figure 5.14. The corrosion of exposed structural components, particularly those of bridges, highways, and offshore structures, undermines the integrity of these structures and asserts economic problems. The importance of finding cost-effective solutions to the corrosion problem cannot be denied. There are three mediums in which corrosion takes place: air, water, and ground soil. Over the years, research on underground corrosion, which involves the soil medium, has been relatively neglected as opposed to research on corrosion in air and water. This is primarily because corrosion underground is not readily visible.

5.10.2 Corrosion Causes, Types, and Protection

(1) General Discussion

Various chemical compounds and microorganisms within the soil matrix trigger and promote the corrosion process and deterioration of steel and concrete structures underground. Natural or man-made events, such as acid rain, hazardous waste dumping, and landfill leaching, pollute the soil and groundwater in many areas. These pollutants provide an aggres-

(a)

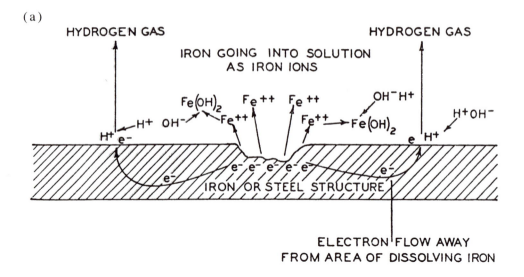

Corrosion Process of Iron or Steel Structural Members

Figure 5.15 Steel corrosion process and mechanism. (a) Corrosion process; (b) example. (After Frye, S.C. [1970], *Proc. Design and Installation of Pile Foundations and Cellular Structures,* Envo Publishing, Bethlehem, PA, pp. 191–208.)

sive environment in which corrosion may become critical. Moreover, continuing metabolic activity of some microbial community results in production of corrosive chemicals. Consequently, steel may be eroded cathodically or physically (pitting). Concrete, which protects the reinforcing steel, may be attacked chemically or invaded by microorganisms that rapidly create an active chemical environment.

(2) *Corrosion Causes*

Basically, corrosion is electrochemical in nature and the presence of oxygen in some form is necessary. For a corrosion cell to function, certain conditions must be met: (1) there must be an anode and a cathode, (2) there must be an electrical potential between the anode and cathode, and (3) there must be a metallic path, electrically connecting the anode and cathode. The anode and cathode must be in a conductive electrolyte, meaning that ions are present. For example, a steel surface which can be classified as heterogeneous and not homogeneous can ideally provide the anodes and cathodes. Seawater or a contaminated soil-water system is an ideal electrolyte solution. All the conditions for corrosion are present.

The difference in potential between the anode and cathode results in the migration of electrons from the anode to the cathode along the metallic connection. At the anode, with the loss of electron positive, ferrous ions combine with the negative OH ions in the water, forming ferrous hydroxide which can also react further to form the familiar rust that we see. The amount of metal removed at the anode is directly proportional to the amount of flowing current. One ampere of current flowing for 1 year can remove 20 lb of steel.

(3) *Bacteria Corrosion*

Bacteria and chemical corrosion are closely related, as bacteria can cause changes in the characteristics of soil-water systems. A typical example of this may be the conversion of H_2S into H_2SO_4 through bacterial action, as shown below:

(b)

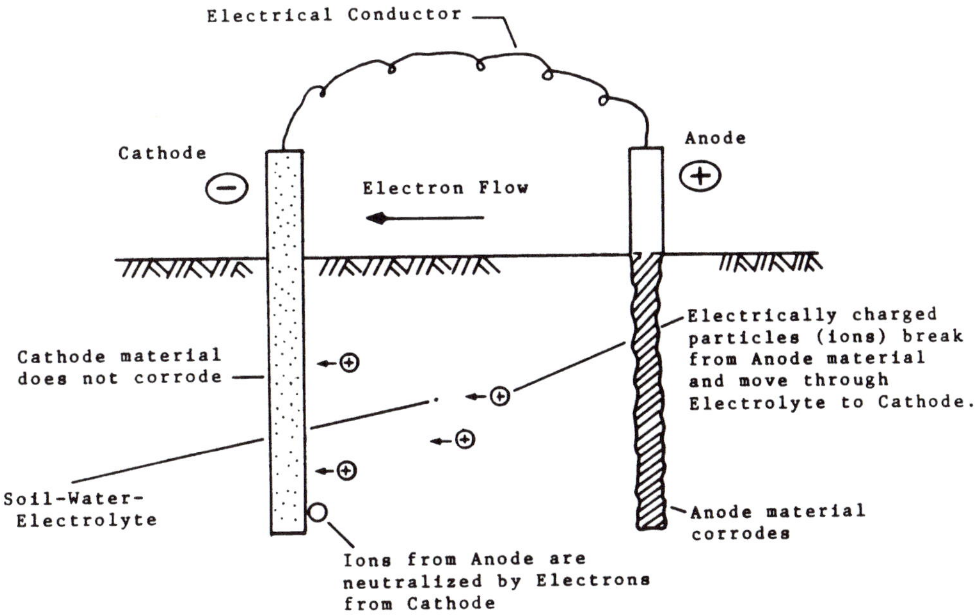

Figure 5.15 (continued)

$$H_2S + 2O_2 \xrightarrow{\text{bacteria}} H_2SO_4 \qquad (5.6)$$

The mechanism of bacterial corrosion has been discussed by a number of investigators in the past. The major points can be summarized as: (1) production of corrosive metabolic compounds by bacteria, e.g., hydrogen sulfide; (2) formation of discontinuous deposits on the structural surface results in differential aeration and concentration cells; (3) distruption of natural and other protective films; (4) breakdown of corrosion inhibitors and coatings; and (5) depolarization of cathodic and anodic reactions (Walch and Mitchell, 1984).

(4) Corrosion Protection and Example

Figure 5.15(a) shows the effect of anodic and cathodic areas on the heterogeneous surface of a piece of steel. A current leaves the anode and flows to the cathode. Conceivably, if an external source of current is impressed on the steel surface greater than that of the corrosion current, all of the steel surface will then become a cathode, and corrosion to all intents and purposes will be nullified. In essence, this is the simplified theory of cathodic protection. Figure 5.15(b) shows one method of using an external anode and the protective flow of current to the cathode. Detailed discussions on other protection methods, performance, and cost are presented by Frye (1970).

In civil engineering, most of the structures erected on a ground surface have extensive underground components. These components may be subject to underground corrosion in time. Underground corrosion is a more complex phenomenon than atmospheric corrosion due to the large number of variables and unknowns involved. In the past, most of the research

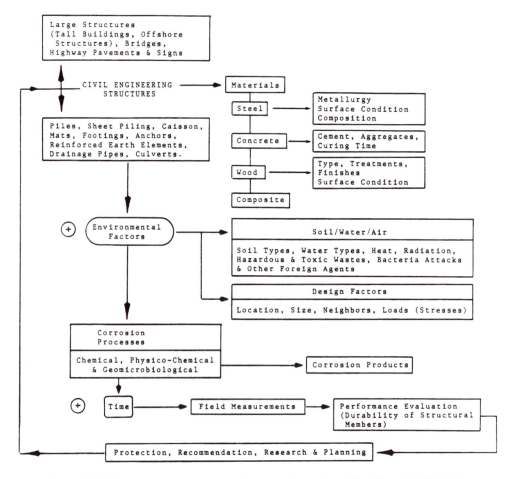

Figure 5.16 Corrosion: cause-process-performance interaction. (After Fang, H.Y. [1987].)

efforts have concentrated on atmospheric corrosion rather than underground environmental corrosion. Clearly, more emphasis is needed on this subject to determine the cause-process-performance factors for underground structures or components of structures in environmentally adverse conditions. These factors are given in a flowchart in Figure 5.16.

Corrosion is a hidden enemy for all underground structural members such as anchors, foundation components, sheet piling joints, steel-bar in rigid pavements, the bridge decks, rivets, and bolted joints, as well as reinforced earth elements. Small holes and crevices are where corrosion-causing bacteria may exist abundantly. Better understanding of bacterial corrosion involves interdisciplinary subjects, such as physicochemical and biological activities that take place in soil. More research is needed with regard to these areas to better assess the conditions and processes that favor bacterial corrosion in soil and to establish systematic, long-term solutions to prevent it.

5.11 EFFECT OF BACTERIA ON BEHAVIOR OF SOIL-WATER SYSTEM

5.11.1 Characteristics of Bacteria

To examine long-term performance of soil-water behavior in the environment, the effects of bacteria on this behavior must be evaluated. In this section, the discussions focus on: (1)

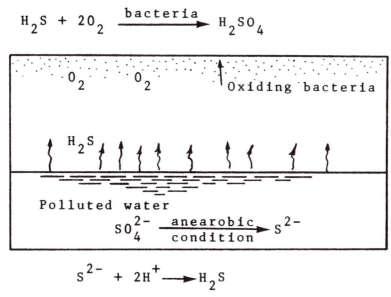

Figure 5.17 Formation of hydrogen sulfide in landfill area resulting from oxidation of hydrogen sulfide to sulfuric acid.

the facts about how bacteria exist in the soil-water system and how they relate to geotechnical engineering, and (2) the mechanism of soil-water-bacteria interaction.

Bacteria are single cell forms. Four major elements — carbon, hydrogen, oxygen, and nitrogen — compose about 98.5%, by weight, of the atoms of the cells.

Phosphorus and sulfur together make up the additional percent. The most abundant molecule in all living cells is water, which usually makes up about 70% of the cell. In general, there are two basic types of bacteria: heterotrophic and autotrophic. The heterotrophic type of bacteria requires energy and carbon directly from soil's organic matter; however, the autotrophic type requires energy from oxidation of mineral constituents (Nester et al. 1978).

5.11.2 Soil-Water-Bacteria Interaction

Most bacteria exist at a shallow depth from ground surface, especially in the A-horizon. It is interesting to know that in 1 g of soil can exist four million bacteria (Figure 3.2[B]). Disturbed soil has more bacteria than undisturbed because the air (oxygen) creates more bacteria. Most bacteria exist in an aerobic condition which is above the groundwater table. The rate and degree of bacteria activity depend on the availability of nutrients, microorganic types, substrate composition, and temperature. In cold weather, bacteria activities are less, but flooding, fluctuating water tables, or disturbing the ground soil, such as construction, will accelerate the aerobic process. In the anaerobic condition, bacteria exist primarily below the groundwater table; in comparison to the aerobic condition, these bacteria activities are a slow process.

There are three basic possible ways that bacteria may cause a soil to change behavior. Bacteria can transfer in an ion exchange which changes the soil-water structure, change the characteristics of pore water as shown in Table 5.6, or change through decomposition (geomorphic process) of soil constituents. Bacteria and chemical corrosion are closely related. Bacteria can cause changes in the characteristics of soil-water systems. A typical example of this effect is presented in Figure 5.17.

Condition Before Loading

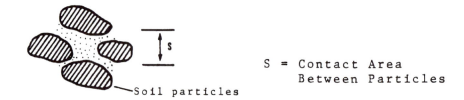

S = Contact Area
    Between Particles

Conditions During Loading

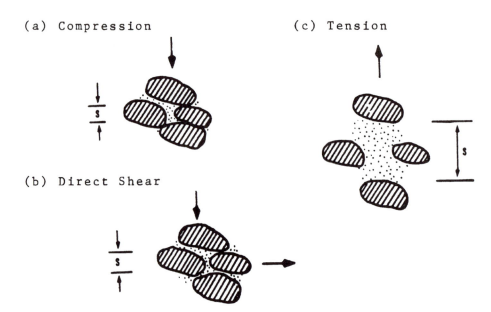

Figure 5.18 Schematic diagram to illustrate the type of test methods as reflected on contact area between soil particles under loading. (a) Compression test; (b) direct shear test; (c) tension test.

5.12 SENSITIVITY OF SOIL TO ENVIRONMENT

5.12.1 General Discussion

The sensitivity of soil to environment is hinged not only to the local environment, but also influenced by naturally inherited mineral structure, such as particle size, bonding characteristics between particles, ion-exchange capacity, etc. The smaller the soil particle the more chance for a soil particle to interact with the environment. The weaker the bonding energy between the particles or higher the ion-exchange capacity, the higher the sensitivity of the particles to the environment. For example, montmorillonite is potentially more sensitive to the environment than illite and kaolinite, because the montmorillonite group has larger surface areas, weaker bonding energy, and higher ion-exchange capacity. In the following section, five case studies are presented with various environmental conditions covering the sensitivity of soil to load, temperature, pore fluid, mineral types, as well as the sensitivity of test results between *in situ* and laboratory experiments.

5.12.2 Sensitivity of Soil to Load

The sensitivity of soil to load has been discussed by many investigators since the concept was introduced by Terzaghi in 1942. However, the types of loads in relation to the degree of soil sensitivity have not clearly been studied.

Figure 5.18 illustrates the influence of the contact area of load between particles in relation to the soil sensitivity to environment. The larger the contact area or surface area of particles per unit volume, the more the chance for soil to react with the environment. Consequently, the type of test will influence soil-environment interaction. To verify which type of loading test is more or less sensitive to the environment, three types of loading tests are studied; namely, unconfined compression test, undrained direct shear test, and tension test. Results of these studies are examined and presented as follows (Fang, 1991b).

From Figure 5.18, indications are that the tension test gives a greater contact area during the loading process than the unconfined compression and direct shear tests. In comparing these test results, the term called '*sensitivity of strength*,' (Δs) is selected to be the indicator of the sensitivity of soil to load. The sensitivity of strength is defined as:

$$\Delta s = S_{OMC} - S_\omega / S_{OMC} \qquad (5.7)$$

where Δs = sensitivity of strength; S_{OMC} = shear strength at optimum moisture content (OMC); and s_ω = shear strength at any given moisture content.

Both shear strength tests are determined by the same type of shear test procedure. Figure 5.19 plots the moisture contents below the OMC (dry-side) and above the OMC (wet-side) vs. sensitivity of strength (Δs). The OMC is determined by the standard compaction test. The strength parameters used are cohesion (c), unconfined compressive strength (q_u), and tensile strength (σ_t) from direct shear, unconfined compression, and tension tests, respectively. In examining Figure 5.19, whether or not above or below the OMC, the tensile strength is more sensitive with respect to the moisture content in comparison with other direct shear and unconfined compression tests. As indicated, the tension test has a larger contact area between soil particles, therefore offering more chance for the soil to react with moisture contact during the loading process.

5.12.3 Sensitivity of Soil to Pore Fluid

In conventional geotechnical engineering, it is assumed that soil parameters such as specific gravity and Atterberg limits are constant for a given soil under any environmental conditions. However, results in various publications indicate that these parameters are not a constant for a given soil and change when the pore fluid or other environmental conditions change. In order to understand the degree of sensitivity of soil to pore fluid, basic characteristics on soil-pore fluid interactions are reviewed and discussed.

Rosenquist (1953) presents considerable investigatory work pertaining to the formation and behavior of Norwegian quick clays undertaken in the 1940s and early 1950s. As a result of these studies, the effect of salt concentration or electrolyte concentration upon clay behavior was considered among many other factors. It is the effect of electrolyte concentration on clay behavior that is of interest herein.

Increasing the electrolyte concentration in the pore fluid has been shown to increase the liquid limit (LL) of the soil. The LL is a measurement of a clay's remolded shear strength. It is a measure of the water content at which, under a given energy, a given type of shear failure occurs. Hence, if the LL increases with increasing electrolyte concentration, it means that for shear failure to occur in the soil, a higher moisture content was required with a higher electrolyte concentration.

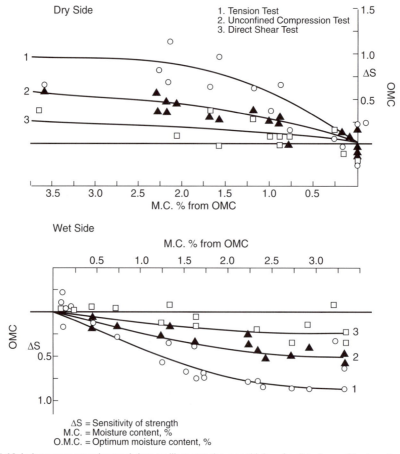

Figure 5.19 Laboratory experimental data to illustrate the sensitivity of soil to type of test methods. (After Fang [1991b].)

Lambe's (1953) study on the effect of polymers on soil properties evaluated their potential as a soil improvement method. It is postulated that the anionic acrylate groups of the polymer strengthened interparticle bonds through the polyvalent cations present. Hence, they are highly effective in causing a flocculation tendency or a soil aggregation.

Matsuo (1957) reported the effect of sodium chloride, potassium chloride, hydrochloride acid, calcium chloride, and magnesium chloride in solution upon the Atterberg limits of three Japanese soils. The study included various concentrations of each of these electrolytic solutions and their effect on the liquid limit, plastic limit, and hence, plasticity index.

Andrews et al. (1967) compared the interaction of three clay minerals with three pore fluids. The pore fluids were water, dimethyl sulfoxide (DMSO), and dimethyl formamide (DMF). The clay minerals used were kaolinite, attapulgite, and sodium montmorillonite. Both DMSO and DMF are acidic solvents. Typical test results on effects on soil constants are summarized in Table 5.7.

Vees and Winterkorn (1967) prepared four clay samples as homoionic clays (Section 4.6). In this way, they could study the effect of cation exchange on the various soil properties including Atterberg limits, cracking, sorption, hydraulic and thermal conductivities, consolidation, and shear tests. Depending upon the primary ion, both the liquid limit and the plastic limit vary. For kaolinite, the PL increases with increasing valence of the exchange ion. Pore fluids' effects on other parameters will be discussed in following chapters.

Torrance (1975) studied the effects of pore fluid chemistry on the behavior of sensitive marine clays of Canada and Scandinavia. Liquid limit tests were conducted on Lada clay

Table 5.7 Effect of Pore Fluids on Specific Gravity and Liquid Limit

Minerals	Pore fluid	ω_L	ω_p	I_p	ω_s	G_s	V_s
Kaolinite	H₂O	62	33	29	29	2.61	0.383
	DMSO	105	50	55	—	2.59	0.386
Attapulgite	H₂O	291	110	181	80	2.61	0.383
	DMSO	309	155	154	—	2.57	0.389
Na-bentonite	H₂O	506	55	451	6	2.51	0.383
	DMSO	140	80	60	—	2.92	0.342

Note: ω_L — liquid limit; ω_p — plastic limit; I_p — plasticity index; ω_s — shrinkage limit; G_s —specific gravity; V_s — specific volume; H₂O — water; DMSO — dimethyl sulfoxide, $(CH_3)_2SO$.

After Andrews, R.E., Gawarkiewicz, J.J. and Winterkorn, H.F. (1967), Comparison of the Interaction of Three Clay Minerals with Water, Dimethyl Sulfoxide, and Dimethyl Formamide, HRR no. 209, pp. 66–78.

using its natural pore fluid as well as electrolytic solutions of various salts. It can be seen that as the concentration of the electrolyte in solution increases, the LL increases.

Griffin et al. (1976) present a series of laboratory leachate tests to evaluate the potential of various clay minerals for attenuating chemical constituents in a permeant. The test utilized leachate from a sanitary landfill permeating kaolinite, illitic, and montmorillonitic clays in a matrix of quartz sand. Prior to permeation, the clay mineral structures were analyzed by X-ray diffraction in order to evaluate the exchangeable ions. The study found that chloride, sodium, and water-soluble organics were relatively unattenuated. Potassium, ammonium, magnesium, silicon, and iron were moderately attenuated. Finally, the heavy metals — lead, cadmium, mercury, and zinc — were strongly attenuated. Conversely, calcium, boron, and manganese were significantly higher in the effluent concentration than in the original leachate. Montmorillonite was more effective in attenuating the chemical constituents, followed by illite and kaolinite. The increase in effectiveness in attenuating chemical constituents is directly related to the cation-exchange capacity of the clay mineral. The principal attenuation mechanism was precipitation for the metals and cation exchange for the other chemical constituents attenuated.

Fuller (1978) conducted a similar study to that conducted by Griffin et al. (1976). Leachates were utilized to permeate 11 different naturally occurring clay soils. The leachate was generated in a homeowner-type septic tank, and at times spiked with various chemical constituents. Soils were analyzed to evaluate their clay mineralogical structure prior to permeation. It was found that the types of clay minerals in the cation-exchange capacity were among the least important factors in the attenuation of chemical constituents from the leachate. The amount of clay in the soil sample, however, was considered to be the most important factor.

5.13 SOIL-WATER-AIR INTERACTION

5.13.1 Behavior of Air in Soil-Water System

Clay minerals establish different equilibria when brought in contact with the water substance in the gaseous or the liquid phase as long as no capillary condensation takes place in the former case. The gas phase consists mainly of the monomer H₂O, while in the liquid phase the entire coherent mass of available water may be considered as one polymeric water molecule. Winterkorn (1958) presents the laboratory experiments on vapor adsorption under a pressure of 30 mmHg. The results show the vapor adsorption vs. temperature for four types of homoionic soils (Figure 5.20). In all cases, the vapor adsorption decreases as temperature increases under a constant pressure.

Figure 5.21 presents a schematic diagram illustrating the progress of gas (air) in soil. There are four steps as listed in Figure 5.21. Rau and Chaney (1988) observed air bubble

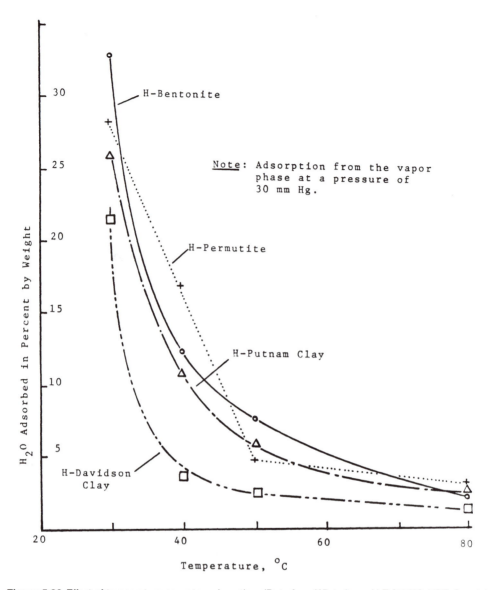

Figure 5.20 Effect of temperature on water adsorption. (Data from Winterkorn, H.F. [1958], HRB Special Report 40, pp. 324–338.)

contents of marine clay. From stage (2) to stage (3), when gas (air) releases in the form of bubbles, the pore fluid volume increases significantly.

5.13.2 Soil-Water-Vapor Interaction

The theory of distillation-condensation-type water movement in the vapor phase has been studied by Derjaguin and Melnikova (1958) and Hallaire (1958). Water may move into, through, and out of soil in the solid, liquid, or vapor phase. Vapor phase movement may be due to gradients in relative humidity of the soil air caused by differences in temperature, moisture content, and in the water affinity of the internal soil surface, to passing air currents, or to the pumping of air into and out of soil as a result of changing barometric pressure in the atmosphere or biosphere. In general, vapor movement by diffusion alone is normally not

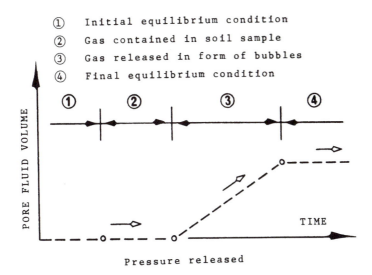

   ①   Initial equilibrium condition
   ②   Gas contained in soil sample
   ③   Gas released in form of bubbles
   ④   Final equilibrium condition

Figure 5.21 Progress of gas release in soil. (After Rau, G. and Chaney, R.C. [1988], ASTM, STP 977, pp. 338–352.)

very important; movement by convection, however, may acquire great significance under certain special conditions as illustrated in Sections 8.8.6 and 18.10.

5.14 SUMMARY

1. A water molecule is a bilateral electric structure. These molecules possess electric polarity. Interaction between a liquid and a solid can occur only on the solid's surface.
2. For evaluation of the soil-water system, we must treat the soil element in three phases: solid, liquid, and gas. The liquid portion can be divided into two parts: gravity water and environmental waters. Environmental waters are strongly influenced by the local environmental conditions. This region is most likely the place for ion exchange and bacterial activities to occur.
3. Soil-water interaction occurs in the various energy fields, and electrochemical properties play an important role during soil-water interaction in the environment.
4. The sensitivity of soil to environment is not only hinged on the environment alone, but on the structural unit of a clay mineral such as particle size, the bonding energy between particles, and ion-exchange capacity. The smaller the particle the more chance for a particle to interact with the environment, and the weaker the bonding energies between particles; or the higher the ion-exchange capacity, the more sensitive it is to the environment.
5. Geomorphic processes (aging process) can be divided into five possible processes which could happen to soil and rock, namely, mechanical, chemical, physicochemical, geomicrobiological, and soil-forming. Each process relating to soil-rock is discussed.
6. In studying the soil-water-gas interaction, it is assumed that water in the gaseous phase consists mainly of the monomer H_2O, while in the liquid phase the entire coherent mass of available water may be considered as one polymeric water molecule.

PROBLEMS

5.1 What are the common dissolved gases found in natural water? Is drinking water a pure water?
5.2 What is hard water and what is soft water? How and why does hard water damage foundation structures? Why do river water and lake water have significant different characteristics?
5.3 If a 0.5-l sample of a water solution is found to contain 2.2 mg of fluoride ion, what is the parts per million (ppm) of fluoride in the water solution?

5.4 What is the dynamic viscosity of pure water at 4°C? What is the absolute viscosity of the same water at 15°C.

5.5 What is the [H$^+$] of a solution with a pH of 6? What is the [OH$^-$] of a solution with a pH of 9? What is the pH of a 0.0075 mol l^{-1} solution of HNO$_3$?

5.6 Explain why some soil is sensitive to moisture as reflected on the shear strength, volume change, and other geotechnical properties and others are not.

5.7 Why is sand more sensitive and clay less sensitive to liquefaction?

5.8 Why is there a loss in strength noted when a soil is soaked but not permitted to swell after compaction?

5.9 A sample of dry sand having a unit weight of 1.62 gr/cc and a specific gravity of 2.70 is placed in the rain. During the rain, the volume of the sample remains constant but the degree of saturation increases to 40%. Determine the unit weight and water content of the soil after being in the rain.

5.10 Four soil samples each have a void ratio of 0.75 and a specific gravity of 2.72, having degrees of saturation of 70, 80, 90, and 100%. Determine the unit weight for each of the four samples.

CHAPTER 6

Shrinkage, Swelling, and Cracking Characteristics of Soil

6.1 INTRODUCTION

Environmental factors significantly affect the soil volume with changing phenomena such as shrinkage, swelling, and cracks. However, all of these phenomena are associated with moisture content of soil. Shrinkage and swelling have been known to geotechnical engineering for years, but cracking behavior is relatively new. To characterize the crack, the sorption, tensile, and fracture of soil must be examined. A schematic diagram illustrates the relationship between skrinkage and swelling as shown in Figure 6.1. The shrinkage process is caused by thermal energy, however, the swelling process is due to multimedia energy; therefore, these two processes both are associated with moisture content but they are not reversible. The causes of shrinkage and swelling are mainly due to the flow path directions as discussed in Section 5.6 and Figure 5.7. Figure 6.2, a schematic diagram, illustrates the interrelationship between cracking and fracture phenomena. Detailed explanations of Figure 6.2 will be presented in Section 6.9. In this chapter, discussion focuses on: mechanisms of shrinkage, swelling, cracking, and fracture as well as sorption and tensile characteristics of soils with special interest in contaminated fine-grained soils.

6.2 SHRINKAGE CHARACTERISTICS

6.2.1 Characteristics of Shrinkage of Soil

Shrinkage is one of the major causes for volume changes associated with variations of water content in soil. Haines (1923), Casagrande (1932), and Hogentogler (1937) have shown that when soil decreases its water content, the volume decreases. Based on Haines and Hogentogler's work, the volume changes relating to soil consistency can be illustrated as shown in Figure 6.3. Relationships among various soil moisture terms are approximate only. In examining Figure 6.3, from points **a** to **b** is normal shrinkage, and the volume changes linearly with decrease of water content. This linear shrinkage is due to surface tension forces of the capillary moisture. When volume change reaches point **b** and the color of soil changes, a small amount of volume change from points **b** to **d** is termed as *residual* shrinkage, also called *curvilinear* shrinkage. Further decrease of water until there is no more change in volume as indicated at point **c** is called the *shrinkage limit* (SL) (ASTM D427-93). The SL can be used as a general index of clay content and will, in general, decrease with increases in clay content.

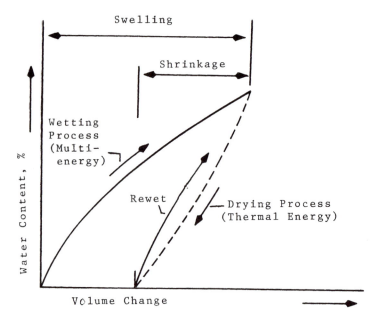

Figure 6.1 Schematic diagram illustrates the relationship between shrinkage and swelling processes.

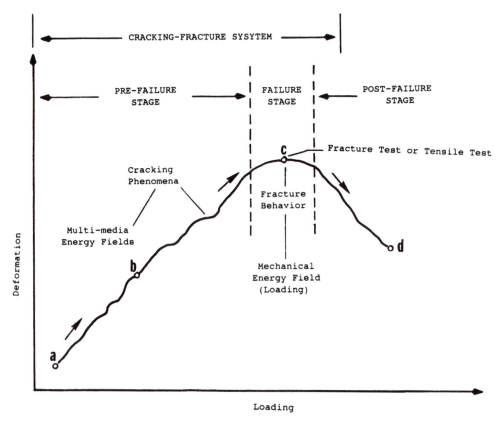

Figure 6.2 Schematic diagram illustrates the interrelationship between cracking and fracture phenomenon.

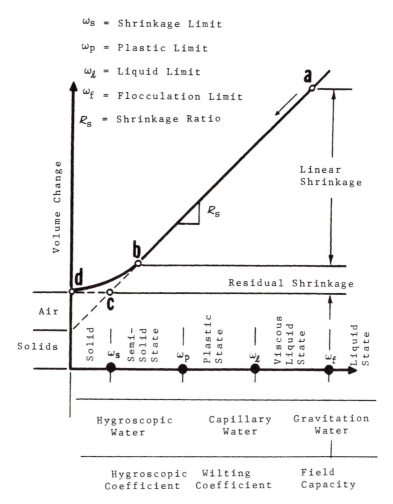

ω_s = Shrinkage Limit

ω_p = Plastic Limit

ω_ℓ = Liquid Limit

ω_f = Flocculation Limit

R_s = Shrinkage Ratio

Figure 6.3 Volume change vs. water content. (Based on Haines, [1923]; and Hogentogler, C.A., Jr. [1937], *Engineering Properties of Soil,* McGraw-Hill, New York.)

The slope of the straight-line portion (from points **a** to **b**) of the volume change and soil consistency curve in Figure 6.1 is called the *shrinkage ratio*, $\mathbf{R}_s$. This value changes within narrow ranges under normal environmental conditions. For example, the $\mathbf{R}_s$ for muck soil is 1.42, black cotton clay (Section 13.4) from India is 2.1, and silty clay from the AASHO Road Test is 1.95.

6.2.2 Factors Affecting Shrinkage of Soil

In general, the shrinkage decreases with increases in clay content. In addition to the clay content, other factors affecting the shrinkage characteristics include dry process, soil particle orientation, unit weight, and pore fluids, as well as exchangeable ions. Rao (1979), reporting on the experimental results of kaolinite and montmorillonitic clays, reveals that the shrinkage limits are affected by the changes in the fabric content, the initial water content, and the consolidation pressure. Figure 6.4 shows the volumetric change vs. pH solution for three basic clay minerals. Figure 6.5 shows the shrinkage limit vs. activity for both natural and homoionic soils of Marshall subsoils.

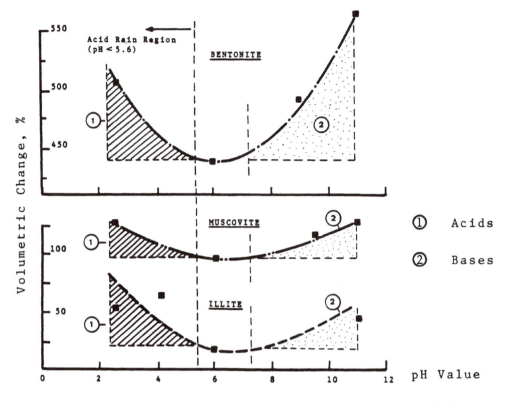

Figure 6.4 Volumetric change vs. pH value for three basic clay minerals. (After Fang [1989a].)

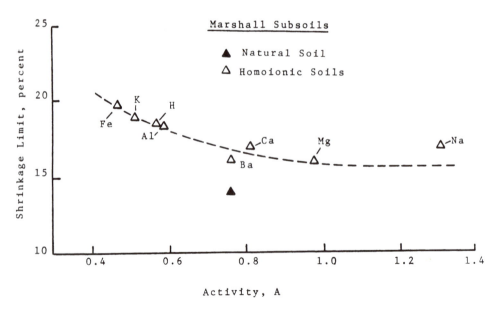

Figure 6.5 Shrinkage limit vs. activity for natural and homoionic soils. (Data from Table 4.8.)

The interrelationship of the shrinkage limit, shrinkage ratio, dry unit weight, and specific gravity of solids can be calculated based on the air-water-solid relationship as described in most standard textbooks. Also, shrinkage measurement procedure has been standardized by both ASTM (ASTM D427-93) and AASHTO (AASHTO T92-68). Soil cracks caused by shrinkage will be discussed in a later section of this chapter. For some soils which exhibited a higher rate of volume changes in the natural state, the rate of volumetric shrinkage is related to the initial moisture content. When the initial moisture content is higher, the rate of the volumetric shrinkage is greater.

6.3 SWELLING CHARACTERISTICS

6.3.1 Swelling Mechanism

When soil loses its water, the volume decreases or shrinks. However, if water is added to the dry soil, the volume of soil increases; this phenomenon is called *swelling*. The mechanism of the swelling process is more complex in comparison with the shrinkage process as discussed in a previous section. Soil swelling is a spontaneous process occurring when conditions are such as to decrease the free energy or increase the entropy of the system. The rate of the process is directly proportional to the available free energy and decreases as the latter is converted; this has two consequences:

1. The rate of swelling decreases with time.
2. The swelling pressure being a measure of free energy also decreases. The rate of swelling is also controlled by the available means of moisture supply and loss by the resistance of the system to the obtaining mode of water conduction.

There are numerous methods for examining and identifying the swelling behavior of soils and classification of swelling soil or expansive clays. Further discussions on origin relating to geology will be discussed in Section 13.4.

6.3.2 Swelling Measurements

These methods can be divided into two major groups: the direct and indirect measurements. The direct measurements use one-dimensional volume changes in conjunction with standard compaction, CBR test, or consolidation test as discussed in standard textbooks. The indirect methods used as a measurement of related soil properties are mineralogy (Seed and Goodman, 1964), centrifuge moisture equivalent (PCA, 1962 and 1992), and dielectric dispersion (Fernando et al. 1975).

The correlations of swelling or shrinking related to soil types and other parameters have been studied by many investigators. Dakshanamurthy (1979) reported that a relationship between maximum volumetric swelling and mean normal stress indicates that maximum axial and radial swelling decrease exponentially with increase in the mean normal stress. This relationship is dependent on the principal stress ratio (ratios of major principal stress to minor principal stress in the triaxial cell apparatus). The swelling ratio (ratios of axial swelling to radial swelling) was found to decrease with time allowed for swelling, reaching an equilibrium value termed the equilibrium swelling ratio which is a function of boundary loading conditions.

6.3.3 Factors Affecting Swelling Characteristics of Soil

Factors affecting swelling characteristics are numerous. Among these factors, the moisture content, unit weight of soil, pore fluid, and exchangeable ions are most important. Baver and Winterkorn (1935) present the exchangeable ions effects on soil swelling and other colloidal properties such as viscosity (Section 5.3), and zeta potential (Section 9.8) of Putnam clay as shown in Figure 6.6. In examining Figure 6.6, the swelling properties of Putnam clay change by varying amounts of exchangeable Li and K ions. It is found that adding Li ion causes much more change in comparison with K ion. Table 6.1 presents the swelling property of

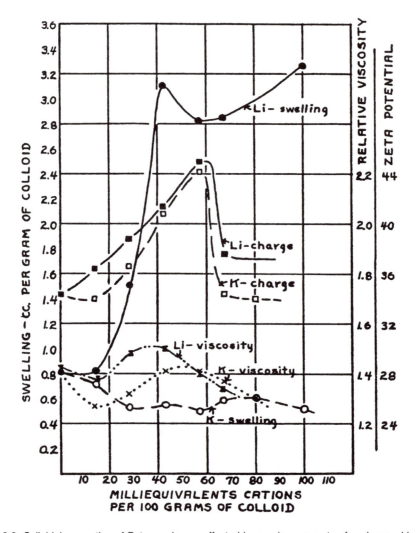

Figure 6.6 Colloidal properties of Putnam clay as affected by varying amounts of exchangeable Li and K ions (After Baver, L.D. and Winterkorn, H.F. [1935], *Soil Sci.,* v. 40, no. 5, pp. 403–419.)

Gila, a predominately montmorillinitic soil, on the macroscopic swelling (gram of absorbed water/gram of soil). The data in Table 6.1 indicate that low salt concentrations and high sodium content in the leachate promote clay swelling (McNeal et al. 1966).

Effects of mineral types on swelling properties are further examined by using three basic clay minerals, including montmorillonite, illite, and kaolinite as shown in Figure 6.7. In examining Figure 6.7, significant results are shown between montmorillonite and kaolinite clays, because montmorillonite is more sensitive to environment than illinite and kaolinite as discussed in Section 5.12. Acar et al. (1989) presents free swelling data vs. dielectric constant and will be discussed in Section 9.7.

Quirk and Schofield (1955) suggested that clay swelling can reduce hydraulic conductivity by causing the blockage of the pore voids. McNeal and Coleman (1966), based on Quirk and Schfield's work, applied varying concentrations of sodium-calcium salt solutions with different exchangeable sodium percentage to seven soils which differed in their clay content and percentages. In comparing the relative sensitivity of the soils to hydraulic conductivity changes, McNeal and Coleman concluded that: (1) soils high in either kaolinite and sesquioxides or in amphorous minerals appeared stable, (2) soils high in montmorillinite appeared

Table 6.1   Macroscopic Swelling of Gila, a Predominately Montmorillonitic Soil

Salt conc (meq/l)	Swelling (g/g)	Salt conc (meq/l)	Swelling (g/g)
SAR = 0		SAR = 50	
800	0.00		
200	0.02	800	0.00
50	0.02	200	0.14
12.5	0.04	50	0.65
3.13	0.06	12.5	2.00
SAR = 15		SAR = 100	
800	0.00		
12.5	0.12	800	0.00
3.13	0.84	200	0.28
		50	1.50
		12.5	2.44
SAR = 25		SAR = 00	
800	0.00		
200	0.05	800	0.00
50	0.08	200	0.91
12.5	0.56	50	2.04
3.13	1.84	12.5	2.70

Note: SAR = sodium adsorption ratio.

After McNeal et al. (1966).

very susceptible to conductivity changes, and (3) soils high in 2:1 lattice clays with moderate amounts of montmorillinite were intermediate in their behavior.

Soil *retention* is partly due to swelling forces and partly to capillary forces. Water is retained also in the matrix of silt and fine sand but to much less a degree than clay soils. Soil retention ability and retention time are important and will be discussed further in Section 7.4.

6.4 SORPTION CHARACTERISTICS

6.4.1 Water Sorption

Soil-water interaction in the environment as discussed in a previous chapter depends on the characteristics of sorption, absorption, and adsorption reaction of soil. These reactions are also influenced by soil types, particle size, and bonding characteristics. As indicated in Figure 4.5, the *sorption* process covers both *absorption* and *adsorption* processes. Vees and Winterkorn (1967) presented a series of laboratory experiments on homoionic modification clays as shown in Figure 6.8.

Figure 6.8(a) shows the water sorption of kaolinite as a function of the type of exchange ion and time. The Th-homoionic clay shows the highest initial sorption rate and the lowest ultimate intake. This is to be expected from its flocculated structure and its low water affinity. For the other ions the picture is not so clear-cut. This is due to the interaction between absorption and adsorption processes with respect to the types of ion-exchange reaction in the soil. Considering the relatively low ion-exchange capacity of the kaolinite, the observable ion effect appears to be markedly influenced by the type of secondary structure induced by the different cations rather than by the individual water affinities of the ions themselves.

Of particular interest is the closeness in rate and final volumes in the water sorption by Na- and Mg-kaolinite, which indicates a similarity in water affinity that has been observed on other occasions. Figure 6.8(b) presents the ultimate volume intake vs. time for attapulgite. The ultimate volume intake is determined by the springiness of the needle-shaped crystallites and of the felt-

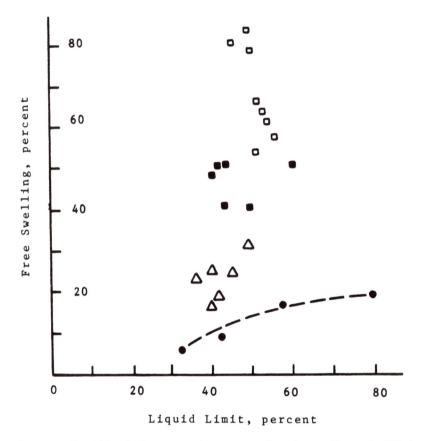

□ Montmorillonite   ⎫ Sensitive
■ Montmorillonite-Illinite Mixtures   ⎬ to
△ Illinite   ⎭ environment
● Kaolinite (less sensitive to environment)

Figure 6.7 Sensitivity of soil type to environment as reflected on swelling vs. liquid limit test results (Data from Tongji University [1979], Soil Properties and Mechanics, Shanghai, China.)

like structure formed by them. This springiness increases with increasing valency of the exchange ions while the water affinity of the clay mineral surfaces decreases. Accordingly, the rate of water intake as well as the ultimate volume increases with increasing exchange ion valency.

6.4.2 Absorption and Degree of Saturation

In geotechnical engineering, the term absorption is used very loosely. The degree of saturation is commonly used as an indication of water absorption character of soil as shown in Figure 6.9. In examining Figure 6.9, indications are that absorption process is a function of time; therefore, the degree of saturation may possibly cover both absorption and adsorption interaction. It is especially true for the fine-grained soils. A dotted vertical line shown in Figure 6.9 indicates the possible dividing line between the absorption and adsorption phenomena. It must be noted that the term of saturation and sorption is similar in geotechnical engineering which includes both absorption and adsorption characteristics.

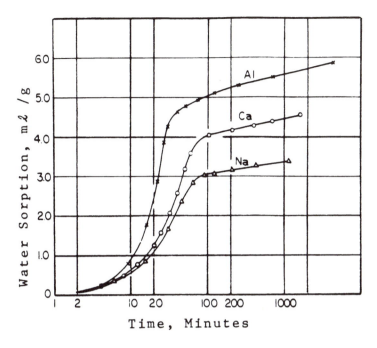

a

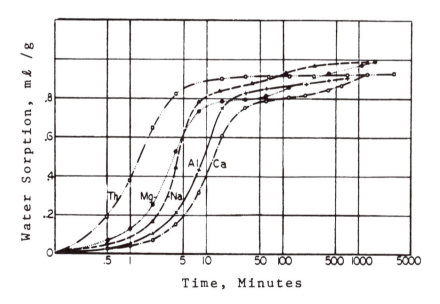

b

Figure 6.8 Water sorption as a function of type of exchangeable ions and time. (a) Attapulgite; (b) kaolinite. (After Vees, E. and Winterkorn, H.F. [1967], Engineering Properties of Several Pure Clays as Functions of Mineral Type, Exchange Ions and Phase Composition, HRR no. 209, pp. 55–65.)

6.4.3 Absorption of Organic Molecules

As discussed in Sections 4.9 and 5.6, in the mechanism of hydrated clay which contains interlayer cations surrounded with water molecules, the cations satisfy the negative charge

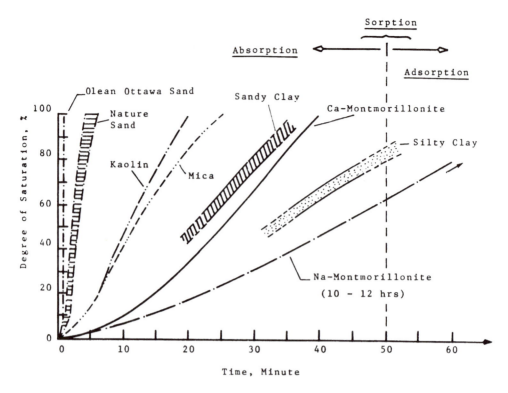

Figure 6.9 Degree of saturation vs. time.

on the clay surface through the hydrogens of the coordinated water molecules. Velde (1992) explained that when organic molecules are placed in an aqueous solution, the water and organic material compete for the interlayer site. As the proportion of the water and organic material is changed in favor of the latter, they are incorporated into the interlayer sites.

6.5 ADSORPTION PHENOMENA

6.5.1 Types of Adsorption

Adsorption is defined as the concentration of a substance on a surface. It is the uniform penetration of molecules of one phase in-between the molecules of a second phase. In order for the adsorption to be appreciable, it is necessary that the adsorption have a large surface area. Therefore, fine-grained soil has more adsorption capacity than larger soil particles. The following factors are to be considered:

- Nature of the adsorbent
- Nature of the substance being adsorbed
- Surface area of the adsorbent
- Temperature
- Pressure

In general, a decrease in the temperature or an increase in the pressure increases the amount of adsorption. There are two types of adsorption known, namely: physical and physicochemical adsorptions, but they are defined loosely.

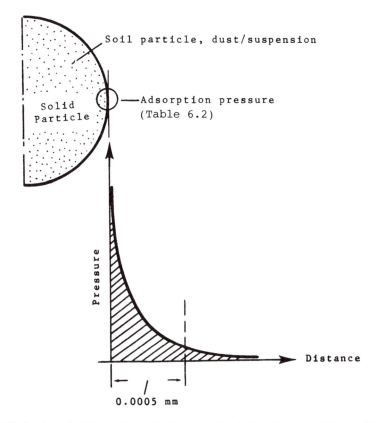

Figure 6.10 A schematic diagram illustrates the intensity of adsorption around the particle surface.

(1) Physical Adsorption

This type of adsorption also is called van der Waals adsorption. It is characterized by low heats of adsorption, in which the surface is gradually coated with a layer of molecules held in place by van der Waals forces. Since van der Waals forces have weak binding energy, the physical adsorption cannot have a sufficiently large energy effect on the adsorbed molecules to affect their chemical properties significantly.

(2) Chemical or Activated Adsorption

Chemical adsorption is much stronger than physical adsorption, characterized by heats of adsorption between 10 and 100 kcal/mol of gas (Porterfield, 1972). This is comparable in magnitude to the heat of formation of a chemical compound. Also, it is believed that chemical adsorption is a combination of gas molecules with surface atoms to form a surface compound. Physical adsorption is common at low temperatures, and chemical adsorption is common at high temperatures, however, physical adsorption at low temperature may pass into chemical adsorption as the temperature is increased.

6.5.2 Adsorption Mechanism

Figure 6.10 illustrates the intensity of adsorption around the particle surface. The mechanism of adsorption phenomena in soil-water system is presented as follows:

Table 6.2 Adsorption Pressure on Soil Surface at Various Mositure and Environmental Conditions

Condition	Adsorption pressure		Ref.
Hygroscopicity	50	kg/cm²	Blanck, 1938
Permanent wilting point	12.50	kg/cm²	
Wilting point (dead water)	6.25	kg/cm²	
Vacuum moisture equivalent	0.55	kg/cm²	
Humid climates	25,000.	psi	Winterkorn, 1955
Solidify water (45°C)	12,000.	psi	

1. The large adsorption forces exerted on water molecules by the surfaces of solid soil particles act similarly to externally applied pressures; that is, they may liquefy solid water or solidify liquid water.
2. Adsorption forces of the order of magnitude that solidify water at 45°C at pressure of 12,000 kg/cm² are not at all uncommon on the surfaces of soil particles (Table 6.2).
3. All physical properties of cohesive soils, such as mechanical, electrical, optical, and acoustical, are functionally connected with the behavior of the water substance in strong pressure; electrical and magnetic fields exist on the surface of the particulate matter in soils.
4. Water in a soil-water system is never pure but holds materials in solution and dispersion. Dissolved materials are mainly salts and acids. The water layers next to the solid particles are under high adsorption pressures, which may be larger than 25,000 kg/cm². The water may be in a solid condition at temperatures above 50°C. The adsorption forces decrease exponentially to about 50 kg/cm² at the hygroscopic moisture content, and hence more slowly to zero for the water content at which the soil-water system behaves essentially as liquid (liquid limit). The course from hygroscopic moisture content to the liquid limit is presented in Table 6.2.

6.5.3 Adsorption Measurements

There are various methods for measuring adsorption of soil, such as batch-adsorption procedures for estimating soil adsorption of chemicals (Roy et al. 1991). These methods are time consuming. For comparison purposes, both absorption and adsorption methods are discussed. Alternative procedures for the estimation of both absorption and adsorption are proposed (Fang, 1993) by using existing ASTM standard procedures, the field moisture equivalent (FME) (ASTM D426-39; AASHTO T93-68), and centrifuge moisture equivalent (CME) (ASTM D425-88(1994); AASHTO T94-54) to indicate the absorption and adsorption phenomena for soils. A brief discussion and justification for how these procedures can be used for these purposes is presented.

(1) Absorption Relating to the FME

By using Field Moisture Equivalent (FME) one can estimate the characteristics of soil absorption behavior. The FME is one of the early standardized test procedures established in 1939 by ASTM. The FME of a soil is defined as the minimum moisture content expressed as a percentage of the oven-dried soil at which a drop of water placed on a smooth surface of the soil will not immediately be absorbed by the soil, but will spread out over the surface and give it a shiny appearance.

(2) Adsorption Relating to the CME

The Centrifuge Moisture Equivalent (CME) is the moisture content of a soil after a saturated sample is centrifuged for 1 hr under a force equal to 1000 times the force of gravity

Table 6.3   Nature of Water Adsorbed on Clays in Relation to the Electrochemical Properties of the Exchangeable Ions

Clay types[a]	Ionic mobility @ 25°C (cm²/volt/sec)	Ionic activation energy (cal/g ion)	Ionic dissociation (relative)	Density of adsorbed water @ 25°C (g/cc)	Structural development of water (relative)
K-clay	0.000051	4037	++	0.981[b]	+
Na-clay	0.000069	4486	+++	0.972	+++
Li-clay	0.000057	4358	++	0.975	++
Na-clay	0.000132	5157	++++	—	—

[a] The first three clays were prepared from <2.0 μm columnated Wyoming bentonite by titration. The last clay was prepared from <0.2 μm electrodialyzed Wyoming bentonite by titration.
[b] These values are for a distance of 10 Å from the clay surface.

After Low, P.F. (1958), Movement and Equilibrium of Water in Soil Systems as Affected by Soil-Water Forces, HRB Special Report 40, pp. 55–64.

(g). This test is used to assist in structural classification of soils. A value lower than 12 indicates permeable sands and silts, while a value greater than 25 indicates impermeable clays with high capillarity. CME values can be as high as 68 for soft marine clays from the Gulf of Mexico and 56 from the Gulf of Maine (Section 14.4).

The study also shows that when FME and CME are both more than 30 and if FME is greater than CME, the soil probably expands upon release of a load and should be classified as an expansive soil (Section 13.6). In addition, as indicated by Winterkorn and Fang (1991), a FME greater than the liquid limit (LL) indicates the danger of autogenous liquefaction of a soil in the presence of free water (Table 6.4). The FME value is similar to the CME, as it is a qualitative indicator property which must be correlated with soil performance in order to have significant meaning.

Table 6.4   Identification of Some Problematic Soil Deposits Based on Soil Constants

Soil constants		Soil behavior
Both FME and CME values	>30	Expansive soil
If FME value	>CME	
Both FME and CME values	>30	Danger of autogenous liquefaction
If FME	>LL	
If LI	≤0.4	Overconsolidated clay

Note: CME = centrifuge moisture equivalent of soils (ASTM D425); FME = field moisture equivalent of soils (ASTM D426); LL = liquid limit of soils (ASTM D423); LI = liquidity index

$$LI = \frac{\text{water content} - \text{plastic limit}}{\text{plasticity index}}$$

PL = Plastic limit of soils (ASTM D424); and PI = Plasticity index = LL – PL.

6.5.4 Adsorption Characteristics in the Environment

(1) Adsorption Characteristics

Ion adsorption is one of the important characteristics of the soil-water system interacting with the environment. The nature of water adsorbed on clays in relation to the electrochemical properties of the exchangeable ions has been reported by Low (1958). Part of the experimental results are summarized in Table 6.3.

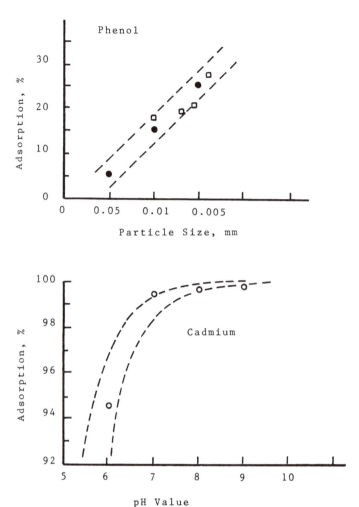

Figure 6.11 Adsorption characteristics of hazardous/toxic substances. (Data from Wang, C.H. [1985], *Water Pollution,* Scientific Publishing, Beijing; and others.)

(2) Polymer Adsorption

In the pollution intrusion process, some smaller organic molecules can be inserted between the swelling layers of clay minerals due to the irregular forms of the polymers. According to Velde (1992), the phenomenon of *polymer adsorption* also follows the electrical double-layer principle as discussed in Section 4.9. The nonlocalized, negative charge on the clay structure is immediately compensated in the zone closest to it by positive ions in aqueous solution. The polymer effects of ionic interaction with charged clay particles in aqueous suspension indicate that increasing the concentration of elements in a solution tends to flocculate the clay particles. Clay-polymer adsorption interaction is similar to clay-organic absorptions; however, the range of clay types is greater in that the swelling property is no longer necessary.

The adsorption characteristics from the vapor phase have been discussed in Section 5.13. It is indicated that dry clay minerals establish different equilibria when brought in contact with the water substance in the gas or in the liquid phase. Results of adsorption from vapor phase for four natural soils are shown in Figure 5.20.

(3) Adsorption of Hazardous Substances

Adsorption characteristics of hazardous/toxic substances are presented in Figure 6.11. It is found that an increase in the pH or decrease of particle size will increase the percent of adsorption for phenol and cadmium.

The adsorbed pentachlorophenol (PCP) is listed as a hazardous substance in natural soils. PCP as a hazardous substance has been used in great quantities as a wood-preserving biocide for at least 50 years. Since PCP-treated wood as aged utility poles, railroad ties, and wooden crates is usually found in landfill, PCP often appears in the leachate of landfills. The adsorption of PCP by soils depends on the chemical and physical properties of the soil, the chemical species that are present in the aqueous phase, and the affinity of the PCP molecules for the surface. Soil properties that influence the extent of adsorption are grain size distribution, organic carbon content, clay mineralogy, and acidity in the soil as reflected by pH, according to studies by Christodoulatos et al. (1994).

Various types of rock also present the characteristics of adsorption behavior. Adsorption of some sedimentary rocks of uranium (U) content is presented in Table 10.7. The behavior of dust and suspension is also closely related to the characteristics of adsorption. Through adsorption, dust is the key to all air pollution problems, as discussed in Section 3.9.

(4) Adsorption and Stabilizing Materials

In ground improvement or soil stabilization, the adsorption of material ions is directly related to the strength of the stabilized materials. Some relatively important stabilization materials such as cement, bitumen, and lime are presented as follows.

1. Soil-cement: In the considerations of surface chemistry and chemistry controlling the efficiency of soil-cement stabilizations, there are certain physical factors having an influence on this process, such as grain size distribution, pore space, and moisture-density relationships. The gross effect of these appears to be a modification of the results to be expected from the electrochemical approach (Section 5.6) and grows in importance as the activity of the clay material increases, resulting in there being an optimum value of the activity for most efficient stabilization.

Some salts (e.g., sulfates of sodium and magnesium) are detrimental by virtue of interacting with the cement to form expanding compounds leading to disintegration of the mixture. Much work has been done on the pedological study of soil-cement stabilization by HRB (1943).

2. Bitumen stabilization: The important factors controlling the bitumen stabilization are (1) ion-exchange capacity, (2) silica-sesquioxide ratio, and (3) adsorption behavior of bitumen and materials selected to be stabilized. The amount of bitumen required depends essentially on the affinity of the bitumen for the soil material. As bitumen predominantly acts as an electromagnetically charged colloid, the amount of bitumen required increases with the number of electronegative charges located on the surface of the material to be stabilized. In the instance of a clay soil and a given particle grading, the percentage of bitumen required increases with the activity of the clay material as discussed by Wooltorton (1955).

3. Chloride, lime, and silicate: Using electrolytes in stabilization, a number of electrolytes, such as calcium chloride, sodium chloride, hydrated lime, ferric chloride, and sodium silicate, can be important, through electrochemical concepts involving ionic exchange, though others may feature a change in hygroscopicity, lowering of the freezing point, possibility of increased maximum densities and shearing resistance, and retardation of evaporation by crystallization within the capillaries.

In addition to its deliquescent properties, calcium chloride can, by leaching, change a highly water-retentive clay, as a sodium clay, into a calcium clay by ion-exchange reaction, thereby reducing the plastic index and causing a volume change. The benefit achieved by this process increases with the increase in the silica-sesquioxide ratio. Hydrated lime may behave similarly with the added advantage of acting as a wetting agent to change the effective sign of highly active clay colloids. In the presence of natural pozzolanic materials, there is also an adsorption effect.

The effect of silicate of soda to soils is to increase water-stable aggregation, resist crushing, and to decrease the plastic limit, plastic index, and swelling. The process involved is apparently that of anionic exchange (Wooltorton, 1955), resulting in the formation of insoluble silicate sheaths around the soil particles and accompanied by adsorption effects.

6.6 CRACKING PHENOMENA AND MECHANISMS

6.6.1 Soil Cracking Phenomena

(1) General Discussion

Soil *cracking* is a natural phenomenon and is frequently observed in many natural and man-made earthen structures. These cracks are a result of an internal energy imbalance in the soil mass caused by nonuniform moisture distribution, temperature distribution, or distribution of compaction energy during construction. Cracking in flood plain clays results from decomposition followed by cyclic expansion and contraction from seasonal wetting and drying. Many preconsolidated clays also exhibit cracking and fissures due to unloading or dessication. These closely spaced, small cracks contribute to progressive erosion or landslides in excavation, slopes, dams, highway embankments, river banks, hydraulic barriers, and other earth structures.

(2) Soil Cracking Types

The types and mechanism of soil cracking have been examined by Fang et al. (1989a, 1994a) and have been classified into four types, namely: shrinkage, thermal, tensile, and fracture cracks. Discussion of these four types of cracks is presented as follows.

(a) Shrinkage Cracking

It is the most common cracking found in earth structures and the phenomenon can be observed on drying mud flats as shown in Figure 6.12. As water is lost from the surface of a soil mass, tensile forces are established in the drying surface layer. Because of the water loss soil also loses its ability to relieve these tensile forces by plastic flow. These stresses are finally relieved by the formation of *shrinkage cracks* that break up the surface layer into pieces of more or less distinct geometric shapes. This geometric shape of the cracks depends on the clay mineral composition, the heating process, pore fluids, and other factors. As shown in Figure 6.12, illite shows finer cracks than muscovite, and oven-dried samples give more irregular cracking patterns than air-dried samples.

If the soil material is homogeneous, the cracking pattern will be hexagonal in accordance with the law of the triple angle, which is a special form of the law of least energy. This phenomenon can also be explained by linear fracture mechanics concepts which will be discussed in the following section. If the soil system is nonhomogeneous, as in the case of the presence of organic matter or different water affinity and greater mobility than the mineral soil components, then the hexagons will tend to become rounded and the organic matter will become concentrated at the surface of the fissures.

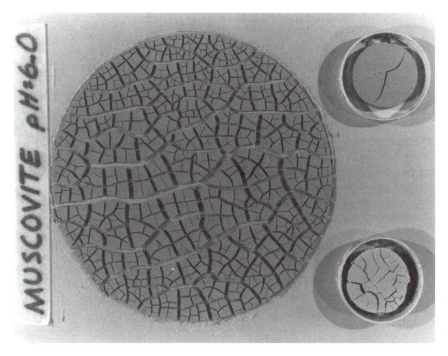

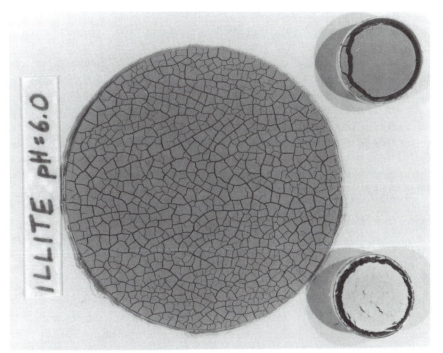

Figure 6.12 Comparison of cracking patterns between illite and muscovite clays.

(b) Thermal Cracking

Thermal cracks are caused by the change of the thermal stresses of material. The stresses are developed when material is heated and then suddenly cooled, a significant change in temperature such as freezing-thawing or wet-dry processes in soil. The thermal cracks for soil are somewhat different than other construction material because the temperature and moisture in the soil mass are so closely related and the state of stress of soil is so sensitive to influence by both temperature and moisture behavior simultaneously.

The plastic deformation occurs in different zones of the soil mass as a result of differential temperature with accompanying volume changes arising during the cooling cycle. If the deformations arising are within the elastic strength range of the soil mass, then no stress remains after temperature equilibrium is reached. As indicated in Figure 6.12, the cracking patterns between oven- and air-dried soil samples of illite and muscovite are different because the rate of heating process in the oven is more intense and less uniform than the air-dried process. Other factors that contribute to the thermal cracks are soil types, unit weight, grain size distribution, void ratio, and geometry of the primary and secondary drying surfaces.

(c) Tensile Cracking

Tensile cracks are caused mainly by overburden pressures including structural loading, rainfall, ice and snow loads, vegetation, and seasonal creep loads. Sometimes, they are also associated with changes in moisture or thermal stress, as discussed previously, and related with fracture loads which will be discussed in the following section.

(d) Fracture Cracking

In a soil mass either man-made or in the natural state, there always exist some cracks due to daily moisture or temperature changes or caused by seasonal groundwater table fluctuation, rainfall or melting snow which will fill into the cracks or voids, and consequently, produce the porewater pressures. These porewater pressures vary with changes of these environmental conditions as does the capillarity tension produced from porewater between the soil particles. When a saturated soil dries, a meniscus develops in each void of the soil structure which produces the tension in the soil-water system and a corresponding compression force in the soil skeleton. This compression stress is just as effective as stress commonly used in soil mechanics in producing soil compression as an external load. Pressure of 200 to 300 kPa can be produced in fine-grained soils. This internal cyclic-type load caused by the combination of shrinkage or thermal stresses and the fluctuation of the porewater pressure between soil particles is called *fracture load* in soil. The cracks produced from the fraction load are referred to as *fracture cracks* in soil.

6.6.2 Cracking Mechanisms

(1) Cracking-Fracture Phenomena and Relationships

For a given material, the cracking and fracture characteristics belong to one system. Cracking represents the prefailure phenomenon of a material and fracture is the behavior of a material at failure condition as illustrated in Figure 6.2. Figure 6.2 shows the interrelationship between prefailure and failure conditions of a soil from a geotechnical engineering point of view. At the early stage of deformation vs. loading curve, cracks develop due to various natural or man-made causes as discussed in previous sections. At this stage, the soil behavior is controlled by multimedia energies such as thermal, electric, and magnetic energies. However, at the failure stage, it is dominated only by mechanical energy which is the loading. Therefore, for evaluation of soil's useful life, it is necessary to investigate the cracking-fracture mechanisms and their interaction.

(2) Cracking-Moisture Relationship

The mechanism and the interaction between cracking and fracture as illustrated in Figure 6.2 are a complex phenomenon. There are numerous factors affecting the cracks and how these cracks interact with the failure condition. In general, all cracks are related to the moisture content in the soil and local environments. As discussed in Section 5.6, soil-water equilibrium conditions differ when in contact with liquid water and water vapor. Even at the same moisture content, the structure and physicochemical properties of a soil-water system may differ considerably, depending upon the flow path from which this moisture content has been attained. Since soil cracking mechanisms are related to the moisture content in all cases, the flow directions or paths are also related to the cracking patterns. There are two basic mechanisms of soil cracking such as the following.

(a) Cracking When Soil is Drying

As wet or moist cohesive soil system loses water, the soil particles move closer and closer together. If the drying proceeds from the surface downward as in a mud flat, the dehydrated surface layer shrinks while the water resistance between the upper and lower layers and in the layers themselves prevent an adjustment to the volume decrease of the surface layer; thereby tensile stresses develop in the surface layer. This cracking pattern produces the greatest stress release with the least amount of work. In other words, cracking and fracture during the drying process are controlled mainly by thermal energy.

(b) Cracking When Soil is Saturated

The entrance of water into a porous dry soil system can also cause cracking/fracture of soil mass. In this case, these causes and phenomena are more complicated than in drying condition, because they are controlled by multimedia energy fields which include thermal-electric-magnetic energies as explained in Section 5.6. The swelling capacity increases when soil becomes saturated. This capacity of the clay minerals may result in internal swelling that decreases the permeability. If internal forces produced by water enter into the soil mass, they may be large enough to cause surface exfoliation of the systems and produce the cracks.

6.6.3 Cracking Pattern and Measurements

(1) Cracking Pattern Measurement Procedure (Lehigh Method)

Characteristics of the cracking pattern is an important phenomenon in the evaluation of environmental geotechnical properties of soil-water system. A simple procedure for preparation of samples and the items to be examined from the cracking pattern test developed at Lehigh University are presented as follows.

(a) Preparation of Soil Specimen

1. All soil samples must pass #40 U.S. standard sieve. Smaller particle sizes or larger surface area per unit volume of soil sample will give more distinct cracking patterns.

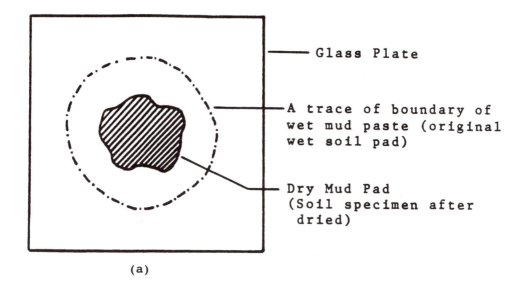

Glass Plate

A trace of boundary of
wet mud paste (original
wet soil pad)

Dry Mud Pad
(Soil specimen after
dried)

(a)

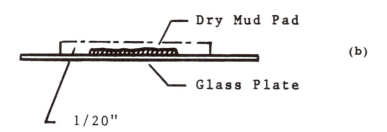

Dry Mud Pad

(b)

Glass Plate

1/20"

Figure 6.13 Preparation of wet mud pad for conducting cracking pattern tests. (a) Top view of dry mud pad. (b) Cross-section of dry mud pad.

2. Approximately 50 to 100 g of soil are needed. The exact amount of soil to be used is not a critical factor.

3. Water content of the specimen must be at the full saturated condition. For simplicity, the water content of the liquid limit (ASTM D423-66) can be used.

4. Soil-water must be mixed thoroughly for at least 5 min. Then spread this wet soil-paste uniformly on a clean glass plate as shown in Figure 6.13. A round-shaped wet-paste (mud pad) sample is formed (Figure 6.13(a)).

5. The maximum height of the wet soil-paste must be less than 1/20 in. as shown in Figure 6.13(b). The thinner the wet mud pad, the more distinct cracking patterns will develop.

6. The drying process for a soil sample can be determined by three common methods used in the laboratory: air dry at room temperature (20°C or 68°F) for 24 hr, oven dry at 110°C for 6 hr, or microwave oven for 3 to 4 min.

It must be noted that when using any of the three drying methods for the determination of moisture content of a soil sample, the results will be very close or practically the same. However, the cracking pattern for the same soil by using these three different methods will be different.

(b) Items to be Examined using the Cracking Pattern Test

1. Geometry of the cracking pattern, crack opening, length, and shape
2. Estimation of percent of shrinkage in comparison with the original (wet condition) soil specimen

3. Describing the soil color after it is dried
4. Pattern changes caused by the heating processes, such as air-dried and oven-dried
5. Pattern changes caused by various pore fluids
6. Pattern changes due to types of clay minerals
7. Soil color changes caused by various pore fluids in the soil voids (The physicochemical or biological interaction between soil particle and pore fluids can be explained by soil-water interaction phenomena.)

Soil cracking pattern tests are a qualitative indicator property which must be correlated with other soil performance in order to have a significant meaning. These performance tests include: (1) tensile test, (2) fracture test, (3) clay mineral structural analysis, and (4) pore fluid chemistry analysis.

(2) Internal Soil Cracking Measurement

In a previous section, we discussed soil surface cracking phenomena and their measurement. However, in many cases such as clay liner, hydraulic barriers, and other earthen structures, they also develop internal cracking due to shrinkage, thermal, tensile, and fracture. The internal cracking characteristics are also important for long-term performance study.

Internal soil cracking measurement may be made by use of *computed tomography* (CT) techniques. CT technique is a relatively new X-ray method for nondestructive testing (NDT) that overcomes some of these limitations, as reported by Armistead and Stanley (1989). CT measures point-by-point density values in thin cross sections of an object, thus allowing three-dimensional imaging of the internal structure when successive transverse sections are compared. In X-ray CT, a collimated X-ray beam passing through the test object is measured by an array of detectors located on the other side of the object. The object is rotated slightly or, in some cases, translated and rotated, and a new set of measurements is made. This process is repeated until the object has been fully rotated. The measured projections are controlled by a computer which is the major advantage of this technique over conventional radiography.

The internal cracking of a laboratory-prepared clay liner (sand-bentonite mixture) specimen is caused by a triaxial compression test. These cracks are not visible at the outer surface of the specimen after the triaxial compression test at failure condition. However, through X-ray CT test, the internal cracking patterns are observed as shown in Figure 6.14. The specimen was prepared at Lehigh University and tested by Advanced Research Corporation in California.

(3) Cracking Behavior of Contaminated Soils

Cracking patterns and mechanisms for contaminated soil differ from noncontaminated soil. By using water (noncontaminated) and acid or base (contaminated) for comparison, soil cracking patterns caused by certain polluted waters because of the differences of dielectric constants and their surface tension in the pore fluids can be observed. Characteristics of these two important factors are discussed in Section 5.3 and 9.7. Typical cracking patterns with various pore fluids are shown in Figure 6.15.

(4) Soil Structures Relating to Cracking Patterns

The laws governing the formation of shrinkage cracks have been discussed in detail by Haines (1923) and Thompson (1942). Effects of freezing-thawing and drying-wetting cycles on soil cracks have been studied by Czeratzki and Frese (1958). Soil structures such as flocculation or dispersion will affect the cracking pattern as illustrated in Figure 6.15. Andrews et al. (1967), Alther et al. (1985), and Evans (1991) pointed out that flocculating and dispersive structures and pore fluids create different types of cracking patterns. As illustrated in Figure 6.16, the flocculation produces larger areas of cracking than dispersive structures.

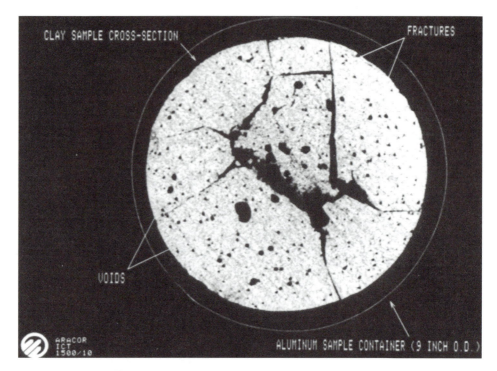

Figure 6.14 Internal cracking pattern by X-ray CT method.

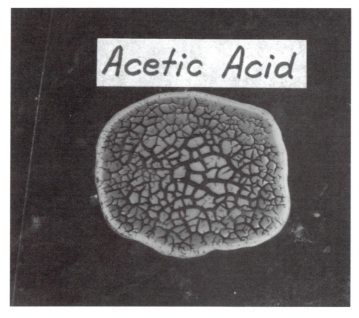

Figure 6.15 Cracking patterns with various pore fluid.

6.7 TENSILE CHARACTERISTICS OF SOIL

6.7.1 Introduction

For quantitive measurement of the cracking behavior of soil as discussed in Section 6.4, determining the tensile strength is one of the simplest approaches. There are numerous test

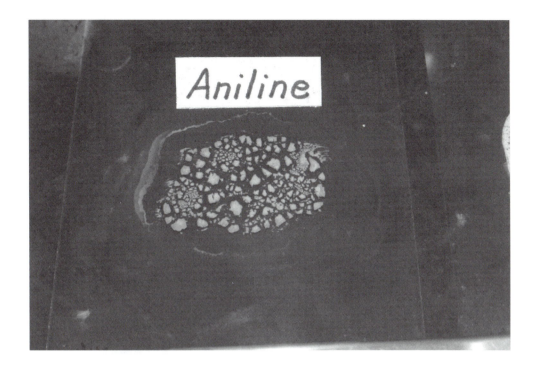

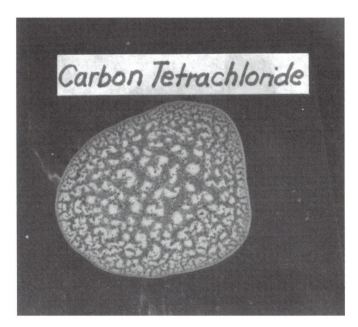

Figure 6.15 (continued)

procedures to measure the tensile strength of soil including the Briquest Gang model, the beam-type test, the split tensile test, the unconfined-penetration test, the triaxial-cell type, and the hollow-cylinder device. The review and evaluation of these methods and data interpretations are given ASTM (Yong and Townsend, 1981).

The test described in this chapter is the *unconfined-penetration* test (UP test) developed at Lehigh University in early 1970 (Fang and Chen, 1971). The UP test is the simplest and least expensive method for measuring the indirect tensile strength of soils or other nonmetallic construction materials. Test results have been compared with the conventional split-tensile

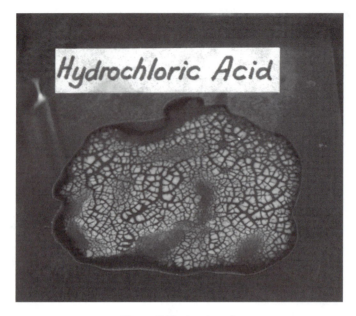

Figure 6.15 (continued)

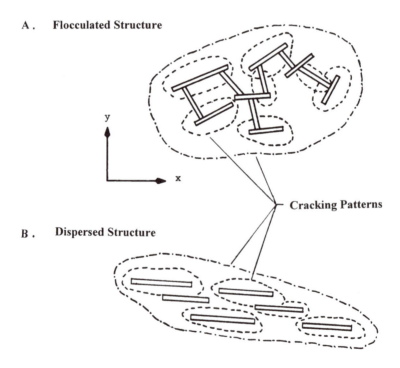

Figure 6.16 Schematic diagram illustrates the effect of soil structure on cracking behavior.

test (ASTM C496-71; AASHTO T198-74) and indicate good agreement. The UP test has an advantage in that it can be conveniently performed in conjuction with routine compaction and CBR tests.

For the UP test, various sample sizes and shapes can be used. In addition, the other available test methods measure the tensile strength across a predetermined failure plane, whereas the UP test always causes failure on the weakest plane which results in the measurement of the true tensile strength of soil.

Table 6.5   Mechanism of Tensile Strength of Soil at Various Stages

Dry soil	Controlled by cohesion, friction, and some tension	

Partially saturated soil	Mainly controlled by tensile strength	Caused by
		   Surface tension of soil moisture
		   Oriented dipoles linking a positive charge on one particle with a negative one on the neighboring particles
		   Formation of an electric field by ions
		   Diffuse double-layer thickness expands
		   Permits a more orderly arrangement of particles
		   Electrolyte concentration reduces

Saturated soil	Tensile strength very small	

6.7.2 Mechanism of Tensile Strength of Soil

1. When soil is dry, it is held together by cementing agents and adhesive films between soil particles. The cementing power of these films is a function of their own physicochemical properties.
2. The adhesive force between soil particles is also dependent on the grain size and its surface areas.
3. If soil gradually becomes wet, the tensile strength develops due to the surface tension and other forces.
4. From a physicochemical point of view, the tensile strength for the partially saturated soil is caused by dipoles linking a positive charge on one particle with a negative one of the neighboring particles. As the diffuse-double layer thickness expands and the electrolyte concentration reduces, the degree of flocculation also decreases which permits a more orderly arrangement of particles.
5. If there is further increase of moisture content which dilutes the concentration of soil particles per contact and reduces the net attractive forces between particles, then as the soil is saturated, there is a very low tensile strength with no engineering significance; therefore, the tensile strength holds together mainly partially saturated soil.
6. When soil is dry, it holds together not only by tensile strength, but with the cohesion and friction forces as well. It should be noted that tensile strength is more sensitive to the moisture content than other strength parameters as discussed in Section 5.12. Table 6.5 summarizes the mechanism of tensile strength of soil.

6.7.3 Measurement of Tensile Strength of Soil

The Lehigh method is also called *unconfined penetration* (UP) method. The laboratory setup of the UP test is shown in Figure 6.17. By using two steel disks (punches) centered on both the top and bottom surfaces of a cylindrical soil specimen, a vertical load is applied on the disks until the specimen reaches failure. Alignment of the test specimen between two disks must be in the same line. The tensile strength of the specimen can be calculated from the maximum load by the formula:

$$\sigma_t = \frac{P}{\left(KbH - a^2\right)} \tag{6.1}$$

where σ_e = tensile strength, psi; P = maximum load at failure, lb; K = constant, depending on specimen-punch size and soil type (Table 6.6); b = radius of specimen, in.; H = height of specimen, in.; and a = radius of disk (punch), in.

(a)

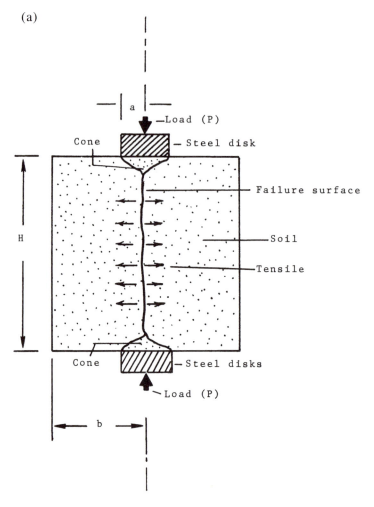

Figure 6.17 Unconfined-penetration (UP) test for determination of tensile strength of soil. (a) Geometric of test specimen; (b) test setup; (c) failure mode of soil samples.

In Equation (6.1), the constant K depends on disk (punch) and specimen size, and the compression-tensile strength ratio. Table 6.6 summarizes the proposed K values for the various types of soil and specimen-punch size.

6.7.4 Tensile Strength Relating to Other Soil Constants

(1) Atterberg Limits and Density

The tensile strength increases as the liquid limit (LL) and plasticity index (PI) increase. For the higher moisture content, as density increases the tensile strength increases slightly; however, at lower moisture content, as density increases, the tensile strength increases sharply. Similar conclusions were reported by other investigators by using split-tension tests (Narain and Rawat, 1970).

(2) Activity and Toughness Index

Tensile strength and activity (A) values are plotted in Figure 6.18(a). It is clearly indicated that the tensile strength increased significantly for all types of soil when air-dried.

Figure 6.17 (continued)

However, the amount and rate of increase in strength depends on the soil type. For a low-plasticity soil, the differences between two moisture conditions are more pronounced than for a high plasticity soil. Figure 6.18(b) shows the tensile strength test results vs. toughness index (I_t) for both moisture conditions. Similar trends are indicated in Figures 6.18(a) and 6.18(b). From these results, it can be projected that there will be greater differences in strength between the two moisture conditions for a low-plasticity soil than for a high-plasticity soil.

Table 6.6   Value for Constant, K, Specimen-Punch
Size, Specimen Height-Diameter Ratio, and
Rate of Loading

Specimen size	K value	
	Soil	Stabilized materials
Proctor mold		
4 × 4.6 in. (10.2 × 11.43 cm)	1.0	1.2
CBR mold		
6 × 7 in. (15.24 × 17.78 cm)	0.8	1.0
Larger than 12 in. (0.30 m)	0.8	
Specimen-punch size: 0.2–0.3		
Height-to-diameter ratio of specimen: 0.8–1.2		
Rate of loading: ASTM recommendation for the axial strain at a ratio of 0.5–2% of height per minute		

(3) Compressive Strength

The ratio of unconfined compressive strength to tensile strength of materials is of interest to all design engineers because of its practical uses. For soils, most tests for this ratio are performed at OMC conditions. A curvilinear relationship has been reported for various soil types (Narain and Rawat, 1970; Krishnaya et al. 1974). Figure 6.19 shows the effect of soil types and molding moisture contents on compression-tension ratio. In examining Figure 6.19 indications for the low-plasticity soil, the molding moisture contents are sensitive to the compression-tension ratio. However, for the high-plasticity soil, the molding moisture content is much less sensitive to the compression-tension ratio.

(4) Cohesion and Friction Angle

The relationship between cohesion and tensile strength is also useful to the practitioner. When soil is dry, the ratio of cohesion and tensile strength influences only the soil types. For low-plasticity soil, a slight increase of the molding moisture content sharply affects the cohesion-tensile ratio, while for a high-plasticity soil molding moisture content has lesser effects on the cohesion-tensile ratio.

The curvilinear relationship between friction angle and tensile strength was observed for all six types of soils It indicates that soils with higher plasticity indices have lower friction angles. It can also be seen that at lower tensile strengths the rate of increase of the friction angle is much higher than at higher tensile strength.

(5) Effect of Exchangeable Ions

Figures 6.20(a) and 6.20(b) show the compressive and tensile strength of specimens of nine ionic modifications of two natural clay soils. Figure 6.20(a) plots tensile strength vs. PI for Hagerstown and Cecil soils. The tensile strength was tested in an air-dried condition. In examining Figure 6.20(a), it indicates that for Hagerstown soil greater variations exist in comparison with Cecil soil. A similar conclusion can be obtained for compressive-tensile strength ratio as shown in Figure 6.20(b).

(6) Soil Classification Based on Tensile Strength

As discussed in Section 5.12, the tensile strength is more sensitive to environmental conditions than compression test results; therefore, it is advantageous to use tensile strength

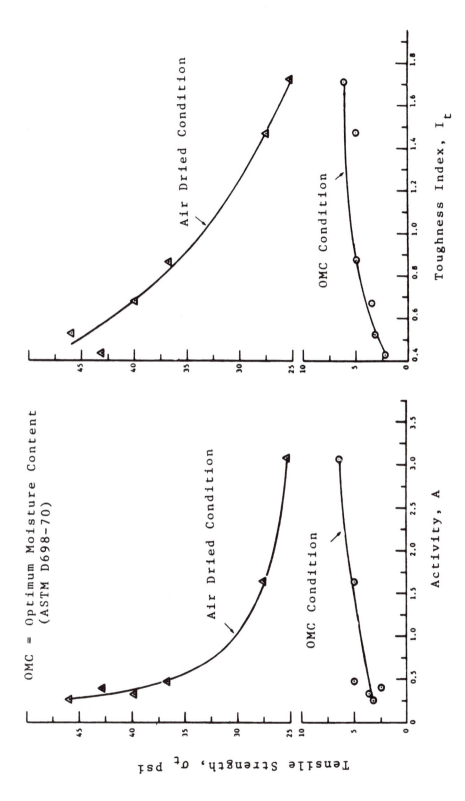

Figure 6.18 Tensile strength vs. soil constants of two moisture contents during tension tests. (a) Tensile strength vs. activity; (b) tensile strength vs. toughness index. (After Fang and Fernandez [1981].)

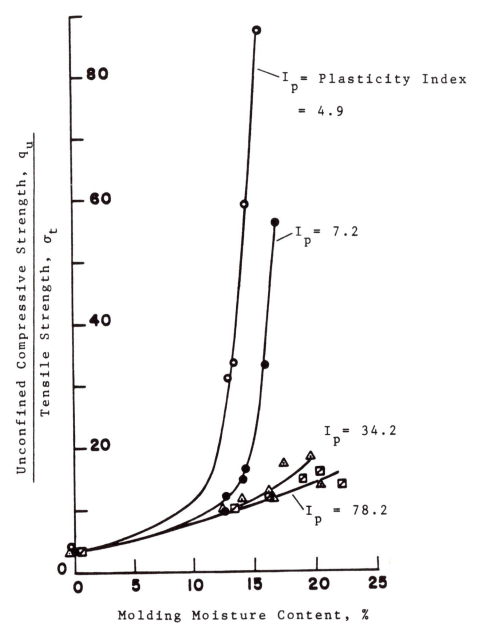

Figure 6.19 Effect of soil types and molding moisture contents on compression-tension ratio. (Data from Fang and Fernandez [1981].)

to classify soil type for soil stabilization or ground improvement in adverse environments. Table 6.7 presents a soil classification system based on tensile strength determined by UP tests.

6.8 FRACTURE CHARACTERISTICS OF SOIL

6.8.1 Introduction

The fracture test is also a quantitive measurement of the cracking behavior of a soil because for a given material, the cracking and fracture characteristics belong to one system.

(a)

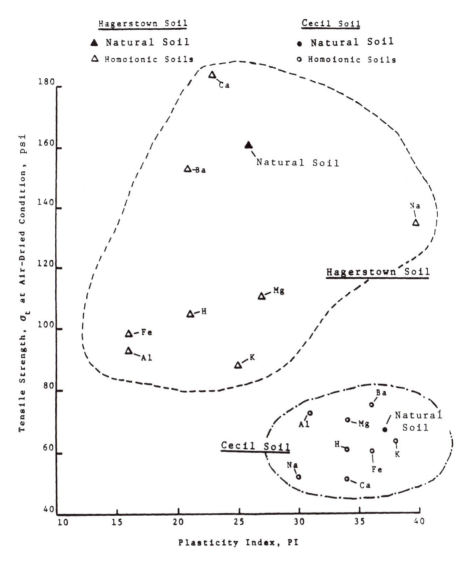

Figure 6.20 Compressive-tensile strength vs. plasticity index on air-dried specimens of natural and homoionic soils. (a) Tensile strength vs. plasticity index; (b) compressive-tensile strength ratio vs. plasticity index. (Data from Winterkorn, H.F. and Moorman, R.B.B. [1941], Proc. 21st Annual Meeting, HRB, pp. 415–434.)

Cracking represents the prefailure phenomenon of a material and fracture is the behavior of a material at failure condition. To illustrate this relationship, a typical deformation vs. loading curve for a given soil is presented in Figure 6.2. This figure shows the correlation between prefailure and failure conditions of a soil from a geotechnical engineering point of view. At the early stage of deformation vs. loading curve, the cracks develop due to various natural or man-made causes. At this stage, the soil behavior is controlled by such energy as thermal, electric, and magnetic. However, at failure, it is dominated only by mechanical energy which is the loading. Therefore, for evaluation of soil's useful life, it is necessary to investigate the cracking-fracture mechanisms and their interaction.

(b)

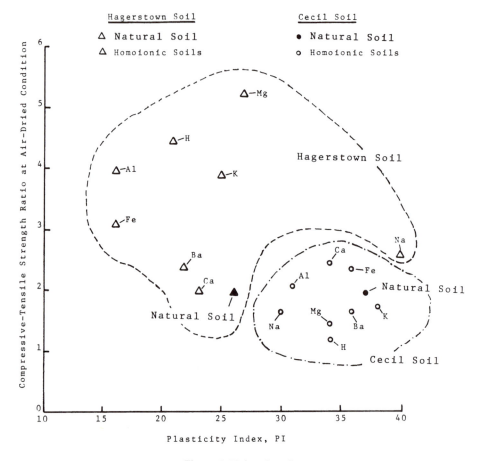

Figure 6.20 (continued)

6.8.2 Application of Linear Elastic Fracture Mechanics (LEFM)

The concept of fracture mechanics theory was developed in 1921 by Griffith. It has acquired a considerable amount of success in predicting failure caused by crack propagation for metals. More recently, it has been applied for predicting fracture behavior in rocks, concrete, pavements, stabilized construction materials, and for compacted fine-grained soils.

Strictly speaking, the theory is limited to linear-elastic materials despite the fact that plastic or nonlinear strains unavoidably prevail in the vicinity of flaws or cracks. Nevertheless, it does provide an ideal and simple way of estimating the amount of energy required to create free surface in the material. The degree of simplicity is achieved by focusing attention on the leading edge of the crack where nonlinear strains exist but are regarded as localized within a zone of negligibly small dimensions. Although the stresses and strains within this zone cannot be analyzed in fine detail, LEFM theory is able to give an adequate description of the gross feature of the stresses and strains near the cracks. In particular, one can calculate the strain-energy release rate, G, which is in a formal sense the force driving the fracture process

Table 6.7 Tensile Strength for Nonmetallic Construction Material Classification

Degree of friability	Tensile strength ranges [Psi (kPa)]	Typical material types[a]
Weak	0–10 (0–69)	Saturated soil; sandy silt sand; cinders; dry fly ash
Very friable	10–30 (69–210)	Compacted soil; stiff natural clay; hardpan
Friable	30–80 (210–550)	Soil-cement; fly ash-line mixtures; lime-stabilized soil; compacted soil-aggregates
Medium friable	80–300 (550–2100)	Cement or bituminous-treated granular materials; aggregate-lime-pozzolan mixture; chemical-treated base materials
Strong	300–1000 (2100–6900)	Mortar; light-weight concrete; lean concrete; bituminous concrete
Very strong	Greater than 1000 (>6900)	Rocks; high-strength concrete; polymer concrete

[a] Typical material types show large variation in the results due to the molding moisture contents, curing time and types, and soil-aggregate interaction.

(Irwin, 1968). At present, plastic and nonlinear strains can be included in the fracture mechanics treatment only in a rather superficial way without incurring serious losses of clarity and simplicity (Shi and Liebowitz, 1968).

Previous work on fracture mechanics has been concerned mostly with crack-toughness testing studies of common metals in an effort to characterize the fracture toughness of a material by the critical value of energy release, i.e., (G_c). In this way, one can compute values of nominal stress necessary for crack propagation for various size-specific cracks. Many of these computations have been carried out in the past for laboratory specimens with simple configurations by assuming material isotropy, homogeneity, etc. However, application of fracture mechanics principle is no longer a simple matter when applied to a complex system such as a multilayered soil.

6.8.3 Laboratory Fracture Tests on Soils

(1) Preparation of Test Specimen

The molds used in preparing the test specimens were developed at the Geotechnical Laboratory of Lehigh University. The mold consists of a steel base with removable sides and interior walls. It can accommodate up to four specimens at the same time. The size of the specimen is $3 \times 3 \times 0.25$ in. as shown in Figure 6.21(a). The length of the notch is 1 in. (2.5 cm). This notch is used for creating the cracks when the fracture load, P, is applied. The interior of the mold was coated with a lubricating agent to facilitate the removal of the specimens. After the specimens were extracted from the mold, they were air dried. A glass plate cover was placed over them to prevent warping.

The size of mold is modified when larger soil particles are involved. The dimension is proportional to the grain-size distribution. For stabilized soil specimens that measured $6 \times 6 \times 0.50$ in. a modified mold was used. The test procedure follows ASTM (E399-90). However, a simple recording device, the mechanical strain device, has been modified for determination of fracture load, P.

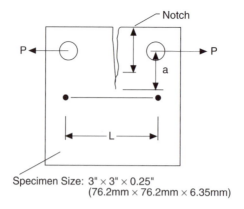

Specimen Size: 3" × 3" × 0.25"
(76.2mm × 76.2mm × 6.35mm)

a

b

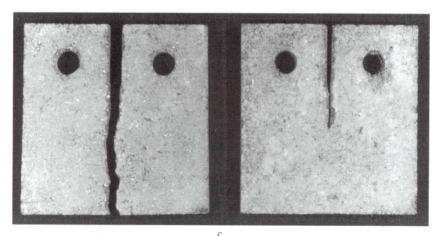

c

Figure 6.21 Determination of fracture load of compacted soil samples by compliance test. (a) Dimensions of soil specimen; (b) soil sample; and (c) failure mode.

(2) G_c and K_c Measurements and Computations

In practice, the value of strain-energy release rate, G, is measured in a laboratory test (ASTM E399-74). It uses a cracked specimen which is pulled apart by a load, P, as shown in Figure 6.21(a). The sketch of the specimen and failure mode is shown in Figures 6.21(a) and 6.21(b). For a gauge length, L, and a crack length, a, one obtains the strain energy release rate, G, for the material under testing as follows:

$$K_c = Y(a/w)\,P(a)^{1/2}\,/Bw \tag{6.2}$$

where a, B, H, L, and w are given in Figure 6.21(a). The term designated as Y(a/w) is a function of a/w, and is the finite width correction (Shi, 1973). G_c and K_c are related as shown in Equation (6.2). The relationship between fracture load and strain energy developed by Irwin (1968) is shown in the following equation:

$$G_c = \frac{1}{2}P^2\,\frac{\partial(\Delta L/P)}{\partial a} \tag{6.3}$$

where G_c = Griffith strain-energy release rate; P = fracture load; ΔL = change in gauge length and a = distance between the line of load application to the crack tip.

The term $\Delta L/P$ in Equation (6.3) is the compliance and the test itself is often referred to as a compliance test. This allows for computation of a stress intensity factor, K, of the newly created fracture surface (crack) as follows:

$$K_c = \frac{G_c E}{1 - v^2} \tag{6.4}$$

where K_c = stress intensity factor; E = Young's modulus of the material; v = Poisson's ratio; and G_c = determined from Equation (6.3).

Equation (6.4) is satisfactory for a specimen whose thickness is large in comparison with the crack size (Shi and MacDonald, 1974). The central idea behind the K factor is that all information relating to crack loading and geometry is contained therein. The K factor changes from problem to problem and is dependent on the stress state near the crack tip.

(3) Fracture Load Tests on Stabilized and Contaminated Soils

Three experimental studies were carried out: (1) soft marine clay, (2) compacted fine-grained stabilized soil, and (3) solidified sludge waste. Material used in the Experiment (3) was made by mixing the dewatered sludge (filter cake) with slag, cement, and water. The slag added was screened (passing 1.25-in. sieve and retaining on 0.25-in. sieve) and the cement was selected as type II. Detailed procedures are discussed in a separate report (Pamukcu and Topcu, 1991). Experiments (1) and (2) are summarized by Fang (1994a).

(a) Soft Marine Clay

Fracture load test on two soft marine clays is shown in Figures 6.22 and 6.23. All samples were passed through sieve #40. Molding water content for all samples was constant at 25%, while varying the molded dry density. All samples were air dried before testing. The fracture

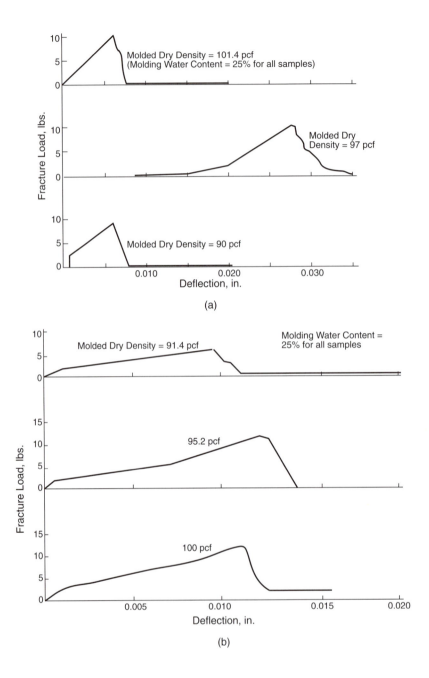

Figure 6.22 Fracture load vs. change in gauge length with various molded dry densities. (a) Gulf of Mexico marine clay; (b) Gulf of Maine marine clay. (After Fang [1994a].)

load and the corresponding change in gauge length were recorded. The crack growth was also determined. In this study, all samples failed rapidly; therefore, the a value in Figure 6.21(a) is assumed as the length between the line of application to the total length of the specimen (= 2.5 in.).

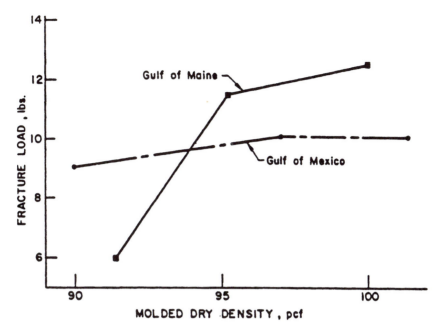

Figure 6.23 Fracture load vs. molded dry density for Gulf of Maine and Gulf of Mexico marine clays. (After Fang [1994a].)

Table 6.8 Summary of Fracture Toughness Data on Three Compacted Fine-Grained Soils

Sample no.	Crack length a, (nm)	Fracture load P, (N)	Fracture toughness MPa.m$^{1/2}$
1	24.13	22.46	0.0979
	27.94	20.91	0.1064
	31.75	16.46	0.0988
	35.56	15.57	0.1134
	39.37	10.68	0.0981
2	26.67	18.24	0.0878
	27.94	19.57	0.0993
	33.02	15.57	0.0994
	35.56	12.90	0.0940
	38.10	9.79	0.0828
3	20.32	22.24	0.0846
	24.13	22.24	0.0969
	24.13	21.35	0.0930
	30.48	15.12	0.0856
	31.75	13.34	0.0801

Note: Computed from Equation 6.2.

(b) Fracture Load on Fine-Grained Soil

Results of compacted fine-grained soil are summarized in Table 6.8. The measured values of fracture load, and the calculated values of fracture toughness, K_c, are given for different crack lengths, **a**, for identically prepared specimens in each group.

(a)

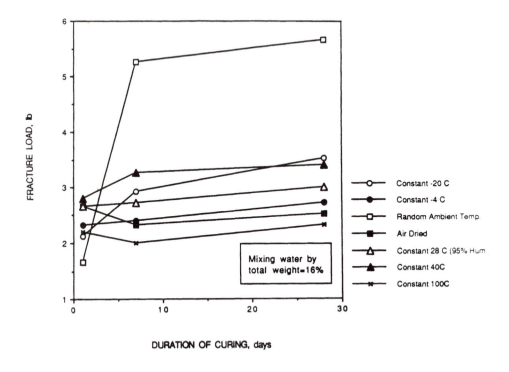

Figure 6.24 Fracture load tests on solidified sludge wastes. (a) Affect of duration of curing; (b) affect of water content for 28-day cured specimens; (c) affect of % of air entrainment reagent; (d) affect of superplastic reagent. (Data from Pamukcu, S. and Topcu, I.B. [1991], Study of Causes of Random Cracking of Solidified Sludge Reused as Capping Material, Final Report to Bethlehem Steel Corp. February, 33 p.)

(c) Fracture Load on Contaminated Fine-Grained Soils

Fracture load tests on solidified sludge wastes are summarized as shown in Figure 6.24 (Pamukcu and Topcu, 1991). The results of the fracture load tests appear to support the assessment of other test results such as unconfined compressive strength and tensile strength. All these tests were measured over a range of water contents and curing temperatures. The principle aim for these measurements was to see if there was a correlation between strength development and fracture load, P, under similar environmental conditions.

Figure 6.24(a) shows the variation of fracture load vs. curing time for the 16% mix water content specimens. It appears that the samples cured in the ambient-temperature (20°C) and relative-humidity (65%) environment are by far more resistant to cracking than the samples cured otherwise. Figure 6.24(b) shows the variation of fracture load with increased water content for 28-day cured specimens. There is a consistant decrease of the fracture resistance for the –20°C, the 100°C cured, and the air-dried specimens with increasing water content.

Figure 6.24(c) shows the influence of increased percentage of air bubbles (percent air entrainment reagent) on the 28-day specimens. However, the largest variations were observed with the frozen and the humidity chamber-cured specimens, whereas there was no significant variation with the results of the ambient condition specimens. Figure 6.24(d) shows the variation of fracture load with increasing percentage of superplasticizer reagent in the mixture. In general, addition of the superplasticizer appears to either produce little change or decrease in the fracture load.

(b)

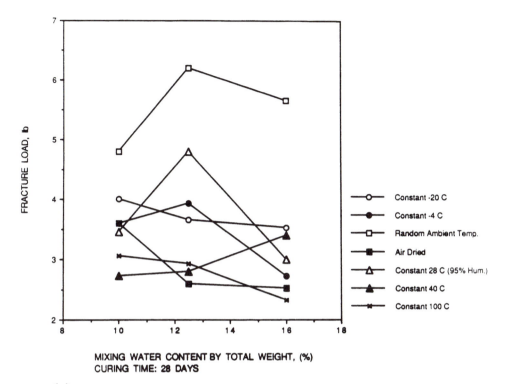

(c)

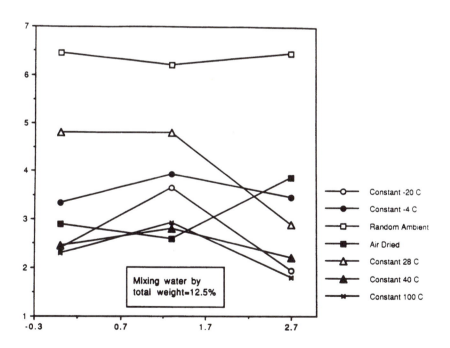

Figure 6.24 (continued)

(d)

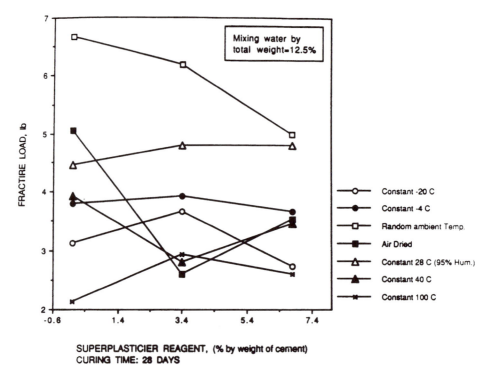

Figure 6.24 (continued)

The effects of environmental factors on fracture load presented in this experiment appear to be inconclusive and more tests are necessary to study the effect of molding water content, humidity, and cyclic freezing-thawing on air entrainment treatment. However, the results of a majority of the tests show that there may be an optimum proportion of the reagent that will result in higher fracture resistance of the solidified mixture.

6.8.4 Applications from Cracking and Fracture Data

The cracking and fracture information of soil also can be used in environmental geotechnical engineering. A brief discussion is presented as follows.

(1) Use as a Guideline in Ground Improvement Program

Section 6.6.1 indicates that in the soil mass either man-made or the natural state, there always exist some cracks due to various environmental factors. Tensile and fracture tests are the quantitative measuring techniques for estimation of the cracking behavior of a soil. Figure 6.25 is a schematic diagram illustrating the cracking-time relationship at various cracking stages. This diagram is used as a basic guideline for evaluation of crack growth in brittle construction materials. In examining Figure 6.25, it is seen that an initial crack of size C (point a) can remain constant for a large number of load cycles before an increase is noted. From points b to c is called the degradation stage. In that period, the ground improvement or repair work must be made, in order to extend the useful life of the material or structure, shortly thereafter the useful life of the structure (point c) will rapidly come about. Also, it can be seen in this figure that shrinkage, thermal-, tensile-, and fracture-caused cracks are

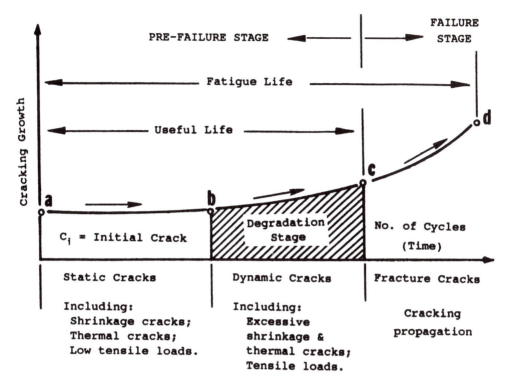

Figure 6.25 Schematic diagram illustrating the cracking-time relationship at various cracking stages.

interrelated with the useful life of any geostructures, especially in progressive failure of slopes, erosion, and landslide problems.

(2) Cracking Patterns used for Identification and Characterization of Soil

Cracking patterns can be used for identification or characterization of soil behavior; also, the color and patterns can be used together for identifying and characterization of contaminated fine-grained soils.

(3) Internal Cracks Used to Evaluation the Soil Disturbance

Characteristics of internal soil cracking can be used for estimation of progressive erosion or predicting potential landslides. This type of cracking measurement can be made by use of computed tomograph (CT) techniques as discussed in Section 6.7.6.

Soil cracking pattern tests are a quantitative indicator property which should be correlated with other soil performance parameters in order to have a significant meaning. These performance tests include tensile, fracture, clay mineral structure analysis, and pore fluid chemistry analysis. Other factors contributing to the soil cracks are soil types, unit weight, grain size distribution, void ratio, and geometry of the primary and secondary drying surfaces.

6.9 SUMMARY

1. Shrinkage is one of the major causes for volume changes associated with variations of water content in soil. If water is added to the dry soil, the volume of soil increases and this phenomenon is called swelling. It must be pointed out that shrinkage and swelling are not a reversible process; however, there is a linear relationship between these two processes with limited ranges.

2. Sorption includes both absorption and adsorption processes. The absorption is a mechanical process and adsorption, however, is a physicochemical process.

3. Soil cracking is a natural phenomenon due to internal energy imbalance in the soil mass and changes in local environments. Four types of cracking are examined, namely: shrinkage, thermal, tensile, and fracture.

4. The characteristics of cracking and fracture belong to one system. Cracking represents the prefailure phenomenon of a material, and fracture is the behavior of a material at failure condition; both are interrelated. Previously, fracture mechanics had been concerned predominantly with crack-toughness testing studies of metals; however, the concept has been expanded to include materials such as concrete, bituminous mixture, and stabilized soils.

5. LEFM is limited to linear-elastic materials despite the fact that plastic or nonlinear strain unavoidably prevails in the vicinity of flaws or cracks. Nevertheless, it does provide an ideal and simple way of estimating the amount of energy required to create free surfaces in the material. LEFM cannot be applied to expansive clays due to the complicated soil-water interactions.

6. The unconfined penetration (UP) test is introduced for determination of tension test. It is simple, easy to perform, and can be used conveniently in conjunction with routine compaction and CBR tests.

PROBLEMS

6.1 A building is to be constructed in a region where the maximum depth of seasonal variation in moisture content is 3 m. The footings are placed 2 m below the ground surface, and at the time of construction the clay layer from this depth down to 3 m was saturated and had a natural moisture content of 32.2%.

Under the most adverse conditions, the moisture content at the 2-m depth could be reduced to 18.0% and that varied almost linearly with depth. The shrinkage limit was found to be 20.0%. Assuming that the building is large enough to prevent evaporation completely beneath its center, what is the maximum differential settlement or volume change that could occur due to shrinkage of the clay layer? (Make any reasonable assumptions you deem necessary.)

6.2 Prove shrinkage limit is a function of unit weight of water, dry unit weight of soil, and specific gravity of solids.

6.3 The shrinkage of a clay soil due to the evaporation of water stops when a certain water content is reached, and when this occurs the clay is still essentially saturated. Explain why?

6.4 Explain the differences between ASTM standard split-tensile test and unconfined-penetration (UP) test for determination of tensile strength of concrete block, rock, and compacted clay.

6.5 Why is there very little tensile strength for saturated soil? Why is the moisture content so sensitive to tensile over the compressive strengths?

6.6 Compare failure modes between split-tensile test and unconfined penetration (UP) test for determination of tensile strength of fine-grained soil.

6.7 Why can the fracture mechanics concept not be applied for evaluation of fracture behavior of expansive clays?

CHAPTER 7

Hydraulic Conductivity and Mass Transport Phenomena

7.1 INTRODUCTION

The amount of water existing in the soil mass significantly influences the engineering behavior of soil because the water content will affect the state of stress of the soil under a load, as is especially true for cohesive soils. Other behavior related to water includes densification (compaction) properties, shrinkage, swelling, settlement, and progressive failure. In this chapter, discussions focus on capillary, hydraulic conductivity, and mass transport phenomena. Applications emphasize the areas of ground improvement engineering and the soil decontamination process. Others such as shrinkage, swelling, densification, settlement, and strength are discussed in Chapters 6 and 11.

7.2 CAPILLARITY

7.2.1 General Discussion

Capillary rise is caused by surface tension of fluid as discussed in Section 5.2. It occurs at the interface between two different materials. For soil, it occurs between the surface of water, mineral grains, and air. Basically, surface tension results from differences in forces of attraction between the molecules of the materials at the interface. The phenomenon of capillarity also can be explained by the Law of Thermodynamics. All surfaces and interfaces of solids and liquids are loci of free energy whose universal tendency to decrease (Second Law of Thermodynamics) leads to the phenomenon of surface and interfacial tension. One consequence of this is the rise of a liquid in a capillary whose walls are wetted by the liquid, and its depression in a capillary whose walls have no affinity for the liquid.

7.2.2 Height and Rate of Capillary Rise

(1) Height of Capillary Rise

There are two types of capillary phenomena, the vertical and horizontal types. The basic equation for capillary rise or depression is

$$h_c = \pm \frac{2T \cos \alpha}{r \gamma g} \tag{7.1}$$

where h_c = height of capillary rise, cm; T = surface tension, dyn/cm; α = angle of wetting; r = radius of capillary, cm; γ = density of liquid, g/cm^{-3}; and g = gravity, cm/sec^{-2}.

Since 2/r equals the ratio of wetted perimeter to cross section of liquid column and also that of wetted area to liquid volume per unit length of liquid column, Equation (7.1) may be written as:

$$h_c = \frac{S\ T}{n\gamma g} \qquad (7.2)$$

where S = internal or capillary surface per unit volume and n = pore space per unit volume.

Equation (7.2) holds reasonably well for nonswelling soil systems in which the thickness of the strongly adsorbed water films is negligible as compared with the pore dimensions, where, of special importance, the potential capillary rise or 'suction' in a soil should be determined experimentally. From basic physics Equation (7.1) for water can be simplified as:

$$h_c = 0.3/d \qquad (7.3)$$

where d = diameter of voids between soil particles, cm

Since the value of d is difficult to obtain, Terzaghi (1942) proposed to let d = D_{10}; then the effective size (Section 4.4) of the soil particle is:

$$h_c = 0.3/D_{10}, \text{ cm} \qquad (7.4)$$

In addition, Terzaghi also proposed that h_c can be estimated from void ratio, e, and D_{10} as:

$$h_c = C/e\ D_{10}, \text{ cm} \qquad (7.5)$$

where C = constant which varies from 0.1 to 0.5.

(2) Rate of Capillary Rise

Often the rate of capillary rise is more important than the ultimate height. In such case one employs the following equation:

$$t = \frac{n\,h}{k}\left[\ln \frac{h}{h-z} - \frac{z}{h} \right] \qquad (7.6)$$

where t = time required for the meniscus of the capillary water to rise to the height z above the free water level; n = porosity; k = coefficient of permeability of the soil; h_c = capillary potential, height of capillary rise and z = distance of capillary meniscus from groundwater level.

7.2.3 Factors Affecting Capillary Rise

(1) Effect of Time and Soil Particle Size

A curvilinear relationship between time on capillary height for sand, silt, and clay are observed. This relationship is closely related with soil particle size as indicated in Figure 7.1. However, local environments play an important role on the capillarity rise which will be discussed in Figure 7.2.

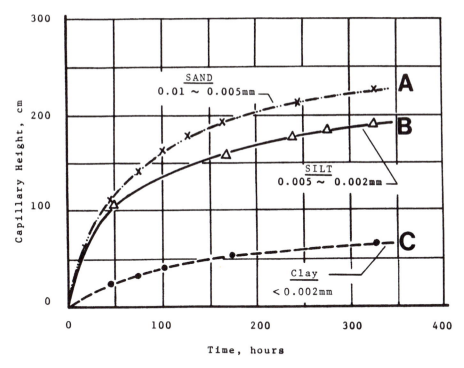

Figure 7.1 Capillary height vs. time for various soil particle sizes. (After Tianjing University [1979].)

(2) Effect of Moisture Content and Types of Pore Fluids

Moisture content, size of soil grains, contact angle, and state of packing affect the value of capillary potential because of their influence on the radii of curvature of the water-wedge surfaces. If the soil particles are not completely wetted by water (unsaturated soil), the contact angle between the menisci and the soil particles will be greater than zero. An increase in the contact angle will tend to increase the curvature of the meniscus and thereby increase the capillary potential of the soil at a given water content.

(3) Effect of Temperature

The surface tension of water is an inverse function of temperature. Hence, a decrease in temperature increases the surface tension and decreases the capillary potential as well as other factors such as dissolved salts, because an increase in the amount of dissolved salt in the soil water increases its surface tension and thereby lowers the capillary potential of the soil.

7.2.4 Capillary Measurements

(1) General Discussion

There are several measuring techniques for estimation of the capillary rise and capillary phenomena in soils, such as the capillarimeter and open-tube method (Lane and Washburn, 1946), capillarity-permeability tests (Taylor, 1948; Lambe and Whitman, 1979), and the ASTM standard method for measuring capillary-moisture relationships for soils by pressure-membrane apparatus (ASTM D3152-77) and by moisture tensions between 0.1 and 1 atm (10 and 101 kPa) (ASTM D2325-81).

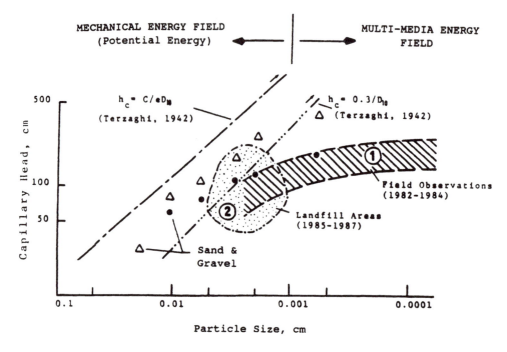

Figure 7.2 Soil-water-environment interaction as reflected on capillary rise.

Other methods for estimation of potential capillary rise are available, such as the Richards soil tensionmeter, *in situ* capillary potential of soils, developed by Russell and Spangler (Spangler, 1951), or 'suction' in a soil suggested by Marshall (1959). The capillary potential or free energy has been measured by the tensionmeter, pressure membrane apparatus, and vapor adsorption method. Winterkorn (1963) presents interrelationships between grain size, capillary rise, internal surface, porosity, and hydraulic conductivity of granular soil systems.

(2) Typical Test Results and Comments

Test results determined by various methods under various environmental conditions are presented in Figure 7.2. In examining Figure 7.2, some comments are made as: Equation (7.2) holds reasonably well for nonswelling soil systems in which the thickness of the strongly adsorbed water films is negligible as compared with the pore dimensions. Equation (7.5) proposed by Terzaghi is based on experimental data provided by Lane and Washburn as shown in Figure 7.2, together with the other experimental data. The experimental results from laboratory open-tube and capillarimeter tests lie between two lines; Equation (7.5) serves as an upper limit and Equation (7.4) serves as a lower limit. Figure 7.2 shows that the smaller the particle size, the larger the h_c. After a period of 300 hr, the h_c levels off for all three grain sizes regardless of the soil particle sizes.

Based on field observations for various soil types from construction projects as shown in Figure 7.2, it is indicated that for cohesionless soil, the h_c is close to the values predicted from Equation (7.4). For fine-grained soil and for most field observations, h_c will not exceed more than several meters. All field observations in both construction and landfill areas are based on the saturated free water surface line located at the soil profile and/or horizons in excavations. For granular soil, such as clean uniform sand or gravel, the h_c, is dominated by the effective grain size, D_{10}. However, for fine-grained soil, the h_c is not only related to the grain size, but also influenced by the grain-size distribution, clay mineral composition, pore fluid, stress history, as well as local environmental factors.

Equations (7.4) and (7.5) proposed by Terzaghi are valid when the D_{10} is greater than 0.02 mm. Also, the rate of the capillarity rise depends on the soil particle sizes. For example, for particle size between 0.01 and 0.005 mm, 7 days are required to reach the capillary height at 2.0 m. However, for particles smaller than 0.002 mm, much more time, perhaps months, is required to reach the same height.

7.3 HYDRAULIC CONDUCTIVITY

7.3.1 Darcy's Law

Flow velocity is the flow through a porous media following *Darcy's* Law as shown in the following equation:

$$Q = kiA \tag{7.7}$$

where Q = the rate of flow; k = a constant known as Darcy's coefficient of permeability; i = hydraulic gradient; and A = cross-sectional area for which flow can pass through.

Equation (7.7) can be rewritten as:

$$v = ki = Q/A \tag{7.8}$$

If the hydraulic gradient, i, is to a unity, then the velocity, v, in Equation (7.8) is equal to k. Also, the velocity, v, is called *approach velocity* or *superficial velocity* (Taylor, 1948). The *average effective velocity*, v_s, of flow through the soil can be computed as:

$$v_s = v\ A/A_v = v\ V/V_v = v/n \tag{7.9}$$

where v_s = average effective velocity, also known as seepage velocity; A = total cross-sectional area of soil specimen; A_v = area of voids in soil specimen; V = total volume of soil specimen; V_v = volume of voids in soil specimen; and n = porosity.

Saturated flow under a hydraulic pressure gradient following Darcy's Law can be written as:

$$k = \frac{V}{A\,i\,t} \tag{7.10}$$

7.3.2 Hydraulic Conductivity Equations

(1) Poiseuille-Hagen Equation

Equation (7.8) is Darcy's Law for hydraulic conductivity in soils, representing essentially laminar flow if k is the coefficient of hydraulic conductivity or permeability. Because of the laminar nature of the flow theoretical calculation of k was attempted by a number of workers on the basis of the fundamental *Poiseuille-Hagen* equation for flow-through capillaries:

$$V' = \frac{\pi r^4}{8\eta} \cdot \frac{\Delta P}{\Delta L} \tag{7.11}$$

in which V' = volume of water yields in time, t; r = internal radius of capillary; η = viscosity of the liquid; and $\Delta P/\Delta L = i$ = hydraulic gradient.

(2) Kozeny-Carman Equation

Best known among the attempts of theoretical derivation of k is perhaps the *Kozeny-Carman* equation:

$$k = \frac{1}{\eta} \frac{1}{k_p t^2 S^2} \frac{n^3}{(1-n)^2} \tag{7.12}$$

in which k_p = pore shape factor; t = tortuosity; S = specific surface per unit volume; n = porosity; and η = coefficient of viscosity of liquid.

(3) Winterkorn Equation (Winterkorn, 1942 and 1963)

The *Winterkorn* equation is based on the Poiseuille-Hagen equation for flow-through capillaries, as Equation (7.11). Considering the porosity of a soil system as due to the presence of (n_c) capillaries of effective radius (r) per unit cross section, we can multiply both sides of the Poiseuille equation with n_c and set the product equal to V/A in the Darcy equation (Equation 7.8). We obtain:

$$k = \frac{(n_c) \cdot V'}{i\,t} = \frac{\pi r^4}{8\eta} \cdot (n_c) \tag{7.13}$$

where n_c = porosity due to capillarity.

If we multiply the numerator and denominator of the right side of the equation (Equation 7.13) with the unit length (ℓ) and let:

$$n = \pi r^2 \, n_c \, (\ell) \tag{7.14}$$

Substituting (n) value from Equation (7.14) into Equation (7.14a) for (k), we obtain

$$k = \frac{n\,r^2}{8\eta\ell} \tag{7.14a}$$

where (k) in Equation (7.14a) is called transmission coefficient and is proportional to the porosity (n) and internal radius of the capillary (r^2). Equation (7.14a) holds reasonably well for porous systems in which there is no detectable interaction between the pore walls and the water substance. In the case of appreciable soil-water interaction, the equation must be written as:

$$k = \frac{(n - n_r)(r - r_r)^2}{8\eta\ell} \tag{7.15}$$

where n_r= pore space and r_r= the portion of the radius occupied by restrained water that does not participate in the flow.

Equations (7.14) and (7.15) can be further modified by considering [$r^2/4$] as an oversimplified expression of [$r^2\pi/2r\pi$]^2, which also corresponds to the square of the ratio of porosity (n), so the pore wall surface (S) can be calculated from the grain size composition or from the results of adsorption experiments. The range of validity of these and similar equations is restricted to well-disturbed uniform soil samples containing no secondary soil particle aggregations.

7.3.3 Hydraulic Head and Energy

To express both potential and kinetic energies of fluid, the term *head* is commonly used to denote the energy per unit of mass. There are three types of heads involving fluid flow in soil; namely, *pressure head* (h_p), *elevation head* (h_e), and *velocity head* (h_v). The h_p is the height to which the water rises in the piezometer above the point under consideration. In geotechnical application, the h_p is needed for computing the porewater pressure and effective stress (Table 7.3). The h_e is the distance from the datum. The pressure head plus elevation head ($h_p + h_e$) is called total head or the piezometric head. The total head is used in Darcy's Law to compute the hydraulic gradient, i, as shown in Equation (7.8). It must be noted that the velocity head (h_v) in soil is of very small magnitude and can be neglected for practical application; however, for estimation of pollution migration, the velocity head in soil should not be omitted.

7.3.4 Hydraulic Conductivity Measurements

Hydraulic conductivity or coefficient of permeability, k, defined in Equation (7.7), can be measured in the laboratory and/or in the field. There are numerous techniques available directly or indirectly to obtain this value. In the laboratory test, three general approaches are (1) variable or falling head, permeameter; (2) constant head permeameter; and (3) computed from consolidation test (Equation 11.7). Using permeameter for measuring the hydraulic conductivity has been standardized (ASTM D2434-68) or described in various standard textbooks on soil testing methods. The pertinent equation for the constant head permeameter is

$$k = V/Ait \tag{7.16}$$

where V is the water yield in time t; A is the cross-sectional area of the specimen; and i = $\Delta P/L$, i.e., the difference in pressure at the two ends of the specimen. For the falling head permeameter:

$$k = 2.3 \frac{La}{At} \log \frac{h_1}{h_2} \tag{7.17}$$

where a = cross-sectional area of the standpipe in which the hydraulic head falls from h_1 to h_2 in time t, while the water volume a($h_1 - h_2$) flows through the soil specimen of thickness L.

Other methods include computing from particle surface area per pore-volume and its porosity made by Winterkorn (1963), based on the experimental results provided by HRB (1959). Also, the k value can be estimated from effective size, D_{10}, as proposed by Hazen (1911). It may also be calculated from seepage, pumping, and pressure tests in the field. Comparison of k from laboratory measurements and calculations of various soil types is presented in Figure 7.3. As seen in the figure, there are large variations among these methods.

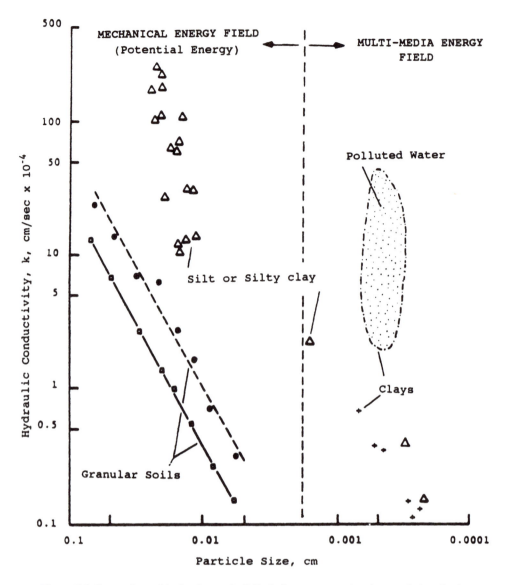

Figure 7.3 Comparison of hydraulic conductivity, k, from measured and computed methods.

7.3.5 Factors Affecting Conductivity Results

(1) General Discussions

In examining Figure 7.3, there are large variations in hydraulic conductivity results, especially for fine-grained soils. An understanding of the influence of pore fluids on clay behavior is essential to the design of many components of foundation structures, especially for hazardous and toxic waste containment systems. There are numerous factors affecting this fine-grained soil, but among the most important factors are exchangeable ion, temperatures, etc. Figure 7.4 presents the comparisons of permeability between natural and homoionic soils including attapulgite and bentonite clays. Significant results caused by exchangeable ions and effects of temperature on coefficient of permeability of slightly organic clay have

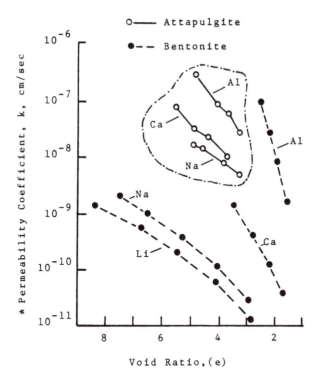

Figure 7.4 Comparison of permeability values between natural and homoionic soils. * Permeability values computed from consolidation test. (After Vees E. and Winterkorn, H.F. [1967], Engineering Properties of Several Pure Clays as Functions of Mineral Type, Exchange Ions and Phase Composition, HRR no. 209, pp. 55–65.)

been obtained. Tests conducted by consolidation tests with two temperature ranges are shown in Figure 7.5.

(2) Hydraulic Conductivity of Contaminated Soil

Pore fluids affect hydraulic conductivity significantly and a brief review is presented herein. Waidelich (1958) used consolidation tests to compute the coefficient of permeability of kaolinite and bentonite clays with various pore fluids, as indicated in Figure 7.6. Green et al. (1980) studied the effect of organic solvents on the permeability of clays. The study found that, in general, the permeability of clays was consistently lower for organic solvents than with water. It was found that the permeability decreased with time and attained equilibrium in several weeks. Permeability results correlated well with the pore fluid dielectric constant (Section 9.7); the lower the dielectric constant, the lower the permeability.

Haxo (1981) presents a study to determine the durability of liner materials for hazardous waste disposal facilities. All tests were conducted on bentonite-sand admixtures as well as natural clay soils. In a constant head permeability test, wastes including acidic, alkaline, lead, and oily substances were placed above a compacted fine-grained soil liner. Test results are simply stated in that the fluids collected after more than 3 years of exposure are essentially neutral and have high solids, mostly salt content. Table 7.1 lists the results obtained for the three soils of highest clay content to water and four organic solvents provided by Schram (1981). Anderson and Brown (1981) had undertaken a study in which two smectite clay minerals were permeated with organic fluids. The permeants were acetic acid, analine,

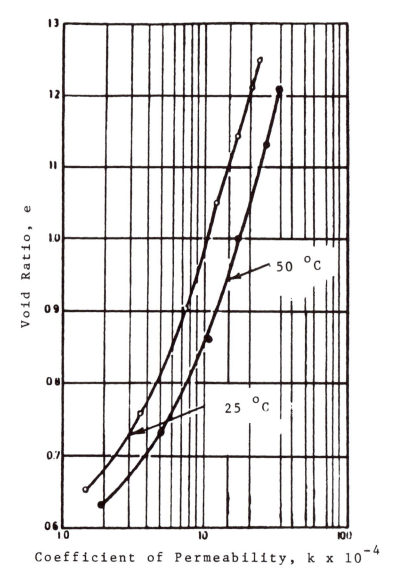

Figure 7.5 Effect of temperature on coefficient of permeability of slightly organic clay. (After Habibagahi, K. [1976], *Geotech. Eng.,* v. 7, no. 2, pp. 95–108.)

acetone, ethylene glycol, heptane, and xylene. Part of the results are presented on Figure 7.7. In summary, significant increases in permeability were obtained with basic neutral polar and nonpolar organic fluids over those values obtained with water.

Coefficient of permeability vs. the amount of pore fluid added is reported by Alther et al. (1985). Four types of pore fluids are used and a set of typical results is presented in Figure 7.8. Figure 7.9 presents a comparison between water and aniline on laboratory-prepared clay liner material. Significant results are obtained. Figure 7.10 shows hydraulic conductivity of compacted montmorillonite to benzene and acetone as reported by Acar and Olivieri (1989). Again, significant results between two pore fluids are found.

Based on the literature review, some generalizations can be made on hydraulic conductivity caused by contaminated pore fluids as:

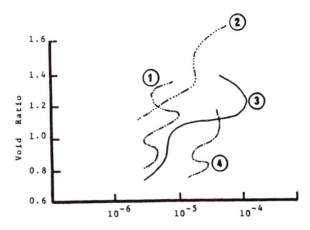

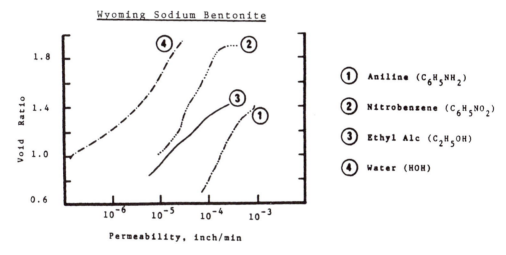

Figure 7.6 Permeability vs. void ratio for kaolinite and bentonite clays with various pore fluids. (After Waidelich, W.C. [1958], Influence of Liquid and Clay Mineral Type on Consolidation of Clay-Liquid Systems, HRB Special Report 40, pp. 24–42.)

1. Hydraulic conductivity increases as the percentages of multivalent cations in pore fluid increases;
2. Hydraulic conductivity increases result when organics are introduced; these increases may be related to the dipole moment and dielectric constants of the solvents.
3. Hydraulic conductivity increases as the salt concentration of a pore fluid increases.

(3) Triaxial-Permeameter for Contaminated Pore Fluid

The measurement of the hydraulic conductivity of fine-grained soils utilizing hazardous and toxic permeants requires equipment specifically designed and constructed for that purpose. A triaxial-type permeameter was developed to serve that purpose at Lehigh University in early 1983 by Evans and Fang (1986 and 1988). Important for studying contaminated pore

Table 7.1 Intrinsic Permeability of Three Clay Soils to Water and Four Organic Solvents

Permeant liquids		Intrinsic permeability (cm²)		
		Lake bottom clay (illite and kaolinite)	Nicholson (vermiculite)	Fanno (montmorillonite and mica)
Nonpolar-aromatic (xylene)	0.03	1.4×10^{-8}	2.9×10^{-9}	1.5×10^{-7}
Nonpolar-aliphatic (kerosene)	—	1.0×10^{-9}	2.9×10^{-9}	1.6×10^{-7}
Polar-glycol (ethylene glycol)	0.48	2.4×10^{-9}	1.4×10^{-10}	6.3×10^{-8}
Polar-alcohol (isopropyl alcohol)	0.25	6.4×10^{-9}	6.8×10^{-10}	9.3×10^{-8}
Water	1	5.1×10^{-10}	4.3×10^{-11}	1.6×10^{-9}

Dielectric constant of permeant.
Dielectric constant of water.

After Schram (1981).

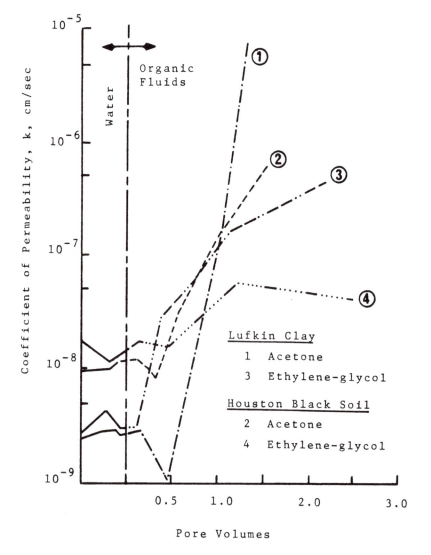

Figure 7.7 Permeability vs. pore volume for some natural soils with various pore fluids. (After Anderson D. and Brown, K.W. [1981], Organic Leachate Effects on the Permeability of Clay Liners, U.S. EPA Report no. 600/9-81-002b, pp. 119–130.)

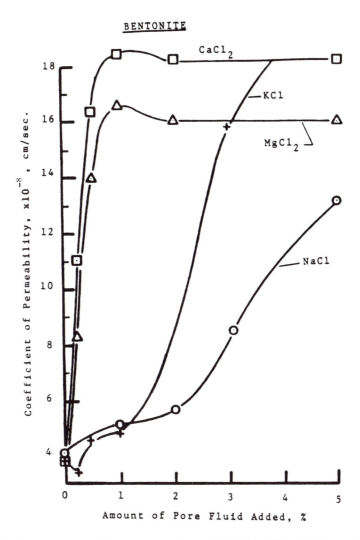

Figure 7.8 Coefficient of permeability vs. amount of pore fluid. (After Alther, G., Evans, J.C., Fang, H.Y. and Witmer, K. [1985], Influence of Inorganic Permeants upon the Permeability of Bentonite, ASTM, STP 874, pp. 64–73.)

fluid, the interaction between the pore fluids in the soils and the resulting soil response must be considered as well as the compatibility between the test equipment and the permeants. Further, a method of modeling the state of stress and permeation relative to field conditions must be incorporated in order to arrive at a system that can adequately determine the effects of hazardous and toxic permeants upon the hydraulic conductivity of fine-grained soils.

The testing system developed for hydraulic conductivity testing with hazardous and toxic permeants has three major components. These are the control panel, the permeability board, and the triaxial cell. The necessary strength and compatibility factors were accomplished through the optimum use of stainless steel and aluminum to provide adequate rigidity while providing the ultimate compatibility through the use of Teflon® at any point in which the permeameter is in contact with the permeant. Equipment design and fabrication considerations in their special applicability for testing of hazardous and toxic permeants are discussed by Evans and Fang (1986).

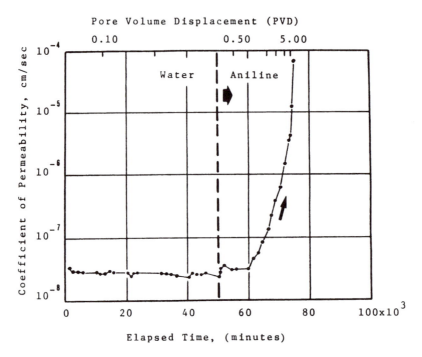

Figure 7.9 Effect of pore fluid on permeability of sand-bentonite mixture. (After Evans J.C., and Fang, H.Y. [1988], Triaxial Permeability and Strength Testing of Contaminated Soils, ASTM STP 977, pp. 387–404.)

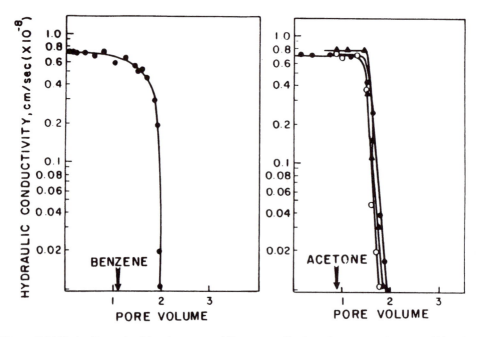

Figure 7.10 Hydraulic conductivity of compacted Ca-montmorillonite to benzene and acetone (After Acar Y.B. and Olivieri, I. [1989], HRB Annual Meeting, 24p.)

7.4 INFILTRATION, PERCOLATION, AND RETENTION

7.4.1 Characteristics of Infiltration

(1) General Discussion

The role of infiltration in the hydrologic cycle was first discussed by Horton in 1933. He defined infiltration as the passage of water through the soil surface into the soil. In a rainstorm infiltration normally begins at a high rate and decreases to a minimum as the rain continues. During the storm, the precipitation may be disposed of in several ways: (1) interception by the canopy of vegetable cover; (2) retention in the depressions upon the land surface; (3) direct infiltration; (4) evaporation; (5) evapo-transpiration; and (6) surface runoff.

(2) Infiltration Capacity

Infiltration capacity (f_p) is the maximum rate at which a given soil condition can absorb rain as it falls. The infiltration capacity will decrease exponentially in time from a maximum initial value to a constant rate. The actual *rate of infiltration* (f_i) is always smaller than f_p except when the rainfall intensity (i) equals or exceeds f_p. It also decreases exponentially with time as the soil becomes saturated and soil particles swell.

The factors that affect both the amount and the rate of infiltration are primarily those which characterize the soil particle sizes and their relative permanency, such as: (1) dynamic characteristics of soil structure and porosity; (2) time in relation to the amount of infiltration; and (3) type of land usage in relation to infiltration.

7.4.2 Infiltration of Water Through the Pavement Surface

The stability of highway embankment is adversely affected by the presence of free water within the pavement structure. This water may come from the sides or beneath the pavement structure in areas having high water tables, or it may infiltrate into pavement components through the pavement surface during the period of rainfall. The portland cement concrete (PCC) and dense grade base course used in the pavement surface are practically impermeable, but they cannot be constructed without joints or some cracks. Studies reported by Ridgeway (1976) and Moulton (1991) give the amount of infiltration through the pavement joints or cracks as follows.

(1) Infiltration of Water Through PCC Pavement

Ridgeway (1976) proposed the following formula for the estimation of the amount of infiltration through PCC pavement and base course.

$$Q = 0.1 \, [N + 1 + (W/s)] \qquad (7.18)$$

where Q = amount of infiltration (ft^3/hr/linear ft of pavement); 0.1 = infiltration rate (ft^3/hr/ft of crack); N = number of lanes; W = pavement width (ft); and s = PCC slab length (ft).

(2) Infiltration of Water Through Base Course

$$Q = 0.2 \, [N + 1 + (W/40)] \qquad (7.19)$$

where 40 = average distance between transverse cracks (ft). Other notations same as Equation (7.18).

Equations (7.18) and (7.19) are based on observations collected on Connecticut highways. They may be applicable in some areas but not in others where infiltration rate or crack spacing is different.

(3) Infiltration of Water Through Both Rigid and Flexible Pavement

Moulton (1991) proposed a uniform design infiltration rate formula for both concrete and asphalt pavements as well as base course.

$$q_i = I_c \, (N_c/W + W_c/WC_s) + k_p \tag{7.20}$$

where q_i = design infiltration rate (ft^3/day/ft^2 of drainage layer); I_c = crack infiltration rate (ft^3/day/ft of crack); N_c = number of contributing longitudinal joints or cracks on either rigid or flexible pavements; W_c = length of the contributing transverse cracks or joints (ft); W = width of aggregate base or subbase subjected to infiltration (ft); C_s = spacing of transverse cracks or joints (ft); and k_p = rate of infiltration (equal to the coefficient of permeability) through the uncracked pavement surface.

Example 7.1

Infiltration into a rigid pavement section (after Moulton, 1991)

Solution

Consider a new PCC pavement consisting of two 12-ft-wide traffic lanes with 10-ft-wide, dense-graded, bituminous concrete shoulders. Transverse pavement joints are placed at 20-ft intervals. The assumption is made that infitration through the uncracked pavement surface is insignificant, i.e., $k_p = 0$. Using Equation (7.20):

I_c = 2.4 cfd/ft
N_c = (N + 1) = 3
N   = number of traffic lanes = 2
C_s = 20 ft
W_c = 44 ft
W   = 44 ft gives
q_i = design infiltration rate
     = 2.4 (3/24 + 44/44(20))
     = 0.42 cfd/sf

In Example 7.1 the conservative assumption is made that the asphalt concrete shoulders have transverse thermally induced cracks at the same intervals as the pavement joints.

Infiltration characteristics of water through the natural ground or the embankment soil are important for analyzing the slope stability. Further discussion on this aspect will be presented in Section 16.5.

7.4.3 Percolation and Retention

Percolation is the water movement within the soil mass. Percolation and infiltration are closely related, but the infiltration cannot continue unimpeded unless percolation provides sufficient space such as voids in the surface layer for infiltrated water.

Retention is a part of the water movement process in the soil-water system. It is closely related to channel precipitation, overland flow, interflow, and groundwater flow. De Wiest

(1965) indicated that a portion of the rain falls directly into the stream channel and is considered as channel precipitation. Initially, almost all of the rain is collected on the ground as surface retention, the sum of interception, depression storage, and evaporation. Later, the ground soil becomes more saturated and more water infiltrates into soil. Finally, overload occurs and becomes surface runoff.

Water retention for soils intermediate in texture is partly due to swelling forces and partly to capillary forces as discussed by Yong and Warkentin (1966). Mineralogical aspects and explanations on the characteristics of water retention are given by Low (1968). Since the characteristics of soil retention are an important controlling factor to soil-water interaction in the environment, further studies have been carried out by many investigators. Yong et al. (1969) reported the temperature effect on water retention and swelling pressure of clay soils. It is observed that increased temperatures cause increased swelling pressures and decreased capillary water retention values in experiments where only one of the components is active. However, where the two components exist together, the two opposing trends cause different effects under temperature changes. Gupta and Larson (1979) developed a method for estimating soil water retention characteristics from particle size distribution, organic content, and unit weight of soil.

7.5 MASS TRANSPORT PHENOMENA IN MOIST FINE-GRAINED SOILS

7.5.1 Creeping Flow

As discussed in previous sections, flow through a fine-grained soil/water system is complex in nature. Understanding it requires knowledge from other disciplines as pointed out in Chapters 5 and 6. Considering such a slow moving flow with the complex nature of pore fluids, the conventional approach based on Darcy's Law as defined in Equation (7.7) may not always be applicable. A description of such a flow which has low velocity is, it is unsteady, nonuniform, or sometimes discontinuous and contains toxic or hazardous substances or is in solid or gaseous forms. The term 'creeping flow' has been introduced by Fang (1986) to describe such behavior. This type of flow will be further discussed in a later section of this chapter.

7.5.2 Nature of Pore Fluid in Soil

As discussed in Section 5.3, natural pore fluid (water) always contains some gases because it contains variable amounts of dissolved oxygen, nitrogen, and carbon dioxide. These gases dissolve in water from the biosphere or atmosphere and can be produced or consumed in the biological processes that occur within the natural water. Other gases are occasionally found in soil-water systems including ammonia (NH_3), chlorine (Cl_2), hydrogen sulfide (H_2S), and radon gas (Rn-222). Also, larger concentrations of methane (CH_4) can be found in the soil-water system if landfill areas are nearby.

Water in a soil system may simultaneously be present as vapor, as a liquid of varying viscosity, and as a solid of varying plasticity. Except in the vapor phase, water possesses a well-developed structure of highly electric character due to its own polar molecular nature, as well as being under the influence of the electrically charged surfaces and of ions in solution. Since the thermal energy of the molecules and ions of the aqueous solutions act counter to the structural-forming factors, water structures and their physical and physicochemical consequences are thermo-sensitive and this leads to tremendous complexity and to interrelationships between all physical and physicochemical factors in soil-water systems.

Table 7.2 Hydraulic Conductivity Vs. Mass Transport Phenomena in Soil-Water System

Mass Transport Phenomena	
Hydraulic conductivity	Mechanical energy field
Due to mechanical (hydrostatic) potential	
Energy conductivity (environmental)	Multimedia energy fields
Hydration energy	
Due to the hydration energy of ions, related to the heat of wetting	
Osmotic energy	
Due to the osmotic energy of ions either held in a kind of Donnan	↓
equilibrium on the solid particle surfaces or free in the aqueous solution	
Capillary potential	
Due to the surface tension of water and the size and geometry of the soil	
pores	
Electric Potential	
Electro-osmosis; electrokinetic phenomena	
Thermal potential	
Thermo-osmosis: thermal-electric effect	
Magnetic potential	
Electromagnetic force	
Vapor pressure potential	

Table 7.3 Pore Pressures Produced During Mass Transport Process in Soil-Water System

Mechanical (hydrostatic) pressure[a]	Mechanical energy field
Environmental pressure	Multimedia energy field
Pressure produced from heat of wetting	
Pressure produced from the kinetic dispersive force	↓
Electric-motive force (emf)	
emf produced from thermal-electric effect	
emf produced from electro-electric force	
Water-vapor force	

[a] Pore pressure (u) commonly used in geotechnical engineering for computing the effective stress (σ). The pressure is due to hydrostatic head only,

such as: $u = h\gamma_\omega$

where u = pore pressure
          h = hydrostatic head
          γ_ω = unit weight of water

For evaluation of the soil-water system, we must treat the soil element in three phases: solid, liquid, and gas. In view of the manifold properties of the water substance and modifications in the soil environment, water in soil responds to the imposition of any energy gradient, be it mechanical (Darcy's Law), thermal, electric, magnetic, or other, and responds to or exhibits coupling effects indicative of the disturbance of the other energy fields. Such response usually results in mass transport phenomena; therefore, flow through a porous, moist system is not simply the hydraulic conductivity, as described by Darcy's Law commonly used in geotechnical engineering. A comparison between hydraulic conductivity and mass transport phenomena for soils and pore pressures produced during the mass transport process in soil-water systems is presented in Tables 7.2 and 7.3. In addition, hydraulic conductivity for a given soil is not a constant, as it will change when the environmental condition changes. In order to examine the mass transport phenomena in the soil-water sysyem, it is necessary to have experimental data; consequently, a new apparatus for measuring conductivity of water movement under various energies is needed.

7.5.3 Soil Energy Conductivity

(1) General Discussion

To describe the mass transport phenomena under various energy gradients, a general expression similar to Darcy's law can be stated as:

$$v = k\,i + k\,i + k\,i + k\,i + \ldots \tag{7.21}$$

where v = flow velocity; k = constants as coefficients of energy conductivity in mechanical (m), thermal (t), electric (e), magnetic energy (n) fields; i = energy gradient in the mechanical energy field (also known as hydraulic gradient based on Darcy's law); i = energy gradient in the thermal energy field (thermal gradient based on Fourier's law); i = energy gradient in the electric energy field (electric gradient based on Ohm's law); and i = energy gradient in the magnetic energy field (magnetic gradient based on Faraday's and Lenz's laws).

Equation (7.21) can be simplified as:

$$\Psi_\varepsilon = K_\varepsilon\,I_\varepsilon \tag{7.22}$$

where Ψ_ε = flow velocity in energy fields; K_ε = coefficient of energy conductivity; and I_ε = energy gradient.

Since flow through a fine-grained soil-water system is not simply the hydraulic conductivity as described by Darcy's Law, the terms *energy gradient* and *energy conductivity* (Fang, 1992a and 1994c) are introduced to replace the hydraulic gradient and hydraulic conductivity commonly used in geotechnical engineering. Equation (7.22) may be called the '*energy conductivity equation*'. The *coefficient of energy conductivity* shown in Equation (7.22) must be determined experimentally. Based on the law of Conservation of Energy, this coefficient for a given material in an isolated system must remain constant under any environmental condition.

(2) Soil Energy Conductivity Measurements

The importance of water movement in fine-grained soil caused by thermal or electric potentials has been discussed by many investigators (Winterkorn, 1955; Mitchell, 1993). More recently, due to hazardous/toxic waste control system construction, this problem has brought up further attention. A simple apparatus for measuring energy conductivity in soil mass has been developed by Fang (1997).

1. The apparatus is portable with the control panel being approximately 20×20 in. Soil specimen size used in the experiment is $5 \times 1\ 1/2$ in.
2. Two cells are both connected at the ends of soil specimen for the purpose of controlling the electric energy.
3. The magnetic energy is created by using two electric currents crossing each other, creating a temporary electromagnetic force in the vicinity of the electric charges (Figure 9.28).
4. The thermal energy is produced through heating elements.

This apparatus can be used for demonstration of thermo-osmosis and electro-osmosis phenomena, as well as the thermo-electric effect in the classroom as a teaching tool. Experimental data generated from such tests can be useful for analysis and design of hazardous/toxic waste control and storage facilities. In addition, a unique feature of the apparatus is that it can also be used for soil decontamination and removal of heavy metals trapped in the soil-water system (Section 9.13)

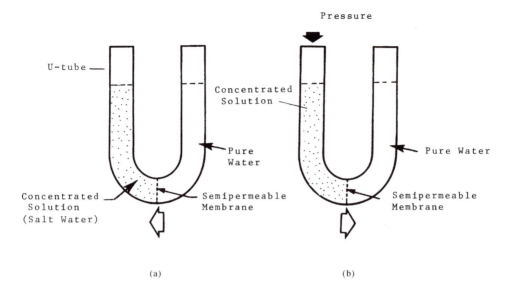

Figure 7.11 Osmosis phenomena demonstration (a) Typical osmosis; (b) reverse osmosis.

7.6 OSMOSIS AND REVERSE OSMOSIS PHENOMENA

7.6.1 Osmotic Phenomena

(1) General Discussion

Osmosis is a phenomenon involving solutions separated by a membrane. The membrane acts as a barrier between the solutions and has the property of allowing certain types of molecules to pass through, while preventing the passage of other species in solution. From an engineering viewpoint, when the concentration of a solution at one point is different from that at another point, there is a tendency for the more dilute liquid to diffuse into the region of higher concentration.

A classic demonstration of osmotic phenomena from basic physics is conducted with a U-tube, as shown in Figure 7.11, in which a semipermeable membrane has been placed to separate the U-tube as indicated by a vertical dotted line. If a salt solution is placed on the right side of the membrane as denoted by **A**, and the pure water is placed on the left-side as denoted by **B**, then the pure water will enter the solution by passing through the membrane (vertical dotted line). Movement of the solute in the opposite direction is inhibited by the membrane. This one-directional movement is indicated by a distinct lowering of the level of the pure water, and a corresponding rise in level of the solution. This simple demonstration is referred to as classical osmotic phemonena.

(2) Osmotic Membranes and Osmotic Pressures

1. Semipermeable membrane and osmotic membrane: As discussed in the previous section, indications are that osmosis is a phenomenon involving solutions separated by a membrane. The membrane acts as a barrier between the solutions and has the properties of allowing certain types of molecules to pass through, while preventing the passage of other species in solution. Such a membrane is called semipermeable membrane. Semipermeable membranes that allow the passage of the solvent but not the solute are called osmotic membranes.

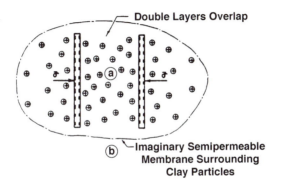

Figure 7.12 Osmotic pressure between two clay particles immersed in water. (After Ladd, C.C. [1960], Mechanisms of Swelling by Compacted Clay, HRB Bulletin 265, pp. 10–26.)

   2. Osmotic pressure: As indicated in Figure 7.11(a), indications are that osmosis can be stopped if a certain opposing pressure or force is exerted on the solution side of the membrane. This pressure is called the osmotic pressure. Pure water has an osmotic pressure of zero. Any dissolved nonvolatile solute will increase the osmotic pressure. Two solutions which have the same osmotic pressure are called isotonic solutions.

7.6.2 Osmotic Pressure in Soils

(1) Osmotic Flow in Soils

Osmotic phenomena can be used to explain certain characteristics in fine-grained soil, as discussed by Hough (1957). The solid particles in fine-grained soils carry a negative electric charge. While this is involved in adsorption, it also results in attracting to the particle surface a number of the positively charged atoms (ions) of the dissociated minerals which are in solution in the surrounding pore water. A fixed number of charges provided by these ions must always remain in the vicinity of a given clay particle in order to provide electrical balance. However, individually these ions may exchange places with others existing in or are introduced into the clay-water system.

(2) Osmotic Repulsive Pressure in Soils

Ladd (1960) used osmotic repulsive pressure to explain the mechanisms of swelling by compacted clay. Osmotic pressures can act in clay since:

   1. There exist differences in solute concentrations (in this case ions are the solute).
   2. The electric field around the negatively charged clay particles acts as a semipermeable membrane.

Combining osmotic phenomena and electric double-layer concept, Ladd proposed the mechanism of swelling soil as illustrated in Figure 7.12. In this case, the only ions in the double-layer will be the exchangeable cations. Because of the exchangeable cations, the concentration of ions in the double layer water is larger than the concentration of the ions in the free water. The exchangeable cations are attracted to the clay particles by the negative electric field arising from the negative charge on the soil particle surface.

The electric field acts as a semipermeable membrane in that it will allow water to enter the double layer but will not allow the exchangeable cations to leave the double layer. One can see that due to the difference in ion concentrations between points **a** and **b**, water would like to flow from **b** to **a**, and an effective stress is required to prevent an increase in interparticle spacing, that is swelling. If effective stress is reduced, then water flows from **b** to **a**, thus decreasing the ion concentration at **a**. At that point, the double layer is expanded until the correspondingly lower osmotic pressure is again in equilibrium with the effective and attractive pressures.

7.6.3 Reverse Osmosis Phenomena

As indicated in Figure 7.11(b), osmosis is the spontaneous passage of a liquid from a dilute to a more concentrated solution across an ideal semipermeable membrane that allows passage of the liquid but not of dissolved solids. The *reverse osmosis* is just the opposite in which the natural osmotic flow is reversed as shown in Figure 7.11(b). The amount of pressure applied to the concentrated solution is sufficient to overcome the natural osmotic pressure of the less-concentrated solution.

The reverse osmosis system is relatively new in comparison with natural osmosis as previously discussed. The phenomenon of the system remains, however, in the laboratory. Reverse osmosis's many applications in the area of environmental engineering have been discussed by Buckley (1975). Further applications into environmental geotechnology have been explored including (1) seawater desalination, (2) oil/water separation, (3) waste water treatment, and (4) soil decontamination.

7.7. SOIL-WATER SUCTION AND DIFFUSIVITY

7.7.1 Soil-Water Suction

The term known as *soil-water suction* or tension was started by Buckingham in 1907 based on the work by Briggs in 1897 and introduced into geotechnical engineering by Schofield (1935). The significance of suction and its measurement techniques were established by Croney and Coleman in 1954 of the Road Research Laboratory, London. After an International Conference on Pore Pressure and Suction in Soil (ICE, 1961), this concept has been well accepted by geotechnical and highway engineers.

The pore water in natural and compacted soils is often in a state of tension; that is, its pressure is negatively relative to atmospheric conditions. This negative pressure is called *suction*. Numerous devices capable of measuring the suction include ceramic resistance blocks, tensionmeters, pressure plates, osmotic tensionmeters, and thermocouple psychrometers. These devices have been examined by Fredlund and Rahardjo (1993) and others.

The unit for soil-water suction is called pF units. On the **pF** scale of measuring suction, the logarithm to base ten of the suction expressed in centimeters of water is equivalent to the pF value. Thus, 10 cm of water equals pF 1, 1000 cm equals pF 3 (approximately 1 atm). Soil in equilibrium with free water has a pF of 0, and oven-dried, almost pF 7.

According to Croney and Coleman (1961), distinguishing between soil suction and negative pore-water pressure can be explained by the following equation:

$$u = s + (\alpha) p \qquad (7.23)$$

where u = negative pore water pressure; s = soil water suction; α = change of negative pore water pressure with constant moisture content; and p = applied pressure.

The value of (α) can be measured directly by a simple loading test on a sample of known suction. Equation (7.23) can be used to estimate the ultimate moisture distribution beneath pavements and structures on saturated and unsaturated ground soils. As pointed out by the Organization for Economic Cooperation and Development (OECD, 1973), properties of soil-water systems must be known in order to predict and control moisture content and movement in ground soils. Elzeftawy and Dempsey (1976) indicated that the absolute magnitude of change in the resilient modulus caused by an increase in the soil-moisture content will not be constant for all soil types, but will vary with the accompanying change in soil-water pressure (suction).

7.7.2 Soil-Water Diffusivity

(1) General Discussion

The *soil-water diffusivity* concept was introduced into geotechnical engineering by Childs and Collis-George in 1950. Their theory for the flow of water in unsaturated soil assumes that Darcy's Law can be written as a diffusion-type water-flow equation in homogeneous soils where gradients of water content rather than gradients of total potential are expressed as:

$$q = D \; i_\omega - k \tag{7.24}$$

where q = water flux; = soil-water content on a volume basis; k = hydraulic conductivity of soil (function of the soil water content); D = soil-water diffusivity (function of the soil-water content); and i_ω = water-content gradient.

The diffusivity, D, is also equal to:

$$D = k/C \tag{7.25}$$

where C = the specific soil-water storage capacity = $\partial\omega/\partial h$, where h is the soil-water pressure head (suction).

The parameters of k, ω, and h in Equation (7.25) can be determined experimentally as described by Elzeftawy and Dempsey (1976) and Fredlund and Rahardjo (1993).

(2) Field Measurements

In the field condition, the water table is established at the bottom of a compacted subgrade soil column, and the movement of water through the soil can be measured under isothermal conditions. A gamma-ray method can be used for nondestructive (NDT) measurement of the water content and a tensionmeter pressure transducer arrangement to measure the soil-water pressure. The unsaturated hydraulic conductivity (Hamilton et al. 1981), diffusivity and soil-water characteristics parameters can be evaluated from these data to predict the movement of moisture through unsaturated ground soils (vadose zone). Some typical relationships of soil-water content to suction and diffusivity for clay, loam, sandy loam, and silty clay soil are presented in Figure 7.13 a and b. Figure 7.14 shows the relationship between diffusivity, suction, and soil-water content of a silty clay. In both cases, when water content increases both diffusivity and suction also increase. Also, Figure 7.15 shows the effects of depth on degree of saturation related to suction of compacted volcanic ash. Decrease saturation of soil, and the pF value increases.

7.8 MOISTURE MIGRATION

7.8.1 Characteristics of Moisture Migration

Migration phenomena of soil is part of the dynamic behavior of soil moisture movement. Dynamic equilibrium between water in the various states, within individual water molecules, will pass back and forth, at stabilized rates, between the vapor and the liquid states, between

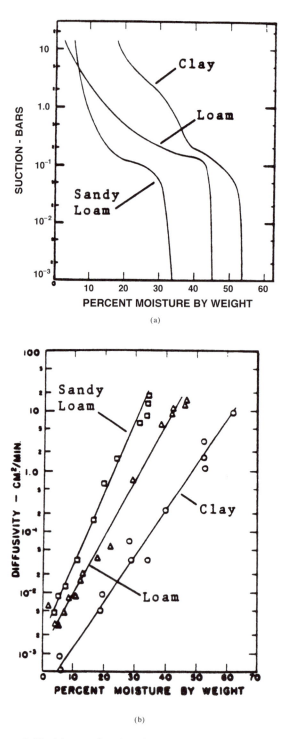

Figure 7.13 Soil suction and diffusivity as a function of water content (a) Soil suction; (b) diffusivity. (After Gardner, W.R. [1958], Mathematics of Isothermal Water Conduction in Unsaturated Soil, HRB Special Report 40, pp. 78–87.)

the strata of capillary and hygroscopic water. Furthermore, within a volume of soil at uniform temperature, water molecules in the various states will diffuse back and forth from one place to another. With this complex, dynamic picture in mind, little imagination is required to

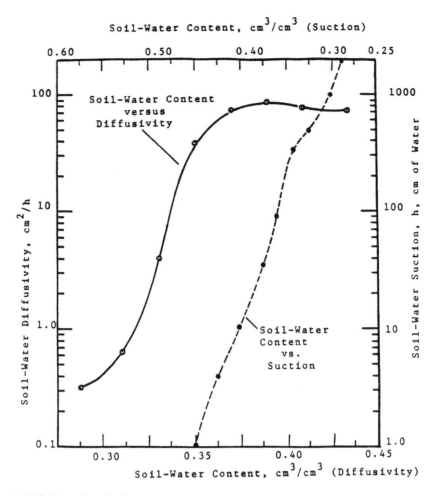

Figure 7.14 Relationship of soil-water content to soil-water suction and soil-water diffusivity for a silty clay soil. (After Elzeftawy and Dempsey, [1976].)

realize that when the temperature level is changed at one location, the equilibria mentioned above are disturbed, and a redistribution of water, both between the several states and between locations now at different temperatures, will be required before new equilibrium conditions are established with the moisture content at the cooler location exceeding that at the warmer. Under such environmental conditions, this redistribution constitutes what may be referred to as moisture migration under thermal gradients. Migration causes can be mechanical, thermal, electrical, and magnetic. Based on the above-mentioned description of moisture migration the following two points must be noted:

1. The migration phenomenon is transient redistribution of water rather than a continuing flow.
2. Water in any of the states in which it exists in the soil may migrate back and forth, if local environmental conditions change (Figure 7.16).

7.8.2 Limits of Moisture Migration

When the amount of migration is studied as a function of the initial uniform moisture content of the soil, it is found that the phenomenon noticeably occurs only over a range of moisture contents bounded by two distinct limits, between which it passes through well-defined limits:

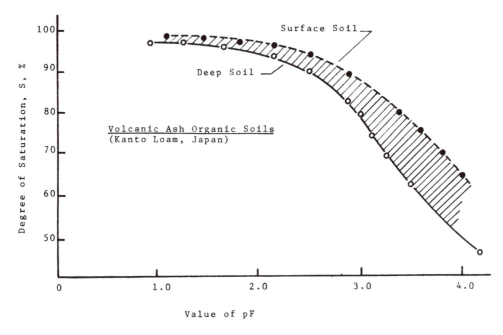

Figure 7.15 Effect of depth on degree of saturation as reflected on pF value. (After Yasuhara and Takenaka, [1977].)

1. The lower limit of moisture migration is at about the hygroscopic coefficient, the maximum at about the wilting coefficient.
2. The upper limit of moisture migration is at about the field capacity.

These lower and upper limits define the range of capillary water. The relationship is illustrated in Figure 7.16. The terms used here such as wilting coefficient, field capacity, and other moistures are explained in Figure 5.1. The upper and lower limitations for nonpolluted water as suggested by Lassen et al. (1952) are indicated in the figure, however, for contaminated water, the range widens.

7.8.3 Factors Affecting Moisture Migration in Soils

(1) Possible Modes

Numerous researchers have studied the causes and factors affecting moisture migration in soils. Jumikis (1958) reported the possible modes affecting moisture migration as indicated in Figure 7.17. Derjaguin and Melnikova (1958) reported the migration phenomena reflected as thermocapillary flow as a function of moisture content of soil. Significantly different results are found for clay loam and sandy loam as shown in Figure 7.18.

(2) Porosity and Specific Surface Area

It is generally recognized that porosity, together with the related parameter-specific surface area, has an important influence on the magnitude and rate of migration and, presumably, on the mode in which it occurs. On the one hand, continuous moisture films are necessary to establish and maintain liquid migration, regardless of the mechanism postulated. Soils with high porosity and a relatively small specific surface area, such as sands, have much higher coefficients of vapor diffusion than do less permeable clay soils.

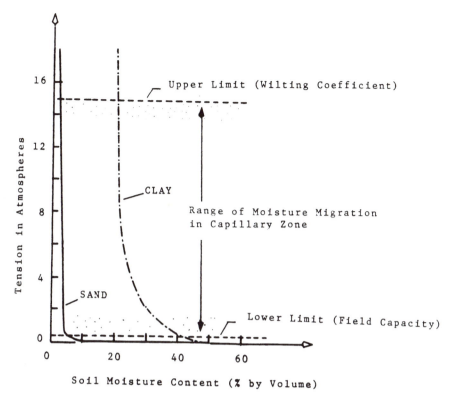

Figure 7.16 Upper and lower limitations for moisture migration in capillary zone. (Data from Lassen, Lull and Frank [1952].)

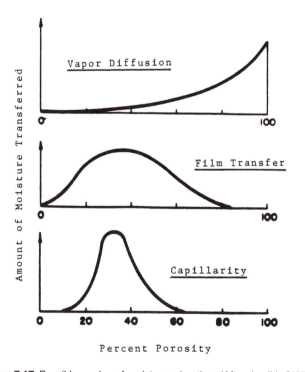

Figure 7.17 Possible modes of moisture migration. (After Jumikis [1958].)

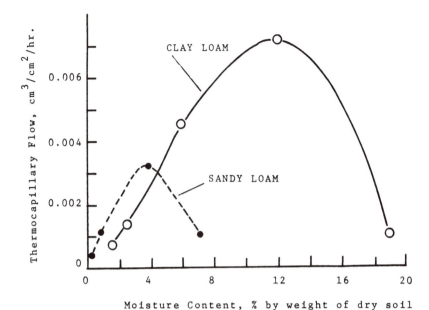

Figure 7.18 Moisture migration (reflected as thermocapillary flow as a function of moisture content. (After Derjaguin and Melnikova [1958].)

Jumikis (1958) has illustrated the ranges of porosity over which the various modes of moisture migration may be effective.

(3) Soil Structure

The structure of the soil is also important. Smith (1943) showed that when soils are fragmented the amount of migration increases appreciably over the values observed for the same soils in their natural, structured condition. For the somewhat granular A-horizon of the soils studied, the increased value was 127% of the original value; for the more structured B- and C-horizons the increased values were 443 and 240%, respectively. Other factors affect migration including ion exchange, leaching, and adsorption.

7.9 DIFFUSION PHENOMENA

7.9.1 General Discussion

Diffusion is the process whereby ionic or molecular particles move in the direction of their concentration gradient under the influence of their kinetic energy. The process of diffusion is often referred to as *self-diffusion, molecular diffusion,* or *ionic diffusion.* In geotechnical engineering, it is referred to as *chemical transport phenomena.* The mass of diffusion substance passing through a given cross-section per unit time is proportional to the concentration gradient. The statement is known as *Fick's* first law. It can be expressed as:

$$F = -D \, dC/dx \qquad (7.26)$$

where F = mass flux; D = diffusion coefficient; C = solute concentration; and dC/dx = concentration gradient, which is a negative quantity in the direction of diffusion.

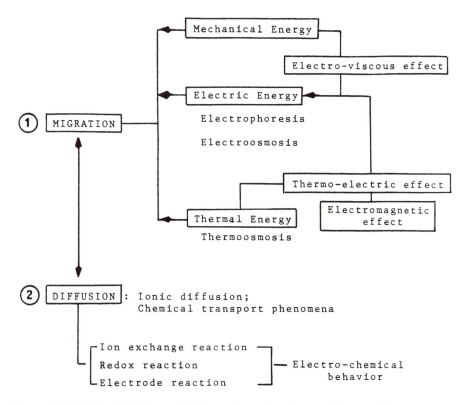

Figure 7.19 Relationship between diffusion and migration characteristics in soil-water systems.

The diffusion coefficients are temperature dependent. In groundwater, the range of diffusion coefficients is from 1×10 to 2×10 m/sec at 25°C. If the soil-water system is polluted, the ranges of coefficient change dramatically.

7.9.2 Diffusion Phenomena in Soil-Water System

Diffusion and migration are closely related in the soil-water system as illustrated in Figure 7.19. The diffusion theory has been applied for evaluation of hazardous and toxic waste movement in the soil-water system by numerous investigators. Tuncan, et al. (1989) developed laboratory equipment for measuring soil behavior under coupled influence of thermal-chemical potentials. Acar and Haider (1990) studied the transport of low-concentration contaminants in saturated earth barriers. Benson and Charbeneau (1991) used the diffusion theory to predict first-passage times for earth landfill liners.

7.10 WATER MOVEMENT IN VADOSE ZONE

The phenomena of water movement in the *vadose zone* (unsaturated zone) of fine-grained soil are similar to the mass transport phenomena and soil energy conductivity as discussed in Sections 7.5 and 7.6. Figure 7.20 presents types of flow and their characteristics in soil-water systems. Water movement in the vadose zone is complex especially in polluted areas. Further discussion will be presented in Section 17.6. Figure 7.21 illustrates the movement of water in unsaturated loess soil. In examining Figure 7.21, there are larger variations existing in the distance between points **a** to **b** influenced by air bubble trapped and swelling behavior in the soil-water system.

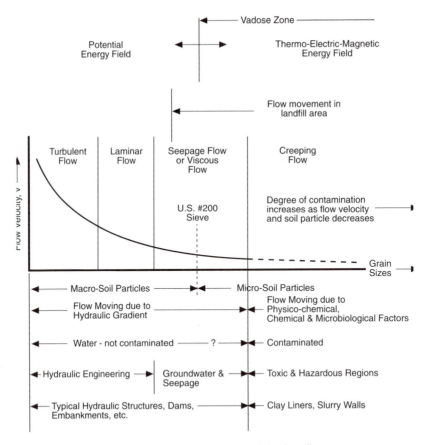

Figure 7.20 Types of flow and their characteristics in soil-water systems.

7.11 SUMMARY

1. Capillary rise, hydraulic, thermal and electric conductivities of soils are discussed and basic concepts of each aspect are examined.

2. Soil is composed of electrically negative mineral surfaces, while water is composed of electric water dipoles and predominately positively charged ions. Soil-water-environment interaction is a mass transport phenomenon and is not simply controlled by hydraulic conductivity as described by Darcy's law.

3. Some limited experimental data are used to illustrate mass transport phenomena of contaminated unsaturated fine-grained soil.

4. A general equation for evaluation of energy conductivity of soil is presented. A simple apparatus for measuring such conductivity is developed.

5. Migration phenomena of soil are part of the dynamic behavior of soil moisture movement. Migration is a transient redistribution of water rather than a continuing flow and moisture in the soil may migrate from place to place if the local environmental conditions change.

6. Diffusion is the process whereby ionic or molecular particles move in the direction of their concentration gradient under the influence of their kinetic energy. The mass of diffusion substance passing through a given cross section per unit time is proportional to the concentration gradient.

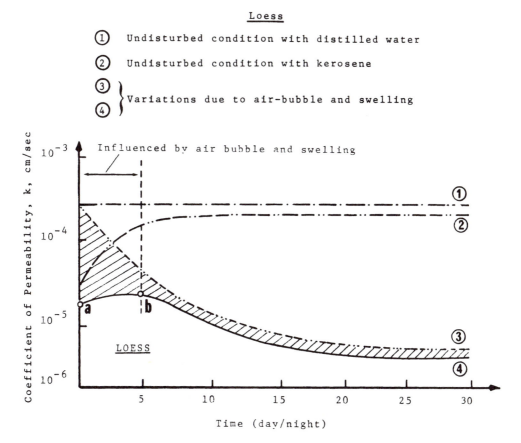

Figure 7.21 Coefficient of permeability vs. time curves. (Data from CAS [1961], Research on Loess, Civil Engineering Research Report no. 13, Chinese Academy of Science, Beijing, 157p.)

PROBLEMS

7.1 Derive from Equation (7.1) the rise of water in the capillary tube, and from Equation (7.2), the capillary rise related to the effective size (D_{10}) of soil particle. What is the maximum height of capillary rise of water in a material having a uniform pore diameter of 0.0001 cm?

7.2 How high does methyl alcohol rise in a glass tube 0.1 mm in diameter with the surface tension being 23 dyn/cm and the specific gravity as 0.8?

7.3 During a constant-head permeability test, some of the tail water evaporates. What effects does this have on the value of the coefficient of permeability, k? Explain. If air is trapped in the soil specimen during the falling head permeability test, would the coefficient of permeability, k, increase or decrease for a clay soil?

7.4 A soil specimen 10 cm in diameter and 5 cm thick was tested in a falling head permeameter. The head dropped from 45 to 30 cm in 4 min and 32 sec. The area of the standpipe was 0.5 cm². (a) What was the capillary rise of water in the standpipe? (b) Considering the effect of capillary rise in the standpipe, compute the coefficient of permeability of the soil in units of cm/sec. (c) The soil sample was material having a specific gravity of solids equal to 2.67 and the sample weighed 617 g dry. Compute the average seepage velocity.

7.5 A horizontal stratified bed of sand 12 ft thick is broken into two equal parts by a silt layer of 2 ft thickness. The coefficient of permeability (k) in the vertical direction of the sand is 2×10^{-2} ft/sec and the coefficient of permeability in the same direction of the silt layer is 3×10^{-4} ft/sec. Establish the value of the average k in the vertical direction for this stratified bed.

7.6 Discuss the inorganic and organic permeant effects on the permeability of clay liners. Why does leachate induce permeability changes?

7.7 A soil has a unit weight of 17.1 kN/m$_3$ and a water content of 6%. How much water in gallons should be added to each cubic yard of soil to raise the water content to 12.5%? (Assume that the void ratio remains constant.)

7.8 In a well-drilling project, a heavy viscous fluid having a unit weight of 150 pcf was required. It was decided to use a suspension of clay ($G_s = 2.82$), sand ($G_s = 2.67$), iron filings ($G_s = 7.13$), and water ($G_s = 1.00$). Assuming that 5 parts of clay are to be used for 1 part of sand (by weight); what weight of filings would be required for 1000 cc volume of suspension?

7.9 Two soils [A] and [B] are placed above each other in a constant-head permeameter tube. The specific gravities and void ratios of these soils are: $G_s[A] = 2.65$, $G_s[B] = 2.69$; $e_A = 0.60$, $e_B = 0.69$. Specimen height for both soils is 40 cm. The hydraulic head from bottom of soil to the top of the water supply tank is 120 cm. If 25% of the hydraulic head is lost by the upward flow through soil [A], compute the critical hydraulic gradient at which the instability occurs.

CHAPTER 8

Thermal Properties of Soils

8.1 INTRODUCTION

Thermal variation in soils has a great influence over several engineering fields. A few examples of the phenomenon are (1) the loss and recovery of bearing capacity under the effect of freezing-thawing; (2) the prediction of pavement temperatures which affects the stabilization of bituminous surface curing, curing of concrete slabs, moisture movements in any type of pavement components, embankments and earth dams, clay liners; (3) the dissipation of heat from buried electric cables or steam pipes, and the thermal exchange in heat pump systems; and (4) thermal conductivity in hazardous and toxic storage controlling facilities.

8.2 CHARACTERISTICS OF HEAT AND HEAT SOURCES

8.2.1 Characteristics of Heat

Heat is a form of energy. The heat capacity per gram of a material is called its *specific heat* which for the *atomic weight* in grams of an element is the *atomic heat*, and for a gram mole of a compound is the *mole* (mol) *heat*. The total heat content (U), of a system possessing the heat capacity (C_s), and being at an absolute temperature (T), is expressed by:

$$U = \int_0^T C_s \, dT \tag{8.1}$$

The concept of heat as a form of motion of the ultimate particulate components of matter dates back to the ancients. Scientific proof of the essential validity of this concept was furnished by Rumford in 1798 and Davy in 1799. The experimentally found relationships between temperature, pressure, and volume of normal gases led to the concept of the 'ideal gas'. The modern theory of the specific heat of gases which takes the electromagnetic phenomena into account is largely due to Albert Einstein (1879–1955), while that for the specific heat of solids has been developed mainly by Peter Debye (1884–1966).

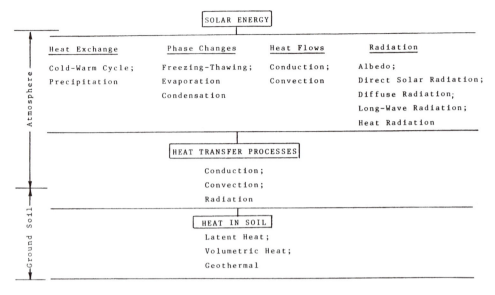

Figure 8.1 Soil-heat interaction near the ground surface.

8.2.2 Heat Sources

Types of heat sources penetrating into the ground soil can be grouped into two major catagories: natural and man-made sources. The natural sources include solar and geothermal. The man-made sources cover the heat released from steam pipes, electrical cable lines, heat pump systems, as well as other man-made sources produced from chemical and nuclear wastes dumped into the natural ground, carbon dioxide produced in landfill areas, and abandoned coal mines which, on occasion, produce underground fires. Some of the heat sources are listed in Table 2.10.

8.2.3 Heat Exchange Near the Ground Surface

The heat balance of ground soil from solar source includes heat exchange, phase change, heat flows, and radiation, as described by Geiger (1965), and climate near the ground relating to the subsurface thermal conditions and various related engineering problems. The actual heat exchange during daytime and at night is quite different as discussed by Geiger based on physical and meteorological concepts. Figure 8.1 shows the heat exchange and soil-heat interaction near the ground surface.

8.3 HEAT TRANSFER PROCESS AND MEASURABLE THERMAL PARAMETERS

8.3.1 Heat Transfer Process

The heat transfer process in the soil-water system is through three basic transfer processes: conduction, convection, and radiation.

(1) Conduction

Conduction is the process in which heat energy is transferred between molecules within a substance, between two substances in physical contact, or by direct molecular interaction. The rate of heat flow or the thermal conductivity (k) of a substance is dependent on the

Figure 8.2 Comparison between conduction and convection of saturated or partially saturated soils. (a) Conduction — heat transfer by direct contact between two soil particles. (b) Convection — heat transfer by moving liquid or air.

capacity of its molecules to give and absorb heat. In general, gases (air) are poor conductors, but are excellent for insulation. Many building insulation materials contain thousands of tiny air pockets and, therefore, can be useful as insulation materials. It also indicates that from a geotechnical point of view for dry soils, an increasing porosity results in an increasing thermal resistivity.

(2) Convection

Convection is defined as the heat transfer between a surface and a moving liquid or gas or the transfer of heat by the movement of the molecules from one point to another. In convection processes, heat always moves from warm to cool. The larger the temperature difference between two substances, the more heat will be transferred. In most soils, the heat transfer process within the soil mass is conduction or convection, or both. Figure 8.2 illustrates a comparison between conduction and convection of saturated or partially saturated soils.

(3) Radiation Process

All materials are constantly radiating thermal energy in all directions because of the continual vibrational movement of molecules as measured by temperature at their surface. The amount of thermal energy a material radiates depends on the temperature of the radiating surface.

The *emittance* (E) of a material is a measure of the material's ability to give off thermal radiation. For example, shiny metals are poor emitters of thermal radiation. This means they radiate very little heat at a given temperature. The value of E, varies from 0 to 1.0 for most construction materials. Not all materials absorb thermal radiation. Some will reflect it and some will transmit it. The capacity of a surface to reflect or absorb thermal radiation will depend primarily upon the density and composition of the surface rather than on its color. For soil, in addition to its density and composition, the moisture content and texture of the soil will also greatly affect the thermal behavior.

8.3.2 Measurable Thermal Parameters

(1) General Discussion

Once the heat is transmitted into the ground soil, the ability of the soil to retain or dissipate heat is dependent on its heat capacity and thermal conductivity. Thermal parameters or constants can be measured experimentally or can be computed from known parameters. Figure 8.3 presents the interaction of soil-heat system and their measurable parameters. Table 8.1

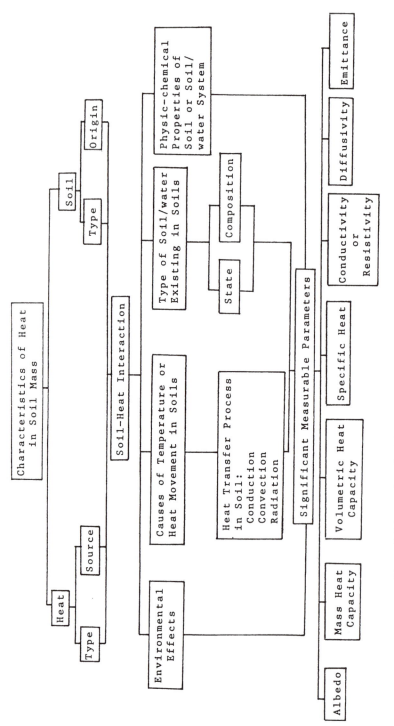

Figure 8.3 Interaction of soil-heat systems and their measurable parameters.

Table 8.1 Summary of Thermal Parameters and Their Units

Aspect	Quantity	Equation	Typical unit	
			SI	c.g.s
Heat retained by soil (heat retention)	Mass heat[a] capacity (C_m)	C_m = Q/Temp.	J/(g)(°C)	Btu/(lb)(°F)
	Volumetric[b] heat capacity (C_v)	$C_v = C_m\gamma^c$	J/(mm³)(°C)	Btu/(ft³)(°F)
	Specific[a,b] heat (C_{sp})	C_{sp} = (C_{ms})substance/(C_{mw})water	Dimensionless	
Dissipation of heat (heat transmission)	Thermal[a] conductivity (k)	k = Q/iAt	W/(m)(°C)	Btu/(ft)(S)(°F)
	Thermal[a,b] resistivity (R)	R = 1/k	(m)(°C)/W	(ft)(S)(°F)/Btu
	Thermal[b] diffusivity (α)	$\alpha = k/C_v$	mm²/S	ft²/hr

[a] Experimentally determined.
[b] Calculated quantity.
[c] γ = unit weight of soil; ρg, where ρ = density; g = gravity.

presents a summary of thermal parameters and their units. In geotechnical engineering, the heat conductivity and resistivity are extremely important; therefore, detailed discussions will be presented in Section 8.6.

(2) Temperature and Heat

Temperature is a measure of the internal motion of an object's constituent molecules. The greater the motion, the greater is the internal energy and the higher is the temperature. There are three different temperature scales namely: Fahrenheit (°F), centigrade (°C) and absolute or Kelvin (°K). On the Fahrenheit scale the freezing point of water is 32°F. The boiling point of water is 212°F. The centigrade scale is fixed at 0°C for the freezing point of pure water, and 100°C for the boiling point of pure water. The Kelvin or absolute scale is similar to the centigrade scale in that the divisions on the scale are the same size. But the zero on the absolute scale is –273°C; the boiling point of water is 373°K. Interrelationship and conversions of all three types of temperature scales are represented by the following equations:

$$T°C = 5/9 (T°F – 32°) \tag{8.2}$$

$$T°F = 9/5\ T°C + 32° \tag{8.3}$$

$$T°K = T°C + 273.18° \tag{8.4}$$

where T = temperature; C = Centigrade scale; F = Fahrenheit scale; and K = Kelvin (absolute) scale.

Heat is thermal energy in transit. If a hot object is placed in contact with a cold object, heat will flow from the hot to the cold object because some of the molecular motion of the hot object will be transferred to the cold object. The unit of heat is the *Calorie* or *calorie* (Cal, cal). One *calorie* is defined as the amount of heat required to raise the temperature of 1 g of water by 1°C.

(3) Heat Capacity and Specific Heat

1. Mass heat capacity (C_m): The *mass heat capacity* is defined as the quantity of heat (Q) required to raise a unit weight of material 1°.

Table 8.2 The Specific Heat Capacity of Some Clay Minerals

Clay minerals	Temp (K)	Specific heat capacity (cal/deg/g)
Vermiculite	303–373	0.205
Halloysite	298	0.228
Kaolinite	298	0.227
Muscovite	298	0.193
Na-Montomorillonite	273–303	0.191

After Low, P.F. (1968), Mineralogical Data Requirements in Soil — Physical Investigations, SSSA Special Publ. Series no. 3, pp. 1–34.

2. Volumetric heat capacity (C_v): The *volumetric heat capacity* is defined as the quantity of heat required to raise a unit volume of material 1°. This quantity can be computed as shown in Table 8.1.
3. Specific heat, c_{sp}: *Specific heat* is the ratio of mass heat capacity, C_m, divided by the mass heat capacity of the water, c_{mv}. Specific heat is a dimensionless quantity. Some typical heat capacities of clay minerals are given by Low (1968) as shown in Table 8.2.

(4) Diffusivity

The *diffusivity* is the quotient of the thermal conductivity and the heat capacity per unit volume; its units in the c.g.s. system are cm²/sec. The diffusivity value may be determined by calculation if the thermal conductivity, specific heat, and density of a soil are known. Based on the work done by Kersten (1949), the following general conclusions regarding the diffusivity of soils may be drawn: (1) changes in temperature do not cause significant changes in diffusivity unless they pass through the freezing point; (2) at low or moderate moisture content, an increase in moisture content of either a frozen or an unfrozen soil causes an increase in diffusivity; (3) an increase in density of a soil causes a slight increase in diffusivity; and (4) mineral composition affects diffusivity to about the same degree that it affects thermal conductivity.

8.4 SOIL-HEAT INTERACTION

8.4.1 Mechanism of Soil-Heat Interaction

Soil-heat interaction near the ground surface and interaction of the soil-heat system and their measurable parameters have been discussed in previous sections. Some of these interactions have been summarized in Figures 8.1 and 8.3. Soil normally consists of solid, liquid, and gaseous phases as discussed in Sections 3.2 and 3.3.

The mechanism of heat transmission differs at different phases. The thermo-geotechnical properties of this polydisperse, multiphase system depend not only on the phase compositions, but also on the dispersion of the phases and on the previous environmental history of the systems. Also, because of the contribution from water movement in the vapor phase and the lightness of the water molecule, the vertically upward thermal conductivity of a moist soil system may be considerably larger than the vertically downward one. Furthermore, in an undisturbed condition, the solid particles, the liquid, and gaseous phases in a grown soil are arranged in a dispersed structure, and while in a remolded condition, distribution of particles and phases is random (flocculated). The actual condition of most soils of interest to the geotechnical engineer lies between these two extremes.

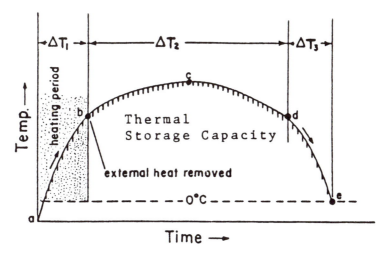

Figure 8.4 The concept of thermal storage capacity (TSC). (After Fang, H.Y. and Chaney, R.C. [1983], Proc. ASCE Engineering Mechanics Specialty Conf. Purdue University, v. 2, pp. 219–231.)

8.4.2 Soil Thermal Storage Capacity

Regardless of the type of heat, once heat is transmitted into the ground, the ability of the soil to retain or dissipate heat is dependent on its heat capacity and thermal conductivity, as discussed. To visualize the *thermal storage capacity* (TSC) and other related parameters, an idealized schematic diagram illustrating temperature-time relationship of ground soil is shown in Figure 8.4. Review of Figure 8.4 shows that from point **a** to point **b** is the heating period corresponding to heat energy being absorbed in the soil at the depth being monitored. The time for the heating period is denoted ΔT_1. The period of time (ΔT_2) from points **b** to **d** corresponds to the period where external heat is no longer being put in at the ground surface but energy still is being received at the depth being monitored. The TSC of the soil can then be defined as the area under the curve described by points **b, c, d** and **e** as illustrated in Figure 8.4. The dissipation stage of the temperature-time curve is from point **d** to **e**. This stage takes place over a time, ΔT_3.

From a practical point of view, the ideal soil for use in thermal storage applications will have a small ΔT_1 and a large ΔT_2. In addition, the temperature levels at points **b, c,** and **d** as shown in the figure will be high and relatively constant. A laboratory experimental program was conducted to explore the feasibility of meeting these requirements in the actual soil. Engineering applications of TSC concept to solar energy collection and storage have also been also studied (Fang and Chaney, 1983).

8.5 THERMAL CONDUCTIVITY AND RESISTIVITY OF SOIL

8.5.1 Thermal Conductivity

As indicated in Table 8.1, the *thermal conductivity*, k, of a particular material is defined as the quantity of heat, Q, which flows normally across a surface of unit area, A, of the material per unit of time, t, per unit of temperature gradient, i, normal to the surface. The flow of heat in the soil also can be illustrated as being similar to the flow of water through porous media which follows Darcy's Law as frequently used in geotechnical engineering. The reciprocal of thermal conductivity, k, is referred to as *thermal resistivity*, R, and is a measure of material's ability to resist heat flow. It can be illustrated by the following equations:

$$k = \frac{Q}{i\,A\,t} \qquad\qquad (8.5)$$

$$R = 1/k \qquad\qquad (8.6)$$

where R = thermal resistivity; k = thermal conductivity; Q = quantity of heat; i = temperature gradient; A = unit area which flows normally across; and t = time.

Thermal conductivity is defined by Equation (8.5). The value of conductivity of a soil depends on the thermal conductivities of its solid, liquid, and gaseous components; and their volume proportions and their arrangement of particles or structure. The structure itself is influenced by the size, shape, and gradation of the components. Generally, solids conduct heat better than liquid, and liquids better than gases. To describe the thermal conductivity of an aggregate of soil particles, it is, therefore, necessary to describe the manner in which the solid particles either make contact with each other or are separated by intervening liquid and gaseous phases.

Figure 8.5 presents both measured and calculated thermal conductivity values for air dry and saturated vs. dry density of Canadian Leda clay as reported by Penner (1963). In all cases, increasing the dry density will increase the thermal conductivity. Figure 8.6 is plotted with logarithm of thermal conductivity vs. porosity of homoionic kaolinite clays (Na, Al, Ca) at constant water contents. In all cases, decreasing the water content and increasing the porosity will decrease the thermal conductivity. Figure 8.7 summarizes the thermal conductivity and resistivity vs. relative density and void ratio of dry Monterey 0 sand. In all cases, increasing the relative density or decreasing the void ratio will decrease the thermal resistivity or increase the conductivity.

8.5.2 Thermal Resistivity

(1) General Discussion

Thermal resistivity is a recipocal of thermal conductivity as indicated in Equation (8.6). The thermal resistivity of the soil involves a number of factors such as density, moisture content, grain size, clay content, etc. Field experience has shown that the resistivity varies over a wide range in different parts of the country. Figure 8.8 shows the relationship between thermal resistivity and conductivity of sand at constant density. The moisture content is varied from dry condition to 100% by weight. Table 8.3 gives thermal resistivities in °C-cm/watt (Rho) of the material commonly encountered.

(2) Practical Aspects of Thermal Resisitivity

Thermal resistivity is an important parameter relating to many environmental geotechnical problems; some comments on practical aspects based on data from Tables 8.3 and 8.4 are presented as:

1. Quartz has, by far, the lowest thermal resistivity of any of the materials listed.
2. Rocks, being relatively dense, homogeneous objects, have a lower thermal resistivity than soils. The more homogeneous rocks (e.g., certain forms of granite) have a lower thermal resistivity than less homogeneous rocks such as limestone and sandstone.
3. Under same conditions, a dense, compact soil should have a lower thermal resistivity than a more porous soil. A wet soil should have a lower thermal resistivity than a dry soil.

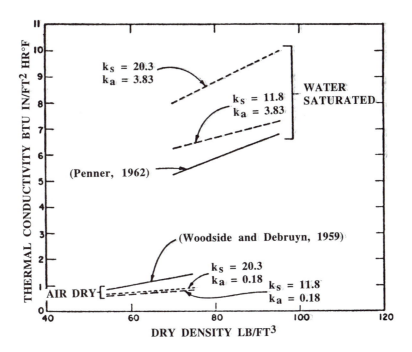

Figure 8.5 Thermal conductivity vs. dry density for Canadian Leda clay (After Penner, [1963], Proc. Int. Conf. of Permafrost, NAS-NRC Publ. 1287.)

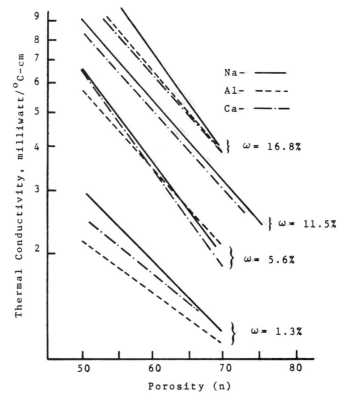

Figure 8.6 Thermal conductivity vs. porosity of natural kaolinite clay (After Reno, W.H. and Winterkorn, H.F. [1967], Thermal Conductivity of Kaolinite Clay as a Function of Type of Exchange Ion, Density and Moisture Content, HRR no. 209, pp. 79–85.)

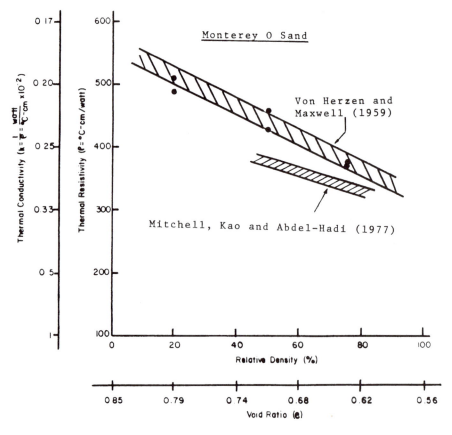

Figure 8.7 Thermal conductivity vs. relative density and void ratio of dry sand. (After Chaney, R.C., Ramavjaneya, G., Kanchanastit, P., and Fang, H.Y. [1983], *ASTM Geotech. Testing J.*, v. 6, no. 4, pp. 220–225.)

(3) Effects of Moisture Content and Density

Effects of moisture content and density on thermal resistivity or conductivity of some soils have been presented in Figures 8.5, 8.6, and 8.8. Their behaviors can be explained as:

1. In a dry or partially saturated soil, the presence of air with its high thermal resistivity greatly increases the overall thermal resistivity of the soil.
2. It is evident that compaction should reduce the thermal resistivity of a material, both by reducing the total void volume and by improving contact between the solid particles. Table 8.4 illustrates the magnitude of this effect.

Examining the data from Table 8.4 clearly shows how the denser and more compact granular system compares with the less dense system. Even a moderate degree of compaction improves the thermal resistivity considerably. A greater degree of compaction can improve the thermal resistivity greatly.

3. Effect of moisture also significantly affects thermal resistivity of soil as indicated in Figure 8.8. The amount of moisture added to a dry, granular material tends to distribute itself in a thin film on the surface of the solid grains. A small amount of this water may go into 'solid solution' (Section 4.9) at the grain surface. The rest remains as a film on the surface. Such a film, which is known as hydroscopic water, should reduce the contact resistance between the grains greatly; and the shape of the thermal-resistivity-moisture content curves (Figure 8.5) indicates that such is indeed the case.

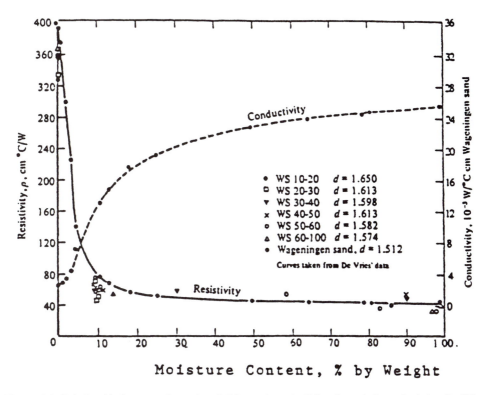

Figure 8.8 Relationship between thermal resistivity and conductivity of sand at constant density (After Van Rooyen, M. and Winterkorn, H.F. [1959], Proc. HRB, v. 38, pp. 576–621.)

Table 8.3   Values of Thermal Resistivity of Soil Components

Material	Thermal resistivity (°C-cm/watt)
Air	4000
Granite	25–58
Limestone	45
Mica	170
Organic matter (wet)	400
Organic matter (dry)	700
Sandstone	58
Water	165

Data selected from Sinclair et al. (1960) and Jumikis (1966).

Table 8.4   Effect of Density and Air Content on Thermal Resistivity of Soil Components

Material	Density (g/cc)	Air by volume (%)	Thermal resistivity (°C-cm/watt)
Quartz	2.640	0	11
Crushed quartz	1.307	50.6	573
Crushed quartz	1.512	42.7	392
Finely crushed quartz	0.824	68.9	1270
Finely crushed quartz	1.600	39.4	228

Data selected from Van Rooyen and Winterkorn (1959) and Sinclair et al. (1960).

8.6 EFFECTS OF HEAT OR TEMPERATURE ON SOIL BEHAVIOR

8.6.1 General Discussion

The effect of temperature and heat on ground soil, it is necessary to understand how temperature or heat affects each of the main components' phases (solid, liquid, gas) separately and also their interaction with each other within the framework of the total system. In this section, we will discuss how temperature or heat relating to the moisture in the soil pores will influence the strength and compressibility of soil and other construction materials.

1. Low temperature range (5 to 100°C)
2. Medium temperature range (100 to 250°C)
3. High temperature range (fusion)

8.6.2 Characteristics of Soil-Heat System

In general, increases in volume of water with increasing temperature at constant pressure and decreasing internal friction of the liquid phase will increase the gaseous phase. This lubricant effect in the soil mass may be quite small, especially if the soil possesses a granular-bearing skeleton. On the other hand, soils without granular-bearing skeletons have more surface area that must be lubricated in compaction and whose interaction with water is dependent on temperature. The existence of such interaction has been known for a long time, as has been the fact that it is a function of type and amount of mineral surface and type and amounts of exchangeable ions as well as of the electrolyte environment in the contacting aqueous phase.

The effect of temperature on this interaction could be explained from physicochemical concepts and was studied in the early 1930s by Baver and Winterkorn (1935) on extracted natural clays over a wide moisture range and a temperature range from 30 to 99°C. In order to explain the results of these and correlated studies, they separated the total interaction (swelling) water at an elevated moisture content into hydration water (Section 4.8), whose binding is associated with a considerable amount of heat of wetting and osmotic water and the relatively unrestrained water between the hydrated cations in what may be called the diffuse-double-layer (Section 9.8). Both types of water are intrinsically related to the dielectric constant of water.

Since the latter decreases with increasing temperature and then more rapidly is the increase in kinetic energy of the exchange ions with temperature increase, a decrease in both the hydration water and the osmotic water in a saturated soil system may be expected with increasing temperature. This means that, at constant water content, increases in temperature decrease the portion that is bound or restrained and increases the portion that is free while also decreasing the viscosity of the latter.

While this temperature effect is quite straightforward in relatively pure clay-water systems within a limited range of temperature, it becomes more complicated even with these when high temperatures may cause marked dispersion or flocculation effects depending on clay mineral and exchange ion type. Even greater complexity is obtained in the presence of nonclay minerals of silt and larger size. The basic phenomenon, however, is a decrease in interaction water with increasing temperature. This decrease will normally result in an increase in the hydraulic conductivity with increasing temperature. However, in the presence of a strong dispersion effect, the conductivity may actually decrease especially if the clay is contained within the interstices of a coarse-granular skeleton.

The facts and considerations just presented should impress on one's mind the great importance of visualizing all the geometric, granulometric, and structural factors within a

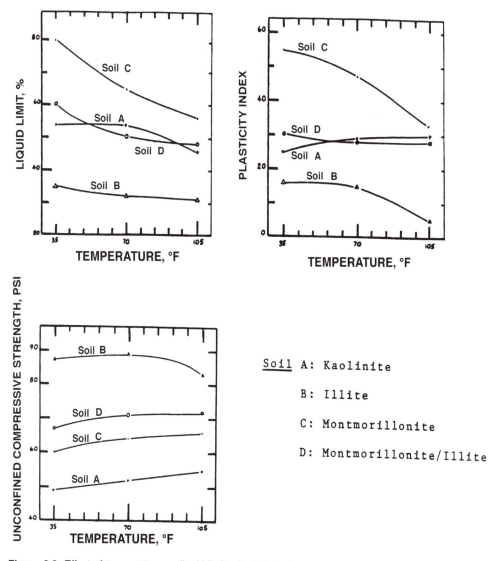

Figure 8.9 Effect of temperature on liquid limit, plasticity index and unconfined compressive strength of four clay minerals. (After Laguros, J.G. [1969], Effect of Temperature on Some Engineering Properties of Clay Soils, HRB Special Report 103, pp. 186–203.)

system that are either restraining or permitting the full expression of the mineral-water-temperature interaction. It is from such a perspective that the data reported in the literature should be assayed. A limited selection of such findings is given as follows.

8.6.3 Effect of Soil Behavior at Low Temperature Range

(1) Soil Constants

Effects of the temperature or heat on the soil parameters as reflected on the liquid and plastic limits of soil have been studied by many investigators. Youssef et al. (1961), in testing Egyptian soft to still dark-brown clays, found that temperature changes caused a change in liquid and plastic limits proportional to the change with temperature of the viscosity of water. Laguros (1969), testing kaolinite, illite, montmorillonitic, and montmorillonitic-illitic clay, found that their liquid and plastic limits decreased with increase in temperature with a maximum effect in the case of the montmorillonitic clay (Figure 8.9).

(2) Unit Weight-Moisture Relationship of Soil

Soil temperature or heat has an effect on compaction, particularly on soils high in clay content. Hogentogler (1936) reported laboratory compaction tests on a clay soil in which unit weight increased 0.47 kN/m³ and the optimum moisture content (OMC) decreased 3% when the temperature of the soil increased from 1.1 to 46°C. In a similar study, Woods (1940) found that the difference in the maximum dry unit weight at temperatures from 1.7 to 23.9°C was 2% for a sandy clay. Burmister (1964) found the dry unit weight of a particular soil compacted at the OMC at 18.3°C to 16.3 kN/m, while it was only 14.1 kN/m at 1.7°C. Laguros (1969) obtained comparable results working with clay soils containing kaolinitic, illitic, montmoril-lonitic, and montmorillonitic-illitic clay minerals, respectively. Further discussion of temperature (above and below 0°C) effects on unit weight-moisture relationship will be presented in Section 11.2 and Figure 11.5.

(3) Compressibility

Gray (1936) reported an increase in compressibility with increase in temperature, the greatest effects being observed in the range of secondary consolidation. Lewis (1950), who tested samples of London clay at temperatures between 5 and 15°C, concluded that a change in temperature had little influence on the coefficient of compressibility, a_v, but was significantly effective on the coefficient of consolidation, c_c. Burmister (1951) described temperature effects that he considered serious enough to render ordinary consolidation test results as useless. Finn (1951), investigating the effect of temperature on the consolidation characteristics of remolded clays during primary compression, found that the c_c, increased radically between 4.4 and 21.1°C, but did not increase significantly between 21.1 and 26.6°C. Lambe (1958) reported decrease in soil volume with increase in temperature observed in tests run at constant pressure with the temperature being the only external variable. Lo (1961) demonstrated that even relatively small changes in temperature can cause a marked change in compressibility in the secondary consolidation range. Similar conclusions on the temperature effect on secondary compression were drawn by Schiffman et al. (1966), and Campanella and Mitchell (1968).

Plum and Esrig (1969) found that the amount of temperature-induced consolidation is related to soil compressibility; the higher the compressibility, the greater the consolidation for a given temperature increase. He also showed the volumetric strain of an illite-water system as a function of overconsolidation ratio, OCR, for a temperature increase from 24 to 50°C. The influence of temperature on primary consolidation behavior of a slightly organic soil is examined by Habibagahi (1976). From a test temperature range from 25 to 50°C, he concluded that the void ratio decreased as temperature increased under constant consolidation pressure.

(4) Porewater Pressures

Campanella and Mitchell (1968) emphasize the role of porewater pressure changes accompanying temperature change. Determinant factors under drained conditions appear to be the thermal expansion of the porewater, the compressibility of the soil structure, and the initial effective stress. Their experimental results show several clay-water systems in which each change in temperature by 1°F changed the porewater pressure by about 0.75 to 1.0% of the initial effective stress. For less compressible materials the porewater pressure change was considerably greater.

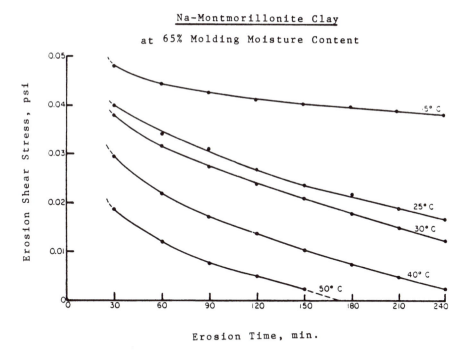

Figure 8.10 Effect of water temperature on erosion shear stress for sodium montmorillonite clay. (After Liou, Y.D. [1970], Hydraulic Erodibility of Two Pure Clay Systems, Ph.D. dissertation, Department of Civil Engineering, Colorado State University, Ft. Collins.)

(5) Modulus and Shear Strength

Murayama (1969), using a rheological model for analysis of the elastic modulus of clay-water systems, showed that the modulus decreased as the temperature increased. Mitchell (1969) studied the relationship between initial stress and strain in stress relaxation tests at various temperatures. Considering the straight-line portions of the curves through the plotted data as representative of the elastic modulus of the soil, he concluded that the modulus decreased with increase in temperature. Figure 8.10 shows the effect of water temperature on erosion shear stress for sodium montmorillonite clay as reported by Liou (1970). It is indicated that for a given time, if the water temperature increases, the erosion shear stress also increases.

Similar conclusions on the temperature effect on shear strength were drawn by many investigators (HRB, 1969). It must be kept in mind that with the exception of the Atterberg limit and compaction tests, the effect of temperature on soil-water systems was determined at constant water content or at free accessibility to a water reservoir. If change in temperature is associated with a change in moisture content, then the total effect is the sum, or the difference, as the case may be, of both temperature and moisture change effects. This, of course, is the reason why field plate load tests yield higher values for moduli and bearing value during the warmer months (Benkelman et al. 1962).

The general effect of temperature on the engineering properties of subgrade soils is summarized in Table 8.5. However, the information contained therein is to be used with caution because of the possible modifying or counteracting effects due to the previously mentioned geometric, granulometric, and soil structure factors.

Table 8.5   General Effect of Temperature on the Geotechnical Properties of Soils

Soil properties	Temperature	
	Low	High
Water retention	High	Low
Liquid limit	High	Low
Plastic limit	High	Low
Maximum compacted density	Decrease	Increase
Optimum moisture content	Increase	Decrease
Void ratio[a]	High	Low
Volume change[a]	Swelling	Shrinkage
Coeff. of compressibility, a_v	Decrease	Increase
Pore water pressure, u	Decrease	Increase
Coeff. of permeability, k	Decrease	Increase
Shear strength	High	Low
Modulus	High	Low

[a] At constant pressure in open saturated system.

8.6.4 Effect of Soil Behavior at Medium Temperature Range

Thermal treatment or *thermal stabilization* is defined as an irreversible and effective increase of the shear strength of soil or rock. Based on experimental work reported by Wohlbier and Henning (1969), the preliminary heat treatment of kaolinite clay, especially of kaolin from Zettlitz, provides effective stabilization, i.e., a permanent increase of the shear strength. Concerning the shear strength of preliminary heat-treated samples of a cohesive soil, the water as a void filling is of special importance. In the case of a material in which the increasing strength is only due to its capillary tension and not to changing structure, the effect of stabilization when reducing this capillarity by adding water is irreversible with a small amount.

The effects of heat treatment on the strength of kaolinite at higher temperature ranges (from 200 to 600°C) are shown in Figure 8.11. In all cases, with the temperature increase, the shear strength also increases. Also, Chandrasekharan et al. (1969), studying the effect of heat treatment on Indian black cotton soil (Section 13.4), found a moderate reduction of water affinity in the initial ranges of heat treatment between 25 and 200°C. The temperature effect on the consistency properties of laterite soil has been reported as not significant because of the relatively inert character of the clay minerals of kaolinite, oxides of iron, and aluminum present.

8.6.5 Fusion

Fusion is a heating process for melting of a solid to a liquid. It also refers to reactions between small atomic nuclei to form larger ones. The solid phase of a soil particle may consist of many different minerals, some of which contain water as an integral part of their crystal lattice structure. Most solid particles also hold surface water layers by means of molecular attraction (adsorption).

A soil mass also contains carbonate and sulfate minerals which, when heated sufficiently, will degas with subsequent alternations in crystalline structure. The clay minerals may also contain a layer of water that bonds to the clay mineral surfaces by the negative charges possessed by the clay mineral structures and the cations. This bound water between mineral structures and the cations requires considerable heat energy to be separated from the solids. When the temperature of the soil body is increased, the free interstitial water is driven off; then the adsorbed water is also driven off. As the temperature further increases, the hydration

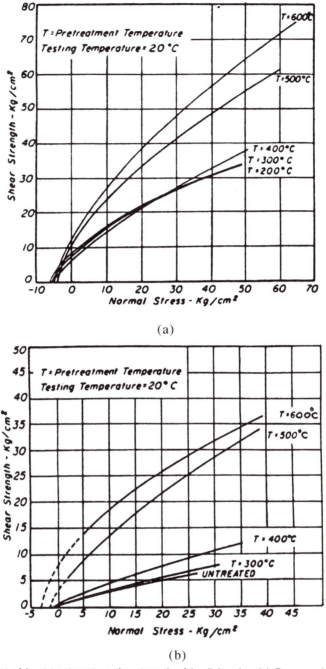

Figure 8.11 Effect of heat treatment on the strength of kaolinite clay. (a) Dry samples and (b) rewet samples. (After Wohlbier, H. and Henning, D. [1969], Effect of Preliminary Heat Treatment on the Shear Strength of Kaolinite Clay, HRB Special Report 103, pp. 287–300.)

water is driven off and new minerals are formed having different crystalline morphologies. Fusion process has been used for soil stabilization and ground improvement. Typical data on melt point of some common clay minerals and fusion range of soil aggregate material are presented in Tables 8.6 and 8.7. The typical mineral compositions are derived from data tabulated by Deer et al. (1962). Detailed discussions on these aspects are given by Post and Paduana (1969).

Table 8.6   Melt Point of Some Common Soil Minerals and Rocks

Mineral	Structural water(s)	Typical atomic proportions of dry melt									Melt point (°C)	
		CaO	K_2O	Na_2O	FeO	Fe_2O_3	TiO_2	MgO	Al_2O_3	SiO_2	Estim.	Observed
Common Soil Minerals (Nonclay)												
Feldspars												
Albite	—	0.004	—	0.114	—	—	—	—	0.200	0.682	1115	1100–1200
Oligoclase	—	0.040	—	0.091	—	—	—	—	0.232	0.637	1200	1200–1300
Andesine	—	0.078	—	0.067	—	—	—	—	0.261	0.594	1305	1300–1355
Microcline	—	—	0.133	0.024	—	—	—	—	0.198	0.645	1355	1150–1530
Orthoclase	—	—	0.105	0.042	—	—	—	—	0.199	0.654	1305	1150–1530
Anorthoclase	—	0.012	0.042	0.080	—	—	—	—	0.207	0.659	1200	1100–1200
Amphiboles												
Hornblende	1.9	0.124	—	0.010	0.101	0.034	0.009	0.150	0.073	0.499	—	1180–1220
Pyroxenes												
Enstatite	0.3	0.008	—	—	0.046	0.006	—	0.353	0.010	0.577	—	1380–1500
Augite	0.5	0.193	—	—	0.094	0.019	0.008	0.142	0.035	0.509	—	1150
Carbonates												
Calcite	44.4 CO_2	1.000	—	—	—	—	—	—	—	—	2550	2570
Dolomite	47.4 CO_2	0.579	—	—	0.025	—	—	0.396	—	—	2300	2100
Sulfates												
Gypsum	20.5	1.000	—	—	—	—	—	—	—	—	2550	—
Zeolites												
Analcite	8.3	—	—	0.147	—	—	—	—	0.254	0.599	1150	—
Iron oxides												
Hematite	—	—	—	—	—	1.000	—	—	—	—	—	1594
Goethite	10.5	—	—	—	—	1.987	—	—	—	0.013	1585	—
Magnetite	—	—	—	—	0.309	0.691	—	—	—	—	—	1594
Alumina oxides												
Gibbsite	33.7	—	—	—	—	—	—	—	0.979	0.021	2025	2020
Corundum	—	—	—	—	—	0.010	—	—	0.983	0.007	2025	2020
Others												
Quartz	—	—	—	—	—	—	—	—	—	1.000	—	1720
Ilmenite	—	—	—	—	0.411	0.071	0.505	0.013	—	—	1390	1400
Halite	—	—	—	—	—	—	—	—	—	—	—	801
Common Clay Minerals												
Kandites												
Kaolinite	14.3	—	—	—	—	—	—	—	0.480	0.540	1810	1785
Halloysite	17.4	—	0.006	—	—	—	—	—	0.456	0.538	1800	1775
Smectites												
Montmorillonite	21.2	0.030	—	—	—	0.029	—	0.041	0.236	0.664	1490	—
Nontronite	22.4	0.029	—	—	—	0.380	—	—	0.071	0.520	1580	—
Micas												
Muscovite	4.8	—	0.106	0.009	—	—	—	—	0.395	0.490	—	1255–1290
Phlogonite	3.1	—	0.098	—	0.048	0.031	0.016	0.231	0.158	0.418	1200	—
Biotite	2.3	—	0.086	—	0.186	0.056	0.038	0.094	0.153	0.387	—	1155–1240
Others												
Ilite	7.2	—	0.071	—	—	0.041	—	0.019	0.279	0.600	1590	—
Vermiculite	19.1	—	—	—	0.012	0.065	—	0.291	0.197	0.435	1580	—
Rapidolite	11.3	—	—	—	0.255	0.034	—	0.185	0.231	0.295	1540	—
Attapulgite	19.9	—	0.006	—	—	0.044	—	0.131	0.129	0.690	1530	1380–1500
Serpentine	13.5	—	—	—	—	0.005	—	0.483	0.012	0.500	1800	—

After Post (1967); Post, J.L. and Paduana, J.A. (1969), Soil Stabilization by Incipient Fusion, HRB Special Report 103, pp. 243–253.

Table 8.7 Fusion Range of Soil Aggregate Material

Soil no.	Material	Approximate melt range (°C)	
		Estimated	Observed
Natural Soils			
1	Black Hills bentonite	—	1340
2	Calcareous soil, Tijeras, NM	1950–2050	—
3	Coastal clayey soil, TX	1600–1620	—
4	Estancia Playa soil, NM	1350–1400	1220–1260
5	Granitic detrital soil, NM	1200–1320	1310–1335
6	Hawaiian lateritic soil	1900–1950	—
7	Hayes, KS soil	1250–1495	1235–1285
8	Magnesian soil NM	1350–1435	1280–1310
9	Podzol soil, WI	1350–1450	—
10	Rio Salado dune sand, NM	1300–1400	1330–1340
Aggregates			
	Basalt (and gabbro)	—	1140[a]
	Clays (North America)	—	1580–1800
	Clays (aluminous)	—	1800–1880
	Clays (carboniferous)	—	1605–1675
	Granite (and rhyolite)	—	800–1100
	Kaoline (southern)	—	1645–1785
	Silica sand	—	1700–1750
Artificial Soils			
	Calcareous soil	1300–1325	1235–1300
	Clay soil	1300–1390	1310–1340
	Organic calcareous soil	1275–1300	1220

[a] Basalt may have a melt range from 1100 to 1400°C.
After Post (1967).

8.7 FROZEN AND THAWING SOILS

8.7.1 Identification and Description of Ice in Soil

Ice is a form of water. The transfer from water to ice can be accomplished at 32°F by the removal of heat energy in the amount of 144 Btu/lb without any change in temperature. The growth of ice as distinct lenses, layers, veins, and masses in soil commonly, but not always, is oriented normally to the direction of heat loss. Ice lenses are the ice formations in soil occurring essentially parallel to each other, generally to the direction of heat loss, commonly in repeated layers. Frost flow is similar to thermal flow. Frost transfer is also carried out by conduction, convection, and radiation.

Identification and description of frozen soils have been discussed in detail by Andersland and Anderson (1978) and Phukan (1985 and 1991). Description of frozen soils based on visual examination and simple manual tests was proposed by ASTM (ASTM D4083-89). This practice is intended primarily for use by soil engineers and technicians in the field, where the soil profile or samples from it may be observed in a relatively undisturbed (frozen) state, or it may be used in the laboratory to describe the condition of relatively undisturbed soil samples that have been maintained in a frozen condition following their acquisition in the field. Table 8.8 presents a general description of frozen soils based on visual examination and simple manual tests proposed by ASTM.

Table 8.8   Description of Frozen Soils (Visual-Manual Procedure) (ASTM D4083-89)

Description of Frozen Soils — Ice Not Visible

Group symbol	Subgroup Description	Symbol	Field identification
N	Poorly bonded or friable No excess ice Well-bonded Excess ice	N_f N_b N_{bw} N_{be}	Identify by visual examination; to determine presence of excess ice, use procedure under *Note 2* and hand magnifying lens as necessary; for soils not fully saturated, estimate degree of ice saturation; medium, low; note presence of crystals or of ice coatings around larger particles

Note 1: Frozen soils in the N group may, on close examination, indicate presence of ice within the voids of the material by crystalline reflections or by a sheen on fractured or trimmed surfaces. The impression received by the unaided eye, however, is that none of the frozen water occupies space in excess of the original voids in the soil. The opposite is true of frozen soils in the V group.

Note 2: When visual methods may be inadequate, a simple field test to aid evaluation of volume of excess ice can be made by placing some frozen soil in a small jar, allowing it to melt, and observing the quantity of supernatant water as a percentage of total volume.

Description of Frozen Soils — Visible Ice Less Than 1 in. (25 mm) Thick

Group symbol	Description	Symbol	Field identification
	Individual ice crystal or inclusions	V_s	For ice phase, record the following when applicable
	Ice coatings on particles	V_c	Location Size
	Random or irregularly oriented ice formations	V_r	Orientation Shape Thickness Pattern of arrangement
	Stratified or distinctly oriented ice formations	V_s	Length Spacing
	Uniformly distributed ice	V_s	Hardness Structure Color
			Estimate volume of visible segregated ice present as percentage of total sample volume

Description of Visible Ice Strata Greater Than 1 in. (25 mm) Thick

Group symbol	Subgroup Description	Symbol	Field identification
ICE	Ice with soil inclusions Ice without soil inclusions	ICE + soil type ICE	Designate material as ICE (*Note 1*) and use descriptive terms as follows, usually one item from each group, when applicable

	Hardness	Structure (*Note 2*)
	Hard	Clear
	Soft (of mass, not individual crystals)	Cloudy Porous Candled Granular Stratified
	Color (Examples)	Admixtures (Examples)
	Colorless Gray Blue	Contains few thin silt inclusions

Note 1: Where special forms of ice such as hoarfrost can be distinguished, more explicit description should be given.

Note 2: Observer should be careful to avoid being misled by surface scratches or frost coating on the ice.

The system for describing and classifying frozen soil is based on an identification procedure that involves three steps: (1) consists of a description of the soil phase; (2) consists of the addition of soil characteristics resulting from the frozen state; and (3) consists of a description of the important ice strata associated with the soil. In addition to the description of the soil profile at a given site it is normally advantageous to describe the local environmental and geologic conditions such as terrain features, vegetation cover, depth and type of snow cover, local relief and drainage conditions, and depth of thaw (Shockley, 1978).

8.7.2 Frost Depth (Penetration) and Heave

(1) Analytical Approaches

The frost penetration depth or penetration in soil can be calculated by means of the heat transfer theory. The theory is based on heat conduction in both steady- and unsteady-state flow of heat. Two general equations for calculating the frost depth are commonly used in geotechnical engineering, the Stefan equation and Berggren equation (Berggren 1943; Aldrich, 1956). The Stefan equation is based on: (1) soil is homogeneous, (2) the volumetric heat of latent fusion is the only heat that must be removed when the soil freezes (i.e. volumetric heat is neglected), (3) the temperature gradient in the frozen zone is linear, and (4) the porewater is not moving. The derivation, limitation, and application of the Stefan equation are given by Jumikis (1955). For the Berggren equation, the basic equation is the heat-diffusion equation. It is similar to Terzaghi's consolidation equation. The rigorous solution of the heat-diffusion equation developed by Berggren was modified by Aldrich and Paynter in 1953 (Aldrich, 1956).

(2) In Situ Measurement

There are numerous methods for measuring frost depth in *in situ* conditions. Among these methods, the electric-resistivity method is most common. A device was developed at the AASHO Road Test (Carey and Andersland, 1957) by which determination of frost depth could be made without disturbing the pavement. The system was based on the knowledge that the electrical resistance of a soil-water system changes rapidly upon freezing due to a large difference in resistivity between the solid and liquid phase of water. Pairs of electrodes buried in the soil at 1-in. intervals of depth were connected to leads that extended to the surface. Measurements of the resistance across these electrodes indicated the depth to which the soil-water system had frozen. Typical frost depth under various cover conditions from AASHO Road Test is presented in Figure 8.12. Further discussion on electric-resistivity method for determination of frost depth will be presented in Section 9.4.

Figure 8.13 presents the relationship between air-freezing index, surface cover, and frost penetration into homogeneous soils developed by Sanger (1963). These relationships are hinged on soil types and properties. The term air-freezing index will be explained in a following section.

(3) Frost Heave and Rate of Heave

Frost heave is the raising of a surface due to the formation of ice in the underlying soil and is indicated by percent heave. This percent is the ratio of the amount of heave to the depth of frozen soil before freezing. The percent of heave is affected by soil types, unit weight

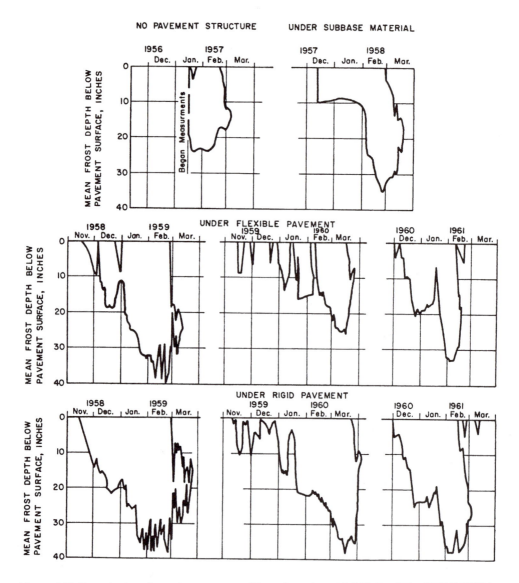

Figure 8.12 Frost depth under various cover conditions. (Data from AASHO Road Test HRB [1962b].)

of soil, and percent of finer than 0.02 mm. In all cases, increase in the percent fine will increase the percent heave for all types of soil. Figure 8.14 presents frost heave vs. dry unit weight of various soil types. Figure 8.15 shows the effect of rate heave on sand added with various amounts of clay minerals and clays. In examining Figure 8.15, in all cases, the average rate of heave increases, as percentage of fines in the soil increased.

The frost heave and the rate of heave both are important for pavement design. The amount of fines in the subgrade soil or base courses is critical for the frost action, frost boil, and other undesirable actions. A frost boil is the breaking of a localized section of a highway or airfield pavement under traffic and ejection of subgrade soil in a soft and wet (soupy) condition caused by the melting of the segregated ice formed by frost action. Table 8.9 presents the guidelines for selection of heave rate or frost susceptibility class in order to reduce the potential damaging of the soil-pavement system.

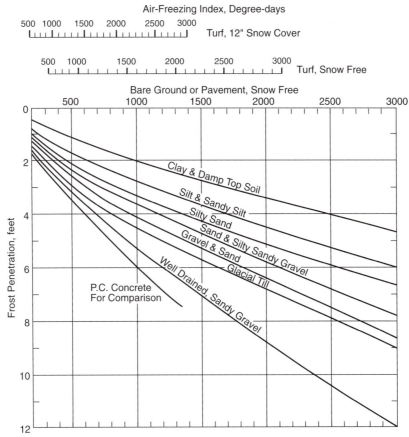

Figure 8.13 Relationship between frost penetration between air-freezing index and surface cover conditions. (After Sanger, F.J. [1963], Proc. Int. Conf. on Permafrost, NAS-NRC Publ. 1287; and TRB [1974], Roadway Design in Seasonal Frost Areas, NCHRP, Synthesis of Highway Practice 26, TRB, 104p.)

(4) Freezing Index

Freezing index (F) is the number of degree-days between the highest and lowest points on a curve of cumulative degree-days vs. time for a freezing season. It is used as a measure of the combined duration and magnitude of below-freezing temperatures occurring during any given freezing season. The index determined for air temperatures at 1.5 m above the ground is commonly designated as the *air freezing index* (F_a), while that determined for temperatures immediately below a surface is known as the *surface freezing index* (F_s). The *mean freezing index* (F_m) is the index determined on the basis of mean temperatures. The period of record over which temperatures are averaged is usually a minimum of 10 years and preferably 30. Determination of freezing index by U.S. Army procedure is illustrated in Figure 8.16.

8.7.3 Environmental Geotechnical Problems of Frozen Soil

(1) Geotechnical Properties and Foundation Problems

Environmental geotechnical problems relating to the frozen soil mainly cover two general topics: (1) the characteristics of soil properties, and (2) the foundation problems including

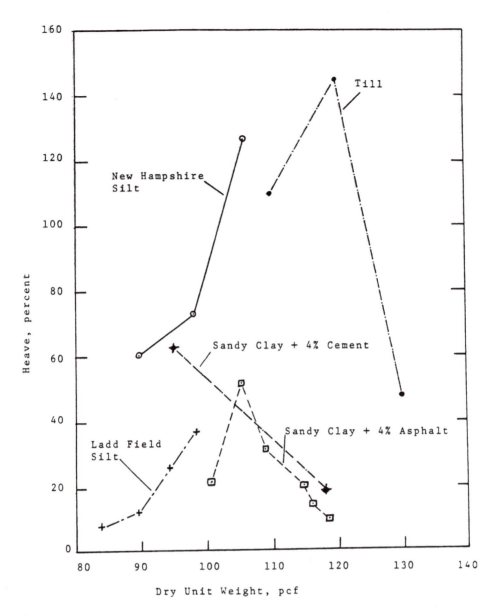

Figure 8.14 Effect of soil type and unit weight on heave (Data from Winn and Rutledge [1940]; and Haley, J.F. and Kaplar, C.W. [1952], Cold-Room Studies of Frost Action in Soils, HRB Special Report 2, pp. 246–267.)

bearing capacity and pile capacity. The foundation problems relating to frozen ground have been discussed by Andersland and Anderson (1978) and Phukan (1985 and 1991) and will be further discussed in Section 12.7.

In a frozen soil, unfrozen adsorbed water can exist in equilibrium with ice over a large temperature range below freezing, as reported by Hoekstra (1969). The adsorbed water freezes gradually as the temperature is lowered, decreasing the thickness of the unfrozen film. The ice phase at each temperature is in equilibrium with an unfrozen water layer whose properties are constantly changing as the temperature is lowered. The reason for the large freezing point depression of the unfrozen water is the presence of exchangeable ions, which are concentrated in thin films upon freezing and specific adsorption forces emanating from a charged surface.

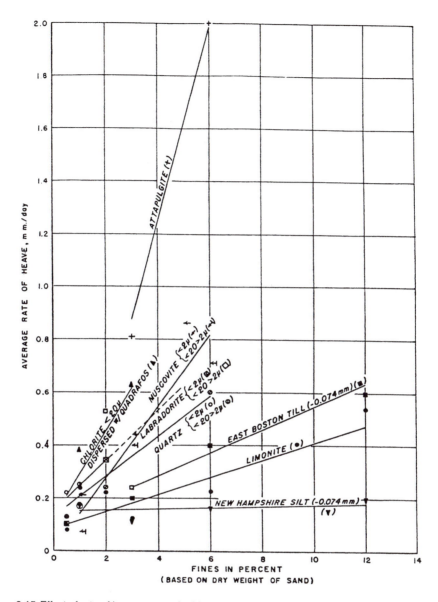

Figure 8.15 Effect of rate of heave on sand added with various amounts of clay minerals and clays. (After Linell, K.A. and Kaplar, C.W. [1959], The Factor of Soil and Material Type in Frost Action, HRB Bulletin 225, pp. 23–44.)

Frozen soils are essentially impermeable to laminar water flow and the mechanism of mass transport is mainly by diffusion in the liquid phase (Section 7.5). Diffusion can occur under the influence of temperature and external stress gradients. Mass transport occurs through the transition layer between the soil particle surface and the ice. Besides mass transport of water, ions also migrate in frozen ground. In general, the engineering properties of frozen soils are temperature dependent in the temperature range from 0 to –10°C, where the phase composition of water changes appreciably. The temperature dependence is related to changes in amount of unfrozen water. The effect of temperature is greater in soil types that contain large quantities of unfrozen water.

The effect of freezing temperature and water content for different soils is presented in Figures 8.17 and 8.18. In examining Figs.8.17 and 8.18, soil types, particles size, and water

Table 8.9   Guidelines for Selection of Heave Rate or Frost Susceptibility Class

Unified classification (soil type)	Symbol	Percent (<0.02 mm)	Heave rate (mm/day)	Frost suscept. classification
Gravels and sandy	GP	0.4	3.0	Medium
Gravels	GW	0.7–1.0	0.3–1.0	Neg. to low
		1.0–1.5	1.0–3.5	Low to medium
		1.5–4.0	3.5–2.0	Medium
Silty and sandy	GP-GM	2.0–3.0	1.0–3.0	Low to medium
Gravels	GW-GM	3.0–7.0	3.0–4.5	Medium to high
	GM			
Clay and silty	GW-GC	4.2	2.5	Medium
Gravels	GM-GC	15.0	5.0	High
	GF	15.0–30.0	2.5–5.0	Medium to high
Sands and gravelly	SP	1.0–2.0	0.8	Very low
Sands	SW	2.0	3.0	Medium
Silty and gravelly	SP-SM,	1.5–2.0	0.2–1.5	Neg. to low
Sands	SW-SM,	2.0–5.0	1.5–6.0	Low to high
	SM	5.0–9.0	6.0–9.0	High to very high
		9.0–22.0	9.0–5.5	
Clay and silty	SM-SC	9.5–35.0	5.0–7.0	High
Sands	SC			
Silts and organic	ML-OL,	23.0–33.0	1.1–14.0	Low to very high
Silts	ML	33.0–45.0	14.0–25.0	Very high
		45.0–65.0	25.0	Very high
Clay silts	ML-CL	60.0–75.0	13.0	Very high
Gravelly and sandy Clays	CL	38.0–65.0	7.0–10.0	High to very high
Lean clays	CL	65.0	5.0	High
	CL-OL	30.0–70.0	4.0	High
Fat clays	CH	60.0	0.8	Very low

After Moulton (1991).

content of the soil play an important role. The amount of unfrozen water increases in the direction of sand > silt > clay.

(2) De-Icing Problems

(a) De-Icing on Pavement

De-icing salt is considered the best option for maintaining clear pavements. Salt is able to dissolve and mix with melted ice, allowing it to drain off the roadway. Experimental studies on effects of de-icing salts on plant biota and soil are given by Hanes et al.(1976). Further discussions of varous aspects of snow removal and ice control on highways and airports are given by TRB (1979).

(b) Ice in River/Streams

A form of ice in a flowing stream is called frazil ice. In a stream, cooling proceeds both by thermal convection and turbulence within the water until the stream is isothermal at 39°F. Ice in streams greatly affects offshore/nearshore structures, especially the broken ice which causes many stability problems for these structures.

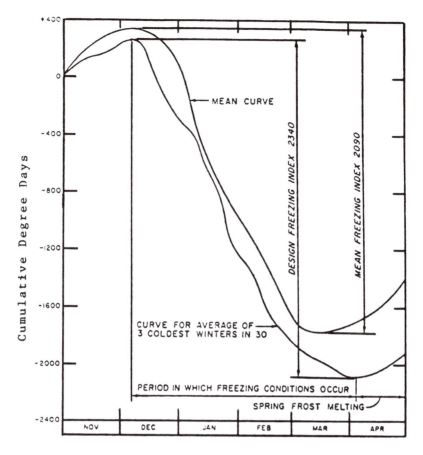

Figure 8.16 Determination of freezing index by U.S. Army procedure (After Linell, K.A., Hennion, F.B. and Lobacz, E.F. [1963], Corps of Engineers Pavement Design in Areas of Seasonal Frost, HRR no. 33, pp. 76–136.)

8.7.4 Characteristics of Thawing Soil

(1) General Discussion

Freezing-thawing cycles are important for various environmental geotechnical problems including pavement performance, bearing capacity of foundation structures, pile capacity, hydraulic barriers, as well as clay liners. Freezing behavior has been discussed in a previous section; some additional information under thawing conditions is added as follows. Among these factors, the thawing index, like the freezing index, is one of the important parameters.

Thawing index (T) is the number of degree days between the lowest and highest points on the curve for cumulative degree-days vs. time for one thawing season. It is used as a measure of the combined duration and magnitude of above-freezing temperatures occurring during any given thawing season. The index determined for air temperature at 1.5 m above the ground is designated as the *air thawing index* (T_a), while that determined for temperatures immediately below a surface is known as the *surface thawing index* (T_s). The *mean thawing index* (T_m) is the index determined on the basis of mean temperatures.

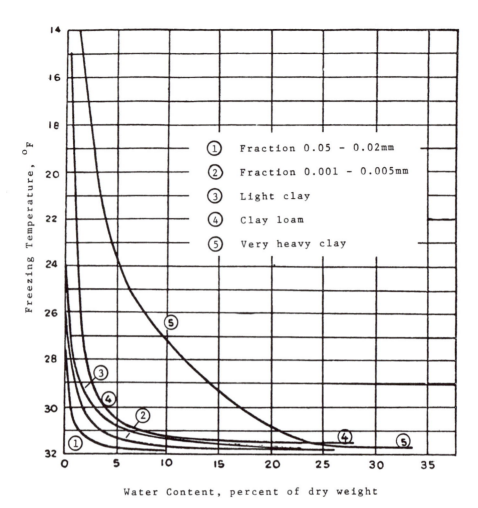

Figure 8.17 Effect of freezing temperature and water content for different soils (After Beskow, G. [1935], Soil Freezing and Frost Heaving with Special Application to Roads and Railroads, Swedish Geological Society Series C, no. 375; and John [1952].)

(2) Properties of Thawing Soil

When temperature increases after passing the freezing point, soil starts thawing. At that time, the strength of soil decreases. Experience shows that shortly after the thawing of the soil beneath the pavement, the soil loses its bearing capacity for a certain period of time. Other important parameters of thawing properties of soil include bearing capacity and settlement of foundation structures.

8.8 CHARACTERISTICS OF SNOW

8.8.1 General Discussion

Snow is also a form of water and directly related to precipitation, as a deposit of ice crystals. There is usually some liquid water present, either as capillary water in the interstices between snow crystals or as molecular water around individual crystals (Section 5.3). The term water content refers to the liquid water in a snow pack. The quality of snow is the

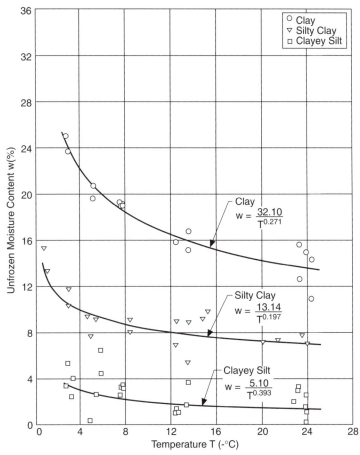

Figure 8.18 Unfrozen moisture content vs. temperature for various types of soil. (After Lovell, C.W. [1957], Temperature Effects on Phase Composition and Strength of Partially-Frozen Soil, HRB Bulletin 168, pp. 74–95.)

percentage by weight which is ice. Snow has a large specific surface area and is at or close to its melting temperature. It is active thermodynamically and very sensitive to local environmental conditions. The basic problems from an engineering point of view are (1) the rates of snowmelt and (2) the result of melt water. In the following section, emphasis is placed on environmental aspects of the snow behavior.

8.8.2 Properties and Problems of Snow

Properties of snow have direct application to a large number of engineering problems, including avalanche mechanics, vehicle mobility, snow plowing, etc. Table 8.10 presents the heat and temperature conductivity of snow. As discussed in Section 8.3, temperature is a measure of the internal motion of an object's constituent molecules. However, heat is thermal energy in transit. Heat may be transferred from the air to the snow pack by conduction or convection. The wind velocity is also closely related to the characteristics of snow, as shown in Figure 8.19 in which (a) shows the wind velocity vs. square root of diameter of packed snow, where a linear relationship is obtained and (b) shows the wind velocity at different altitudes vs. density of packed snow; curvilinear relationships are found in both cases.

Table 8.10 Heat Conductivity and Temperature Conductivity of Snow

Snow Density (g/cm³)	Heat conductivity (cal/cm-sec°C)	Temperature conductivity (cm²/sec)
0.12	0.00010–0.00020	0.0016–0.0032
0.16	0.00021–0.00047	0.0026–0.0058

After Keranen (1929); cited by Winterkorn, H.F. (1953), Macromeritic Liquids, ASTM STP 156, pp. 77–89.

8.9 SUMMARY

1. Characteristics of heat, heat sources, and heat transfer processes are discussed. The heat transfer process includes conduction, convection, and the radiation process in the soil-water system and is examined.
2. Measurable thermal parameters such as heat capacity, specific heat, and diffusivity are discussed and summarized in tabulated form and typical data for each case are also presented.
3. Effects of temperature and heat on geotechnical properties of soil are examined. These properties include soil constants, unit weight-moisture relationship, compressibility, shear parameters, porewater, etc. Heat treatment and fusion processes on soil, rock, and stabilized soil are evaluated.
4. The influence of seasonal factors on the performance of the soil-pavement system is discussed in terms of the pavement surface deflections as well as in terms of the strength of pavement components. It is shown that there is a significant difference in both deflection and strength between spring and fall for both traffic and nontraffic pavements.
5. The engineering properties of frozen soils are very dependent on temperature in the range from 0 to –10°C. The temperature dependence is related to changes in the amount of unfrozen water. The effect of temperature is most pronounced in soil types that contain large quantities of unfrozen water. The amount of unfrozen water increases in the direction of sand > silt > clay.
6. Applications relating to the environmental geotechnical areas such as underground cable and de-icing of pavement are discussed.

PROBLEMS

8.1 Distinguish clearly between heat and temperature.
8.2 Does organic soil absorb more heat than clean uniform sand? Explain.
8.3 Can clay store more heat than sand? Explain.
8.4 If air temperature is 80°F, what is the temperature in degrees Celsius (°C) and absolute (K)?
8.5 Convert the thermal conductivity of soil from cal/cm-sec-°C and from watt/m-°C into kcal/m-hr-°C.
8.6 Define the following terms: (a) Fusion, (b) specific heat capacity, (c) diffusivity, (d) emittance, (e) thermal storage capacity, (f) freezing index, (g) albedo, (h) heat of sublimation, (i) thawing index.
8.7 What are the reasons for observing seasonal variations in thermal resistivity in some soils and not in others?
8.8 What limitations must be observed in the operation of buried cable systems to insure stable soil thermal conditions?
8.9 What range of effective thermal resistivity values apply in various parts of the country? How should these values be adjusted for different depths of burial and for changes in climatic conditions?
8.10 Very fine sands and silts are the soil most susceptible to the detrimental effects of frost action. Explain why.

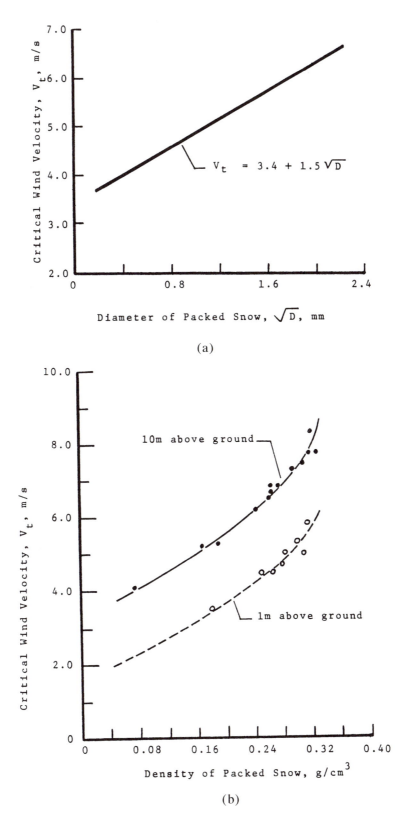

Figure 8.19 Characteristics of snow, (a) Critical wind velocity vs. diameter of packed snow; (b) critical wind velocity vs. density of packed snow. (After CAS [1978].)

CHAPTER 9

Electrical Properties of Soils

9.1 INTRODUCTION

Electrical energy existing in the soil-water system plays an important role as discussed in Section 2.9 and illustrated by Figure 2.3. All other energy such as mechanical, thermal, or magnetic hinges on the characteristics of electric energy. Electrical characteristics in the soil-water system have multiphases due to the following reasons:

1. Both soil and water have inherited electrical characteristics as discussed in Chapters 4 and 5.
2. Electrical energy is closely related to thermal and magnetic energies, and difficult to separate in the natural environment.
3. Electrical-chemical interaction in the soil-water system is sensitive to local environments and, in many cases, these interactions are not clearly understood. The magnitude and behavior of electrical properties in the soil-water system cannot be measured effectively.
4. Some soil-electricity interaction in the environment is also not clearly understood and a theoretical approach sometimes is oversimplified. Differences between theory and experimental observations are significantly different.

Understanding the characteristics of electrical properties of the soil-water system benefits our knowledge for: (1) further understanding the causes of progressive failures, erosion, and landslides; (2) development of more effective techniques for ground improvement; and (3) improvement of more effective approach for soil decontamination by electrokinetic and electromagnetic processes. There are various types of electric sources in the soil-water system including both naturally inherited and man-made sources. In this chapter, discussion focuses on characteristics of man-made electric current applied to fine-grained soils.

9.2 CHARACTERISTICS OF PARTICLES AND ELECTRICITY

9.2.1 General Discussion

The fundamental characteristics of particles have been discussed in Section 2.2. In this section, further discussion is placed on particles interacting in the soil-water-electricity system.

Ordinary atoms are neutral, i.e., they do not carry an electrical charge. However, under certain circumstances atoms can become electrically charged. Such charged atoms are called *ions*. Some elements form positive ions, called *cations,* and some form negative ions, called *anions*. Since atoms are able to form ions, atoms also contain basic units of a negative charge which can be lost to form positive ions or gained to form negative ions. These units of charge are the same units of charge which characterize electricity. Such units of negative charges are known as *electrons*. The positive charge is necessary to neutralize the negative electrons in a neutral atom. Such positive subatomic particle present in the atom is called *proton*. If a subatomic particle carries no electrical charge (neutral), it is known as a *neutron*. Basic types of particles which serve as building units of matter are presented in Table 2.1.

9.2.2 Electric Field and Electric Charge

An electric field is a condition in space set up by electrical charges to which other electrical charges react. The electric charge is the input energy. There are two kinds of charges, positive and negative. The positive charge in ordinary matter is carried by protons, the negative charge by electrons. Charges of the same sign repel each other; charges of opposite sign attract each other. The net charge is equal to the total positive charge minus the total negative charge. According to the *principle of conservation of charge*, the net electric charge in an isolated system always remains constant. The basic unit of electric charge is the electron charge (e$^-$) where:

$$e^- = 1.602 \times 10^{-19} \text{ C (coulomb)} \tag{9.1}$$

The *electron volt* (eV) is a unit of energy. If a particle carrying a charge (e$^-$) accelerates from rest through a potential difference of 1 V, it acquires a kinetic energy of 1 eV, where

$$1 \text{ eV} = 10^{-6} \text{ MeV}$$

$$= 1.60 \times 10^{-19} \text{ J (joule)} \tag{9.2}$$

9.3 MEASURABLE PARAMETERS IN SOIL-WATER-ELECTRICITY SYSTEM

9.3.1 General Discussion

From basic physics, we learn that if one *joule* (J) of work is required to move one coulomb (C) of positive charge from one point to another, the potential differential between the two points is a volt (V). The measure of electric current is the amount of charge that passes a given point per unit of time. If 1 C of charge passes a point in 1 sec, the current is defined to be 1 ampere (A). Table 9.1 summarizes measurable electrical parameters, definitions, computational formulas, and units. Some common terms frequently used in this text are briefly explained as follows.

9.3.2 Electromotive Force and Resistance

(1) Electromotive Force

An *electromotive force* (emf) is defined as any force that can cause electric charges to move and thereby give rise to an electric current. A battery is a typical example of a source of emf. The emf also can be generated from thermal or electromagnetic energy. The unit of emf in general is cm-volt. Also, it must be noted that 'electromotive force' is a misleading term, since emf refers to a quantity related to energy and not to force. It is for this reason,

Table 9.1 Summary of Measurable Electrical Parameters, Their Definitions, Computational Formulas, and Units

Parameter	Definition or formula		Units
Current (I)	charge (Q)/time (t)	I = Q/t	1 (A) = 1 (C/sec)
Voltage (V)	work (W)/charge (Q)	V = W/Q	1 (V) = 1 (J/C)
Resistance (R)	voltage/current	R = V/I	1 ohm = 1Ω
			= 1 volt/ampere
Power (P)	work (W)/time (t)	P = W/t	1 (W) = 1 (J/sec)
		= VI	
		= I^2 R	
Specific resistance or resistivity (ρ)	R = ρ L/A^a		ohm-meter, or Ω–m
Conductance (g)	g = 1/R		mho = 1/ohm = Ω^{-1}
Specific conductance or conductivity (γ)	g = γ A/L^a		mho/m, Ω^{-1}/m
Capacitance (C)^a	charge (Q)/voltage (V)	C = Q/V	1 (F) = 1 coulomb/volt

Note: A = ampere; C = coulomb; J = joule; sec = second; F = farad.

[a] A = cross-sectional area; L = length of the conductor; and C = capacitance (see text).

as a common practice, reference is made only to (emf) or (EMF), or represented by the symbol (E), and not to the misnomer 'electromotive force' (Tilley and Thum, 1971).

(2) Resistance

Property of an electric circuit tending to prevent the flow of current and at the same time causing electric energy to be converted into heat energy is called *resistance* (R). The ohm (Ω) is the practical unit of resistance and is defined as shown in Table 9.1.

9.3.3 Conductance, Capacitor and Capacitance

(1) Conductance

Conductance (g) is a reciprocal of resistance (R) and may be defined as being that property of a circuit or of a material which tends to permit the flow of an electric current (I). The unit of conductance is the reciprocal of *ohm* (1/ohm or (mho)).

(2) Capacitor

A *capacitor* is a system that stores energy in the form of an electric field. A capacitor consists of a pair of parallel metal plates separated by air or other insulating material. The potential difference (V) between the plates of a capacitor is directly proportional to the charge (Q) on either of them, so the ratio (Q/V) is always the same for a particular capacitor.

(3) Capacitance

Capacitance (C) is the ratio of (Q/V). The unit of capacitance is the *farad* (F), where 1 F = 1 coulomb/volt. Since the farad is too large for practical purposes, the microfarad and picofarad are commonly used, where

$$1 \text{ microfarad} = 1 \text{ F} = 10^{-6} \text{ F}$$

$$1 \text{ picofarad} = 1 \text{ F} = 10^{-12} \text{ F} \tag{9.3}$$

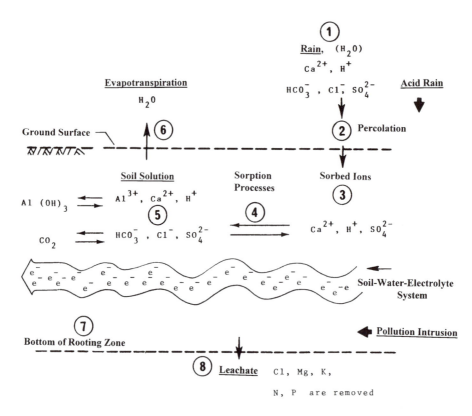

Figure 9.1 Effect of electrochemistry of soil-water interaction in the environment.

9.4 FUNDAMENTALS OF SOIL-ELECTROCHEMISTRY

9.4.1 General Discussion

The basic concept of *electrochemistry* was established by Michael Faraday in 1834, who discovered the fundamental law of electrolysis. However, the application of this concept in engineering is a relatively new interdisciplinary subject. It is a combination of chemistry, physicochemistry, engineering, and their interactions with electricity and environment.

Progress for development of soil-electrochemistry also hinges on the development of other related subjects such as clay mineralogy, ion-exchange capacity of soil, electrode reaction, electrolysis, electrolytes, and redox reaction. Some of these aspects have been discussed in Chapters 4 and 5, and some further discussions will be presented in this chapter.

9.4.2 Soil-Electrochemical Interaction and Its Measurement

Soil-electrochemical interaction is complex. A schematic diagram illustrating the interaction with the environment is presented in Figure 9.1. The chemical formulas described in the figure are obtained from Reuss (1978). Some of the elements indicated in Figure 9.1 have been discussed in previous sections such as rain and acid rain (Section 5.4), infiltration and percolation (Section 7.3), sorbed ions (Section 4.6), sorption processes (Section 6.4), and soil solution (Section 5.3), and others will be explained as follows.

(1) Evapotranspiration

Evapotranspiration process is part of the energy exchange. Eddy motion is the principal mechanism by which water vapor is removed from the vicinity of the plant or evaporating surfaces. In general, evaporation is considered to be a process of mass transport analogous to heat transfer. The moisture movement in the rooting zone and soil-root-leaf interaction is caused by electroviscous and thermo-electric effects as discussed in Sections 5.6 and 5.7.

(2) Leaching

Leaching is caused by rainwater or drainage processes due to some clay minerals removed by a natural process. During the leaching process, common mineral elements such as calcium (Ca), magnesium (Mg), potassium (K), nitrogen (N), and phosphorus (P) are removed. As a majority of calcium is lost, nitrogen is leached from soils in the form of nitrates (NO_3). The quantity of phosphorous lost by drainage is small, but comparatively large quantities of sulfur (S) may be found in drainage waters.

If additional electric current is applied into the soil-water system and/or the polluted water intrudes into the soil-water system as indicated in Figure 9.1, then the complication of soil-water-electricity interaction is significantly increased. Further discussions on this aspect will be discussed in the following sections.

9.4.3 Electrode and Cell

An electrode consists of two pieces of metal which connect to an electron pump (e.g., battery) and are immersed in a solution containing ions; the cations are attracted toward the negative piece of metal and the anions toward the positive metal. In this condition, the pieces of metal are called *electrodes*. The electrode that attracts the cations is called the *cathode* (–) and the electrode that attracts the anions is called the *anode* (+). The arrangement of such a system is referred as a *cell* (Figure 9.2). The process of subjecting a solution to the conditions which will produce electrode reaction is called *electrolysis*.

9.4.4 Electrolysis

The process of subjecting a solution to the condition which will produce electrode reaction is called electrolysis. The fundamental law of electrolysis was discovered by M. Faraday in 1834. He stated his discovery in the form of two laws.

(1) Faraday's Laws

1. First law: "the chemical power of a current of electricity is in direct proportion to the absolute quantity of electricity which passes."
2. Second law: "several substances were placed in succession and decomposed simultaneously by the same electric current... the results were comparable, the tin (Sn), lead (Pb), chlorine (Cl), oxygen (O), and hydrogen (H) evolved being definite in quantity and electrochemical equivalents to each other."
3. Restatement of Faraday's laws (first and second) into one law is as follows (Selwood, 1954): "The number of equivalent weights of any substance deposited or evolved at an electrode is proportional to the quantity of electricity passing through the electrolyte, i.e., substance being electrolyzed."

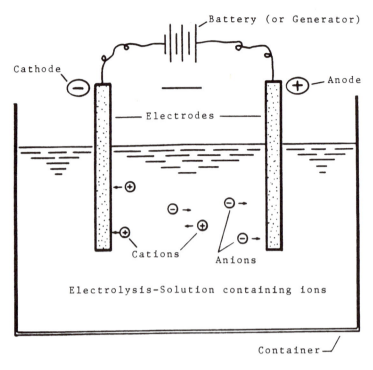

Figure 9.2 Typical setup of electrolysis apparatus.

(2) Faraday Unit

The quantity of electrical charge associated with 1 mol of electrons is called the *faraday*. The relationship between faraday and other units commonly used is

$$1 \text{ faraday} = 1 \text{ mol of electrons}$$
$$= 96{,}500 \text{ coulombs} \qquad (9.4)$$

9.5 ELECTROLYTES AND ELECTRICAL REACTIONS

9.5.1 Characteristics of Electrolytes

Substances that form ions in water solution are called *electrolytes*. Some substances that ionize completely are called *strong* electrolytes. Others ionized only slightly are called *weak* electrolytes. A substance that forms an aqueous solution that does not conduct electricity is called a *nonelectrolyte*. Many molecular substances such as sugar ($C_{12}H_{22}O_{11}$) and alcohol (C_2H_5OH) are nonelectrolytes. Some important electrolytes are discussed as follows.

(1) Salts and Base

Most of the salts are strong electrolytes. For example, $NaCl$, NH_4NO_3, KBr, $CaCl_2$, and Na_2CO_3 are strong electrolytes. Salts such as $AgCl$ do not produce highly conductive solutions because they have a very low solubility. However, the small amount that is in solution is completely dissociated into ions.

All soluble metal hydroxides (OH) are strong electrolytes. For example, $NaOH$, KOH, and $Mg(OH)_2$ are strong electrolytes. But bases such as NH_3 and the amines tend to be weak.

The strongest base that can exist in water is OH^- because any stronger amide ion is an example of such a base. Some acids are strong electrolytes, other are weak. Most acids fall somewhere between the two extremes.

(2) Natural Electrolytes — Acid Rain and Acid Drainage

As discussed in Section 1.6, indications are that acid rain is the major part of soil-water interaction in the environment, because acid rain together with acid drainage are the natural electrolytes (Section 5.4) which will influence the ion-exchange reactions and consequently, will affect the behavior of the soil-water-electrolyte system.

9.5.2 Electrode and Redox Reactions

(1) Electrode Reaction

Electrode reactions are actually chemical reactions involving electron transfer. A reaction in which a species loses or gains electrons at an electrode and is converted to a new species is called an electrode reaction. An electrode reaction only occurs when a simultaneous reaction involving the gain of electrons occurs at the other electrode in the system.

(2) Redox Reaction

All chemical elements can accept or donate electrons. If an element loses electrons by a substance it is called *oxidation* or if it gains electrons it is called *reduction*. Oxidation and reduction always occur together because a substance can only donate electrons if another substance can accept them, and vice versa. A reaction involving oxidation and reduction is called an oxidation-reduction reaction and is also referred to as a *redox reaction*.

(3) Redox Potential

The *redox potential* is an intensity measurement. It measures the availability rather than the quantity of electrons. Because the redox potential is governed by the potential rate of electron donation vs. the rate of electron acceptance, it is also a rough estimate of the amount of electron donors present. Many inorganic soil chemical reactions are biological reactions of carbon, nitrogen, and sulfur which are redox reactions. Most redox reactions also transfer H ions. The significance of (H) vs. (e) in soil chemical reactions depends largely on the presence or absence of oxygen. Oxygen is generally available in most waters for plant roots and microbes.

9.5.3 Characteristics of E_h and pe and Their Relationship

Electrode, electron, reversible and equilibrium potentials, and electron activity are all closely related terms for the equilibrium potential of the electron. Electron potential or electrode potential is generally associated with the term, (E_h), while electron activity is associated with the term (pe).

The E_h is dependent on the pH value and the partial pressures of H and O. In general, the diagram of (E_h) vs. (pH) is useful for evaluation of characteristics of contaminated soil-water systems; because all biological and soil systems contain water, their oxidation-reduction region is the region of water stability (Figure 1.4). The (pe) vs. (pH) diagrams can be used to portray electrochemical equilibria. The (pe) values range from 20 to 78 at highly oxidizing conditions. The relation between (pe) and (E_h) at 25°C has been proposed by Bohn et al. (1985) as:

$$E_h = 0.059 \text{ (pe)} \tag{9.5}$$

In a soil-water system, the organic oxidation and methane fermentation, nitrification, and denitrification reactions are of this type and are mediated by bacteria (Section 5.9). Also, pollutants such as cadmium (Cd), mercury (Hg), arsenic (As) and lead (Pb) are frequently found in polluted soil-water systems. Cadmium is used extensively in the manufacturing of batteries, paints, and plastics. In addition, it is used in manufacturing of various structural elements such as bolts, nuts, and plates. Mercury is one of the most problematic heavy metals found in many soil-water systems. Most of the recent research concerning the toxicity of mercury has involved the methgylated mercury compounds. Figure 9.3 shows the mercury (Hg) and arsenic (As) compounds existing in the soil-water system under various environmental conditions as reflected by E_h values. In examining Figure 9.3(a), when pH changes from 8.00 to 8.75, the E_h value will change from 330 to 420 mV. Other elemental pollutants, such as arsenic (As), exist in soil-water systems and under various pH values produce even more complex forms of the arsenic compounds as reported by Wang (1985).

9.6 ELECTRIC PROPERTIES OF SOIL, WATER, AND ROCK (WHEN ADDITIONAL ELECTRICITY IS APPLIED)

9.6.1 Mechanism of Soil-Electricity Interaction

The purpose of applying additional electric current into the soil mass is to identify and/or characterize the water, soil, and rock properties, because reactions of electricity will reflect their behavior. The interaction between soil and electricity depends on soil particle size, mineral structures, mineral surface conditions, and characteristics of pore fluid, as well as ion-exchange capacity and properties of electrolytes as discussed in Sections 4.6 and 4.8. From Figure 9.4 it is indicated that the ion movement direction, in general, follows the direction of electric current; and the influence area (electric energy field) is related to the magnitude of electrical charge and characteristics of soil-water system. Also, there are two distinct cases in the soil-electricity interaction as:

1. Dry soil condition: When soil is dry, the electric resistivity, in general, is very high, because there is very little interaction between electric charge (input energy) and ions existing in the soil.
2. Moist soil condition: For moist soil, the electricity increases and resistivity decreases significantly due to the moist film around soil particles which serve as a bridge linking the electric charge (e⁻) and the ions existing in the soil.

9.6.2 Properties Measured by Electric Current and Voltage

Measurable electrical parameters such as electric current, charge, resistivity, or conductivity are frequently used for determination of both laboratory and *in situ* soil-water properties. Some typical experimental results based on these parameters for practical use in geotechnical engineering are presented as follows.

(1) Soil-Water Properties

Electric current (I) is the ratio between electric charge (Q) and time (t) and can be used to identify soil-water characteristics. Figure 9.5(a) shows the electric current vs. time for soil, water, and potassium chloride (KCl). In all cases, electric current decreases as time duration

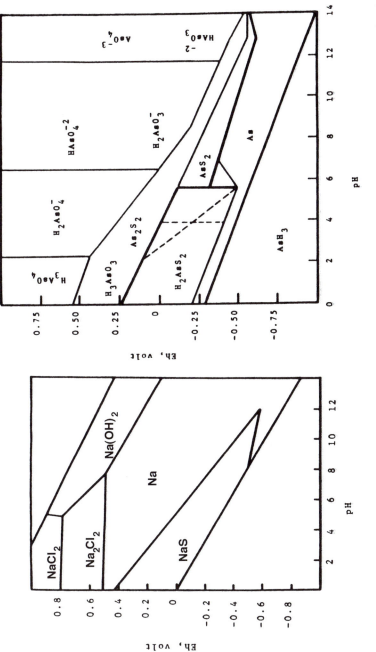

Figure 9.3 Forms of heavy metal existing in soil-water systems under various environmental conditions as indicated by redox reaction (a) Mercury (Hg); (b) arsenic (As). (After Wang, H.T. [1985], *Water Pollution*, Scientific Publishing, Beijing.)

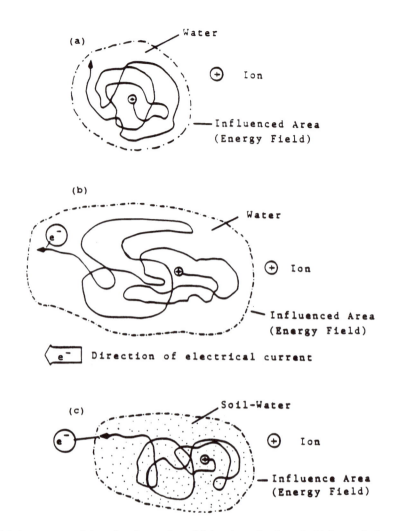

Figure 9.4 Ion movement in soil-water system. (a) Ion in water (no electricity applied — Brownian movement); (b) ion in water when electricity applied; (c) ion in soil-water system when electricity applied.

increases. Figure 9.5(b) shows the current decreases when voltage increases for all cases with various soil and water conditions.

(2) Locating Leak Sources of Geomembrane Liners

Using voltage for locating the leak sources of geomembrane liners was developed by Darilek and Parra (1988) of the Southwest Research Institute, San Antonio, TX. The project was sponsored by U.S. EPA's Hazardous Waste Engineering Research Laboratory.

9.6.3 Electric Charge and Surface Charge Density

(1) Characteristics of Electric Charge in Soil-Water System

The net charge is the summation of the positive and negative charges. In soil, the negative charge is generally larger than the positive charge, except in higher acidity regions, where most of the soil carries a negative charge. Soil cation-exchange capacity exists in most soils under certain pH conditions because cation-exchange capacity is usually around where pH

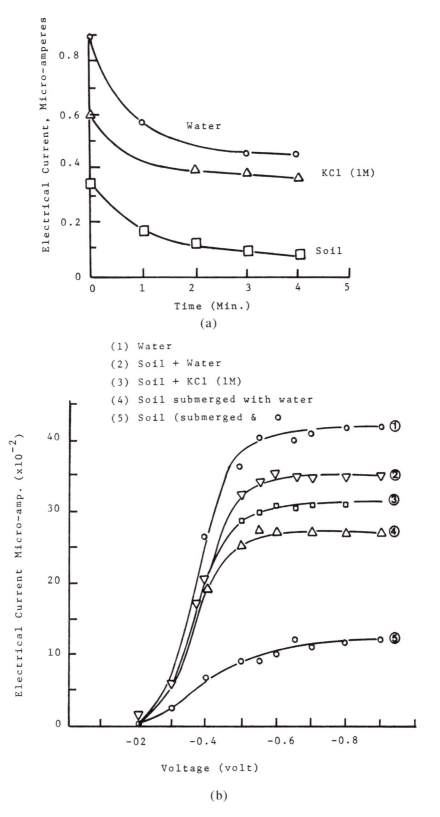

Figure 9.5 Characteristics of soil as measured by electric current. (a) Electric current vs. time; (b) electric current vs. voltage. (After Yu, T.Y. and Chiang, H.Y. [1984], *Electro-Chemical Properties of Soils,* Scientific Publishing, Beijing.)

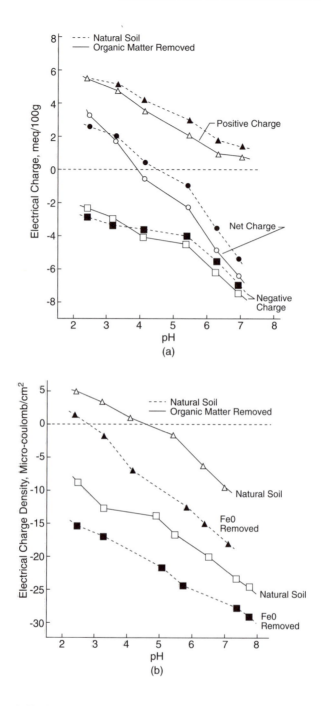

Figure 9.6 Effect of pH value on electrical charge of red laterite clay. (a) Electrical charge; (b) net charge; (c) charge density. (After Yu, T.Y. [1964], *Electro-Chemical Properties of Soils,* Scientific Publishing, Beijing.)

= 7.0. The pH can influence exchangeable negative charges in soil, but would not influence permanent negative charge (Schofield, 1949). In most soil, pH less than 5.5 would not be influenced by pH greatly. Figure 9.6 shows the effect of pH value on electrical charge of red laterite clay from the southwest of China (Yu, 1964).

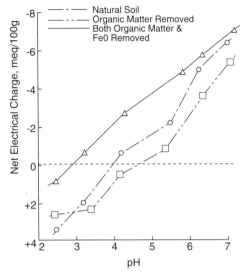

Figure 9.6 (continued)

(2) Effect of Electric Surface Density on Soil-Water System

Table 9.2(a) shows the experimental data on negative electric charge for soil particles larger than 1 mm at pH = 7.0. Table 9.2(b) presents the summary of pH effect on surface electric density of red laterite clay.

9.6.4 Electric Resistivity of Soil, Rock, and Water

(1) General Discussion

Using electric resistivity is a most common method for measuring the characteristics of soil-water systems as well as engineering properties of soil-rock. The applications can be grouped into the following general areas:

1. Environmental effects on soil or water characteristics, such as temperature and moisture content.
2. Mechanical properties of soil, such as unit weight, porosity, Young's modulus, and strength parameters of soil or rock.
3. Frost depth (frozen or unfrozen soils), or location of groundwater level.
4. Locating subsurface materials (such as gravels, sand and soil deposits) as well as sinkholes or other cavities.
5. Locating buried geostructural members such as pipes, piles, as well as hazardous/toxic wastes.
6. Locating leaks in geomembrane liners.

(2) Soil and Rock

Using electric resistivity for measuring soil rock properties requires injecting electric current into the soil (or ground) using a pair of surface electrodes. Resistivity is related to the moisture content and temperature of materials. When moisture content and/or temperature increases, the resistivity decreases. Resistivity also is related to the physical or mechanical properties of soil or rock. A correlation between Young's modulus of rock samples vs. electric

Table 9.2(a) Surface Electric Density of Red Laterite Clay Related to Mineral Composition

| Soil types and locations | pH | | a | Mineral composition |
	H₂O	KCl		
Reddish-brown lateritic soils (Canton)	5.0	4.8	5.2	Kaolinite group
Red clay from granite (Canton)	4.8	4.2	12.0	Kaolinite group
Red clay (Central)	4.8	4.1	22.0	Kaolinite, illite, and mixtures
Loess (Central)	7.0	6.1	40.0	Illite, montmorillonite, and mixtures
Tropical soil (Hainan Island)	7.2	6.3	53.5	Montomorillonite and kaolite groups
Chernozems (Inner Mongolia)	—	—	91.0	Montmorillonite

[a] Negative electric charge, meq/100 g.

Table 9.2(b) Surface Electric Density of Red Laterite Clay Related by pH Value

| Oxisol (4th century) red laterite clay | | Reddish-brown laterite soils (Latosols) | |
pH	Surface density	pH	Surface density
4.9	−13.3	2.4	+4.6
6.5	−19.6	5.4	−1.4
7.3	−22.9	6.3	−6.1
7.7	−24.0	7.0	−9.0

FeO Removed from Natural Soil

pH	Surface density	pH	Surface density
5.1	−21.3	2.4	+1.8
5.7	−24.2	4.2	−6.7
7.3	−27.4	6.3	−14.9
7.7	−28.8	7.1	−17.7

Note: (1) Unit for surface electric density: micro-coulom/cm² and (2) (−) negative electric charge; (+) positive electric charge.

After Yu, T.Y. (1964), *Electro-Chemical Properties of Soils,* Scientific Publishing, Beijing.

resistivity for thermally shocked rock samples is reported by Keller (1974). Figure 9.7 shows the frost depth determination by electric resistance measurements from the AASHO Road Test at Ottawa, IL. Figure 9.7(a) shows the ground temperature vs. electric resistance. Figure 9.7(b) depicts electric resistance vs. depth of subsurface soil layer. The electric resistance can be used to identify whether or not the ground soil is frozen.

(3) Water Quality Measured by Electric Resistivity/Conductance

One common practice for the use of electric resistivity or conductivity is the determination of soil-water salinity and acidity. To determine the electric conductivity, two electrodes of constant geometry, including constant distance of separation, are placed in the solution. When an electric potential is imposed, the amount of current varies directly with the total concentration of dissolved salts. At constant potential, the electric current is inversely proportional to the solution's resistance and can be measured with an electric resistance meter.

Electrical conductance is the reciprocal of electrical resistance as shown in Table 9.1. The measured electrical conductance is the net result of the solution's salt concentration and the electrode geometry. The effects of electrode geometry are embodied in the cell constant,

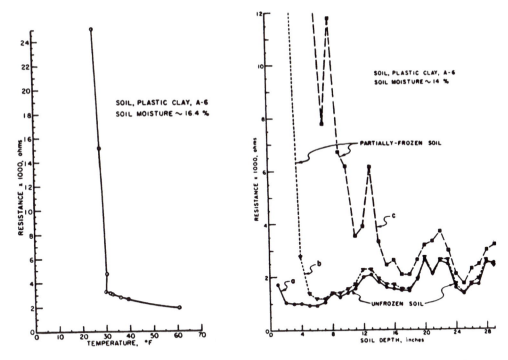

Figure 9.7 Frost depth determination by electric resistance measurements at AASHO Road Test, Ottawa, IL. (a) Temperature vs. electrical resistance and (b) soil depth vs. electrical resistance. (After Carey, W.N., Jr. and Andersland, O.B. [1957], *Highway Res. Abstr.*, v. 27, no. 4.)

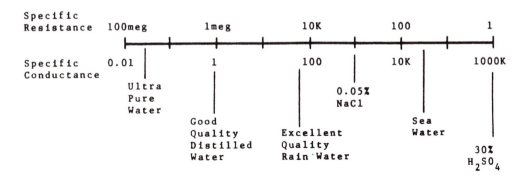

Figure 9.8 Resistivity and conductivity of various types of water. Units: specific resistance — ohm-cm; specific conductance — micromhos/cm (μmhos/cm $\times 10^2$ = s/m).

which is related to the distance between electrodes divided by their effective cross-sectional areas. The cell constant is commonly measured by calibration with potassium chloride (KCl) solutions of known concentration. Water quality classification based on electric resistance or conductance have been proposed by ASTM as shown in Figure 9.8.

9.6.5 Electrical Properties of Contaminated Soil

Engineering properties of contaminated soil-water system are closely related to the electric properties of the same system. In general, it measures the electric surface charge, dielectric constant, and redox reaction. Electric charge vs. pH values have been summarized in Tables 9.2 and 9.3 and Figures 9.5 and 9.7. Notice from these tables and figures that the pH solution in the soil-water system plays a significant role, because contaminated pore fluids will change

Table 9.3   Dielectric Constant and Other Physicochemical Properties of Water and Some Organic Compounds

Compound	Formula	ε	μ	γ	DM
Water	H_2O	80.4	0.894	1.0	1.89
Acetic acid	$C_2H_4O_2$	6.1	1.05	1.28	1.74
Acetone	C_5B_6O	20.7	0.79	0.33	2.90
Aniline	$C_6H_5NH_2$	6.9	1.02	4.40	1.55
Benzene	C_6H_6	2.28	0.88	0.65	0
Carbon tetrachloride	CCl_4	2.24	1.50	0.97	0
Chlorobenzene	C_6H_5Cl	5.71	1.11	0.80	1.69
Cyclohexane	C_6B_{12}	2.02	0.78	1.02	
Dimethyl formanide	$(CH_3)_2HCON$	26.6	0.802	0.950	3.85
Dimethyl sulfoxide	$(CH_2)_2SO$	48.9	1.98	1.1008	3.90
Ethanol	C_2H_6O	24.3	0.79	0.20	1.69
Ethylene glycol	$C_2H_6O_2$	38.66	1.11	21.0	2.28
Heptane	C_7H_{16}	1.0	0.68	0.41	0
Hexane	C_6H_{14}	1.89	0.66	0.33	0.08
Methanol	CH_4O	33.62	0.79	0.54	1.66
Nitrobenzene	$C_6H_5NO_2$	35.74	1.20	2.03	4.22
P-dioxane	$C_4H_8O_2$	2.21	1.44	1.03	0.45
Phenol	C_6H_6O	13.13	1.06	12.7	1.45
Tetra chloroethylene	Cl_2CCl_2	2.30	0.72	1.60	0

Note: ε = dielectric constant (20°C); μ = viscosity, centipoises (20°C); γ = unit weight, g/cm³ (20°C); and DM = dipole moment, debyes.

the ion-exchange reactions and consequently, will influence the electrical properties in the soil-water system. For measuring environmental geotechnical behavior of contaminated soil-water system, the redox reaction as an effective measuring technique is discussed in Sections 5.8 and 9.4. Some test results have been presented in Figure 9.3.

9.7 DIELECTRIC CONSTANT OF SOIL-WATER SYSTEM

9.7.1 Characteristics of the Dielectric Constant

The concept of dielectric constant proposed by Quinke in 1859 as a measure of a material's ability to perform as an insulator is discussed in Section 5.3. Because this constant is a useful measurable parameter for evaluation of contaminated soil-water system, further discussions are needed. Fundamental characteristics, measuring techniques, typical data, and correlation with other soil parameters will be presented in this section.

In a soil-water system, the water molecule possesses an electric dipole moment. Because of this behavior, the water molecules tend to orient themselves in a strong electric field in a manner that the positive poles face the negative electrode and the negative poles face the positive electrode. Also, the negative electron cloud around the oxygen nucleus is deformed or polarized in an electric field. Thus, an induced dipole is formed which adds its moment to that of the permanent dipole. The random kinetic movement of the water molecules tends to counteract the aligning effect of the electric field on the permanent dipoles, while it has practically no effect on the induced dipoles.

The alignment of the dipoles in an electric field, taking place against the disordering effect of the kinetic energy of the molecules, decreases the effective strength of the field as a function of the strength of the dipoles and their concentration per unit volume and of the counteracting kinetic energy. A measure of the bulk effect is the dielectric constant of a substance.

9.7.2 Dielectric Constant of Substance

(1) General Discussion

To illustrate the dielectric constant (ε) of a substance, Equation (9.6) may be used. If a substance is placed between two point charges, c_1 and c_2, located at a distance (r) from each other, it reduces the force between these charges from

$$c_1 \, c_2/r^2 \text{ to } c_1 \, c_2/r^2(\varepsilon) \qquad (9.6)$$

For certain soil-water interactions, the use of the dielectric constant leads to easier description and understanding. Dielectric constant has wide ranges, from 1.0 for heptane to 80 for water. Table 9.3 summarizes dielectric constant and other physicochemical properties of water and some organic compounds.

(2) Dielectric Constant Measurement

Dielectric constant is not directly measured. It is computed from the following equation as:

$$\varepsilon = C_s \, d/A \qquad (9.7)$$

where ε = dielectric constant; C_s = capacitance, F (Section 9.8); d = length of a specimen (= 0.7 cm); A = cross-sectional area of specimen (= 2.85 cm^2); and = 8.85×10^{-14} F/cm.

For a typical soil specimen, where d and A are known (Tuncan, 1992):

$$\varepsilon = 2.775 \, C_s \qquad (9.8)$$

(3) Electrical Dispersion Measurement

Electrical dispersion is the modification of the dielectric constant as proposed by Arulanandan and Smith (1973). The technique is based on the variation with alternating current frequency of the apparent dielectric constant and electric conductivity (mhos/cm) of saturated clays in the radio frequency range (10^6 to 10^8 Hz). This phenomenon is referred to as dielectric dispersion. A typical dispersion curve for a saturated illite clay is shown in Figure 9.9. By increasing the frequency is seen the decrease in the dielectric constant and increase in the conductivity. It has been shown that the major factors influencing the dielectric dispersion of the fine-grained soils in the radio frequency are the compositional properties of the different phases and the heterogeneous nature of the system (Smith and Arulanandan, 1981).

9.7.3 Dielectric Constant of Water and Other Pore Fluids

(1) Dielectric Constant of Water

Based on experimental data obtained from Eisenberg and Kanzmann (1969) cited by Winterkorn (1974), the course of the static dielectric constant of water between 0 and 100°C is well expressed by the following equation:

$$\varepsilon = 87.740 - 0.40008(t) + 9.398 \times 10^{-4}(t^2) - 1410 \times 10^{-6}(t^3) \qquad (9.9)$$

where ε = dielectric constant and t = temperature,°C.

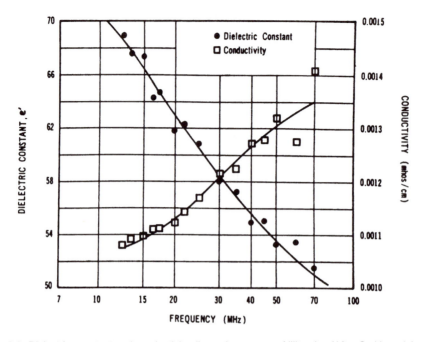

Figure 9.9 Dielectric constant and conductivity dispersion curves of illite clay (After Smith and Arulanan-dan [1981].)

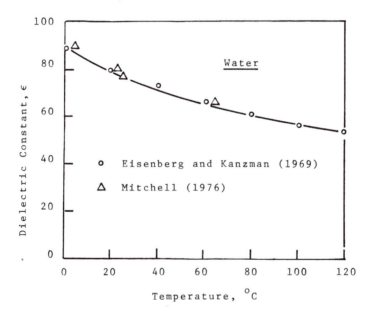

Figure 9.10 Dielectric constant of water vs. temperature.

Figure 9.10 plots dielectric constant vs. temperature for water. The dielectric constant of water is about 80 and is highest among common liquids as indicated in Table 9.3. The high dielectric constant of water is the reason for its great solvent power for strong electrolytes. Thus, the electric attraction forces existing between two ions in a vacuum are reduced to about 1/80 for the same distances in the case of interposed water. This reduction is sufficient to make the dispersive tendency of the kinetic molecular energy prevail.

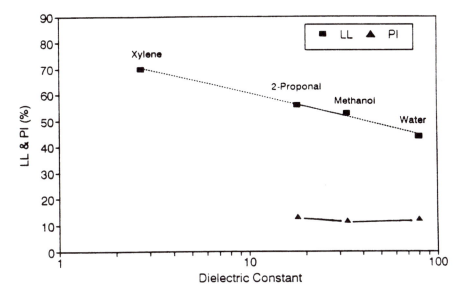

Figure 9.11 Atterberg limit of solvents as a function of dielectric constant. (After Kaya and Fang, [1995].)

(2) Dielectric Constant Versus pH Value

As discussed in Section 5.4 and Figure 5.6, indications are that pH is only measuring the concentration of the H ion in the solution. However, the dielectric constant is measuring both types of ions and their concentrations in the solution. Therefore, for contaminated soil-water system, the dielectric constant is a more effective indication in comparison with pH measurements. Based on limited data obtained from laboratory testing conducted at Lehigh University, indications are that (a) when pH value ranges between 6 to 9, the H ion dominates in the solution, the pH indicator is effective; however, (b) when solution contains multi-ions, then measurement by dielectric constant is more effective. Further testing on these two parameters under various environmental conditions is needed in order to better understand the mechanisms of these two tests.

9.7.4 Effect of Dielectric Constant on Soil Behavior

(1) Soil Constants

The effect of dielectric constant on soil behavior is significant, because the dielectric constant is influenced by both ion concentration and types of ions in the soil-water systems. Acar and Olivieri (1989) show the effect of organic fluids as reflected by dielectric constant on the Atterberg limits of Georgia kaolinite clays in that the dielectric constant of the solvent decreases, the liquid limit (LL) of the soil increases, while plasticity index (PI) remains constant. During the mixing of kaolinite with the solvents, it was observed that soil becomes aggregated by forming small domains.

Kaya and Fang (1995), using the Atterberg limit of solvents as a function of dielectric constant as shown in Figure 9.11, found these observations are totally compatible with the observation of Mesri and Olson (1971) and Fernandez and Quigley (1985), in that the dielectric constant affects the flocculation and channelization in the soil structure. Decrease in dielectric constant causes aggregation and channelization within a soil mass, thus causing an increase in the flow area.

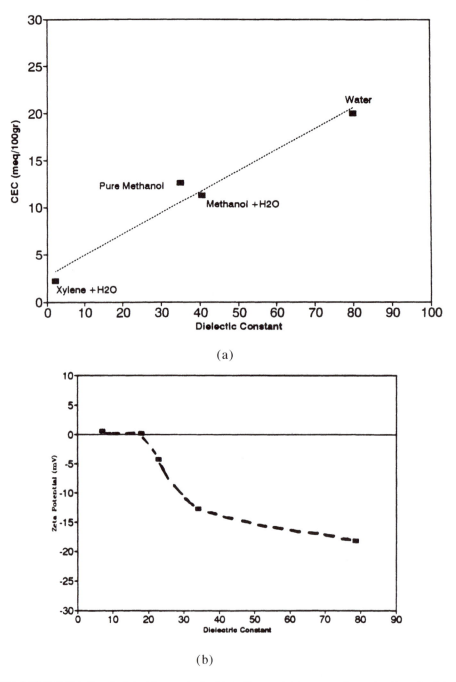

Figure 9.12 Relationship of dielectric constant with cation exchange capacity and zeta potential. (a) Cation exchange capacity; (b) Zeta potential. (After Kaya and Fang, [1996].)

(2) *Dielectric Constant Related to CEC and Zeta Potential*

The cation exchange capacity (CEC) of kaolinite soil is determined by sodium extraction. To completely dissolve the sodium acetate in the solvents, all solvents were prepared with 25% water (75% solvent and 25% water). The result of CEC of kaolinite is presented in Figure 9.12(a). From Figure 9.12(a), it is obvious that CEC linearly increases with an increase in the dielectric constant of the pore fluid.

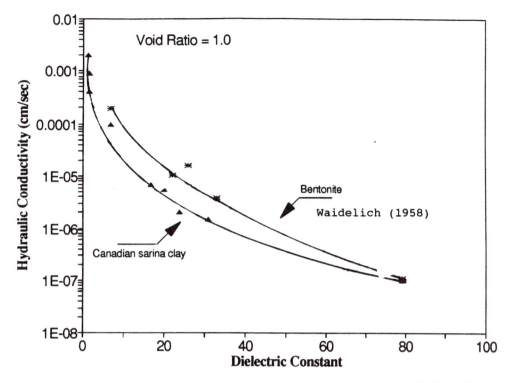

Figure 9.13 Dielectric constant vs. hydraulic conductivity. (After Fernandez and Quigley [1985] and Kaya and Fang [1996].)

Zeta (ζ) potential of soil is an important parameter in the electric energy field. Detailed discussion on this aspect will be presented in Section 9.9.4. However, in this section, the relationship between dielectric constant and zeta potential is discussed. Zeta potential vs. dielectric constant is presented in Figure 9.12(b). In examining Figure 9.12(b), it can be seen that the dielectric constant of the pore fluid decreases and reaches zero value within the experimental error range which can be attributed to the fact that surface charge density of the soil particles decreases as proton surface charge density gets smaller and smaller.

(3) Effect of Dielectric Constant on Conductivity and Compressibility

For fine-grained soil, there is no definite relationship between hydraulic conductivity and void ratio (Figure 7.6) which is especially true for contaminated soils. Using the same data from Figure 7.4, it is found that there is a good relationship between hydraulic conductivity with dielectric constant as shown in Figure 9.13. Discussions on identification of contaminated soil by dielectric constant and conductivity are given by Kaya and Fang (1977). Based on Waidelich's data (1958), dielectric constant vs. compression index (c_c) and swell index (c_s) of three clay minerals with seven types of pore fluids also can be obtained.

(4) Effect of Dielectric Constant on Swelling

Basu (1972) presents a relationship between swell and dielectric constant of wet soil cluster, indicating that a good correlation between increasing dielectric constant of the wet soil cluster and increasing percentage of swell was obtained for these various soil types. It is concluded that dielectric constant is a measure of a soil's water absorption characteristics and can be used for the evaluation of swell potential in soils.

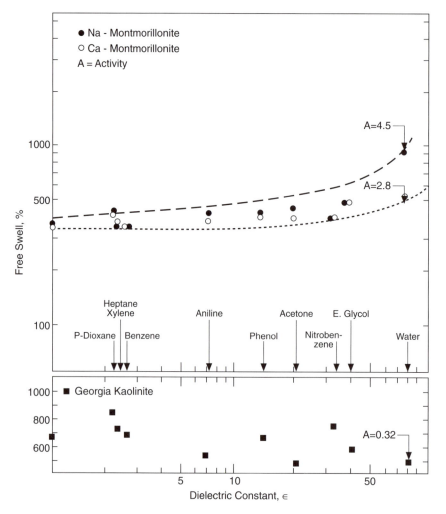

Figure 9.14 The effect of organic fluids as reflected on dielectric constant on free swell of montmorillonite clay. (After Acar, Y.B. and Olivieri, I. [1989], HRB Annual Meeting, 24p.)

Brown and Thomas (1987) have reported on the swelling of kaolinite, illite, and smectite. They found that there is a significant relationship between the dielectric constant of the solvent and swelling. An increase in swelling with a decrease in the dielectric constant was attributed to an increase in the flocculation of the clay particles with a decrease in the dielectric constant. Acar and Olivieri (1989) reported the effect of organic fluids as reflected on dielectric constant on free swell of montmorillonite clay (Figure 9.14). Also, significant relationships among various organic fluids have been found.

9.8 ELECTRIC CONDUCTIVITY AND ELECTROKINETIC PHENOMENA

9.8.1 General Discussion

Fluid flow through a soil-water system caused by electric potential is called *electric conductivity*. The behavior of such movement of fluid under electric potential is called *electrokinetic phenomena*. The phenomena can be divided into two major groups as shown in Table 9.4:

Table 9.4 Classification of Electrokinetic Phenomena of Soil Particles in Clay-Water Micelle

Electrokinetic Phenomena Between Soil Particles

Electro-osmosis — flow movement between soil particles due to electric potential; only flow (such as water) is moving, the soil (solid) particle is not moving (see Figure 9.15a)

Electrokinetic Phenomena in Clay Suspension

Physical (or mechanical) behavior of particle
  Particle movement in clay suspension due to external force (emf) acts on particle related to the physical or mechanical behavior of particle (see Figure 9.15b); there are three possible phenomena
  Streaming potential — due to hydraulic gradient or due to difference in potential (Δp)
  Sedimentation potential — due to gravity settling
  Uplift potential — due to capillarity action
Physicochemical behavior of particle
  Electrophoresis — charged particles in the clay suspension are attracted or repelled from one another; negatively charged particles move toward the anode (see Figure 9.15c)

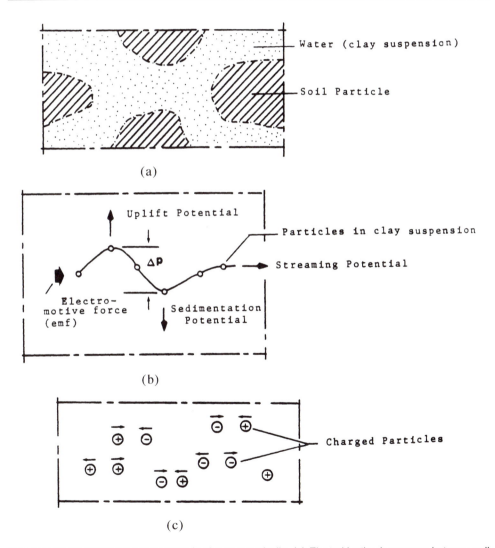

Figure 9.15 Electrokinetic phenomena in clay-water micelle. (a) Electrokinetic phenomena between soil particles (electro-osmosis); (b) physical (mechanical) behavior of soil particles; (c) physicochemical behavior of soil particles (electrophoresis).

1. Phenomena between Soil Particles: In fluid flow through soil particles, only the water moves; the soil particles remain stagnant. This phenomenon is referred to as *electroosmosis* as illustrated in Figure 9.15(a).
2. Phenomena in Clay Suspension: Soil particles move in a suspension between other particles as described in Figures 9.15(b) and 9.15(c). Description in Figure 9.15(c) is referred to as *electrophoresis* and it will be further discussed in Section 9.12.

The movement of water through capillaries and pores as a result of the application of electric potentials was first described by Reuss in 1808. He found that a direct electric current could move water through porous clay and glass diaphragms toward the negative pole. The classical theory to explain this phenomenon is due to Helmholtz in 1879 and was simplified by Smoluchowski in 1903. It was established originally for the simple condition of flow in a glass capillary and then developed in a way analogous to that leading from the Poiseuille equation to a Darcy-type equation together with assumptions discussed in the following section.

9.8.2 Assumptions and Equation of Helmholtz-Smoluchowski Theory

1. The liquid carries a charge opposite to that of the rigid wall and the two charge systems form an electric double layer (Section 4.9) along the wall.
2. The thickness of the electric double layer (Figure 4.15) is extremely, but not infinitesimally, small.
3. The layer of liquid molecules in contact with the wall is immovable, while the rest of the molecules in the liquid within the electric double layer are movable.
4. Within the electric double layer, laminar flow occurs whose velocity rises from zero next to the solid wall to a maximum at the center of the countercharge layer in the liquid. Hence, the velocity remains constant for the remainder of the cross section.
5. The externally applied electric potential acts on the charges of the double layer.
6. The wall is an insulator and the contained liquid possesses electric conductivity.

Analysis of the validity of the Helmholtz-Smoluchowski theory of the electric double layer is given by Gouy (1910), Stern (1924), Bolt (1955), and Gray and Mitchell (1967). The basic equation is presented as:

$$\zeta = \frac{4\pi\sigma\,d}{\varepsilon}$$ (9.10)

where ζ = zeta potential; σ = surface charge density; d = displacement distance; and ε = dielectric constant.

It should be noted that Equation (9.10) states that zeta potential (ζ) is proportional to the surface charge density (σ) of the particle.

9.8.3 Discussions of Helmholtz-Smoluchowski Theory

(1) Limitations of Helmholtz-Smoluchowski Theory

1. The Helmholtz-Smoluchowski theory is based on an electric double-layer thickness of colloid chemistry. It holds quite well for pores that are not too small and for walls whose electric double layer is not too diffuse (Winterkorn 1963).
2. The theory can be applied only to the narrow range of soil types such as undisturbed or remolded silty clay (Tschebotarioff, 1951 and Casagrande, 1952).
3. The theory cannot be applied to well-disturbed clay soil commonly used for the liners in hazardous and toxic waste control systems.

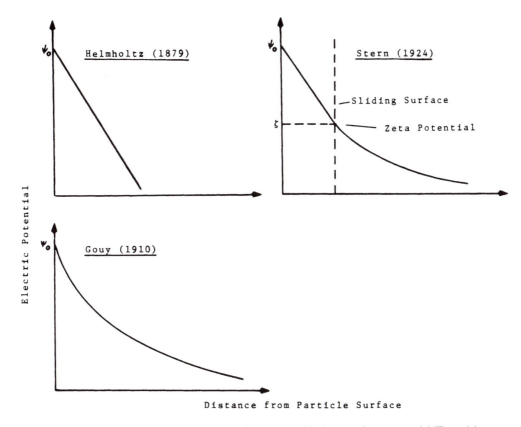

Figure 9.16 Comparisons of characteristics of electric double-layer and zeta potential (ζ) models.

4. It also cannot be used for heavy organic content soils or with animal and vegetable origin, such as the diverse material found in landfill areas (Fang, 1995b).

(2) Modifications and the Gouy-Chapman Theory

Further developments on Helmholtz-Smoluchowski theory are provided for by many investigators. Major modifications focus on the relationship between electric potential (ψ_c) and distance from particle surface (Figure 4.12). These modifications by three investigators are summarized in Figure 9.16.

There are numerous theoretical equations based on assumptions described in Figure 9.16 to describe the clay behavior of the clay-water system; the Gouy-Chapman theory (1910 and 1913) is most commonly used in geotechnical engineering. The Gouy-Chapman model tells us that the tendency toward flocculation or aggregation is usually increased by a decrease in the double-layer thickness corresponding to a decrease in the interparticle repulsion force. Conversely, the tendency toward dispersion is increased as the double-layer thickness increases. Hence, if the double-layer thickness is increased, the electrical repulsion between particles is increased, which causes the particles to disperse. The tendency toward dispersion is caused therefore by an increase in the double-layer thickness.

The Gouy-Chapman theory of ion distribution was incorporated into soil mechanics from colloidal chemistry to help explain and understand the behavior of clay-water systems. As discussed in Section 4.9 and Figure 4.12, a clay mineral has a net negative charge on the colloidal surface. This net negative charge causes the attraction of positive ions or cations in solution. Further information on double-layer thickness will be discussed in the following section.

9.8.4 Electric-Double-Layer Thickness and Zeta Potentials

(1) Electric-Double-Layer Thickness

Some outstanding research was done by Quinke between 1859 and 1861. His significant contributions to electrical properties of soil include: streaming potential, electrical double-layer concept, and dielectric constant. Two other topics are discussed in separate sections. However, the electric double-layer concept will be further discussed herein.

Based on laboratory experiments, Quinke proposed the concept of an electrical double layer. In a system containing a solid in a liquid (suspension), at the interface there are two electrical layers of opposite signs, one forming a coat adhering to the solid and the other contained in the fluid. For an idealized soil particle and its associated ions, if an external difference in potential is applied, it can be observed that soil and a thin layer of attracted cations, while in the diffuse system of counterions and the water associated with it, will move toward the cathode. This system is referred to as the diffuse double layer. In simpler terms, the double layer consists of the clay particles, adsorbed cations, and water molecules in one layer, while the other layer is the diffuse swarm of counterions. The theory explaining this system is called diffuse double-layer theory. The double-layer theory neglects the effects of the potential energy in the oriented water molecules that surround the clay particles as pointed out by Low (1951). However, this theory is useful to explain some basic phenomena in a clay-water-electrolyte system. The equation for the double-layer thickness based on Gouy-Chapman theory is given as (van Olphen, 1977):

$$t = \sqrt{\frac{\varepsilon K T}{8\pi\, n e^2\, v^2}} \tag{9.11}$$

where t = double-layer thickness; ε = dielectric constant; K = Boltzman constant; T = temperature; n = electrolyte concentration; e = elementary charge; and v = ionic valence.

From Equation (9.11), Lambe (1958) has presented a list of variables in the soil water system which affect colloidal stability. These are (1) electrolyte concentration, (2) ionic valence, (3) dielectric constant, (4) temperature, (5) size of hydrated ion, (6) pH, and (7) anion adsorption. Based on the double-layer thickness and its effect on the structure of clay minerals in combination, various particle associations are defined. It states that flocculated or aggregated structures are those structures in which the particles tend to come together. Dispersed or deflocculated structures are those structures in which the clay particles tend to move apart.

Based on the Gouy-Chapman model (Equation 9.11), the effect of pore fluid properties upon the double-layer thickness has been discussed in Section 4.9 and Table 4.3. However, some additional information on pore fluids related to the thickness of double layer are shown in Figs.9.17 and 9.18. In examining Figure 9.17, two pore fluids are used and, in both cases, the concentration of pore fluids increases as thickness of double layer decreases. Figure 9.18 shows the electric potential (ψ) vs. thickness of double layer with three pore fluids as reflected on dielectric constant (ε). A more detailed definition of these structures as given by Van Olphen (1977) and relating to the contaminated soil-water system is given by Evans (1991).

(2) Zeta (ζ) Potential

The zeta (ζ) potential is the electric potential developed at solid-liquid interface in response to movement of colloidal particles as shown in Figure 4.12. Under the influence of

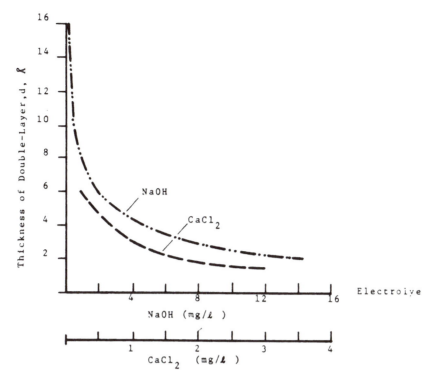

Figure 9.17 Relationship between thickness of double layer and types and concentration of pore fluid (electrolyte). (After Marshall [1949].)

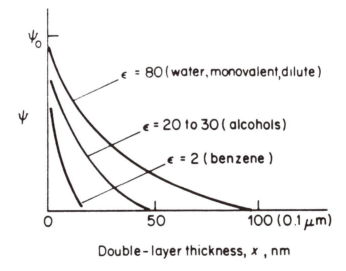

Double-layer thickness, x, nm

Figure 9.18 Effect of pore fluid as reflected by dielectric constant on the double-layer thickness. (After Fernandez and Quigley [1985].)

an applied potential, the particle and a fixed film containing the ions between the particle surface and the boundary line (dotted line) move toward the negative electrode. The thickness of the double-layer affects the magnitude of the potential. When electrolyte concentration increases, the potential decreases. In general, it has been shown that the stability of colloidal

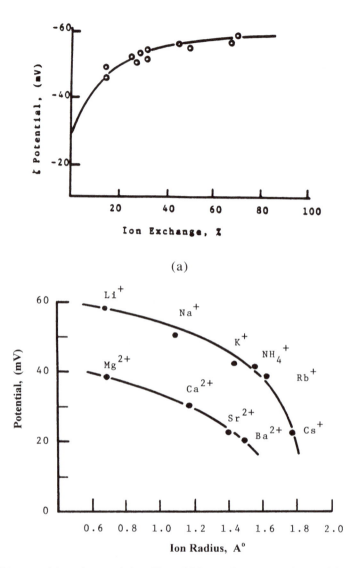

Figure 9.19 Zeta potential vs. characteristics of ions. (a) Ion-exchange capacity and (b) ion radius. (After Jenny and Reitemeier [1935].)

particles is a function of the potential. In Section 9.9.2 various model studies have been used to illustrate analytical solutions of zeta potential. Factors influencing zeta potential include ion types, temperature, adsorption, particle size, and shape of particles.

Analytical solution for zeta potential can be computed from Equation (9.10). It also can be measured by experimental techniques. A detailed discussion on measurements of zeta potential can be obtained from a standard physicochemical textbook. Numerous factors affect zeta potentials. Some relatively important factors are presented as shown in Figures 9.19 to 9.21. Figure 9.19 shows the effects of the percentage of ion exchange and the radius of ion on the zeta potential. Figure 9.20 shows the pH effects on zeta potential on four basic clay minerals. Figure 9.21 shows the effect of pH on one soil with various environmental conditions.

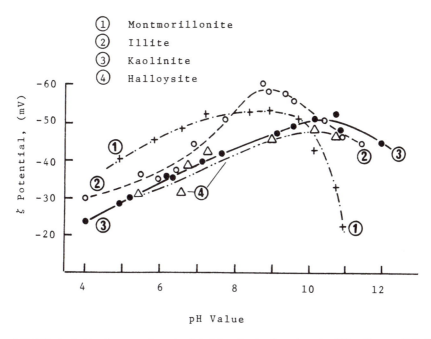

Figure 9.20 Effect of pH value on zeta potential of four basic clay minerals. (After Bergna, H.E. [1950], Trans. 4th Int. Congress of Soil Society, v. 3, pp. 75–80.)

9.9 GROUND IMPROVEMENT BY ELECTROKINETIC PROCESS

9.9.1 Characteristics and Mechanisms

The limitations for using electrokinetic process for dewatering (drainage) have been discussed by Tscheborarioff (1951). It indicates that only a narrow range of soil particle sizes can be used effectively, as shown in area [2] of Figure 9.22. Also in Figure 9.22, in area [1], the flow is controlled by multimedia energy fields, not by electric energy alone. In area [3], the mechanical energy (Darcy's Law) dominates the flow movement. In other words, additional electric energy is not necessary. As indicated in the figure, only silt or silty clays are suitable for electrokinetic process.

9.9.2 Dewatering and Soil Stabilization

The mechanisms of the electrokinetic process used for dewatering and decontamination are different as illustrated in Figure 9.23. Figure 9.23(a) is for the dewatering process. When fluid (water) is removed from soil voids by the electric current, then the voids between soil particles will decrease. Consequently, the density of the soil is increased. Figure 9.23(b) is the mechanism for soil decontamination which will be discussed in Section 9.11. Also, the electrochemical process for soil stabilization will be discussed in Section 9.12.

9.9.3 Factors Affecting the Electrokinetic Process

As indicated in Figure 9.22, the electrokinetic process is limited within the narrow ranges of soil particles and soil types. In addition to soil particle size, there are other factors such

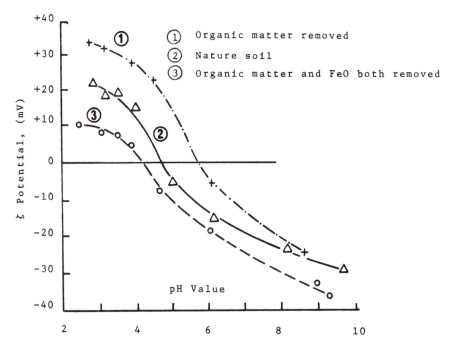

Figure 9.21 Effect of pH value on zeta potential of laterite soil with various environmental conditions. (After Chiang and Shen [1962]; cited by Yu and Chiang [1984].)

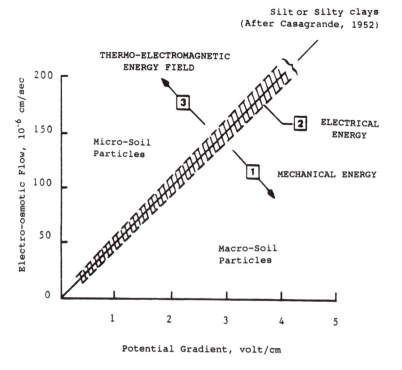

Figure 9.22 The classical theory assumptions are limited to soil within a narrow range of soil particles.

as electrode materials, ion-exchange capacity, ion concentration, etc. as reported by Lockhart (1983), as well as the rate of electric flow, voltage, direction of flow, and types of pore fluid as presented by Khan et al. (1989).

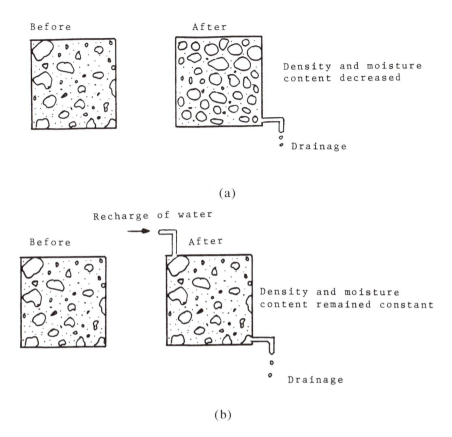

Figure 9.23 Mechanisms of dewatering and decontamination by electrokinetic process. (a) Dewatering (drainage) process and (b) soil decontamination process.

In addition to dewatering by electrokinetic process, other applications including soil stabilization of unstable saturated silt, loess, quick clay, and saline-alkali soils can be accomplished by the same process. Electrokinetic process often is tied-in with electrochemical method which will be discussed in Section 9.13. Furthermore, this process has been applied to various foundations such as pile foundation, reinforced earth, and soil anchor for the purposes of increased density or decreased moisture content around the foundation structure. A state-of-the-art review on soil stabilization and dewatering is given by Casagrande (1983). Some additional information on electrokinetic treatment applications in environmental geotechnical engineering is given by Zyl and Shackelford (1988).

9.10 SOIL DECONTAMINATION BY ELECTROKINETIC PROCESS

9.10.1 General Discussion

Methods for cleaning up hazardous/toxic waste sites have changed since 1980 when the Comprehensive Environmental Response, Compensation and Liability Act (CERCLA), or Superfund, was enacted. Early remedial action for contaminated soil consisted primarily of excavation and removal of the contaminated soil from the site and disposal at a landfill. More emphasis is now placed on *in situ* treatment of soil. The electrokinetic decontamination process is commonly used.

Quinke, in 1859, was probably the first one to study a decontamination process by use of an electrokinetic process. Four types of pore fluids were used for his experiment including:

water, acid, salt, and turpentine oil. He found water was always directed toward the cathode (−). However, turpentine oil flow was found to be directed toward the anode (+). From these simple observations, he concluded that the soil particles were electrically charged and carried negative charges in the case of water and positive charges in the case of turpentine oil.

9.10.2 Mechanism of Soil-Decontamination

As illustrated in Figure 9.23(b), the mechanism between dewatering and decontamination by electrokinetic process is different. To remove unwanted contaminated water or pollutants trapped between soil particles, the spaces between soil particles must remain open. In many cases, due to swelling, shrinkage, air bubbles, soil-chemistry, soil bacteria, and other soil-environmental interactions as illustrated in Figure 9.1, the characteristics between soil particles are unpredictable. A recharge of a nonpolluted water system as shown in Figure 9.23(b) should be installed during the electrokinetic soil decontamination process.

Numerous factors affect the soil decontamination by electrokinetic process. One of the major factors is the electrochemical properties of soil, as illustrated in Figure 9.1. Table 9.5 shows the significant differences between leachate and effluent. The major cause is due to electro-osmosis chemistry and water quality, as reported by Segall et al. (1980). Behavior of some of these soil-electrochemical interactions is not clearly understood. In many cases, the electrokinetic process is an effective method for soil decontamination, and in other cases it is not. (Further discussions are presented by Segall and Bruell (1992) and Yeung and Datla (1995).)

9.11 ELECTROPHORESIS AND ELECTROMIGRATION

9.11.1 General Discussion

Electrophoresis is a part of the electrokinetic phenomena as discussed in Section 9.2 and illustrated in Table 9.4 and Figure 9.15(c). In the clay-water micelle, most of the soil particles are generally electrically charged and they will migrate toward the electrode of opposite charge when placed in an electric field. This migration phenomenon, illustrated in Figure 9.15(c), is called *electrophoresis* or *electromigration*. This value can be computed from Equation (9.10) or measured in laboratory conditions. The common unit for this is the micro/sec/cm. In recent years, many technical instruments for measuring electrophoresis have been developed, such as capillary electrophoresis (CE) reported by Hewlett-Packard (1995).

9.11.2 Experimental Measurements

There are numerous laboratory measuring techniques and apparatuses available. Ultraviolet (UV) and ultraviolet/visible (UV/Vis) absorbance detection currently is the most commonly used detection technique in capillary electrophoresis (CE). Using a diode array detector, absorption at a large number of wave lengths can be measured simultaneously, giving spectral information. Since the observed spectrum is characteristic for any analysis, this information can be used for identification of unknown compounds. Multiwave length detection and peak identification are capabilities that make the diode array detector a good tool and an ideal detector for CE.

Effect of various ions on Putnam soil is shown in Figure 9.24. In examining Figure 9.24, the characteristics of ion-exchange capacity play a significant role. Effects of electrophores

Table 9.5   Comparison of Water Quality Before and After
            Electrokinetic Process Is Applied

Characteristics	Before Leachate	After Effluent	
Total dissolved solids	29,000	40,500	
pH	7.8	13.4	←
Heavy metals			
Mercury	<0.001	0.500	
Cadmium	<0.100	<0.200	
Nickel	<0.500	<0.500	
Chromium	<0.050	0.100	
Copper	<0.100	0.760	
Zinc	0.220	6.2	
Lead	0.250	2.2	
Arsenic	0.030	4.5	
Total hardness, as $CaCO_3$	4,270	59 ⎫	
Calcium	318	120 ⎬	
Magnesium	1,690	14 ⎭	
Manganese	3.0	0.11	
Iron	0.2	20	
Sodium	6,150	16,300	
Potassium	350	510	
Alkalinity, as $CaCO_3$	960	33,780 ⎫	
Hydroxide	0	5,950 ⎪	
Carbonate	0	9,790 ⎬	
Bicarbonate	1,171	0 ⎭	
Chlorides	13,670	1,120	
Sulfates	640	820	
Sulfides	<0.05	<0.05	
Total phosphorus	0.23	20	
Nitrates, as N	2.5	<0.1	
Nitrites, as N	<0.1	<0.1	
Organic nitrogen, as N	4	15	
Ammonia nitrogen, as N	67	128	
Total organic carbon	30	2,000	←
Redox potential, in millivolts	−58	−395	

Note: Volatile loss on ignition (600°C) of dredged material is
       8.75%. (Data from Segall et al. (1980)

for various ionic radii have been reported by Millar and Turk (1943). Typical data for various soil types and clay minerals are summarized in Table 9.6.

9.12 ELECTROCHEMICAL PROCESS

9.12.1 Fundamentals of Electrochemical Process

The *electrochemical* method is an extension of the electro-osmosis process by introducing additional ions into the existing soil-water system for the purpose of increasing the ion-exchange activities in the system. The increasing ion-exchange activities will serve two purposes as discussed in Section 9.2: (1) improve the ground stability, and (2) speed up removal of unwanted water trapped in the soil-water system.

As discussed in Sections 4.8 and 5.4, the ion activity in the soil-water system depends on the amount and types of ions in the existing soil. If insufficient ions exist in the ground soil, additional ions such as aluminum (Al^{3+}), calcium (Ca^{2+}), magnesium (Mg^{2+}), etc. may

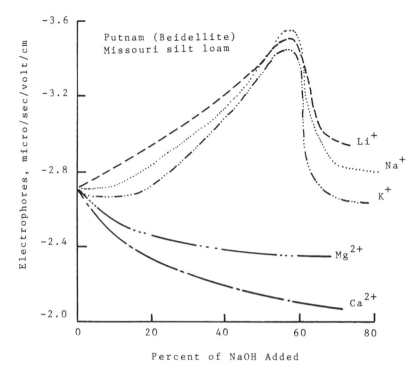

Figure 9.24 Effects of NaOH added on electrophores of Putnam soil. (After Baver and Hall [1937].)

Table 9.6   Electrophores for Various Types of Soil and Clay Minerals

Ion type/ Soil type	Electrophores (micro/sec/volt/cm)							Ref.
	H	Li	Na	K	Mg	Ca	Ba	
Sharkey	3.20		3.80			1.90		Mattson (1928)
Bentonite	2.75		4.30	3.80				Mattson (1929)
Cecil		2.40	2.60	2.30	2.20			Anderson (1929)
Fallon		4.30	3.80	2.10	2.10			Anderson (1929)
Putnam	2.70		3.60	3.50	2.30	2.10		Baver (1929)
Decomposed matter in soil	1.04	1.42	1.42	1.33		0.66	0.44	Baver and Hall (1937)
		3.45	3.31		3.18	3.27		Millar and Turk (1943)
Hayllosite (<2 μm)	2.50		3.50					Beavers and Marshall (1950)
Illite (<2 μm)	3.30		4.00					Beavers and Marshall (1950)

be added into the soil. Electrochemical treatment of the saturated unstable soil deposits brings more improvement in the physical properties than that of electro-osmosis alone. The principal changes affected by this treatment are

- Decrease in moisture content
- Increase in unit weight or density of soil
- Decrease in hydraulic conductivity
- Increase in cohesion and shear strength

9.12.2 U.S. Bureau of Reclamation (USBR) Method

To achieve a successful and economical application of electro-osmotic or electrochemical treatment for ground improvement of the fine-grained saturated soil, the following field and laboratory determinations should be made as suggested by USBR, Earth Laboratory Report No. EM-512.

(1) Approaches

1. Obtain a representatively undisturbed sample (20 to 25 kg) of the material to be treated.
2. Determine *in situ* density and moisture content.
3. Determine physical and physicochemical properties of soil and porewater prior to and after electrical or electrochemical treatment, i.e., determine density, moisture content, porosity, Atterberg limits, ion-exchange capacity, etc. of the original (untreated) and treated soil.
4. Use petrographic study using X-ray diffraction (Section 4.4) and microscopic analysis to identify the principal constituents of the soil and to determine the proportions of clay-type (kaolinite, illite, or montmorillonite) materials.
5. Determine the relationship between voltage and hydraulic head or gradient and the coefficient of osmotic permeability.

(2) Laboratory Condition: Computation of Osmotic Permeability and Specific Resistance

(a) Osmotic Permeability

$$K_e = q_e \ L/E \ \text{cm/sec/volt/cm} \tag{9.12}$$

where K_e = osmotic permeability; $q_e = Q_e/tA$; = unit flow/unit time per unit area of specimen; Q_e = volume of expelled liquid from soil at zero hydraulic gradient by known (emf) at observed time; L = length of laboratory specimen, cm; E = electromotive force (emf), volts; t = observed time, sec; and A = cross-section area of specimen, cm².

(b) Specific Resistance (ρ)

$$\rho = EA/IL \ \text{ohm cm} \tag{9.13}$$

where ρ = specific resistance of the saturated soil; E = voltage; A = area, cm²; I = amperage, and L = length of specimen, cm.

(3) In situ Conditions

Application of the laboratory test results (Step [2]) for selected field resistance (R) and volume of effluence (Q) as follows.

(a) Field Resistance (R)

$$R = \frac{\rho}{\pi \ell} \ 2.303 \ \log \frac{d}{r} \ \text{ohms} \tag{9.14}$$

$$I = \frac{\pi E \ell}{2.303 \ \rho \ \log d/r} \text{amperes} \tag{9.15}$$

$$E = RI \ \text{volts} \tag{9.16}$$

(b) Volume of Effluent (Q)

$$Q = \frac{2\pi K_e \ell H_e}{2.303 \ \log d/r} \ \text{cm}^3 \tag{9.17}$$

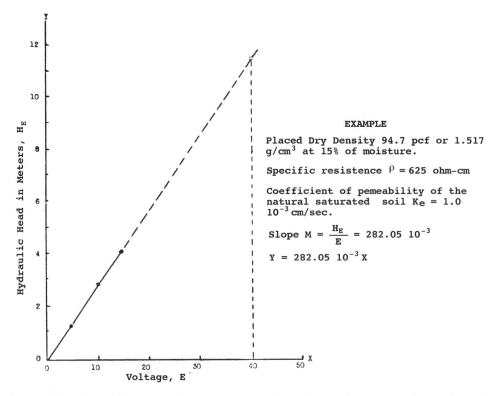

Figure 9.25 Relationship between voltage and developed hydraulic head for a given coefficient of osmotic permeability. (After Karpoff, K.P. [1976], *New Horizons in Construction Materials,* Envo Publishing, Bethlehem, PA pp. 265–273.)

where ρ = specific resistance, ohm cm; ℓ = length of electrodes, cm; d = distance between electrodes, cm; r = radius of electrode, cm; K_e = coefficient of permeability saturated soil, cm/sec; and H_e = hydraulic head in cm developed by selected voltage obtained from laboratory test relationship curve between voltages and hydraulic heads developed by these voltages (Figure 9.25).

Whenever conventional drainage methods (Section 7.3) are ineffective in stabilizing a soil, the electrochemical process may be used to change the principal properties of the soil. This can be done by using a perforated aluminum anode and iron drainage pipe as cathode with the addition of about 2.5% $CaCl_2$ or 5% $AlCl_3$ solution through the anode and with a possible variation in power from 5 to 50 kW/hr/m³.

Electro-osmotic treatment should be applied to soils in saturated conditions. The osmotic effect is in direct proportion to the concentrated pore water volume. The electric current passing through unsaturated soil remains constant until the soil dries out at the anode, after which the rate of expulsion of 'water' rapidly decreases.

A review of dewatering and soil stabilization by electrokinetic process is given by Casagrande (1983). In situ dewatering of contaminated water by electrokinetic process is reported by Shang and Lo (1996).

9.13 MULTIENERGY EFFECT ON SOIL-WATER SYSTEM

9.13.1 General Discussion

As discussed in Sections 2.9 and 9.4 the electrical energy is the center of all energy fields. Therefore, soil-water interaction is under the multienergy effect including thermal, electric, and magnetic energies (Figure 2.3). Linkage between various energy fields such as electro-viscous, thermoelectric, and electromagnetic illustrated in Figure 9.26 will be discussed in this section.

9.13.2 Electroviscous Effect

Electroviscous effect in the soil-water system is a multimedia energy field problem. When water flows under a hydraulic gradient (mechanical energy) into soil voids between negatively charged mineral surfaces (thermoelectric energy fields) as shown in Figure 9.26(a), the exchangeable cations are swept downstream and a streaming potential develops. This is a phenomenon for which Elton (1948) postulated that the streaming potential exerts an electrical retarding force on the exchangeable cations. This retarding force is transmitted to the sur-rounding water by viscous drag and supplements the normal viscous drag and retarding force. According to Michaels and Lin (1955), this phenomenon is the electro-osmotic flow in the reverse direction. The result is equal to the hydrodynamic flow minus the counterelectro-osmotic flow as:

$$k * - k = \frac{\varepsilon^2 \zeta^2}{32 \pi^2 \eta \lambda} \frac{e}{1+e} \tag{9.18}$$

where k* = permeability in the absence of counterelectro-osmosis; k = apparent permeability when counterelectro-osmosis occurs; ε = dielectric constant; ζ = zeta potential; λ = specific conductance (surface + solution); and e = void ratio.

Street (1959) visualizes the streaming potential as being responsible for a counterelectro-osmotic pressure rather than a counterelectro-osmotic flow.

9.13.3 Thermoelectric Effect

(1) General Discussion

A phenomenon of thermoelectric interaction was discovered by J. T. Seebeck in 1822. His discovery of a novel method for the direct transfer of heat into electric energy is now known as the *Seebeck* or *thermoelectric effect*. The device for measuring such effects is called a thermocouple. Figure 9.26(b) illustrates that the thermo-osmosis and thermal-electric effect are very closely related, and it is difficult to separate them in a soil-water system. In early 1940 Winterkorn and associates (Winterkorn, 1947) conducted laboratory experiments on clays which will be presented in the following sections.

(2) Laboratory Experiments

Winterkorn and associates conducted a simple experiment. If one applied a hot plate at one end of a well-insulated cylindrical specimen of moist soil and a cold plate at the other end, a number of thermal, electrical, and mechanical phenomena can be observed. Among the latter is the movement or shift of water within the specimen from the hot side to the cold. This phenomenon can be called thermo-osmosis if we take the term osmosis in its original Greek meaning of 'pushing', without implying a specific theoretical picture for the mechanism or mechanisms involved in this 'pushing'.

The actual problem of the combined heat and mass transport that takes place under the stated experimental conditions is quite complex, as discussed in Highway Research Board reports (HRB, 1958). While a number of mechanisms have been proposed with particular emphasis on transportation of the vapor in the case of soil-water-air systems and in the film

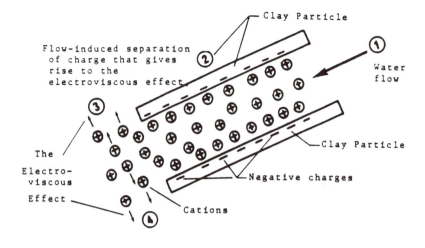

(a)

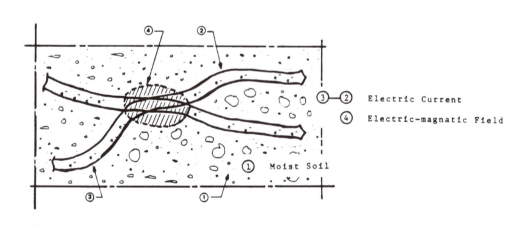

(b)

(c)

Figure 9.26 Schematic diagrams illustrate the multienergy effect on soil-water system. (a) Electroviscous effect; (b) thermoelectric effect; and (c) electromagnetic effect.

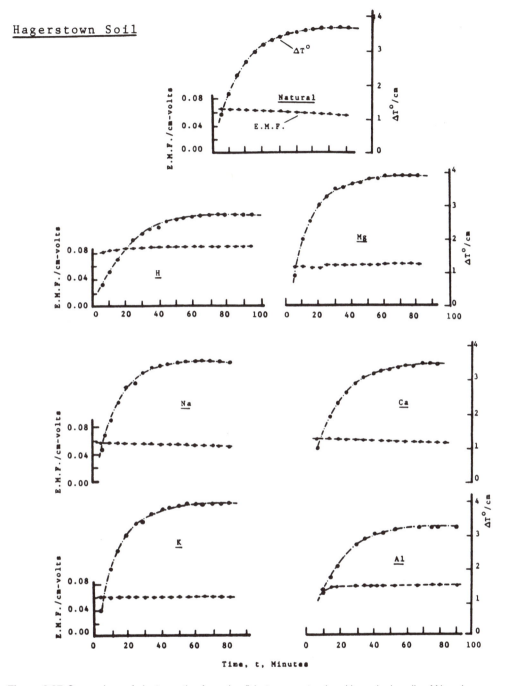

Figure 9.27 Comparison of electromotive force (emf) between natural and homoionic soils of New Jersey Hagerstown soil. (After Winterkorn, H.F. [1958], Mass Transport Phenomena in Moist Porous Systems as Viewed from the Thermodynamics of Irreversible Processes, HRB Special Report 40, pp. 324–338.)

phase in saturated soil-water systems, any one mechanism cannot be exclusive. It appears that nature employs, though to a different extent, every mechanism that scientists have been able to develop on the basis of the experimental evidence. Some laboratory testing data on thermal electric effects of natural and homoionic Hagerstown soil from New Jersey are presented in Figure 9.27.

9.13.4 Electromagnetic Effect

(1) General Discussion

As discussed previously, indications are that thermal-electrical energy fields are closely related in the natural soil-water system. Their distribution in the system is in a random pattern due to complex soil-water interactions in the environment. As indicated in Figure 9.4, ion movement in the soil-water system is also in a random motion. A new phenomenon has been observed that the fluid flowing through a polluted, unsaturated, fine-grained soil can be controlled by thermal-electromagnetic process as reported by Fang (1991a, 1997). A laboratory measuring apparatus has been developed to demonstrate such behavior. A comparison between existing methods (electrokinetic process) and this proposed process for soil decontamination indicates that the process has significantly greater capacity. The energy source for the electromagnetic energy is illustrated in Figure 9.26(c). The *magnetic energy* is created by using two electric currents crossing each other, then a temporary *electromagnetic force* is created in the vicinity of the electric charges.

(2) Characteristics of Electromagnetic Force

Basic physics (Beiser, 1988) indicates that two electrical charges at rest exert forces on each other according to Coulomb's law. When the charges are in motion, the forces are different, and it is customary to attribute the differences to magnetic forces that occur between moving charges, in addition to the electrical forces between them. The total force on a charge (Q) can be divided into two parts: an electric force that depends only on the value of (Q) and a magnetic force that depends on the velocity (v) of the charge as well as on the charge (Q).

In actual condition, there is a single interaction between charges — an electromagnetic interaction in which a magnetic field (B) is present wherever a magnetic force acts on a moving charge. The magnetism of (B) is equal numerically to the force on a charge of 1 C (coulomb) moving at 1 m/sec perpendicular to (B). Electromagnetic induction is a current produced in a conductor whenever it cuts across magnetic lines of force.

Electromagnetic waves consist of coupled electric and magnetic fields. They are perpendicular to each other. From an engineering viewpoint, the magnetic force is very small in comparison with the electric force. However, from theoretical and practical aspects, it changes from one-dimensional into a three-dimensional condition. Consequently, the influencing area or energy field increases significantly.

Ferromagnetism is an important characteristic in the electromagnetic energy field and it also is an important parameter for use in soil decontamination by using magnetism's unique properties such as ferromagnetism to identify or characterize various types of pollutants such as heavy metal trapped in the soil-water system.

(3) Laboratory Experiments

A simple apparatus is shown in Figure 9.28. The apparatus is portable with the control panel being approximately 20×20 in. Soil specimen size used in experimentation is 5×1.5 in. Two cells, both 3.5×1.5 in., are connected at the ends of the soil specimen for the purpose of controlling the electric energy. The magnetic energy is created by using two electric currents crossing each other, creating a temporary electromagnetic force in the vicinity of the electric charges as illustrated in Figure 9.28 (Fang, 1997).

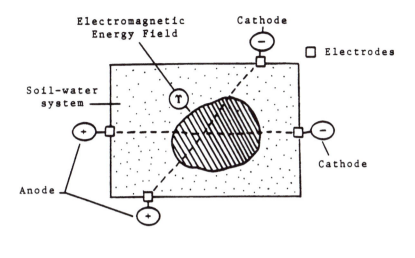

(a)

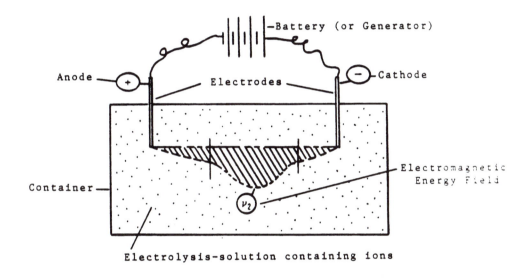

(b)

Figure 9.28 Electromagnetic process in soil-water system. (a) Top view and (b) cross section.

9.14 SUMMARY

1. Electrical energy is the center of all energy fields including mechanical, thermal, electrical, and electromagnetic energy field. Soil-electricity interaction is explained based on a particle energy field theory.
2. The importance of dielectric constant for evaluation of contaminated soil-water system is discussed together with experimental data to support such conditions. Dielectric constant is a better parameter to indicate the degree of contamination than the commonly used pH value.
3. Mechanism of soil-electricity interaction provides the rational basis in developing a better understanding of the electrochemistry associated with the electrokinetic process. The relationship of the chemistry generated by the electrodes in the soil-water-electricity system is not yet fully understood. The potential use of electrokinetic processing of soils can be enhanced if the basic knowledge of those issues is developed.
4. Mechanisms and techniques of ground improvement, dewatering, and soil stabililization by the electrokinetic process are discussed. Comparison of mechanisms between dewatering and soil decontamination by electrokinetic process and electromagnetic wave excitation is discussed.

PROBLEMS

9.1 Why is electric resistivity high in the dry soil and why does resistivity decrease when soil gets wet?

9.2 Electromotive force (emf) is a measure for force or energy? Explain.

9.3 Why is the geometry of the electrode important for measuring electrokinetic phenomena in the soil-water system?

9.4 Why are electrodes made of metal not suitable for measuring electrokinetic phenomena in a contaminated soil-water system?

9.5 What are the advantages for using graphite for electrodes in the contaminated soil-water system?

9.6 What are the characteristics of the dielectric constant? What is the relationship between pH and dielectric constant? How would one measure the dielectric constant for contaminated soil-water?

9.7 Dielectric constant can indicate both ion types and concentrations in the solution. Why do so few people use the dielectric constant for indicating contamination? Explain.

9.8 Explain the mechanisms between ground improvement and soil decontamination by the electrokinetic process.

9.9 What are the limitations for soil decontamination by the electrokinetic process?

9.10 Discuss the characteristics and mechanisms of soil decontamination by the electromagnetic process.

CHAPTER 10

Radiation Effects on Water, Soil, and Rock

10.1 INTRODUCTION

10.1.1 General Discussion

There are approximately 1500 different nuclei known to be unstable. In contrast, only 280 stable nuclear isotopes are presently known. The characteristics of these unstable radioactive elements are summarized in Table 10.1. An unstable nucleus suddenly and spontaneously changes and is accompanied by the emission of one or more rapidly moving particles. Such a nucleus is said to be radioactive and their changes are called radioactive decay. Geotechnical problems interacting with a radiation energy field can be grouped into four general areas:

- Control of radioactive and toxic radon gas
- Disposal or management of nuclear wastes
- Utilization of nuclear energy for large construction operations
- Utilization of gamma-ray for use as a nondestructive testing method

To tackle these problems, we must understand basic knowledge of atomic and nuclear physics including atomic, nuclear, and molecular structures, radioactive decay processes, and soil/rock interaction in the radiation energy field.

10.1.2 Review of Basic Nuclear Physics and Chemistry

The structure and properties of the nuclei of atoms are the concern of nuclear chemists and physicists as well as engineers who are interested in nuclear waste disposal. For our purpose, we can consider that atomic nuclei are aggregates of the nuclear particles, protons and neutrons. The proton and the neutron are called nucleons. Other particles such as the electron, the alpha particle, and gamma photon are associated with the properties of nuclei.

The *atomic number* (Z) of an element is equal to the number of protons (+e) in the nucleus of one of its atoms. Although all the atoms of an element have the same number of protons in their nuclei, the number of neutrons may be different. Each variety of nucleus found in a given element is called an isotope of the element or as:

$$_Z^A X \tag{10.1}$$

where X = chemical symbol of element; A = mass number of isotope; = number of proton + neutrons in nucleus; Z = atomic number of element; = number of protons in nucleus.

Table 10.1 Characteristics of Unstable Radioactive Elements

Series no.	Element	Symbol	Atomic no.	Atomic weight	Decay types α	Decay types β	Half-life Years	Half-life Days	Half-life Min	Half-life Sec
1	Uranium	U	92	238	x		4.5×10^9			
2	Thorium	Th	90	234		x		24		
3	Protactinium	Pa	91	234		x			1.2	
4	Uranium	U	92	234	x		2.5×10^5			
5	Thorium	Th	90	230	x		7.6×10^4			
6	Radium	Ra	88	226	x		1620			
7	Radon	Rn	86	222	x			3.8		
8	Polonium	Po	84	218	x				3.1	
9	Lead	Pb	82	214		x			27	
10	Bismuth	Bi	83	214		x			20	
11	Polonium	Po	84	214		x				1.6×10^{-4}
12	Lead	Pb	82	210		x	22			
13	Bismuth	Bi	83	210		x		5		
14	Polonium	Po	84	210	x				138	
15	Lead	Pb	82	206			Stable			

Data from Sears et al. (1987); Beiser, A. (1988), *Applied Physics,* McGraw-Hill, New York, pp. 238 and 280.

The mass of an atom is always less than the sum of the masses of the protons, neutrons, and electrons of which it is composed. The energy equivalent of the missing mass is called the binding energy of the nucleus. The greater its *binding energy,* the more stable the nucleus. The nuclei can be transformed into others of a different kind by interaction with each other. Since nuclei are all positively charged, a high-energy collision is necessary between two nuclei if they are to get close enough together to react. In any nuclear reaction, the total number of neutrons and the total number of protons in the products must be equal to the corresponding total numbers in the reactions.

10.2 CHARACTERISTICS OF RADIOACTIVITY

10.2.1 Radioactive Decay

Certain nuclei are unstable and undergo radioactive decay into more stable ones. Five types of radioactive decay, namely: gamma decay, alpha decay, beta decay, electron capture, and position emission, together with decay event and the reasons for instability, are summarized in Table 10.2. Among these five decay processes, the alpha decay, beta decay, and gamma decay are closely related to the engineering problems and further explanations will be presented in Section 10.3.

10.2.2 Half-life, Mean Lifetime, and Activity

During the radioactive decay process, the radioactive substance separates from its parent, the number of radioactive nuclei remaining as time elapses decrease. The term *half-life* is used to describe such changes. This half-life of a radioactive sample is defined as the time at which the number of radioactive nuclei has decreased to one half the number at t = 0. Using radon gas (Rn-222) as an example to illustrate the meaning of half-life, Figure 10.1 shows the radioactive decay process for radon gas. From Table 10.1, the half-life decay time for radon is 4 days (3.825 days). At the end of 4 days, the number of radon nuclei remaining is half the number present at the beginning of the interval. For the same reason, at the end of 8 days, the number of radon neclei remaining is 25% of the number present at the beginning of the interval.

Table 10.2 Characteristics of Radioactive Decay

Type of decay	Decay event	Reason for instability
Gamma decay	Emission of gamma-ray reduces energy of nucleus	Nucleus has excess energy
Alpha decay	Emission of alpha-particle reduces size of nucleus	Nucleus too large
Beta decay	Emission of electron by nuclear proton	Nucleus has too many neutrons relative to number of protons
Electron capture	Capture of electron by nuclear proton changes it to a neutron	Nucleus has too many protons relative to number of neutrons
Position emission	Emission of position by nuclear proton changes it to a neutron	Nucleus has too many protons relative to number of neutrons

Based on Beiser, A. (1988), *Applied Physics,* McGraw-Hill, New York, pp. 238 and 280 and others.

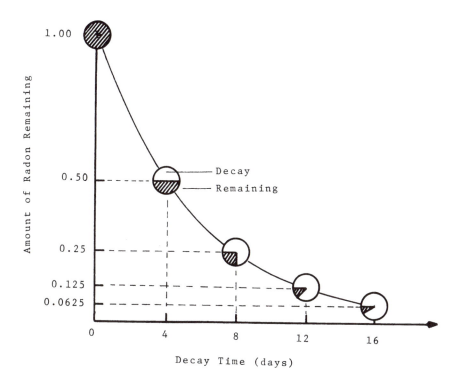

Figure 10.1 Radioactive decay curve for radon gas (Rn-222). Half-life = 4 days (3.825 days).

The *mean lifetime* or *average lifetime* of a nucleus or of an unstable particle is related to its half-life (Sears et al. 1987). In particle physics, the life of an unstable particle is usually described in terms of mean lifetime rather than the half-life.

The term *activity* is used frequently to describe the number of disintegration per unit time. The commonly used unit for the activity is the curie (Ci). The SI unit of activity is the bequerel (Bq).

10.2.3 Units of Radioactivity

(1) Curie (Ci)

A *curie* of radioactivity represents 3.7×10^{10} decay events per second. Originally, one curie (1 Ci) meant the number of disintegrations per second taking place in 1 g of radium (Ra-226), but the definition has been broadened and standardized to mean exactly 3.7×10^{10}

disintegrations per second of any radioactive material. A picocurie is 1 trillionth of a curie and commonly used for measuring radon gas. Standard laboratory sources are usually near 10^{-6} or 1 Ci (microcurie):

10^{-6} curie = 1 Ci (microcurie)
10^3 curie = 1 kCi (kilocurie)
10^6 curie = 1 MCi (megacurie)

(2) Bequerel (Bq)

One Bq is one disintegration per second (1 Bq = 1 sec^{-1}).

(3) Atomic Mass Unit (amu, AMU, u)

The atomic mass unit is the unit used for expressing the masses of atoms and molecules. 1 amu is defined to be the mass of a carbon atom whose nucleus consists of six protons and six neutrons such as ^{12}C.

$$1 \text{ amu} = 1.66 \times 10^{-27} \text{ kg}$$
$$= 1.49 \times 10^{-10} \text{ J}$$
$$= 931 \text{ MeV}$$
$$\text{Proton mass} = m_p = 1.0073 \text{ amu}$$
$$\text{Neutron mass} = m_n = 1.0087 \text{ amu}$$

Example 10.1

Determination of the molecular mass of water (H_2O)

Solution

$= 2 \times$ (atomic mass of hydrogen)
   $+ 1 \times$ (atomic mass of oxygen)
$= (2 \times 1 \text{ amu}) + (1 \times 16 \text{ amu}) = 18 \text{ amu}$

Other related units of radiation are *Rad* and *Rem*. The Rad is a unit that is used to specify the amount of radiation energy absorbed by an object. 1 Rad = 0.01 J/kg. The Rem is a unit that is used to measure radiation dosage in living tissue.

10.3 RADIOACTIVE DECAY PROCESS

10.3.1 General Discussion

As discussed in a previous section, radioactivity is the property possessed by certain substances to become transformed spontaneously into other substances by emission of alpha particles or beta particles from their nuclei. Radioactivity is unaffected by temperature, pressure, or chemical form. In the various radioactive decay processes, particles (He-4 nuclei), electrons, and rays (high-energy photons) are emitted. When these particles enter a piece of matter, energy is transferred to the material through collisions with the atoms in the material. These interactions lead to the ejection of electrons from the atoms and therefore produce ions in the material. This ionization, in turn, gives rise to chemical reactions and to a general heating of the absorbing material. There are several various reasons as summarized in Table 10.2; however, some additional explanations are presented as follows.

10.3.2 Conservation Laws and Radioactive Decay Process

(1) Conservation Laws in Radioactive Decay

The conservation laws in radioactive decay are based on two facts about the radioactive decay process as: (1) the total number of protons and neutrons are exactly equal before and after the decay takes place; (2) the total electrical charge is the same before and after the decay takes place. For example: the mass number of Ra-226 (radium) equals the sum of the mass number of Rn-222 (radon) and He-4 (helium) [222 + 4 = 226]. In the decay of Ra-226, there are 88 protons present before decay and 86 + 2 = 88 after decay.

(2) Radioactive Decay Process

(a) Alpha (α) Radiation

The *alpha* particles have extremely high ionizing action within their range; they are composed of two protons and two neutrons. As they are expelled from the nucleus, the rapidly stripped electrons are neutralized. When electrically stable, alpha particles become helium (He) atoms. When an alpha particle passes through matter, it proceeds in a straight line, leaving a high density of ions along its path (Marion, 1976).

(b) Beta (β) Radiation

In the *beta* decay process, an electron is emitted from the nucleus. The removal of a negative charge from the nucleus means that the (positive) nuclear charge increases by one unit (by +e). Beta particles exhibit relatively low energy levels; they can travel further and faster than alpha particles but, like alpha particles are rapidly attenuated by a thin layer of solid material. Beta radiation consists of negatively charged particles moving at speeds ranging from 30 to 99% of the speed of light.

(c) Gamma (γ) Radiation

Gamma-rays are often emitted by nuclei following radioactive alpha or beta decay. A gamma-ray is a bundle or photon of very high frequency electromagnetic radiation. They are quanta of energy which are very similar to X-rays in their velocity and nature. However, whereas X-rays originate in the electron shells of the atoms, gamma-rays emanate from the nucleus. Gamma-rays have the highest energy level of the radioactive decay emissions with fast traveling speed and greater penetrating power. The unit of gamma radiation is the *photon*.

10.3.3 Energies of Radiation

Since alpha and beta particles have mass and gamma radiations do not, it is possible to establish a single system of expressing energies through use of *Einstein's energy-mass equivalence* formula:

$$E = mc^2 \tag{10.2}$$

where E = energy, g·cm/sec or ergs (more convenient energy unit for nuclear particles is electron volt, eV); m = mass of the particle in grams; and c = velocity of light (2.998×10^{-10} cm/sec).

The electron volt (eV) is the energy necessary to raise one electron through a potential difference of one volt. The conversion factor is $1 \text{ eV} = 1.602 \times 10$ erg. The energies of alpha and beta particles and of gamma photons range from several thousand up to several million electron volts. For this reason, energies are usually expressed as million electron volts (MeV).

10.4 URANIUM AND RADIUM INTERACTION WITH ENVIRONMENT

10.4.1 Uranium and Radium

(1) Uranium (U)

Uranium is an elementary metal, symbol U. The metal has a specific gravity of 18.68 and atomic weight 238.2. The melting point is about 1860°C. It is hard but malleable. Chemically, uranium has a number of isotopes, and its atomic weight varies from 234 to 239. It has the highest atomic weight of any of the materials occurring normally in nature, being 92 on the period scale. Uranium is highly unstable and can be made to disintegrate with explosive violence. Uranium has many uses. Uranium compounds, especially the uranium oxides, are used for making glazes in the ceramic industry and also for paint pigments.

(2) Radium (Ra)

Radium is a radioactive element discovered in 1898 by the Curies. Radium is only commercially obtainable from uranium ores and is believed to be formed from the disintegration of uranium or ionium. The ratio of radium (Ra) to uranium (U) ore is always about 1:3,000,000. Radium is marketed in the form of bromides or sulfate in tubes and is extremely radioactive in these forms.

Other unstable elements include *thorium* (Th), *polonium* (Po), *lead* (Pb), etc. as listed in Table 10.1. Figure 10.2 is a graphical presentation based on data from Table 10.1 by plotting atomic weight vs. atomic number of the uranium series.

10.4.2 Interrelationship Among Uranium, Radium, and Radon Gas

As indicated in Table 10.1 and Figure 10.1, radon gas is derived from uranium (U) and radium (Ra). It is an alpha type of decay process as illustrated in Equation (10.3).

$$\begin{array}{cccc}
\text{Radium (Ra} - 226) & \text{Radon (Rn} - 222) & \mathbf{pn} & \\
88 \text{ protons} & -\alpha \text{ decay} \rightarrow \quad 86 \text{ protons} & + \quad \mathbf{np} & (10.3) \\
138 \text{ neutrons} & 136 \text{ neutrons} & \text{He} - 4 &
\end{array}$$

Equation (10.3) can be rewritten into a nuclear equation as:

$$^{226}_{88}\text{Ra} - \alpha\, \text{decay} \rightarrow ^{222}_{86}\text{Rn} + ^{4}_{2}\text{He} \qquad (10.4)$$

In examining Equations (10.3) and (10.4), the radium nucleus (Z = 88), mass weight (A = 226) has 88 protons and 138 neutrons. When Ra-226 emits an alpha particle (He-4), two protons (p) and two neutrons (n) are carried away. Therefore, the residual nucleus has 86

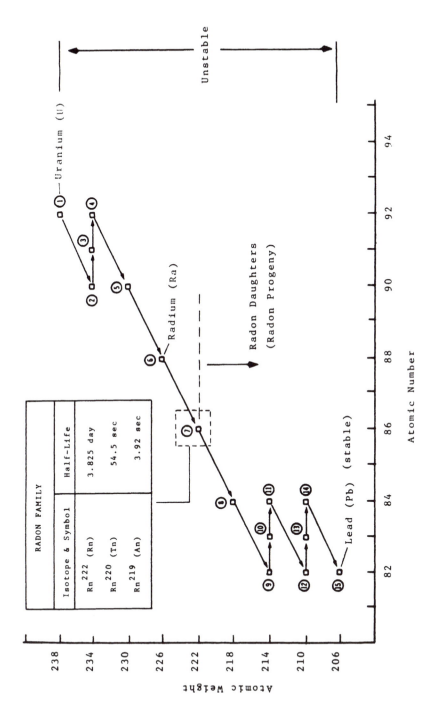

Figure 10.2 Atomic weight vs. atomic number of the uranium series. (Data from Table 10.1).

Table 10.3 Comparison of Physicochemical Properties of Noble Gases

Name	Helium	Neon	Argon	Drypton	Xenon	Radon
Symbol	He	Ne	Ar	Kr	Xe	Rn
Atomic number	2	10	18	36	54	86
Atomic weight	4.00	20.18	39.95	83.80	131.30	222
Valence electrons	ls^2	$2s^22p^6$	$3s^23p^6$	$4s^24p^6$	$5s^25p^6$	$6s^26p^6$
Radius of atom, Å	—	1.60	1.92	1.97	2.17	—
Melting point, °C (under pressure)	−272	−249	−189	−157	−112	−71
Boiling point, °C	−269	−246	−186	−152	−107	−62
Ionization energy, eV[a]	24.6	21.6	15.8	14.0	12.1	10.7
Color	Yellow	Red	Red and blue	Yellow-green	Blue-green	Colorless

[a] Data from Porterfield (1972); and Allinger, N.L., Bigelow, N.J. and McAllister, H.C. (1976), *An Introduction to General, Organic and Biological Chemistry,* Wadsworth Publishing, Belmont, CA.

protons and 136 neutrons. The product of radium decay (the daughter) is a different element: radon (Z = 86). The atomic electron structure changes, following the decay event, to accommodate the new nuclear change by releasing 2 of its 88 electrons. These two electrons eventually attach themselves to the emitted alpha particle and form a neutral atom of He-4. Thus, the original neutral radium (Ra) atom decays and two neutral atoms are formed, one radon (Rn) and one helium (He).

10.5 THE NOBLE GAS FAMILY

10.5.1 Characteristics of Noble Gases

There are six generally known *noble gases,* namely: helium (He), neon (Ne), argon (Ar), krypton (Kr), xenon (Xe), and radon (Rn). They are all gases, with very low boiling and melting points. All noble gases are inert gas which lacks the usual or anticipated chemical or biological action. The stable chemical compounds of them have been formed at Lawrence Radiation Laboratory, University of California at Berkeley since 1962 (Bartlett, 1962).

The important differences between noble gases and common gases is that the noble gases exist as an atom (nonatomic molecule). All others are gaseous elements such as hydrogen (H), nitrogen (N), oxygen (O), and chlorine (Cl) occurring as diatomic (two-atomic) molecules.

The general properties of these noble gases are presented in Table 10.3. All noble gases have a valence electron arrangement in which all the (s) and (p) orbitals in the valence shell are filled with electrons.

Based on data from Table 10.3, a plot between atomic weight and atomic number vs. ionization energy of noble gases is prepared as shown in Figure 10.3. It is indicated that radon (Rn) has the highest atomic number (Z = 86) and atomic weight (A = 222) but has lowest ionization energy (10.7 eV).

The ionization energies of the heavier elements such as He, Ne of the noble gas family are not particularly high, being about the same as oxygen (0) and lower than fluorine (F). It is therefore theoretically possible that Kr, Xe, and Rn could form compounds. In 1962, Dr. Bartlett of Lawrence Radiation Laboratory succeeded in preparing the first such compound. Soon after this discovery, a crystalline solid was made by heating Xe and F together at 400°C. Many compounds of Xe and a few compounds of Kr and Rn have subsequently been reported.

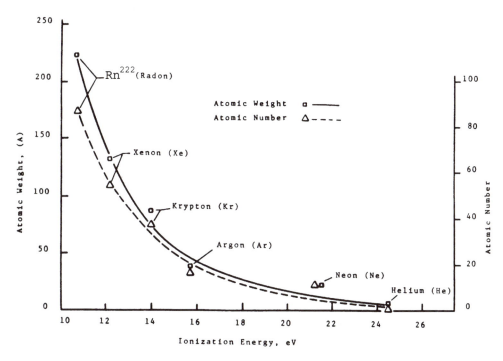

Figure 10.3 Atomic weight and atomic number vs. ionization energy of noble gases. (Ionization energy data from Allinger, N.L., Bigelow, N.J. and McAllister, H.C. [1976], *An Introduction to General, Organic and Biological Chemistry,* Wadsworth Publishing, Belmont, CA.)

Table 10.4 Fundamentals of Particle Characteristics Relating to Radon Family

Radon family	Element	Proton	Neutron	State
Parents	Uranium (^{238}U)	92	146	Solid
	Radium (^{226}Ra)	88	138	Solid
Radon	Radon (^{222}Rn)	86	136	Gaseous
	Thoron (Tn) (^{220}Rn)	86	134	Gaseous
	Actinon (An) (^{219}Rn)	86	133	Gaseous
Daughter	Polonium (^{218}Po)	84	134	Solid

10.5.2 Radon and Radon Family

Among all the six known noble gases, radon (Rn) is one that has the most unique characteristics among them and least understood in the field of multimedia pollution problem. It is not only of interest to physicists, chemists, and biologists, but practical engineers as well. It is a highly interdisciplinary subject, and intercooperation is needed. Table 10.4 presents the fundamentals of particle characteristics relating to the radon family including the parents uranium (U-238) and radium (Ra-226), brothers thoron (Rn-220) and actinon (Rn-219), and daughter polonium (Po-218).

Radon (Rn) is the only type of noble gas that is directly related with uranium (U) and radium (Ra). Figure 10.4 illustrates the interrelationship of these elements in the thermal energy field.

As indicated in Equation (10.1), radon is alpha particle decay. Krypton (Kr-93) is beta-decay, and its half-life is about 1.29 sec. Neon (Ne) is the gamma decay and illustrated as follows:

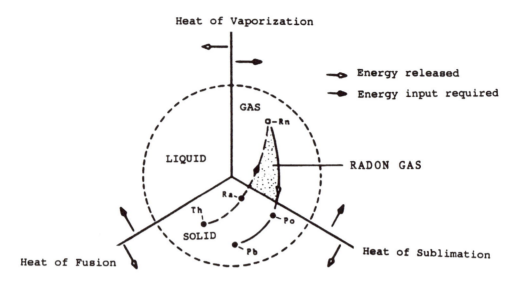

Figure 10.4 Interrelation of U, Ra, and Rn in thermal energy field.

Neon (Ne - 20)   Neon (Ne - 20)

10 protons γ decay   10 protons   +   (10.5)

10 neutrons   10 neutrons   gamma ray

10.6 ENVIRONMENTAL GEOTECHNICAL ASPECTS OF RADIATION

10.6.1 General Discussion

Three basic radiation elements are of interest to environmental geotechnology as mentioned previously. The geohydrological and hydrogeological characteristics of these three basic elements are presented in Table 10.5. In examining Table 10.5, there are three belt divisions, namely: oxidation, oxidation-reduction, and reduction belts. The oxidation belt which is close with ground surface is more active with environments than the other two belts. Also, the water-radiation interaction is most important and will be discussed in further detail in the following section.

10.6.2 Water-Radiation Interaction

Radioactive elements existing in the groundwater are related to the sources of groundwater as indicated in Table 10.6. The surface water contains small amounts of U and Ra in comparison with groundwater from a nearby uranium mine. It is especially distinct for Rn. Figure 10.5 shows the discharge and drawdown vs. time for U and Rn. In examining Figure 10.5, indications are that for the first few hours, the Rn content in the water is only a small amount, then it will increase significantly.

Radioactivity affects water chemistry significantly. Figure 10.6 presents the effect of water chemistry on uranium content in water. Figure 10.6(a) shows uranium content vs. hydrogen carbonate ion in water, and a linear relationship was found. Figure 10.6(b) shows uranium content vs. oxygen and dihydrogen sulfide. A curvilinear relationship is indicated. Figure 10.7 shows the radon content vs. uranium content in water. Figure 10.8 presents the relationship

Table 10.5 Geohydrological/Hydrogeological Characteristics of Uranium, Radium, and Radon Gas

Belt division	Develop depth (m)	Groundwater dynamic movement condition	Water type	E_h (eV)	pH	Fe^{2+}	Fe^{3+}	$\frac{Fe^{2+}}{Fe^{3+}}$	Uranium (U) (g/ℓ)	Radium (Ra) (g/ℓ)	Radon (Rn) (amu)
						me/ℓ					
Oxidation belt Strong	5/70	Strong exchange capacity	HCO_3-Ca or HCO_{3-}	529–384	≤7.2	0 2.0	0.2–38.0	0–0.4	$n \times 10^{-7}$–$n \times 10^{-5}$	<1.96×10^{-12}	5050–204
Weak	35–125		Ca-Mg	384–250	7.0–7.4	0.1–9.0	0.5–11.0	0.4–0.85	$n \times 10^{-6}$–$n \times 10^{-3}$	1.96×10^{-12}–8.27×10^{-12}	204–30
Oxidation reduction belt	50–275	Slow exchange capacity	HCO_{3-} Ca-Mg, HCO_{3-} SO_4-Ca-Mg	250–0	7.2–7.7	0.5–18.0	0.5–2.5	0.7–3.0	$n \times 10^{-6}$–$n \times 10^{-4}$	8.27×10^{-12}–5.12×10^{-11}	60–0
Reduction belt		Very slow or no exchange capacity	HCO_{3-} Ca-Mg, HCO_{3-} Ca-Mg-Na	0–201	7.0–7.5	1.5–18.0	0–1.5	>1.5	$n \times 10^{-7}$		40–0

Selected data from Kao and Sun (1983). Cited by Fang, (1989b) and (1990).

Table 10.6 Comparison of Uranium (U), Radium (Ra), and Radon (Rn) Contents in Groundwater

Content/ Sources of groundwater	Uranium (U) (g–l)			Radium (Ra) (g/l)			Radon (Rn) (amu[a])		
	Min	Max	Ave	Min	Max	Ave	Min	Max	Ave
Surface water (river)	2×10^{-8}	5×10^{-5}	6×10^{-7}	2.5×10^{-13}	4.2×10^{-12}	2×10^{-13}	0	0	0
Under sedimentary rock	2×10^{-7}	8×10^{-6}	5×10^{-6}	1.0×10^{-12}	6.0×10^{-12}	2×10^{-12}	1	50	15
Under granite rock	2×10^{-7}	3×10^{-5}	7×10^{-6}	1.0×10^{-12}	7.0×10^{-12}	2×10^{-12}	10	400	100
Under uranium mine	5×10^{-5}	9×10^{-2}	6×10^{-4}	8.0×10^{-12}	2.0×10^{-11}	8×10^{-11}	50	50,000	10,000

[a] amu = a.m.u. = atomic mass unit.
Data from Fang, H.Y., Lou, G.Y. and Chu, T.G. (1987), Proc. 1st Int. Symp. on Environmental Geotechnology, v. 2, pp. 219–231.

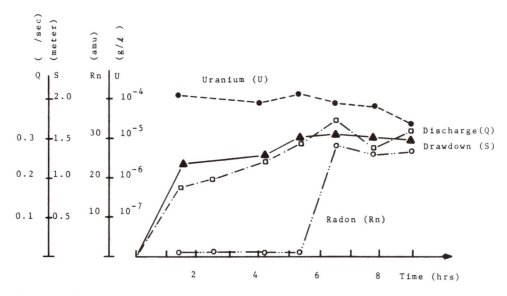

Figure 10.5 Discharge and drawdown vs. time for uranium (U) and radon (Rn) content. (After Liao [1983].)

between the concentration of Ra-226 and total dissolved minerals in groundwater. Figure 10.9 presents the ratio of Rn in water and in air vs. temperature. A curvilinear relationship was found.

10.6.3 Rock/Soil-Radiation Interactions

(1) Rock

Rock mass has a tendency to adsorb gas. Table 10.7 presents laboratory test data of five rock types soaked with uranium-contaminated water for 24 and 48 hr. Results are measured of the uranium content and E_h values. There are significant differences between the 24- and 48-hr soaking periods. Table 10.8 presents radioactive elements in various types of rocks in both igneous and sedimentary rocks. Three conditions include U, Ra and Rn. Significant differences exist among these conditions.

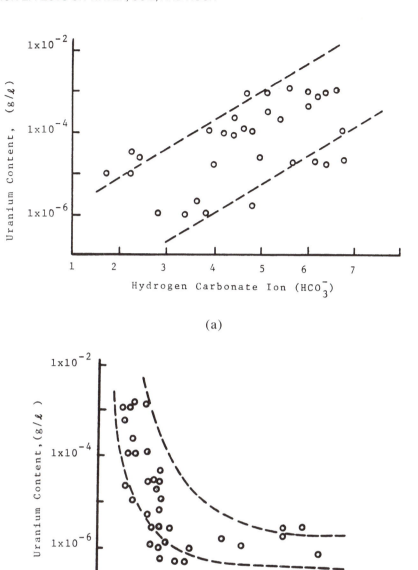

Figure 10.6 Effect of water chemistry on uranium content in water. (a) Uranium content vs. HCO_3 content in water; (b) Uranium content vs. O_2 and H_2S in water. (After Liao [1983]. Cited by Fang [1989b].)

(2) Soil

Most of the radon measurements conducted in a subsurface soil are on soil-gas extracted from tubes of various lengths placed in an auger hole about 10 cm in diameter and extending to 180 cm in depth, as reported by Rose et al. (1988). Each length of tubing is covered with

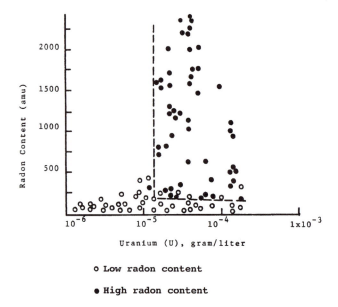

o Low radon content

• High radon content

Figure 10.7 Uranium content vs. radon content in water. (After Liao [1983]. Cited by Fang [1989b].)

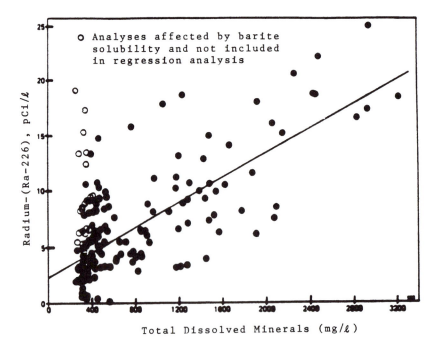

Figure 10.8 The relation between the concentration of Ra-226 and total dissolved minerals in ground-water. (Location: Champaign, Illinois.) (After Gilkeson, R.H. and Cowart, J.B. [1987], *Radon in Groundwater,* Lewis Publishing, Chelsea, MI, pp. 403–422.)

a piece of window screen at its lower end. Sand is packed into the section of auger hole for about 4 cm above the screened end of tubing. Sections of the auger hole between tubing inlets are packed with moist kaolinite clay to separate the sample points. Both the sand and clay are analyzed for radon content. Figure 10.10 presents typical summertime results for Rn-222 and Rn-220 in a soil layer in a Central Pennsylvania limestone region.

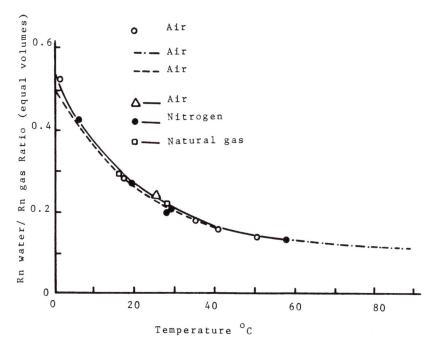

Figure 10.9 Effect of temperature on radon/water distribution (After Cameron, C.P. [1987], A Review of Radon Emanation and Mobilization in Minerals and Rocks, Waterways Experiment Station, Report no. MSSGL-87-27, 42p.)

Table 10.7 Adsorption Characteristics of Some Sedimentary Rocks Relating to Uranium Content

| | Adsorption test: samples soaked in water | | | |
| | 24 hr | | 48 hr | |
Characteristics of rock samples	**Uranium content (g/l)**	**E_h (mV)**	**Uranium content**	**E_h (mV)**
Carbonic conglomerate and breccia	4.5×10^{-6}	419	3.5×10^{-8}	394
Breccia contains little carbonic matter	1×10^{-5}	419	7×10^{-6}	399
Carbonic-organic argillaceous	6.5×10^{-6}	419	6×10^{-6}	414
Carbonic arenaceous and carbonic calcareous	1.5×10^{-5}	439	1×10^{-5}	404
White color calcite-breccia	2×10^{-5}	490	1.9×10^{-5}	470

Note: All test samples are same weight = 100 g. Volume of water contains uranium, all = 0.5 l. Before the test (1) uranium content in water for all samples = 2×10^{-5} g/l; (2) electrical potential (E_h) for all samples = 498 mV.

Selected data from Kao and Sun (1983). Cited by Fang (1989b) and (1995b).

Table 10.8 Radioactive Elements Content in Various Types of Rocks

Rock types	**Uranium (%)**	**Radium (%)**	**Radon (%)**
Igneous rock			
Granitic	4.0×10^{-4}	1.40×10^{-10}	7.6×10^{-6}
Basic	1.2×10^{-4}	0.47×10^{-10}	1.7×10^{-10}
Ultrabasic	0.7×10^{-4}	0.20×10^{-10}	6.5×10^{-10}
Sedimentary rock			
Limestone	1.5×10^{-4}	0.50×10^{-10}	
Sandstone	4.0×10^{-4}	$0\text{-}1.50 \times 10^{-10}$	6.5×10^{-4}
Clayshale	3.0×10^{-4}	1.00×10^{-10}	
Clay	4.3×10^{-4}	1.30×10^{-10}	

After Luo, G.Y., Chu, T.Q., Wu, H. and Liu, G.C. (1991), Proc. 2nd Int. Symp. on Environmental Geotechnology, v. 2, pp. 229–241.

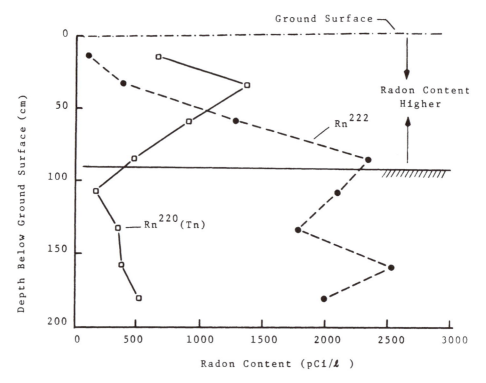

Figure 10.10 Typical summertime results for Rn-222 and Rn-220 in soil layer in University Park, Pennsylvania (After Rose, Washington, and Greeman [1988].

10.7 RADIOACTIVE AND TOXIC RADON GAS

10.7.1 Nature of Radon Gas

(1) General Discussion

Radioactive and toxic radon gas (Rn) is produced naturally in the ground by the normal decay of uranium (U) and radium (Ra) and widely distributed in trace amounts throughout the earth's crust. It is a colorless, odorless, and tasteless element as indicated in Table 10.3. Most radon gas is concentrated in the oxidation belt which is at a relatively shallow depth from the ground surface. Under normal conditions, the amount of radon gas seeping into the atmosphere or residential buildings is very little and will not be harmful to human health.

As discussed in Chapter 1, in recent years, many natural farm lands, forest, and wetlands have been destroyed by conversion into residential and industrial developments: consequently, such construction activities change the dynamic equilibrium of the ecosystem, which in turn triggers and accelerates radon gas emanation and mobilization in the environment. This change is the major reason for the problem of indoor concentration which has significantly increased in recent years.

Recent findings indicate that radon gas is not the totally inert element as previously thought (Fang, 1989b). It is influenced to some degree by local environments such as temperature, humidity, pH value, ion exchange reaction, redox reaction, etc. Also, radon gas can be found in ground soil, water, air, and other media becoming a complex multimedia pollution problem.

Radon gas itself is not a major concern to human health. However, if radon continues to undergo radioactive decay, it produces new naturally radioactive materials. These new materials, which are solid and not gaseous, can stick to surfaces, such as dust particles (Section 3.8) in the air, and thereby these contaminated dust particles can be inhaled and deposited on the surface of the lung of human or animal. Also, it can stick on trees, flowers, or vegetation, and becomes part of the pollution cycle in our day-to-day environment.

The original nucleus is named the parent nucleus and the residual nucleus, the daughter. These descendents of radon gas are called radon daughters or radon progeny. Radon daughter (Po) can also adhere directly to the lung lining. Several radon daughters emit alpha radiation which has a high alpha energy resulting from radioactive decay in the tissue of the lung lining. Recent studies (NAS, 1988) indicate that the result of extended exposure to high levels of radon gas or radon daughters increases the risk of lung cancer.

Radon gas seeps into basement structures through earth floors, sump pumps, floor drains, wall cracks, slabs, joints, etc. (Lafavore, 1987) and can be trapped in a comparatively small air volume. The radon in a house has less air with which to mix; therefore, indoor concentrations are usually higher than those found outdoors. Insulated houses with less opportunity for dilution have higher indoor radon levels than do drafty houses.

(2) Radon Gas and Human Health and Environments

Whether or not radon will cause lung cancer is a hot topic to be discussed. Numerous publication have discussed the various levels of hazards affecting the population's exposure to low levels of ionizing radiation (NRC, 1980). In 1988 the National Research Council (NRC) published the health risks of radon and other internally deposited alpha-emitters (NRC, 1988 and Cothern and Rebers, 1990). Figure 10.11 presents radon concentration and risk of lung cancer vs. air change rate in residential buildings.

Radon is a naturally occurring radioactive gas that has been identified as a high lung cancer risk in studies. Based on the experiences of miners, some experts have estimated that natural radon accumulating in houses could cause about 15,000 lung cancer deaths per year. In 1988, the Environmental Protection Agency (EPA) issued an official statement which claimed that radon is a national problem and is promoting a national program calling for reduction of residential radon.

In 1994, the *Journal of the National Cancer Institute* presented the results comparing the radon exposure of 538 nonsmoking Missouri women who have lung cancer with the radon exposure of 1183 matched subjects who did not have cancer. "An association between lung cancer and the exposure to domestic levels of radon was not convincingly demostrated," the researchers concluded. "The magnitude of the lung cancer risk from radon levels commonly found in U.S. dwellings appears low."

Although researchers for years have been unable to prove that residential radon actually poses a lung cancer risk, some research has concluded that there is a danger, and other studies, like the one in the *Journal of the National Cancer Institute*, show no evidence of risk. Regardless of whether or not radon has a direct effect on human health, it is a part of the air pollution problem. To control air pollution requires knowledge of environmental geotechnology; therefore, it becomes a part of the pollution remedial action and will be further discussed in following sections.

10.7.2 Engineering Properties of Radon Gas

(1) General Discussion

Radon gas is one of the six generally known noble gases as indicated in Figure 10.1. It has the highest atomic number and atomic weight but has the lowest ionization energy in comparison with the other noble gases. Radon gas is derived from radium (Ra), as shown in Table 10.3, with an alpha type of decay process; that is, as Ra-222 emits an α-particle (He-

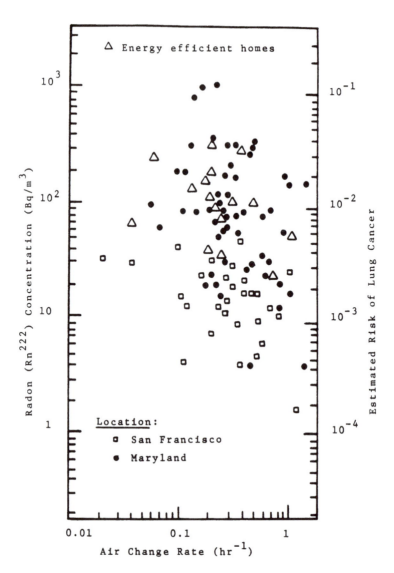

Figure 10.11 Radon concentration and risk of lung cancer vs. air change rate in residential buildings
(After Nero [1989].)

4), two protons and two neutrons are carried away. Therefore, the residual nucleus has 86
protons and 136 neutrons (Table 10.3).

   Radon gas also can be dissolved in groundwater. Rain can absorb radon as it falls through
the lower troposphere — the region of highest atmospheric concentration of radon. When
radon-contaminated rainfall seeps into groundwater, surface waters can become contaminated
with radon.

10.7.3 Radon Emanation in Rock or Minerals

(1) General Discussion

   Radon gas may be widely distributed in the earth's crust, but it is concentrated in certain
types of rock formation. Since radon is derived from uranium and radium as shown in

Figures 10.2 and 10.4, these elements originate from igneous rock (parent rock) such as granite, basalt, gabbro, diorite, etc. During the long period of time required for uranium's change into radium and for the radium to change into radon, many processes are possible such as transport and deposition, hydraulic action, earthquake, etc., moving some of the uranium or radium from one place to another. As radon moves from the parent rock formation to the temporary locations in the rock formation, these temporary locations become host rock formations. One common feature of these host rocks is that they possess large porosities or cavities, such as limestone, dolomite, conglomerate, breccia, mudstone, etc. Among these hosts in eastern Pennsylvania, limestone and dolomite are common storage rocks for radon gas.

(2) Giletti and Kulp Approach (1955)

The mechanisms of radon gas release from rocks are important. In 1955 Giletti and Kulp proposed three possible mechanisms, namely: (1) α-recoil approach; (2) diffusion; and (3) transportation. A brief discussion of each method is presented.

(a) α-Recoil Approach

α-recoil derives its name from the process by which a radon (Rn) atom recoils from a decaying parent radium (Ra) atom. On decay, radium (Ra-226) emits an alpha (α)-particle (He-4) (Equation 10.2) to form radon (Rn-222). Kinetic energy of this α-particle is sufficient to cause the newly formed radon atom to be recoiled some distance in the opposite direction. This distance (ℓ) is known as the 'α-recoil length'. This distance has been shown to vary between 10^{-6} and 10^{-3} cm depending on the medium through which the recoil occurs (Andrews and Wood, 1972). Further developments on this approach are reported by Fleischer and Mogro-Camero (1978) and Fleischer (1983).

(b) Diffusion Process

Diffusion process is another possible route for radon atom release from rock. As a particle migrates with respect to a still ambient medium (solid, air, or liquid) and with the recoil process (case [a]), the diffusion mechanism has also associated characteristic lengths for radon movement which again vary with the medium through which diffusion occurs. The diffusion distance is indicated as:

Solid medium — 10^{-9} cm
Liquid medium — 2^{-9} cm
Gaseous medium (air) — 200^{-9} cm

Further discussions on diffusion behavior for radon (Rn-222) are given by Andrews and Wood (1972) and Davis et al. (1987).

(c) Transportation

Transportation is particle movement by the action of a moving ambient medium (liquid or air). Because of the low primary porosities and low intrinsic permeabilities of crystalline rocks, air and water are not expected to readily flow through the bulk of these rocks. As a result, then, transportation of radon by air or by water through crystalline rocks is not a mechanism that should exert significant control on radon loss (Davis et al. 1987).

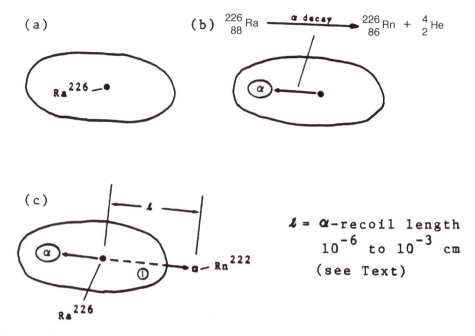

Figure 10.12 Schematic diagrams illustrating α-decay process and α-recoil mechanisms. (a) Ra existing in the rock/mineral; (b) decay process; (c) α-recoil length.

(3) α-Recoil Approach (Fleischer and Mogro-Camero, 1978)

The α-recoil derives its name from the process by which a radon atom recoils from a decaying parent radium atom. On decay, radium emits an α-particle (He) to form radon (Rn-222). Kinetic energy of this α-particle is sufficient to cause the newly formed radon atom to recoil some distance in the opposite direction as shown in Figure 10.12. This distance, known as the α-recoil length, has been shown to vary between 10^{-6} and 10^{-3} cm, depending on the medium through which the recoil occurs. This approach has been further refined by Fleischer (1983).

(4) α-Recoil-Environmental Approach (Fang, 1990)

The recoil approach as indicated in case (3) does not consider the environmental stresses during the decay process. Why these stresses are important are explained as:

1. The α-particle does not travel far and is relatively weak in comparison with the beta and gamma decay process.
2. The α-recoil length is dependent on the crystal lattice structure of each mineral and the bonding energies between particles.
3. The degree of decay or weathering of rock/minerals is related to the α-recoil length. The force produced from α-recoil (Figure 10.12) alone may not be strong enough to push a radon atom out of the rock mineral.

The forces produced during the decay process include heat of wetting, kinetic dispersive force, and electromotive force (emf) and are discussed in Sections 5.7 and 5.8. These forces are influenced by local environments, such as temperature, pore fluids, and pressures. A schematic diagram of a modified mechanism illustrating how radon gas is released from rock minerals is presented in Figure 10.13. Figure 10.13(a) shows the α-recoil length where the rock mineral is not decayed, and Figure 10.12(c) shows that if the rock mineral is decayed, the recoil length is longer as shown in Figure 10.13(b).

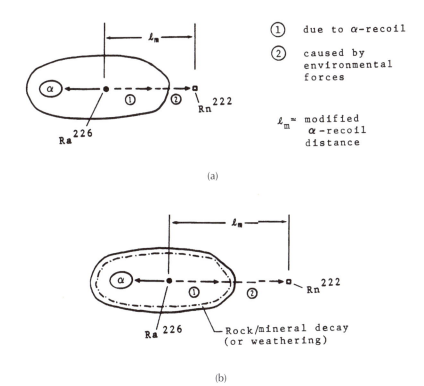

Figure 10.13 Modified α-recoil approach — α-recoil/environmental mechanism. (a) Rock/mineral not decayed; (b) Rock/mineral decayed. (After Fang, H.Y. [1990], 83rd Annual Meeting and Exhibition, Air and Waste Management Association, Paper no. 90-153.10, Pittsburgh, PA, June 19 p.)

10.7.4 Interaction Mechanisms Between Radon and Environment

Regarding the interaction mechanism between radon and the environment, Fang (1990 and 1991a) has proposed the following explanations based on particle-energy-field theory. Brief explanations are presented as follows.

(1) Interaction Based on Radon's Molecular Structure

(a) Low Ionization Energy

As indicated in Figure 10.3, radon gas has the highest atomic number and atomic weight but lowest ionization energy of the noble gases (Allinger et al. 1976). The ionization energy is the binding energy of an atom and is the minimum energy required to free an electron from an atom, or the binding energy of a nucleus as the energy required to separate the nucleus into free protons and neutrons (Marion, 1976). In other words, the smaller the binding energy between particles, the easier for particle interaction with the environment.

(b) Unbalanced Excessive Thermal Energy

The energy flow in the changes of the state of matter of radon gas is shown in Figure 10.4, which indicates that radon's parent, radium (Ra), is a solid, but radon (Rn-222) itself is in a gaseous state and the radon daughters polonium (Po-218) and lead (Pb-214) are also solid. Consider the energy input required to transfer from solid to a gaseous state and then the energy released as the gas is converted into a solid again! Since the half-life of radon gas is 3.825 days (Figure 10.1), during that short period of time, the state of matter in the thermal energy field creates temporary unbalanced excessive thermal energy. This excessive thermal energy will trigger and accelerate the separation of radon particles, since radon gas itself has a low binding energy.

(2) Radon Particle Direct Interaction with Water Molecules

The H^+ and OH^- ions in water are strongly hydrated, as discussed in Chapters 5 and 7. The heat of hydration of 1 mol H ion at 25°C is about 279 kcal. This means that the protons (H^+) are strongly associated with H_2O molecules and form $(HO_3)^+$ ions and larger charged complexes. Even so, the protons may easily move along H-bonds from one water molecule or complex to another. Also, one proton might jump on the back of an $(H_2O)_n$ complex and another jumps away to the other side. It is necessary to remember that the radius of a proton is smaller than one ten thousandth that of the H_2O molecule. Because of this small size protons can jump and react very quickly and can get into narrow places that are not accessible to other ions or molecules. The neutralization reaction of H^+ and OH^- ions for one molar solution has been estimated to take only about 10^{-11} sec. Since radon (Rn-222) contains 86 protons, it may take less than a second for radon to interact with water molecules.

(3) Interaction through Dust-Suspension

Fang (1990) proposed a new approach suggesting that dust and suspension (Sections 3.10 and 3.11) are the bridge between radon and the environment. This proposal indicates that:

1. Radon interacts with the environment through common types of medium, such as dust in the air and suspensions in the water or in the soil-water systems.
2. Interaction between radon and dust-suspension is primarily through physical types of adsorption action (Section 6.5).

The above interaction mechanisms may explain why radon gas is not a totally inert element and can interact with the environment. Also, it is believed that dust and suspension are the key factors in controlling air-water-ground soil pollution problems.

10.8 RADON MITIGATION METHODS

There are two major methods to control this unwanted gas at the present time which present: (1) the passive approach, and (2) the active approach to the problem. Currently, the passive approach is recommended by the U.S. Environmental Protection Agency (EPA, 1986) and Department of Environmental Resources (DER, 1985b). This approach measures radon in the atmosphere, then uses conventional construction methods to either keep radon gas out of a house or chase indoor radon gas out of a house. Both approaches are presented as follows.

10.8.1 Radon Mitigation by Passive Approach

(1) U.S. EPA/DER Method

Special equipment is needed to detect it. The two most popular commercially available radon detectors are the "Charcoal Canister" and the "Alpha Track Detector". Both of these devices are exposed to the air in the radon problem home for a specified period of time and sent to a laboratory for analysis. The test period for Charcoal Canisters is about 3 to 7 days.

Table 10.9 Radon Risk Evaluation Chart

pCi/l	WL	Estimated number of lung cancer deaths due to radon exposure (out of 1000)	Comparable exposure levels	Comparable risk
200	1	440—770	1000 times average outdoor level	More than 60 times non-smoker risk / 4 pack-a-day smoker
100	0.5	270—630	100 times average indoor level	20,000 chest x-rays per year
40	0.2	120—380		
20	0.1	60—210	100 times average outdoor level	2 pack-a-day smoker
10	0.05	30—120	10 times average indoor level	1 pack-a-day smoker
4	0.02	13—50		5 times non-smoker risk
2	0.01	7—30	10 times average outdoor level	200 chest x-rays per year
1	0.005	3—13	Average indoor level	Non-smoker risk of dying from lung cancer
0.2	0.001	1—3	Average outdoor level	20 chest x-rays per year

After U.S. EPA (1986).

Table 10.10 Radon Gas Accepted Level from Various Organizations

Organizations	Accepted level (pCi/l)
Canada	
Health and Welfare	10
Sweden	
New construction	2
Buildings undergoing renovation	5
Existing buildings	11
U.S.	
American Society of Heating Refrigerating and Air-conditioning Engineers (ASHRAE)	2
U.S. EPA	4
U.S. DOE	5
Florida EPA	6
National Council on Radiation Protection	8

After Lafavore (1987); and others.

The minimum test period for the Alpha Track Detectors is 2 to 4 weeks. The measurement resulting from these two types of devices are reported as: working levels (WL) or the concentrations of radon gas as picocuries per liter (pCi/l). Brief discussions on these two terms are presented as follows.

(a) Working Levels (WL)

WL are measurements of energy release. According to EPA (1986), 1 WL is the amount of radon daughters the decay of which will result in the emission of 1.3 billion volts of electron energy. If converted to heat, it might raise the temperature of a cup of water about half a degree. The cumulative exposure is expressed in working level months (WLM). Based on the uranium miner's experience, 1 WLM is 170 hr, the amount of time the average mine worker spends in a mine over the course of a month exposed to 1 WL.

(b) Picocuries per Liter (pCi/l)

As discussed in Section 10.2.3, a picocurie is 1 trillionth of a curie. One picocurie per liter of air is about two radon atoms per minute disintegrating in every quart of air in a room.

Table 10.9 presents the radon risk evaluation chart provided by U.S. EPA (1986). Table 10.10 presents the radon gas accepted level suggested by various organizations. In the U.S. the accepted level ranges from 2 to 8 pCi/l.

(2) Discussions — Advantages and Disadvantages

Present general remedial actions constitute the so-called passive approach. The passive approach tries to keep radon gas out of the house. This approach has the following advantages and disadvantages. The advantage for the passive approach is promoted by U.S. EPA and Department of Energy (DER). The disadvanges are listed as follows:

1. There is no guarantee that radon gas will stop seeping into the house. Radon gas will seep in wherever there is air.
2. It is costly and a heavy burden for the individual house owner and causes many public complaints (in some cases they go to court for legal action).
3. There is no standard on how radon concentration will affect human health.
4. Present methods are only concerned with residential buildings. No attention is paid to radon in drinking water, churning of water in showers, toilets, sinks, and washers.
5. Radon collecting on vegetation, trees, grass, and flowers which directly or indirectly affect animals, milk, fruit trees and humans is also not considered.

10.8.2 Radon Mitigation by Active Approach

Radon mitigation by active approach (dewatering) is proposed by Fang (1990 and 1995b). Dewatering techniques are a simple, low-cost, straightforward approach for controlling sub-surface-surface drainage patterns to dilute and/or redistribute radon gas in the environment before seeping into the atmosphere or buildings. A schematic flow diagram illustrating how this proposed method works for radon mitigation design is summarized in Figure 10.14.

In general, all subsurface-surface drainage or dewatering systems can be used for the purpose of redistribution and/or diluting radon-contaminated water by means of intercepting ditches or trenches. Since all drainage and dewatering systems are part of routine construction procedures, all qualified contractors should know how these procedures function, therefore, no design details are given. However, all analysis and design procedures can be easily obtained through computer-aided design (CAD) processes. It must be noted that drainage and dewatering systems are closely related to the subsurface soil-rock properties, topographical features, groundwater levels, local climatological conditions, building locations, foundation types, etc. Careful planning and cooperation among developer, contractor, and designer are necessary to have effective, economical systems.

10.8.3 Monitoring Techniques

(1) General Discussion

Monitoring systems serve two purposes: (a) to determine sources where there is radon gas leaking, and (b) to check whether or not this radon mitigation method works. There are many commercially available devices measuring radon gas and radon daughters in the air, with special focus on indoor measurements. Various types of devices currently in use have been discussed in a number of references. The general mechanism is designed to allow radon

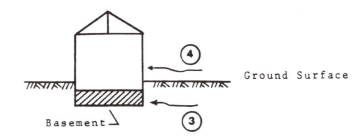

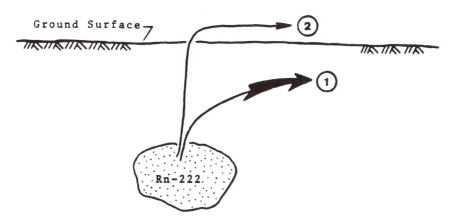

Figure 10.14 Systematic approaches for development of radon gas mitigation design procedures. 1. Remove large portions of radon gas in subsurface soil layers by dewatering or/and alteration of subsurface drainage pattern channelling subsurface water (radon contaminated water) away from site in question. 2. Some radon gas escaping from ground soil into the air can be removed by alteration of surface drainage pattern by channelling surface water. Some radon gas in the atmosphere will mix with dust and float in the air and join with natural dynamic equilibrium of the ecosystem. 3 and 4. Since large portions of radon gas are removed by steps 1 and 2, the remaining radon gas content will be significantly reduced. It still has the possibility that some will enter into buildings. In such cases, routine house cleaning can be used effectively.

gas to diffuse into it, then the device records the natural decay of the captive radon into its daughters through imperfections or tracks on special plastic plates inside the detector. As discussed in the previous section, measuring the radon content in the air generally results in a wide range of measured values. However, it will tell us if there is existing radon gas.

(2) Commercially Available Methods (EPA Approved Methods)

Several methods are used to measure radon daughter concentrations. One method is the Kuznetz method which was developed for use in mines. A sample of air is drawn through a filter that collects air particles with attached radon daughters. At present, the Kuznetz method is frequently used to screen houses to determine if radon gas is a significant problem requiring further sampling (DER, 1985a).

Another method to measure radon daughter concentration is the Radon Progeny Integrating Sampling Unit (RPISU). A RPISU pumps air through a small filter which removes air particles with radon attached by a radiation detecting chip which is later read through a special device to determine the absorbed radiation. The absorbed radiation is expressed as a radon daughter concentration by calibration. In all measurements, the best information results from long sampling periods, minimally a few months and up to a year to account for seasonal variations.

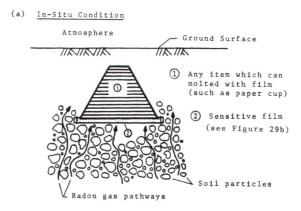

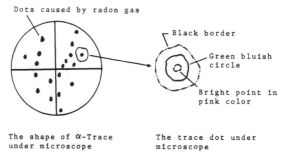

Figure 10.15 Subsurface radon gas measurement device. (a) *In situ* installation; (b) laboratory microscope evaluation (After Fang, H.Y., Luo, G.Y. and Chu, T.G. [1987], Proc. 1st Int. Symp. on Environmental Geotechnology, v. 2, pp. 219–231.)

(3) *Radon Measurements in Subsurface Soil Layer*

Radon measurements in subsurface soil layer have been discussed in Section 10.6. There are several models of equipment, such as EDA Model 200 and Pylon Model NR-1025, for subsurface radon content in soil measurements as reported by Rose et al. (1988). Other radon measurements in groundwater are reported by Graves (1987), and soil, water and rock are reported by Gundersen and Wanty (1992).

(4) *Photochemical Method*

Another *in situ* radon measurement in subsurface soil layers is presented by Fang et al. (1987). The method is based on photochemistry principles. This method can measure radon gas in the subsurface soil layer or in the air. The device is made of a sensitive photographic film. The film can catch the track of radioactive gas reflected on it. The device is available in various sizes and can be installed at various locations and depths in subsurface soil layers. The amount of radioactive gas can be determined under a microscope in the laboratory as shown in Figure 10.15. Figure 10.15(a) shows the *in situ* condition, and Figure 10.15(b) is laboratory microscope evaluation from sensitive photograph film.

Figure 10.16 illustrates the method for locating radon gas by the trace method. It must be noted that the information sought is the relative amount of radon gas not the absolute

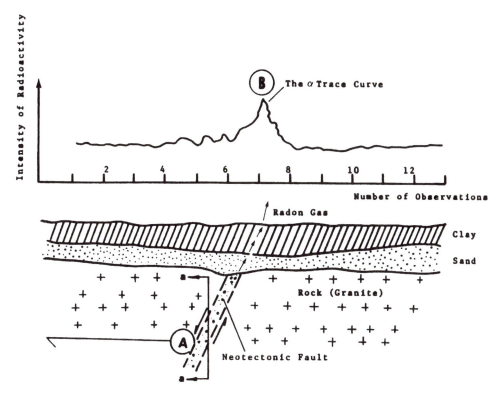

Figure 10.16 Method for locating radon gas — the trace method.

amount. The methods described in Figures 10.15 and 10.16 are low cost, simple to operate, and will serve the purpose to determine whether or not the proposed method described in Section 10.8.4 for radon mitigation works, even though we do not know the absolute amount of radon content in the location.

As indicated in Figure 10.14, the large portion of radon gas existing in the subsurface soil layer can be removed by alteration of subsurface drainage patterns and dewatering. There is always some radon gas escaping from ground soil into the atmosphere. In such a case, radon gas is removed by alteration of surface water drainage patterns, or pumping out radon-contaminated subsurface water. Another suggestion includes routine house cleaning. Damp and dirty basements will contain more radioactive dust than dry and clean ones. Construction processes contribute larger amounts of radon into the atmosphere. The less disturbance of the natural environments, the less radon content in the air. If construction operation is necessary, then careful planning is needed.

10.9 NUCLEAR WASTE DISPOSAL

10.9.1 Review of Background

As reviewed by Rahn (1986), high-level nuclear waste, generated during the reprocessing of spent reactor fuel, is a source of high radioactivity. These highly radioactive components in waste can be sources of groundwater contamination. Groundwater has been contaminated at some federal government facilities. For example, at Hanford, WA, high-level radioactive liquid wastes are stored in about 150 underground tanks overlying basalt and gravel. In 1986,

20 leaks were reported at Hanford. The most serious occurred in April 1973, when 450 m^3 of high-level waste soaked into the ground (Farney, 1974). This included 14,000 Ci of strontium (Sr-90) and 40,000 Ci of cesium (Cs-137). About 500,000 m^3 of low-level waste at Hanford has also been deliberately discharged into the ground soil by cribs and trenches. A large observation well system has been installed by government hydrogeologists to study the movement of contaminated groundwater; ruthenium (Ru-106) and H-3 were observed to be the more mobile elements (Graham, 1981).

At the Idaho National Engineering Laboratory (INEL), the National Reactor Testing Station reported that large volumes of radioactive water have been injected by a recharge well. Since 1952, radioactive wastes have been discharged into the Snake River aquifer which consists of interbedded basalt and gravel. Monitoring of the aquifer shows that tritium (heaviest isotope of hydrogen) has traveled at least 12 km from the well and plutonium (Pu-239) has traveled at least 2.5 km. Both Sr-90 and Cs-137 seem to be adsorbed and remain tied to the aquifer skeleton within approximately 2 km of the injection well (NRC, 1984). Tritium, 66 Ci of Sr-90, 120 Ci of Cs-137, 120 Ci of Co-60 (cobalt), and other wastes have also been discharged into seepage pits since 1954. Most of the dissolved Cs-137 and Sr-90 have been adsorbed by the sediment and basalt layers between the ponds and the water table, which is 137 m below the ground surface.

A major problem confronting the nuclear industry is the disposal of spent-fuel assemblies. The total spent fuel stored in power plants at the close of 1980 was about 27,000 assemblies weighing 7000 tons, having a volume of a cube 15 m in length, and emiting 2 billion curies of radioactivity. The spent-fuel assemblies contain artificial (transuranic) elements. During the past years at least 14 transuranium elements with atomic numbers greater than uranium (93 through 106) have been synthesized. Much of this work has been done by a group at the University of California at Berkeley. For further discussion on this aspect see standard textbooks on nuclear physics.

10.9.2 Sources and Types of Nuclear Wastes

(1) Basic Sources of Nuclear Wastes

There are four basic sources where nuclear wastes can be generated, namely:

1. Mill tailings from uranium (U) mines and mills with their piles of powdered rock containing large amounts of radium (Ra) and radon (Rn)
2. Assemblies of fuel rods stored under water near nuclear reactors
3. Nuclear wastes from commercial use, such as waste from hospitals and private research laboratories
4. Nuclear wastes from the manufacture of nuclear weapons

(2) High-Levels of Radioactive Wastes (HLRW)

High-level waste generated during the reprocessing of spent reactor fuel contains thousands of curies per cubic meter. Spent fuel is another source of high radioactivity as are the transuranic elements. These highly radioactive components in waste can be the sources of groundwater contamination.

As discussed in previous sections, indications are that high-level nuclear radiation comes from spent-fuel wastes. Reprocessed waste includes some 4×10^5 m^3 of highly radioactive solutions containing U-235, Pu-239, Sr-90, Ce-137 (cerium), and other radionuclides. Most high-level radiation and transuranic waste is stored at the U.S. Department of Energy facilities at the Hanford Reservation, WA, the Idaho National Engineering Laboratory, and the Savannah River Plant, GA.

Table 10.11 Nuclear Waste Solidification

	Existing solids	Total radionuclide inventory in "production" solids[a] (MCi)	Engineering development status	Backup research data on process properties, etc.
Hot solidification only	Crystalline = ceramic) radiophases	100	Successful fully engineered system (Hanford)	Small amount
	Cryptocrystalline radiophases (calcines)	100	Successful continuous operation for 8+ years (Idaho)	Small amount
	Glass (+ crystalline) radiophases	30[b]	Hot tests run in U.K and U.S.; hot plant in France	Very large amount
	Metal matrix composites	1	Hot pilot plant in Belgium	Moderate amount
Final disposal complete	Cement-based composites	1	Fully operational in U.S. (Oak Ridge)	Small amount
	Natural mineral matrix composites formed *in situ* and pumped into porous sandstone	100	Hot full-scale disposal in U.S.S.R.	Large amount

[a] Order of magnitude only.
[b] Including a total of 24 MCi in France.
After Roy (1980).

(3) Low-Level Radioactive Wastes (LLRW)

Low-level radioactive waste (LLRW) is very loosely defined. In general, LLRW come from commercial use from hospitals and private research laboratories. Most LLRW has a low radioactivity of about 35 Ci/m^3. However, some LLRW is extremely radioactive and may contain relatively large quantities of fission products with half-life longer than 25 years (Rahn, 1986). A typical example shows that 1 g of uranium (U-238) has a radioactivity of only 3×10^{-6} Ci and strontium (Sr-90) has a radioactivity of 140 Ci/g and a half-life of only 28.1 years.

10.9.3 Management of Nuclear Wastes

For disposal or management of nuclear wastes, the following four major components must be considered:

1. Storage: To reduce the intensity of radioactivity by storing wastes in a safe place because radioactivity decreases in intensity by a factor of 10 to 100 in 50 years
2. Solidification: Develop a technique for converting nuclear waste to a solid form for the purpose of reducing the mobility and dispersibility; Table 10.11 summarizes nuclear waste solidification techniques by Roy (1980)
3. Isolation: Isolating wastes from human society by storing them in isolated places such as deep sea floor deposition, bottom of mine shafts, distant and deep places in deserts, or space such as the moon
4. Emplacement: Safe transportation and insertion in a repository

To achieve an effective radioactive waste management system, three basic criteria must be evaluated as pointed out by McCarthy (1975), White (1980), and Chaney et al. (1986).

1. The waste form itself must be inert and insoluble in the repository environment.
2. The canister and overpack material must confine any radionuclides that do leach out of the primary waste form.
3. The rock formation should be impermeable.

Interactions between nuclear waste and surrounding rocks have been discussed by McCarthy (1978) and White (1980). The major interest in rock interaction with nuclear waste is attempting to determine what type of rock formation will interact less or be less sensitive to waste and remain in an impermeable state. Evidence at present indicates favoring a silicate-type rock.

10.9.4 Disposal Locations and Disposal Techniques

(1) General Discussion

The Federal Nuclear Waste Policy Act of 1982 and the Department of Energy (DOE) in 1983 give general guidelines for the location of an underground site to serve as the nation's first permanent storage facility for high-level nuclear waste. The permanent storage of high-level radioactive waste will be in geologic repositories specifically mined for that purpose. Major problems with long-term confinement of high-level radioactive waste are the threat of human intrusion and the possibility of transport of chemicals in solution through groundwater migration (Davis, 1984; TRB, 1986).

(2) Disposal Locations

In 1985 the three leading site candidates chosen by Department of Energy (DOE) for HLRW dump sites include: Deaf Smith County in the Texas panhandle; Yucca Mountains, about 160 km northwest of Las Vegas, NV; and the federal government's Hanford Works near Richland, WA.

Prior to 1962, all LLRW was disposed at six federal facilities (NRC, 1984). After 1962 six commercial facilities were opened, namely: Maxey Flats, KY; Sheffied, IL; West Valley, NV; Beaty, NY; Richland, WA; and Barnwell, SC. They handle most of the nation's commercially generated LLRW, which amounts to about 75,000 m³/year. The federal government buries about an equal volume of LLRW from research and defense activities at five major U.S. Department of Energy facilities and several minor sites.

The three states which accept commercial LLRW have become increasingly reluctant to continue as the nation's only nuclear wastes dumps. In 1980, the U.S. Congress passed the Low-Level Waste Policy Act which mandated that individual states establish their own sites or join interstate compacts on a regional basis by 1986.

(3) Radioactive Waste Storage

As pointed out in Section 10.3.3, radioactive waste storage is one of the major tasks at the present time. It includes the following steps: interim storage as liquid; conversion to solid; interim storage as solid; transportation to, and storage or disposal in, a federally operated surface or subsurface repository. Storage or disposal must prevent contact of the nuclides with atmosphere, biosphere, or hydrosphere for a period of at least 1000 to several hundred thousand years. Numerous methods for the storage or disposal of HLW have been proposed. Winograd (1974) proposed that radioctive waste storage be in an arid zone.

(4) Waste Form and Leaking Routes

Most radioactive waste is in liquid form. The main leaking routes are infiltrated into trenches. Burial trenches are typically 15 m wide, 10 m deep, and up to 200 m long. After waste containers are emplaced, about 2 m of a clay cap is placed over the waste. Compaction, settlement prediction, and other geotechnical aspects are closely followed similar to the landfill's clay liner analysis and design procedures.

Liquid form of waste is called leachate which may migrate into groundwater flow aquifer. According to Robertson (1984): "During the period when the six commercial disposal sites were chosen, there were no uniform regulations providing comprehensive site-specific geo-hydrologic criteria to be applied to the selection and operation of disposal sites. The Atomic Energy Commission had some general guidelines and performance standards for low-level sites but allowed states to set their own standards if they assumed responsibility for regulation sites."

(5) Comments on Existing Waste Sites

Some comments on waste sites as discussed by Rahn (1986) and others are presented as follows:

1. The West Valley, NY site is underlain by clay-rich glacial till. Glacial till is unsorted drift, or the mixture of rock fragments and fine materials left by melting glaciers. Water infiltrates into trenches, causing the 'bathtub effect' whereby water seeps out on the ground.
2. The Maxey Flats, KY site is shale having low permeability but containing numerous fractures which control groundwater flow. Plutonium and other isotopes have migrated laterally through a thin fractured sandstone bed.
3. The Sheffield, IL site was apparently intended to be a clay-rich glacial till, but it was discovered later that the till contained gravelly sand layers with high permeability. Tritium has been found in some of the monitoring wells surrounding the site. Tritium is moving at about 8 m/year through the glacial outwash. Outwash is stratified accumulation of water-deposited drift. The material is laid down by the meltwater streams issuing from the face of the glacier ice.
4. The Barnwell, SC site was placed in coastal-plain sediments consisting of sandy clay which is somewhat more permeable than the tills or shale of the other sites. U.S. Geological Survey reported that tritium has migrated vertically and laterally through permeable sand layers and is believed to be moving at 27 m/year.
5. Beatty, NV and Richland, WA both are in arid regions and do not suffer the problems of the sites in more humid locales. The average annual rainfall at both sites is 10.1 and 16.5 cm, respectively. Both sites are excavated into mixed coarse-grained unconsolidated sediments.

Nuclear waste interaction with groundwater and other related geotechnical engineering properties is presented by Rahn (1986), McCready (1986), and in Section 10.8. Flooding has been a problem at all sites. The protection of waste sites is also important.

10.10 UTILIZATION OF NUCLEAR ENERGY FOR CONSTRUCTION APPLICATIONS

10.10.1 General Discussion

The development of nuclear explosives has made a new low-cost energy-saving high-power energy source. In late 1940s, the U.S.S.R. demonstrated large-yield conventional explosives useful for excavation, mining, and dam construction. In 1957, the U.S. Army Corps of Engineers and Lawrence Radiation Laboratory, Livermore, CA established a program to investigate possible industrial applications for nuclear explosives. Their investigations included excavation of roadway cuts, aggregate production by open pit mining, drainage diversion or interception by craters and landslide removal, and encouragement or stabilization. Some of their findings that have been reported will be discussed in the following section.

Table 10.12 Structural Damage Thresholds for Buildings and Equipment Near Underground Nuclear Explosions

Type of structure	Type of damage	Peak surface velocity threshold, major damage (cm/sec)	Corresponding scaled distance in granite[a] (ft/kt$^{1/3}$)
Residential (old)	Plaster cracking	10	2500
Residential, concrete block	Cracking	20	1250
Cased drill holes	Vertical displacement, horizontal offset	40–50	750–550
Mechanical equipment (pumps, compressors, generators, etc.)	Skids bent, shafts misaligned	100	375
Prefab metal buildings on concrete pads	Cracked pads, distorted steel	150	250
Rigid steel tanks, 50 gal to several thousand gallons	Buckling	≥300	125
Utility poles	Falling	≥300	125

[a] Distances are for structures resting directly on a hard crystalline rock such as granite, with the explosion detonated in the same medium, and do not include a safety factor. They are, therefore, a minimum and apply only to this ideal case. For most industrial application, distances would probably be substantially increased (doubled or more) to provide a safety factor and to take into account possible presence of different and less favorable geologic conditions.

After Cauthen (1964); cited by Hansen, S.M. and Toman, J. (1966), Aggregate Production with Nuclear Explosives, HRR no. 107, pp. 37–53.

10.10.2 Research Findings and Safety Considerations

(1) Research Findings

Research findings include: (1) the effect on crater dimensions of variation in the spacing between charges; (2) the effect of spacing on the irregularity or cusping in the crater; and (3) the shape of the lip relative to the lip obtained with point charges, on both the sides and the ends of the crater. Research done at Lawrence Radiation Laboratory, Livermore, CA and U.S. Army Corps of Engineers has come to the following conclusions as reported by Nordyke and Circeo, (1966):

1. Use of a spacing equal to approximate a single crater radius results in a smooth-sided crater with apparent dimensions about 10 to 20% larger than expected on the basis of single charge data.
2. Use of a spacing of about 1.25 times a single crater radius results in a ditch with dimensions approximately equal to those expected from single charges.
3. Use of spacing of 1.5 times a single crater radius results in a crater which is somewhat smaller than a single crater radius and quite irregular in cross section.

(2) Benefit/Cost Ratio and Safety Considerations

Nuclear excavation costs in hard rock are estimated to be competitive with conventional excavation in the 10-kt range, rapidly decreasing to a few cents per cubic yard in the megaton range. Advancement in the development of low-fission thermonuclear explosives is expected to reduce radioactivity levels 100-fold below 1962 Sedan levels.

Safety considerations are extremely important. Any use of nuclear explosives will require careful attention to problems of safety. The Atomic Energy Commission and the Lawrence Radiation Laboratory have provided some guidelines as shown in Table 10.12.

10.11 OTHER APPLICATIONS

In recent years the need has been for a quick and accurate method of determination of material properties such as unit weight and moisture content of materials used in the construction of highway pavement and its components. Using nuclear devices for nondestructive tests, the calibration of the devices is critical. The best calibration curve for unit weight determination is obtained from the field data.

The laboratory calibration curve based on all materials tested gives the best results for moisture determinations. The major advantages for using nuclear devices are to save time, because the nuclear density tests are a nondestructive test method (NDT); the laboratory maximum densities cannot be corrected for coarse particles as suggested by ASTM Standards. ASTM suggests two common procedures:

1. Soil density and moisture content measurements: Numerous researchers report using nuclear measurements of soil properties (HRB, 1966). ASTM provides a standard test procedure (ASTM D3017-88 [1993]) for measuring moisture content and density for soils and other highway materials.
2. Asphalt contents in paving mixture: Use the sources of gamma energy for measurement of pavement cracking, moisture content, and density of various types of construction materials and earth structures (Fwa and Tan, 1992).

10.12 SUMMARY

1. Environmental radiation aspects of geotechnology including nuclear waste, radon gas, nuclear energy for construction, and utilization of nuclear energy as nondestructive test methods are discussed.
2. Basic nuclear physics is reviewed including nuclear structure, nuclear physics and chemistry, nuclear reactions, radioactivity, and radioactive decay process.
3. The nature of the noble gases family and radioactive and toxic radon gas is discussed including origin, characteristics, distributions, and engineering properties. The mechanism of radon gas emanation in rock and minerals is examined. A relatively low-cost radon mitigation method is proposed. This method will work to protect residential houses, as well as minimizing other effects to protect natural environments.
4. Nuclear waste disposal techniques are discussed including sources and types of nuclear wastes. Management of nuclear wastes and procedures is also examined.
5. Utilization of nuclear energy for construction applications and benefit/cost ratio and safety considerations is discussed. Also utilization of nuclear energy as a nondestructive (NDT) test used in highway construction is examined.

PROBLEMS

10.1 Define the following terms: nucleic acids, tritium, transuranic elements.
10.2 What parts of its structure are the major responsibility for an atom's mass and for its chemical behavior?
10.3 What happens to the mass number (A) and atomic number (Z) of a nucleus that emits a gamma-ray photon? What happens to its mass?
10.4 Discuss the causes and control of groundwater contamination by radioactive waste.
10.5 Discuss the characteristics of high-level and low-level nuclear wastes.
10.6 Radium (Ra) spontaneously decays into the elements helium (He) and radon (Rn). Why is radium itself considered an element and not simply a chemical compound of helium and radon?

10.7 Why is radon gas considered to be of the noble gas family? Define a radon daughter and radon progeny.

10.8 Discuss the benefit/cost ratio of using nuclear energy for construction.

10.9 What are the reasons for using gamma-rays as a nondestructive (NDT) test for determination of material properties?

Compressibility and Stress-Strain-Time Behavior of Soils

11.1 INTRODUCTION

For analysis and design of foundation structures, two important soil parameters, namely, compressibility and stress-strain of soil, must be examined. The compressibility or volume changes include shrinkage, swelling, compaction (densification), and consolidation (settlement). The stress-strain-time characteristics of soil include strength, failure criteria, and creep. Shrinkage and swelling have been discussed in Sections 6.2 and 6.3. The other parameters are discussed in this chapter.

11.2 SOIL COMPACTION (DENSIFICATION)

11.2.1 General Discussion

Compaction or densification is the simple and low-cost mechanical process for ground improvement and is a major part of geotechnology. The main purpose of this process is to change the soil voids by means of mechanical devices in order to reduce the settlement, hydraulic conductivity, and to increase the shear resistance or bearing capacity of soil.

The simplest way to illustrate this process is through the unit weight (density)-moisture relationship of soil from the compaction tests (Proctor, 1933). The compaction test procedures have been standarized by both American Society of Testing and Materials (ASTM) and American Association of State Highway and Transportation Officials (AASHTO). There are two major types of test procedures: the Standard AASHTO (ASTM D698-91; AASHTO T99-70) and the Modified AASHTO procedures (ASTM D1557-91; AASHTO T180-70). The major difference between these two methods is that the modified AASHTO gives a higher compactive effort than the standard procedure. A typical relationship for a silty clay is shown in Figure 11.1. It must be pointed out that the terms of unit weight and density are exchangeable in geotechnical engineering.

In examining Figure 11.1, the dry unit weight can be computed from wet unit weight and the corresponding moisture content. The point **c** shown in the curve is called *maximum dry unit weight* (γ_0) and the corresponding moisture content is called *optimum moisture content* (OMC). The moisture content below the OMC is referred to as *dry-side* and above the OMC is referred to as *wet-side*. The curve on the right side of the figure is called the *zero-air-void curve* and is defined as the weight of solids per unit volume of a saturated soil mass.

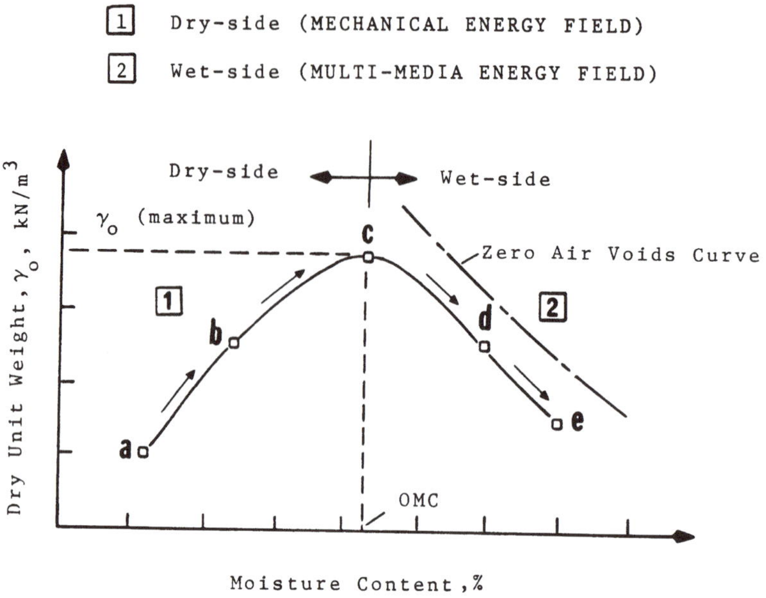

1 Dry-side (MECHANICAL ENERGY FIELD)
2 Wet-side (MULTI-MEDIA ENERGY FIELD)
Dry-side
Wet-side
γo (maximum)
Zero Air Voids Curve
c
b
a
d
e
1
2
OMC
Dry Unit Weight, γo, kN/m3
Moisture Content ,%

Figure 11.1 Unit weight vs. moisture content of a silty clay.

As reported by Shook and Fang (1961), several types of unit weight-moisture curves are indicated depending on soil types. For common soil such as A-6 soil used for earth dams and highway embankments, a single-peak-shaped curve is indicated in Figure 11.2[2]. If the granular material such as subbase or crushed stone base material is used in the highway embankment, there is a double peak or $1^1/_2$ peak as indicated in Figure 11.2[1], and if heavy clay content is contained, then the curve will be oddly shaped such as Figure 11.2[3].

11.2.2 Compaction Theories

To understand the mechanism of the soil densification process, it is necessary to examine the compaction theories and concepts. There are several theories and concepts proposed to explain the compaction process of soil and moisture-unit weight relationships. Brief discussions of some theories and concepts are presented as follows.

(1) Proctor Theory (1933)

Proctor assumes that the soil mass is composed of gravels, sands, silts, and clays, and compaction is the act of forcing the fine grains into the voids between the larger grains. It is contended that the water coats the surface of the soil grains and serves as a lubricant which reduces the frictional resistance between the soil particles and permits the compacting force to become more efficient in arranging the fine soils into the voids between the larger particles. If the moisture content is not sufficient to produce adequate lubrication, the unit weight of the compacted soil will be relatively low because the compacting force is not enough to overcome the frictional resistance between the soil grains.

(2) Hogentogler Theory (1937a)

Hogentogler proposed that compacted soils undergo stages of wetting; when the moisture contents are expressed as percentages of the combined volumes of soil solids and moisture, the moisture-unit weight relation becomes a series of straight lines with different slopes. Four

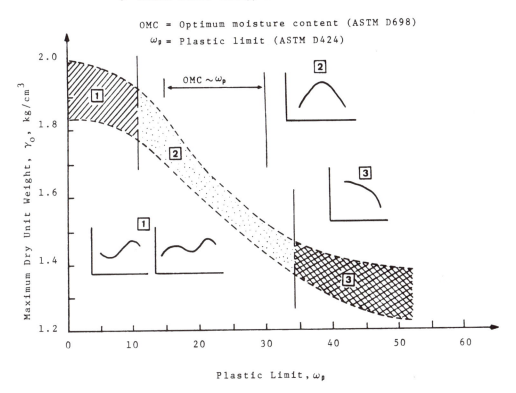

Figure 11.2 Maximum dry unit weight vs. soil types as reflected by plastic limit.

stages are proposed as hydration (points **a** to **b** in Figure 11.1), lubrication (points **b** to **c**), swell (points **c** to **d**), and saturation (points **d** to **e**).

(3) Lambe Theory (1958)

Lambe used a physicochemical concept to explain the unit weight (density)-moisture relationship of soil. Low density, as shown in point **a** of Figure 11.1, is due to insufficient water for the diffuse double-layer which gives a higher concentration of electrolytes and reduces the interparticle repulsion causing a tendency toward flocculation of the colloids. The flocculation structures mean a low-degree of particle orientation and low density. Increase the moisture content from **a** to **b** and the double layer expands, and electrolyte concentration reduces and permits a more orderly arrangement of particles which gives higher density. Further increase of moisture content from **b** to **c** induces further expansion of the double layer, further reducing of the electrolyte concentration, and continued reduction in the net attractive forces between particles. For the higher compactive effort which gives greater input of work, the more nearly parallel are the clay particles.

(4) Compaction Theory Based on Particle-Energy-Field Theory

The particle-energy-field theory (Chapter 2) introduced herein is to explain the density-moisture relationships of soil in the environment. As indicated in Figure 11.1, there are two distinct characteristics of soil existing in the density-moisture relationship, the dry side and the wet side. Since the behavior of soil in these two stages is significantly different, further explanations of these relationships are needed and particle energy field theory may help to clarify this relationship as follows:

1. On the dry side, the soil particle arrangement is controlled by mechanical energy as described by the Proctor theory, forcing the smaller soil particles into the voids between the larger particles by compaction effort. There is no significant physicochemical interaction between soil-water interaction on the dry-side condition.
2. On the wet side, the soil-water interaction between particles will be influenced by physico-chemical behavior of soil as described by Lambe theory.
3. In the case of contaminated soil, the physicochemical-biological effects affect the soil-water interaction as discussed in Sections 5.7 and 5.8. In both cases (2) and (3), interactions are both influenced by multimedia energy fields such as thermal, electrical, and electromagnetic energies.

11.2.3 Factors Influencing Compaction Test Results

Factors influencing compaction test results have been discussed by numerous investigators (Johnson and Sallberg, 1962; Hilf, 1991). Some additional comments on environmental factors affecting test results are presented as follows:

1. To avoid irregular and meaningless unit weight-moisture curves in laboratory and field tests on highly cohesive soils, it is imperative that the moisture be evenly distributed throughout the secondary soil aggregates. This may take from 1 to 7 days in the case of highly cohesive soils to which water is added in the dry condition. This time may be reduced by the use of live steam.
2. Because the viscosity of water and the degree and intensity of the interaction of mineral surfaces with water are functions of temperature, the moisture-density relationships are also temperature dependent. In general, the temperature effect increases with increasing specific surface (surface per unit volume) and therefore with clay content, and is also a function of the clay mineral and exchangeable ions and the electrolytes in the aqueous phase. As discussed in Section 8.4. it is indicated that compaction characteristics of frozen ground are significantly different in comparison with nonfrozen soil. Figure 11.3 shows the effect of freezing temperatures on compaction of silty fine sand. Significant effects when temperature goes below freezing point are indicated.
3. Compaction procedures significantly affect maximum unit weight and OMC. Four test procedures are used to illustrate the effects on compaction test procedures including recompaction on same soil and compaction on fresh soil sampled with various moisture distribution conditions. Results of this soil testing are presented in Figure 11.4.
4. Effects of exchangeable ions on characteristics of compacted Putnam soil as reflected on penetration of cone results are presented in Figure 11.5. Compaction tests were performed based on Standard Proctor test (ASTM D698-91). Results of maximum dry density, optimum moisture content (OMC), and penetration of cone results are obtained. In examining Figure 11.5, encircled points represent data at OMC. The experimental data indicated that the potassium (K) soil received higher maximum dry density and lower OMC.
5. As to laboratory and field compaction methods, unit weight vs. moisture content relationship in the laboratory and field conditions are significantly different, because the environmental conditions in both conditions are different. In the laboratory, the testing condition is specified by ASTM or AASHTO such as temperature and pore fluid. However, in the field condition, the local environmental factors such as temperature, rainfall, pollution migration, etc. cannot be controlled properly in comparison with the laboratory condition. Also, compaction results are influenced by compaction equipment and procedures. However, equipment and procedures used in both laboratory and field are different and therefore, results obtained are also different. Even within the laboratory or in the field, equipment or procedures are also different. Figure 11.6 presents results from two types of dynamic methods used in the laboratory, namely, shaking table and drop hammer. In this study, the effects of gravel content in the soil mixture are also evaluated.

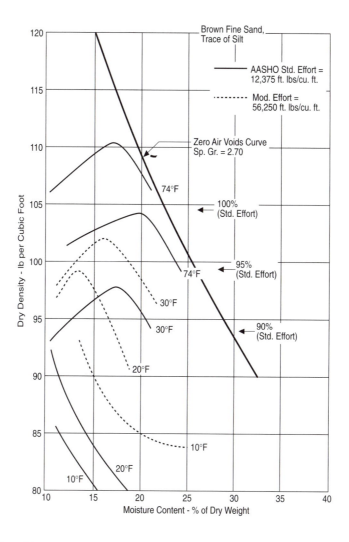

Figure 11.3   Effect of freezing temperatures on compaction of a silty fine sand. (After Hofmann, [1961]; cited by TRB [1971].)

11.3 DYNAMIC CONSOLIDATION (DYNAMIC DENSIFICATION)

11.3.1 General Discussion

Dynamic densification, or *dynamic consolidation,* is a mechanical process to consolidate loose soil deposits at great depths. The process used at the present time is not new. The largest construction project using this technique was during World War II in early 1940 when an airfield was built in Kunming, in southwest China, for U.S. Flying Tiger B-29 bomber landings. The method is frequently used around the world, especially in China, but the lack of publicity has encouraged little scientific study of the process. An early model of the dynamic densification equipment used in 1957 in China shows the total pounder weight about 36 to 72 kN dropped from 10 m.

In 1970, the Menard group (Menard and Broise, 1975) gave a scientific approach for the analysis of the dynamic densification process in which they included vibration behavior during the process in correlation with basic geotechnical properties of soil. Since then, this method

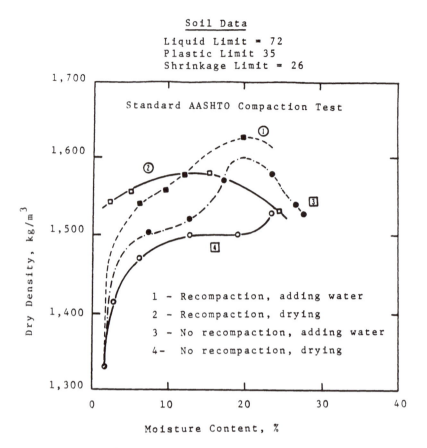

Figure 11.4 Effects of recompaction and moisture distribution on unit weight-moisture curve. (After Tamez, E. [1957], Some factors Affecting the Dynamic Compaction Test, ASTM STP 232, pp. 54–66.)

has been widely used in many large-scale construction projects for densification of granular material at great depths and it also has been extended and shows some success on clay deposits. The dynamic densification equipment currently used in construction has a total pounder weight reaching as much as 356 kN (40 tons) and the height of drop has reached to 40 m. Due to the improvement of crane capacity, the impact energy from pounder to the ground surface can reach to 10.8×10 kJ (4000 ft-ton).

11.3.2 Load-Deformation Relationship and Effective Depth

Figure 11.7 illustrates the typical soil-load interaction in the laboratory study during dynamic consolidation (densification). In examining Figure 11.7, a weight (W_x) is freely dropped from a given height (H_x). The radius (r) and depth (δ) of crater and the effective depth (D_e) are also indicated in Figure 11.7.

The *effective depth* (D_e) is defined as shown in Figure 11.7 as how deep the pounder (weight), dropped freely from a certain height, will affect the fill material below the ground surface. Menard and Broise (1975) proposed that the effective depth or depth of influence is equal to the following.

(1) Menard and Broise Method (1975)

$$D_e = [W_x\, h_x]^{0.5} \tag{11.1}$$

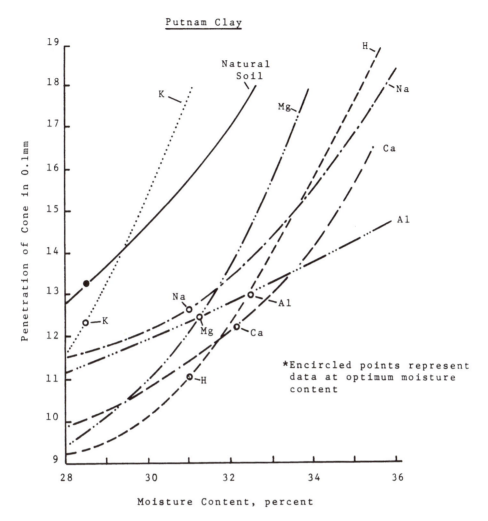

Figure 11.5 Effect of exchangeable ions on characteristics of compaction soil as reflected on penetration of cone results (After Winterkorn and Moorman [1941].)

(2) Leonards et al. Modified Method (1980)

$$D_e = 0.5 \, [W_x \, h_x]^{0.5} \tag{11.2}$$

(3) Lukas Method (1980)

$$D_e = (0.65 \text{ to } 0.80) \, [W_x \, h_x]^{0.5} \tag{11.3}$$

where D_e = effective depth or depth of influence, m
W_x = weight of pounder, metric tons
h_x = height of free drop, m

(4) Charles et al. Method (1981)

Based on field studies that incorporate the strength of the tamped soil the following is proposed as shown in Equation (11.4):

$$D_e = 0.4d \, [(E/A)(1/d)(1/s)]^{0.5} \tag{11.4}$$

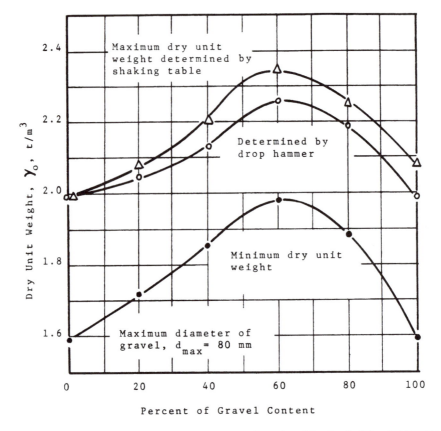

Figure 11.6 Percent of gravel content relating to the dry unit weight of soil. (After Shi [1981].)

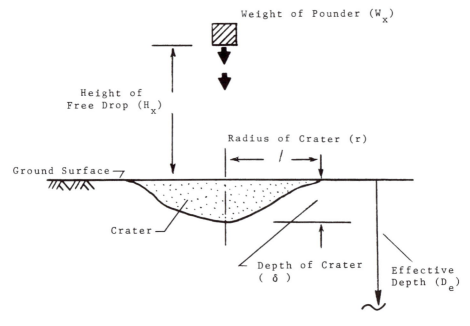

Figure 11.7 Load-deflection relationship under dynamic consolidation.

where   d = diameter of the tamper
       E = applied impacts energy
       A = area of impact
       s = undrained shear strength of soil

The applied energy (Equation 11.4) is expressed in terms of the energy per unit area of the impact surface of the tamper. Luongo (1992) has presented data on impact energy parameters from case histories of various field projects. A minimum applied energy of 200 t-m/m^2 has been suggested by Hayward Baker Co., an experienced consulting firm who conducts field dynamic compaction tests frequently. Inyang (1992) proposed the concept of *specific energy* to quantitatively describe the magnitude of applied energy. In such case, the specific energy applied can be estimated as follows:

$$SP = N(E)\ (H)\ (A)\ (D) \tag{11.5}$$

where   SP = specific energy expressed as energy per unit volume
       N = number of impacts applied at the same point
       E = kinetic energy applied
       A = area of the impact surface of the hammer
       D = thickness of the soil layer(s) compacted

(5) Fang-Ellis Method (1983, 1995a)

Equations (11.1) to (11.3) have been used for the *in situ* ground improvement process and field control. The equations do not consider the type of fill material, size of pounder, and degree of saturation of ground soil. These additional parameters are considered by Fang and Ellis (1983) cited by Fang (1995a) in evaluation of the effective depth in the laboratory investigation. To understand the densification process, it is necessary to examine the mechanism and mechanics of soil-pounder interaction. As shown in Figure 11.7, when the pounder is applied to the soil mass, deformation will result from immediate elastic and inelastic deformation of the soil structure. Pore water drains from the soil resulting in reorientation of soil particles. This process depends upon the soil properties, drainage conditions, stress history, and environmental conditions. The *coefficient of effective depth* (ψ) shown in Figure 11.8 is a function of degree of saturation and radius/depth ratio during dynamic compaction where the effective depth, D_e, can then be computed from Equation (11.6).

$$D_e = \psi[W_x\ h_x]^{0.5} \tag{11.6}$$

where ψ = coefficient of effective depth

The dynamic consolidation process has been frequently used in large-scale construction projects for densification of deep granular soils and more recently for the densification in landfill sites and desert regions.

11.4 CONSOLIDATION

11.4.1 General Discussion

Consolidation is one of the important processes in geotechnical engineering. It accounts for settlement of structures founded on compressible soil deposits and its understanding, together with pertinent laboratory tests, permits the calculation of total settlement, as well as

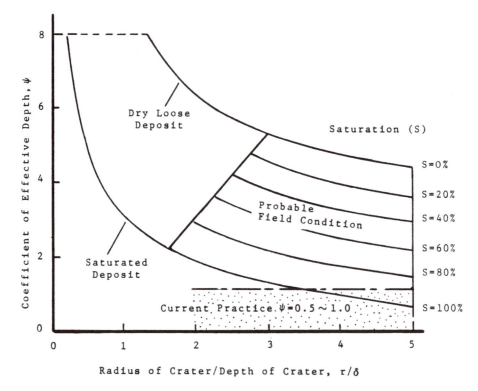

Figure 11.8   Chart for determination of coefficient of effective depth. (After Fang and Ellis [1983]; cited
by Fang [1995a].)

of its time rate. Extensive research on this subject under various aspects both theoretically
and experimentally has been carried out since their significant value was introduced in the
early 1920s. Taylor (1948) carried out comprehensive laboratory studies on various aspects
including drainage systems, secondary compression, and standardization of the test procedures. Figure 11.9, a schematic diagram, illustrates the settlement vs. time curve. Also shown,
the Terzaghi consolidation theory concerns the mechanical energy only. Other causes such
as physicochemical and biological are not included in the Terzaghi theory. More recently,
the consolidation theories other than the Terzaghi concept have been introduced into environmental geotechnology include electrokinetic, thermal, and electromagnetic approaches.
Using physicochemical approach as the consolidation mechanism of soft clays is also suggested (Akagi, 1994).

11.4.2 Terzaghi Consolidation Theory

When a load is applied to a saturated compressible soil mass, the load is usually
carried initially by the water in the pores because the water is relatively incompressible
when compared with the soil structure. The pressure which results in the water because
of the load increment is called *hydrostatic excess pressure*, because it is in excess of that
pressure due to the weight of water. If the water drains from the soil pores, the hydrostatic
excess pressure and its gradient gradually decrease and the load increment is shifted to
the soil structure. In other words, the transfer of load is accompanied by a change in
volume of the soil mass equal to the volume of water drained. This process is known in
soil mechanics as *consolidation*. A theory relating to loading, time, and volume change

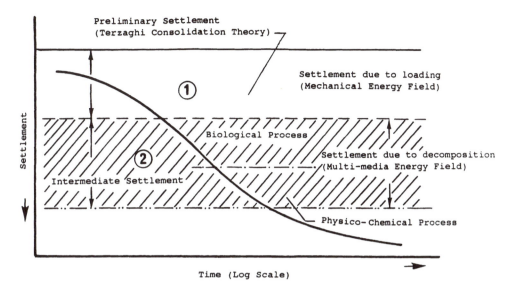

Figure 11.9 Schematic diagram illustrating settlement vs. time for a degradable material.

was proposed and has become known as the *Terzaghi theory of consolidation*. One of the major assumptions in the theory is that volume change and the outflow of porewater occur in one dimension only. For this reason it is sometimes referred to as the Terzaghi one-dimensional consolidation theory.

The major conclusions of the Terzaghi consolidation theory are that it produced two main equations: (1) the settlement equation, and (2) time rate. The *degree of consolidation* is a link between these two equations. Other parameters associated with the theory include: *compression index*, c_c, *coefficient of compressibility*, a_v, and *coefficient of consolidation*, c_v. From a standard one-dimensional consolidation test (ASTM D2435-90), the *coefficient of permeability*, k, can be computed from the following equation:

$$k = c_v \, m_v \, \gamma_\omega \tag{11.7}$$

where c_v can be obtained from a routine consolidation test and m_v is known as the *coefficient of volume compressibility*. It is equal to:

$$m_v = a_v / 1 + e \tag{11.8}$$

The m_v value in Equation (11.8) is used to indicate the compression of the clay per unit of original thickness due to a unit increase of pressure. The unit of m_v is the same as a_v (cm/kg), where e is the average void ratio for the increment during the consolidation test. Zeevaert (1983) reported that the k value presented in Equation (11.7) is a linear relationship with the coefficient of consolidation, c_v, when m_v varies from 0.1 to 0.001 cm/kg. In addition, a comparison of k, computed from Equation (11.7) and from conventional laboratory permeability tests of various soil types, is shown in Figure 11.10. To express the expansion which may occur upon unloading of a soil sample one commonly uses the *swell index*, c_s, as shown in Figure 11.10. The c_s values are always much smaller than the c_c values for a virgin compression. The typical value for Boston blue clay ranges from 0.07 to 0.09 and for soft silt from Shanghai region (Section 13.10) from 0.02 to 0.04.

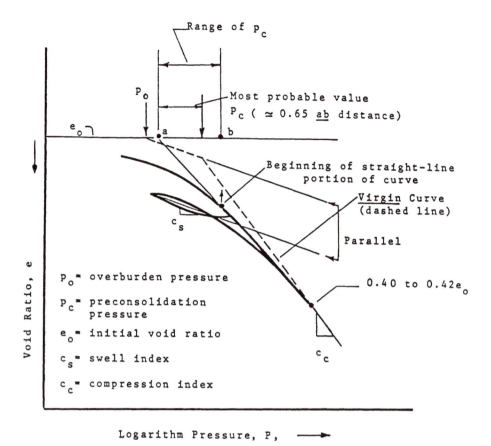

Figure 11.10 Typical void ratio vs. log pressure curve.

11.5 NORMALLY AND OVERCONSOLIDATED CLAYS

11.5.1 General Discussion

As discussed in previous sections, due to the past stress history of a natural soil deposit, soil deposit may be overly consolidated. If maximum intergranular pressure is larger than the present *overburden pressure*, p_0, then this intergranular pressure is called *preconsolidation pressure*, p_c, or called past pressure. There are numerous reasons and factors causing this preconsolidation including geological features, change in pore pressures, change in soil structures, and change in environmental conditions as discussed by Brumund et al. (1976) and Holtz and Kovacs (1981). Causes of preconsolidation are summarized in Table 11.1.

11.5.2 Preconsolidation Pressure and Overconsolidation Ratio

(1) Preconsolidation Pressure

If preconsolidation pressure is caused by loading conditions, then it can be estimated from a void ratio, e, vs. logarithm pressure, p curve, as shown in Figure 11.10. The techniques for estimating, p_c, have been proposed by many investigators as discussed in standard textbooks. The range of the p_c value is indicated in Figure 11.10.

Table 11.1 Causes of Preconsolidation of Soil

Causes	Types	Energy field
Loading	Overburden pressures	Mechanical energy field
	Preloading	
	Past structures	
	Geological loading (glaciation)	
Environmental (chemical, physicochemical, geomicrobiological)	Leaching process	Multimedia energy field
	Ion exchange	
	Temperature and heat	
	Freezing-thawing	
	Wet-dry	
	Acid rain	
	Groundwater fluctuation	

(2) Overconsolidation Ratio (OCR)

The ratio between preconsolidation pressure, p_c, and overburden pressure, p_0, is called *overconsolidation ratio*, OCR, as:

$$OCR = p_c/p_0 \qquad (11.9)$$

where OCR = overconsolidation ratio
p_c = preconsolidation pressure
p_0 = overburden pressure

If OCR = 1.0, then the soil deposit is called *normally* consolidated clay.
If OCR > 1.0, then it is called *overconsolidated* clay.
If OCR < 1.0, it is called *underconsolidated* clay.

Degree of overconsolidation can be classified as: (1) lightly overconsolidated clay: the range of OCR greater than 1.0 and less than 2.5; (2) heavily overconsolidated clay: if OCR value is beyond 8.0. OCR value reaches as high as 35. This OCR value can be used to indicate the stress history of subsurface soil condition as shown in Figure 11.11. In examining Figure 11.11 it can be seen that there are three distinct subsurface layers with various OCR values which show the stress history of Shanghai soft clay region.

(3) Shear Characteristics of Overconsolidated Clay

Based on the typical stress-strain curve, volume change, and porewater pressure of normally consolidated and overconsolidated clays discussed in the standard text (Ladd, 1971; Holtz and Kovacs, 1981), it can stated that: (1) the overconsolidated soil specimen has a greater strength than the normally consolidated clay; (2) the maximum shear of overconsolidated clay occurs at a much lower strain than for the normally consolidated specimen; (3) for the porewater pressure, the normally consolidated specimen develops positive porewater pressure. However, for the overconsolidated specimen, after a slight initial increase the porewater pressure goes negative; (4) the volume change for overconsolidated clay is expansion during shear, while normally consolidated clay compresses or consolidates during shear; and (5) the Mohr failure envelopes for total and effective stresses are different for the normally consolidated and overconsolidated clays. For normally consolidated clay, the internal friction angle, ϕ (total stress), is less than ϕ', (effective stress), and often it is about one half of ϕ'. However, for the overconsolidated clay the ϕ' is less than ϕ.

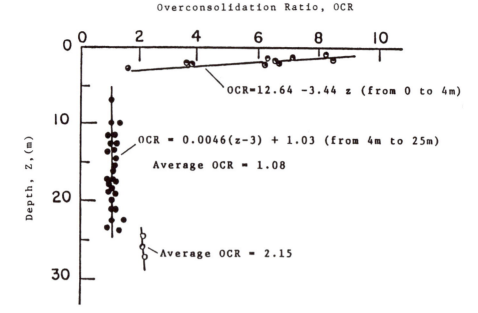

Figure 11.11 Overconsolidation characteristics of alluvia sediments. (Data from Tongji University.)

11.5.3 Factors Affecting Consolidation Test Results

(1) Temperature or Heat

Temperature or heat effects on compressibility have been discussed in Section 8.12. Additional information on volume changes during the consolidation process due to heat is reported by Tsuchida et al. (1991) and Towhata et al. (1993).

(2) Pore Fluid

Pore fluids have significant effects on the compressibilty of soil. As reported by Waidelich (1958), the consolidation tests were run with the three clays and the seven organic and inorganic liquids including water. The data show that the respective indices for the clays with nonexpanding lattices (kaolinite and white clay) vary with the dipole moment. For the kaolinite, the larger the dipole moment of the liquid used, the larger is the slope of the compression curve as reflected by the compression index (c_c). For the white clay (halloysite), however, the larger the dipole moment of the liquid used, the smaller is the slope of the compression curve.

(a) Nonexpansive Clay

The liquids aniline ($C_6H_5NH_2$) and water (H_2O) are two exceptions common to both kaolinite and white clay. It is possible that the aniline engaged in base exchange with the two nonexpanding soils causes the formation of anilinium soils. Water is the other exception to the observation. Water is one of the most common liquids and, yet, very little is really known about it. Methyl alcohol (CH_3OH) is another exception in the case of the white clay. This liquid, having a molecular structure closely resembling water, acts similar to water on occasion. As discussed by Waidelich, the dipole moment can be seen to be a localized force coupled within the molecule itself. The particles of a nonexpanding clay do not absorb liquids within the lattice as an expanding clay will.

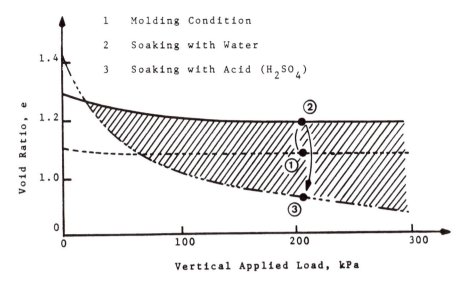

Figure 11.12 Effect of acid on the compressibility of laterite soil. (Data from Sun [1989].)

(b) Expansive Clay

Montmorillonite clay, which possesses an expanding lattice, is affected by the dipole moment to a lesser extent than the clays with a nonexpanding lattice. Rather, the dielectric constant, which is an integral function of the dipole moment and molecular structure, seems to have a more dominating influence on clay minerals.

Leonards and Girault (1961) reported the effective pore fluid on the compressibility of undisturbed samples of Mexico City clays. In this study, the naturally occurring pore water was replaced with carbon tetrachloride (CCl_4) by circulating the pore fluid through the sample until the effluent was pure carbon tetrachloride. Figure 11.12 shows the effect of acid and water on the compressibility of laterite soil. Significant changes of void ratio are obtained.

(3) Exchangeable Ions

Figure 11.13 also shows the coefficients of compressibility, c_v, as functions of the exchange cations and of consolidation range and the coefficient of consolidation, c_v, for the different homoionic soils as a function of their void ratios. Since the compressibility indicates the intensity of the forces holding the water films, this figure gives a picture of the water-fixing ability of the Putnam clay as a function of the exchange ions.

Based on data indicated in Figure 11.13, the water-holding forces decrease in the series: (1) void ratio at 1.1: Na > H > Mg = Ca > K = natural soil; (2) void ratio at 1.5: Na > Ca > Mg > natural soil > K. The great difference between the effect of the Na- and K-ions is impressive.

11.6 PREFAILURE DEFORMATION AND FAILURE CONDITION OF SOIL

11.6.1 Prefailure Characteristics of Soil

When the load is applied to a soil mass, the soil deformation is elastic or nonelastic until the deformation becomes unacceptably large; then it is said that the soil mass has *failed*. Therefore, the strength of the soil is an important criterion in estimating how soil can be sustained under such applied load. This load may be referred to as failure load.

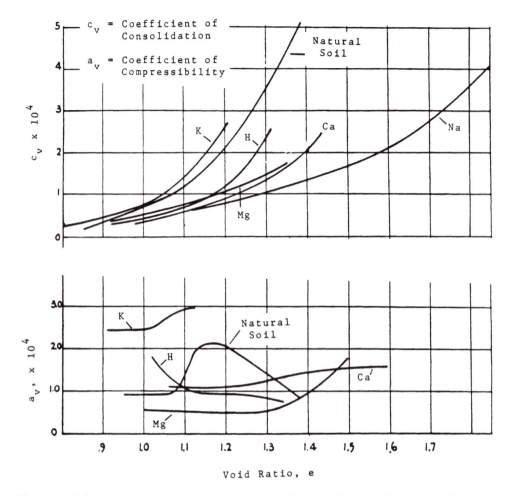

Figure 11.13 Comparison of c_v and a_v between natural and homoionic soils for Putnam clay. (Data from Winterkorn, H.F. and Moorman, R.B.B. [1941], Proc. 21st Annual Meeting, HRB, pp. 415–434.)

However, this failure load is related to the behavior of prefailure conditions of the same soil. At present time, most engineers are interested only in the failure load of the soil, not the prefailure soil conditions. Unfortunately, many premature and progressive failures frequently occur, without proper explanations, because the prefailure and failure conditions are interrelated. In other words, the stability of soil mass is not only affected by the applied load, but affected by prefailure conditions or the genetic or past stress history of soil itself.

Study of prefailure conditions includes creep, progressive failure, stress-softening-hardening, and volume changes. In most studies, the characteristics of prefailure conditions are based on mechanical energy (loading) alone. Unfortunately, in many actual cases, prefailure behavior is controlled or influenced by multi-energy fields.

11.6.2 Relationship Between Prefailure and Failure Conditions

Figure 11.14, shown in schematic diagram, illustrates the three basic failure conditions when soil is subjected to applied load, namely: (1) prefailure stage, (2) failure stage, and (3) postfailure stage. The characteristics of each stage are controlled by various energy fields. Failure (point cde) and postfailure (point ef) conditions are controlled by mechanical energy.

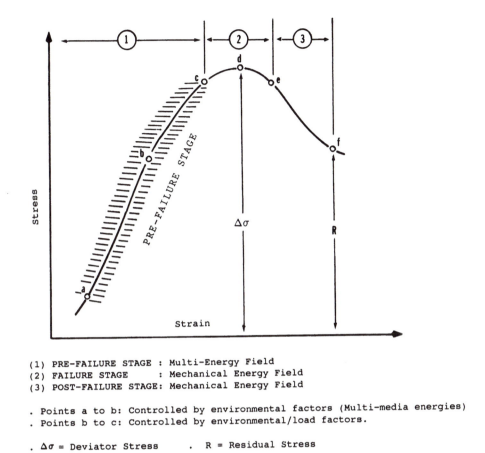

(1) PRE-FAILURE STAGE : Multi-Energy Field
(2) FAILURE STAGE   : Mechanical Energy Field
(3) POST-FAILURE STAGE: Mechanical Energy Field

. Points a to b: Controlled by environmental factors (Multi-media energies)
. Points b to c: Controlled by environmental/load factors.

. $\Delta\sigma$ = Deviator Stress   . R = Residual Stress

Figure 11.14 Schematic diagram illustrates the three basic failure conditions when soil is subjected to an applied load. (After Fang [1994b].)

The (R) value indicated in Figure 11.14 is called *residual stress*, and ($\Delta\sigma$) is called *deviator stress*. The strength of soil at failure is closely related to the prefailure condition of the same soil. The characteristics of peak point (4) are influenced by the behavior of points (ab) and (bc). Natural characteristics and phenomena of soil at prefailure stage are complex. Basically, it includes two stages.

(1) Prefailure Stage (Point a to b in Figure 11.14)

The stress-strain characteristics of soil at prefailure stage are controlled by multimedia energies which are caused by environmental factors. The major factors include: thermal, electric, and electromagnetic energies.

(2) Prefailure Stage-Environmental-Load (Point b to c)

The stress-strain characteristics of soil are shared with both mechanical and multimedia energy fields. Basically, it includes two substages: (1) soil-water interaction at various water contents without load; and (2) soil-water interaction with load. In addition, the soil-water interaction at prefailure stage hinges not only on local environment, but also the natural inherence of clay mineral behavior. That failure condition can be controlled by prefailure condition.

11.7 STRESS-STRAIN-STRENGTH CHARACTERISTICS OF SOIL

The stress-strain relationship for any material is used for analyzing the stability of structures. However, in a conventional approach, the failure stage of the stress-strain relationship is of concern. Numerous mathematical models have been developed to predict soil behavior including elastic, plastic, viscoelastic, elastic-plastic, and hardening and softening stress-strain behavior. These various proposed theories, mathematical models, and their limitations have been discussed by many investigators.

11.7.1 Failure Criteria

Numerous failure criteria have been proposed for the stability analysis of soil mass, but most of them are borrowed from basic mechanics. Since soil is a complicated material, some stress-strain-time behavior is highly nonlinear. However, for practical uses the linear elastic model and Mohr-Coulomb criterion and their shear equation (Equation 11.10) are commonly used. In 1925, Terzaghi introduced that the effective stress concept to include the porewater pressure effects during the shear tests has modified.

$$\left. \begin{aligned} s &= c + \sigma \tan \phi \\ s &= c' + \sigma' \tan \phi' \end{aligned} \right\} \tag{11.10}$$

where S = shear strength
   c = cohesion
   c' = effective cohesion
   ϕ = internal friction angle
   ϕ' = effective internal friction angle
   σ = normal stress on shear plane
   σ' = effective normal stress on shear plane

11.7.2 Types of Shear Testing

(1) Standard Shear Testing

To determine the shear parameters, c, c', ϕ, and ϕ', in Equation (11.10), the shear tests must be performed either in the laboratory or in the field. Currently, various test methods are available including direct shear, triaxial shear, hollow cylindrical shear, and torsion vane. Among these methods, the direct shear (ASTM D3080-90) and triaxial shear (ASTM D2850-70) are most commonly used. There are three types of loading and drainage conditions under which shear tests may be performed: *unconsolidated undrained* test (UU or quick test), *Consolidated undrained* test (CU test), and *Consolidated drained* test (CD or slow test).

The unconfined compression test (ASTM D2166-91) is considered as a special case of the unconsolidated-undrained (UU) test with confined pressure (σ_3) equal to zero. The deviator stress ($\Delta\sigma$) at failure is called unconfined compressive strength and denoted as q_u. For soft clay, when internal friction angle is small or equal to zero ($\phi = 0$), the shear strength (S) or cohesion (c) is equal to half of the unconfined compressive strength, (q_u), or $S = c = \frac{1}{2} q_u$.

(2) Determination of Undrained Shear Parameters from q_u and σ_t

Unconsolidated undrained (UU) shear parameters, cohesion, c, and internal friction angle ϕ, are commonly determined by triaxial and direct shear tests. These test methods are generally time consuming, expensive, and particularly poorly suited to testing stabilized material because of the large particle size and high strengths involved. This frequently necessitates the use of larger test specimens which results in the need for larger test equipment and higher test loads. A simple method for overcoming these difficulties is proposed by Fang and Hirst (1973). The proposed method for determining these shear parameters, c and ϕ, of soils and stabilized materials requires knowledge of only the unconfined compressive strength (q_u) and tensile strength (σ_t).

The method described herein is based on the modified Mohr-Coulomb failure envelope suggested by Chen and Drucker (1969). To establish the failure envelope requires that at least three points on the envelope be given. One point is determined from tensile tests as described in Section 6.7. The second point is determined from the unconfined compression test. This information provides two of the three points necessary to define the failure envelope. Experimental data indicate that cohesion, c, is related to the tensile strength, σ_t, of the material. Then let:

$$c = \frac{\sigma_t}{\xi} \tag{11.11}$$

If soil is in a dry condition, the ξ is relatively constant. For rocks, the ratio is equal to two times the tensile strength. For most soil this value varies from 2.0 to 5.0. The unconfined compression and tensile tests are both simple and easy to perform. No additional equipment is needed, and the tests can be conveniently performed in conjuction with routine CBR and compaction tests.

11.7.3 Factors Affecting Shear Strength

(1) Temperature and Electrolyte Concentration

Temperature and heat effects on shear strength are discussed in Section 8.6. In this section the combined temperature and electrolyte effects on shear are presented. Skempton and Northey (1952) reported a study of remolded shear strength in the effect of leaching. It was found that high values of sensitivity could only be obtained with samples that were originally formed from clay-water slurries with a high salt concentration at that position and then subsequently leached. It was found that heavily overconsolidated clays and freshwater lacustrine clays exhibited sensitivities in the low to medium range. High sensitivities were only found in marine clays which have evidence of subsequent leaching and reduction in salt concentration.

Leonards and Andersland (1960) present shear strength tests evaluating the effects of temperature and electrolyte concentration which yielded results contrary to those predicted by the Gouy-Chapman theory. An increase in salt concentration according to colloidal theory reduces the thickness of the double layer, thereby decreasing the repulsive force between particles. This should result in a more flocculated clay structure and hence a higher shear strength. As the salt concentration increased, the undrained strength decreased as indicated in Figure 11.15.

(2) Contaminated Pore Fluids

Influences of industrial and hazardous/toxic wastes on the geotechnical properties of soils have been reviewed by Evans et al. (1983) for various soil types. Using organic fluids effects on the strength, deformation and permeability of soil-bentonite mixtures have been studied

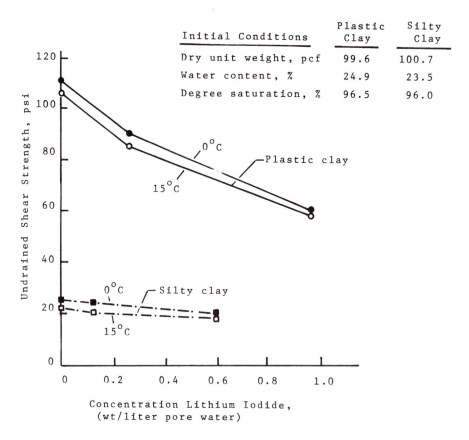

Initial Conditions	Plastic Clay	Silty Clay
Dry unit weight, pcf	99.6	100.7
Water content, %	24.9	23.5
Degree saturation, %	96.5	96.0

Figure 11.15 Effect of salt concentration on undrained shear strength. (After Leonards, G.A. and Andersland, O.B. [1960], Proc. ASCE Research Conf. on the Shear Strength of Cohesive Soils, pp. 793–818.)

by Evans et al. (1985). Naik (1986) studied the effects of pore fluid by using various pH solutions mixed with laboratory-prepared clay liner material (sand-bentonite mixture). A typical result is shown in Figure 11.16.

(3) Ion-Exchange Reaction

Vees and Winterkorn (1967) present two sets of laboratory shear tests both involved in the ion-exchange reactions.

(a) Tests on Systems in Equilibrium with Normal Pressure

Figure 11.17 shows the shear resistance of homoionic modifications of kaolinite and attapulgite as a function of normal pressure, which is also the consolidated pressure. The effect of the exchange ions on the shear resistance for the attapulgite is Al > Ca > Na. Hence, it is due to the degree of springiness imparted to the crystals and not to the degree of interaction with water which would give the reverse sequence. In the case of the kaolinite the sequence is Th > Al = Na > Ca. This also indicates a predominant influence of the tendency to form flocculated structures.

Figure 11.18(a) shows the same shear data for kaolinite as a function of the void ratios reached under the respective consolidation pressures. This time, the logarithm of the shear

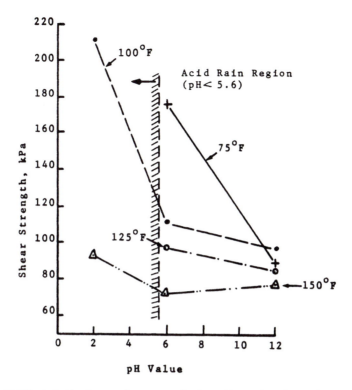

Figure 11.16 Effect of pH value on shear strength of sand-bentonite mixture. (After Naik [1986].)

strength increases linearly with decreasing void ratio. At low void ratios, the structural influence is most effective as indicated by the order of the cation effect: Th > Na > Al > Ca. With increasing void ratio, the effect of the Th-ion decreases rapidly and falls below those of Na and Al at a void ratio of 1.4 and 1.45, respectively.

Figure 11.18(b) shows the corresponding picture for attapulgite. Because of the predominant influence of the springiness of the attapulgite system, the three curves are more closely clustered together. At low void ratios, the sequence of the cation effect is Al > Ca > Na; it is completely reversed at the high void ratios.

(b) Tests on Samples Preconsolidated at Normal Pressures

Tests on samples preconsolidated at normal pressures of 4 kg/cm$_2$ and sheared at normal pressures of 0.5, 1, and 2 kg/cm$_2$ are shown in Figure 11.19(a). For the kaolinite, the data show a sequence of cation effect at high normal pressures similar to the virgin curves in Figure 11.17. The sequence is inverted, however, in the case of the attapulgite. Because of this and the fact that the data showed a greater pressure dependency than is to be expected for quick tests on previously overconsolidated specimens, the void ratios of the samples were calculated from their moisture contents determined after the test.

Figure 11.19(b) shows the same shear data for plots of the shear strength data as a function of the void ratios reached under that recalculated from their actual water contents during testing. For the kaolinite the sequence of the cation effect was the same at the low void ratios as for the virgin samples; however, the magnitude of the shear resistance was at a lower level than Figure 11.18. This could be due to an ordering effect of the overconsolidation. Obviously, the overconsolidation left its effect on the soil structure which is reflected in the shear strength data. Similar conclusions are given for the attapulgite specimens.

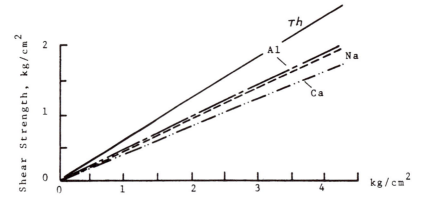

Normal Pressure equal to Consolidated Pressure

(a)

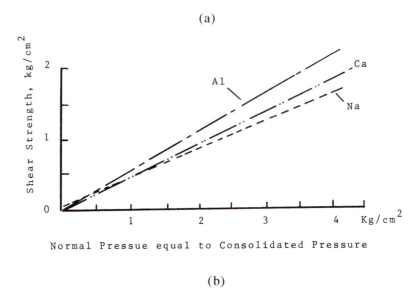

Normal Pressue equal to Consolidated Pressure

(b)

Figure 11.17 Shear resistance of homoionic kaolinite and attapulgite samples as a function of normal pressure. (a) Kaolinite; (b) attapulgite. (After Vees, E. and Winterkorn, H.F. [1967], Engineering Properties of Several Pure Clays as Functions of Mineral Type, Exchange Ions and Phase Composition, HRR no. 209, pp. 55–65.)

11.7.4 Sensitivity, Creep, and Relaxation of Soil

(1) Sensitivity

Most clays lose a portion of their strength when remolded, as discussed in Section 5.12. To measure this phenomenon, the term sensitivity was introduced (Terzaghi, 1944). Winterkorn and Tschebotarioff (1947), based on physicochemical concepts, attempted to explain the possible causes of this phenomenon. The *sensitivity* used in geotechnology is the ratio of undisturbed strength to remolded strength. For most clays ranges from 2 to 4 are recorded, peat from 1.5 to 10, and for marine sediments from 1.6 to 26. From the point of view of their sensitivity to remolding, clays may be classified based on the degree of sensitivity, as suggested by Skempton and Northey (1952) and modified by Bjerrum (1954). The sensitivity value less than 2.0 is insensitive, and value between 4 and 8 is classified as sensitive material. A value between 16 to 32 is classified as very sensitive.

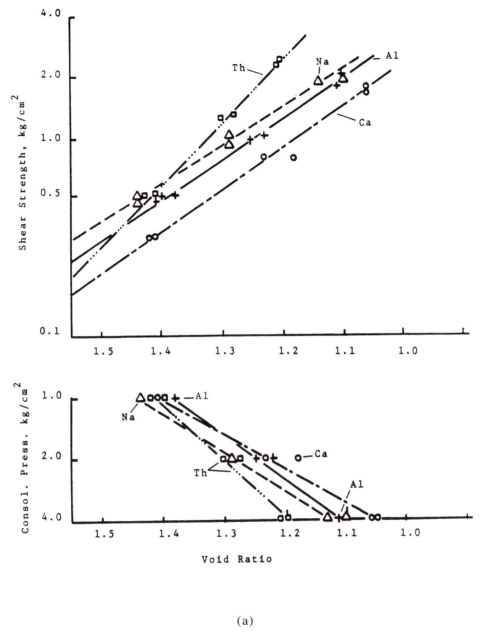

(a)

Figure 11.18 Shear resistance of homoionic soils as function of void ratio. (a) Kaolinite clay; (b) attapulgite clay. (After Vees, E. and Winterkorn, H.F. [1967], Engineering Properties of Several Pure Clays as Functions of Mineral Type, Exchange Ions and Phase Composition, HRR no. 209, pp. 55–65.)

For measuring the strength for both undisturbed strength to remolded strength, commonly the unconfined compressive strength is used (Equation 11.12). Because the unconfined compressive strength is not sensitive enough when dealing with environment as discussed in Section 5.12, the use of tensile strength as an indicator to estimate the sensitivity of fine-grained soil has been suggested (Equation 11.13). The sensitivity based on tensile strength vs. soil types as reflected by plasticity index (PI) is shown in Figure 11.20. The other characteristics of tensile strength have been discussed in Section 6.7.

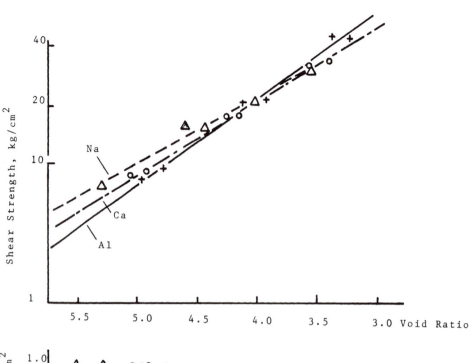

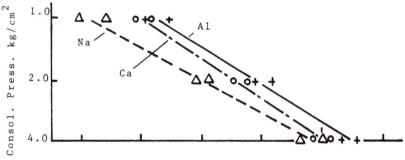

(b)

Figure 18 (continued)

$$S = q_u\,[1]/q_u\,[2] \qquad\qquad (11.12)$$

$$S' = \sigma_t\,[1]/\,\sigma_t\,[2] \qquad\qquad (11.13)$$

where S   = sensitivity
  S′   = sensitivity (proposed)
  $q_u[1]$ = unconfined compressive strength at undisturbed condition
  $q_u[2]$ = unconfined compressive strength at remolded condition
  $\sigma_t[1]$ = tensile strength at dry condition
  $\sigma_t[2]$ = tensile strength at optimum moisture content (OMC) condition

(2) *Creep of Soils*

Creep may be defined as the continued deformation of a material when subjected to constant stress. The study of the creep phenomenon requires the investigation of the change

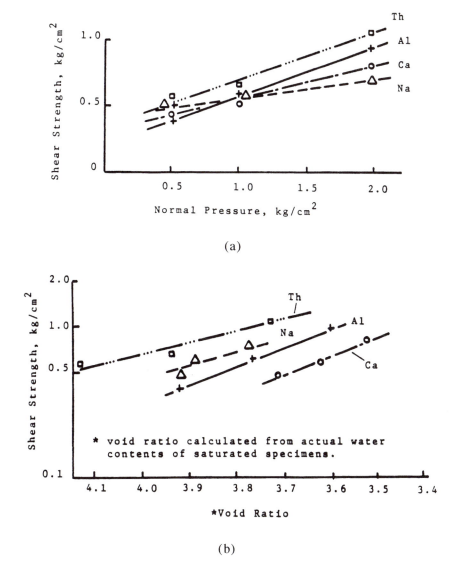

(a)

(b)

Figure 11.19 Shear resistance of overconsolidated homoionic kaolinite specimens. (a) Function of normal pressure; (b) function of void ratio. (After Vees, E. and Winterkorn, H.F. [1967], Engineering Properties of Several Pure Clays as Functions of Mineral Type, Exchange Ions and Phase Composition, HRR no. 209, pp. 55–65.)

in deformation of an element with respect to time when the material (soil) is subjected to a constant stress. Many materials creep at room temperature, whereas others require an elevated temperature before significant deformation will occur under a constant stress that may be insignificant compared to the stress the material can withstand under short time loading conditions. The term creep is also used in a more general sense to indicate any inelastic deformation that occurs with time.

Deformation may be characterized by an instantaneous elastic and inelastic strain followed by a period during which the rate of deformation decreases. Figure 11.21 presents the creep phenomena of various soil types. In all cases, when time period increases, more variations will have occurred largely due to local environments. Detailed discussions on soil creep is given by Singh and Mitchell (1968) and Mitchell (1993).

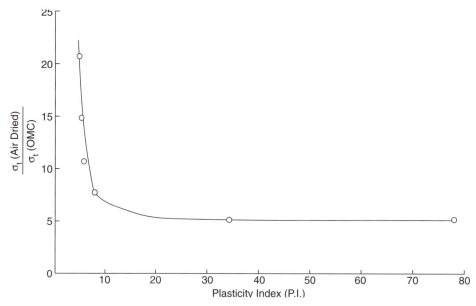

Figure 11.20 σ_t (air dried)/σ_t (oMc) ratio vs. plasticity index. (Data from Fang and Fernandez [1981].)

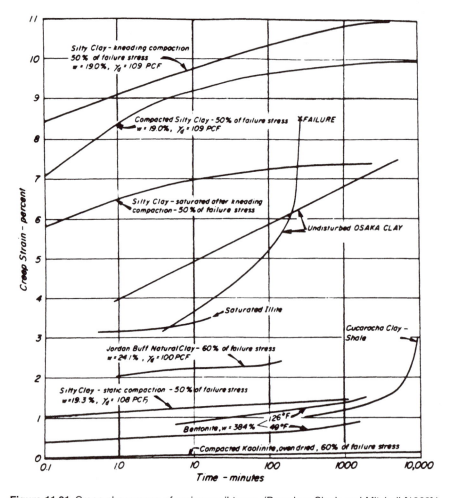

Figure 11.21 Creep phenomena of various soil types. (Based on Singh and Mitchell [1968].)

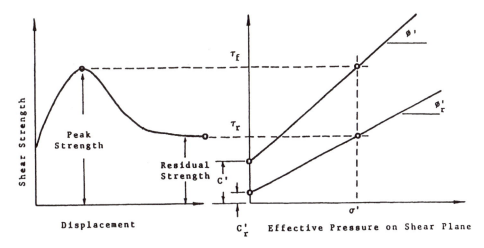

Figure 11.22 The concept of residual shear strength.

(3) Relaxation

Relaxation is the continuous decrease of stress occurring when a material is subjected to constant strain. The same general properties of a material that influence the creep of a material will influence its relaxation. Both creep and relaxation are important phenomena for evaluation of long-term soil behavior, especially for contaminated fine-grained soil. Unfortunately, very little experimental data are available about this aspect.

11.8 RESIDUAL SHEAR STRENGTH

11.8.1 The Characteristics of Residual Shear Strength

(1) General Discussion

For analysis of shear characteristics of overconsolidated soils relating to stability problems, ordinary shear tests are not suitable because they give too high a shear value. Skempton (1964) showed that the strength remaining in laboratory samples after large shearing displacement corresponded closely with the computed strength from actual landslides; therefore, he proposed a *residual strength* concept as shown in Figure 11.22 for the analysis of slope stability of overconsolidated clays as:

$$S_r = \sigma' \tan \phi_r' \qquad (11.14)$$

where S_r = residual shear strength
σ' = effective pressure on shear plane
ϕ_r' = residual effective friction angle

In examining Figure 11.22, the peak strength, τ_f, the corresponding effective friction angle, σ', and effective cohesion, c', are used for conventional slope stability analysis. However, for overconsolidated clays, the ϕ_r' and c_r' are suggested for the slope analysis. The c_r' value is usually very small or zero.

The characteristics of residual strength from laboratory tests indicate that for a given soil sample, if the molding moisture contents are the same for both conditions, the final residual strength for both conditions remains the same. Residual shear strength for both overconsol-

Table 11.2 Residual Shear Strengths of Clay Shales Determined by Annular Shear, Repeated Direct Shear, and Rotational Shear Tests

Shale	ϕ' (km/cm²)	Residual shear strength, ϕ'_r		
		Annular shear (deg)	Repeated direct shear (deg)	Rotational shear (deg)
Taylor I	6·0	10·6 (ϕ'_r not reached)	9·6	9·8
(Laneport Dam, TX)	12·0	10·8	10·3	
	3·0	8·6		
	1·5	8·3		
Strawn	6·0	8·7	9·3	7·2
(Proctor Dam, TX)	12·0	9·5	9·9	
	3·0	9·3		
	1·5	9·4		
Kincaid	1·5	5·5		
(Cooper Dam, TX)	3·0	4·7		
	6·0	4·8	5·5	Material too gritty
	12·0	5·5	5·3	to conduct test
	3·0	5·6		
	1·5	5·5		
Bearpaw	6·0	6·9	6·3	6·4
(Fort Peck Dam, MT)	12·0	6·7	6·2	
	3·0	7·1		
	1·5	8·1		
Pierre	6·0	3·0	3·0	3·1
(Oahe Dam, SD)	12·0	3·6	3·3	
	6·0	3·5		
	3·0	3·1		

After Townsend, F.C. and Gilbert, P.A. [1973], *Geotechnique*, v. 23, no. 2, pp. 267–271.)

idated and normally consolidated clays is the same even though the peak strength and initial moisture content for both cases are different.

(2) Methods for Determination of Residual Strength

The residual shear strength in Equation (11.14) can be obtained from slow drained direct shear tests as suggested by Skempton (1964). Since then several test methods and procedures have been developed including modified shear box, triaxial compression test, and ring shear. Among these methods, the ring shear is most effective (Bishop et al. 1971). There are various types of ring shear available such as Russian ring shear (CRRI, 1976), CRRI ring shear (CRRI, 1976), and others. Because each test apparatus and test procedure is different some discrepancies or variations are reported by Townsend and Gilbert (1973), the CRRI (1976 and 1979), and others.

11.8.2 Residual Strength Data Interpretations

Townsend and Gilbert (1973) summarize the residual shear strength of clay shales from five dam sites (Table 11.2) including Texas, Montana, and South Dakota. Three types of test methods are used including angular shear, repeated direct shear, and rotational shear. The effective pressure, ϕ' along the shear plane varied from 1.5 to 12 kg/cm². The results of effective residual friction angle, ϕ'_r, indicate that there is no significant difference among these methods.

Comparison of residual strength data was determined by three laboratory tests (CRRI, 1976 and 1979), including repeated direct shear, triaxial compression, and ring shear. Four

Table 11.3 Comparison of *In Situ* and Laboratory Test Methods for Determination of Residual Strength Parameters

(1)		(2)		(3)		(4)	
C_r (kg/cm^2)	ϕ_r	C_r (kg/cm^2)	ϕ_r	C_r (kg/cm^2)	ϕ_r	C_r (kg/cm^2)	ϕ_r
0.07	2°34′	0.10	2°00′	0.09	2°20′	0.03	3°00′
0.08	3°12′	0.18	3°06′	0.09	3°14′	0.07	3°00′
0.12	3°18′	0.09	4°50′	0.12	2°20′	—	—

Note: (1) At natural sliding surface; (2) laboratory-repeated direct shear test; (3) *in situ* shear measurement on large shear surface; (4) back calculation.

After CRRI (1976 and 1979), Collected Papers on Landslides, v. 1 and 2, Chinese Railway Research Institute, Lanzhou, China.

soil types are used. For all cases the results from ring shear give the higher residual friction angles. The rate of shear for determination of residual strength was studied by CRRI. It is indicated that the smaller the rate of shear, the higher the residual friction angle (ϕ_r). The rate of shear varies from 0.08 to 0.12 mm/min and ϕ_r varies from 3°36′ to 7°31′. The rubber membrane used in the triaxial apparatus for determination of residual strength also shows some significance. As the lateral pressure increases, the correction values for the rubber membrane increase. Table 11.3 presents the comparison of *in situ* and laboratory test results for determination of residual strength parameters.

Environmental effects on residual shear strength have been studied. Influence of colloidal, clay mineral composition, and chemicals in the soil all affect the residual shear strength significantly (Kenney, 1967).

11.8.3 Correlation of Residual Strength to Other Soil Parameters

Because the peak or residual shear parameters are relatively time consuming and expensive, for practical uses some simple experimental equations and correlations for estimating these strength parameters have been proposed by numereous investigators which are summarized as follows.

(1) Correlation with Clay Content and Atterberg Limits

(a) Clay Content

Skempton (1964) shows the ultimate ϕ' and is related to the clay content (<2 μm) for various soil types from England. De Beer (1969) presented a simple relation between the effective residual friction angle, ϕ_r', and the plasticity index (PI) as shown in Figure 11.23.

(b) Liquid Limit

Equation (11.15), proposed by Jamiolkowski and Pasqualini as cited by CRRI (1979), is based on the weathering of blue and gray clays from Italy. The coefficient of correlation, r^2, for the equation is equal to 0.76.

$$\phi_r' = 453.1 \ (\text{LL}^{-0.85}) \tag{11.15}$$

where ϕ_r' = effective residual friction angle
LL = liquid limit

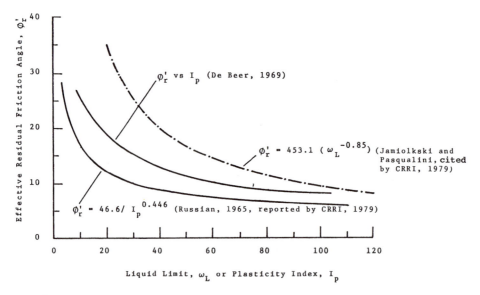

Figure 11.23 Graphical comparison of results computed from various published equations.

(c) Plasticity Index

Equation (11.16) was proposed by a Russian engineer in 1965 as reported by CRRI (1979). Figure 11.23 presents the graphical forms of various equations for comparison purposes:

$$\phi_r' = 46.6/\ PI^{0.446} \tag{11.16}$$

where PI = plasticity index

(2) Correlation with Liquidity Index (LI)

(a) Residual Cohesion

The correlation of residual cohesion, c_r, with liquidity index (LI) and liquid limit (LL) is presented in Table 11.4. For a given liquidity index, the c_r value increases as the liquid limit increases.

(b) Residual Friction Angle

Equation (11.17) proposed by the CRRI (1979) was based on 236 laboratory undrained multirepeated shear tests and 59 field observations on 91 types of sliding soils. Further discussions on this aspect will be presented in Section 16.6. In examining Equation (11.17), the LI is more of an influence on residual shear stress than other parameters.

$$\text{Log } \phi_r = 2.4278 - 1.2279 \text{ Log LI} - 0.1173 \text{ Log LI} \tag{11.17}$$

(3) Correlation with Applied Normal Stress

A relationship between residual strength and normal stress was proposed by a Russian engineer cited by CRRI (1979) as:

Table 11.4 Correlations of Residual Cohesion, C_r, with Liquid Limit, ω_L, and Liquidity Index, I_L

C_r/I_L	Liquid limit (ω_L, %)			
	20–30	30–40	40–50	50–60
0–0.1	0.03–0.05	0.13–0.15	0.14–0.15	0.20
0.1–0.2	0.02–0.04	0.08–0.09	0.13–0.14	0.19–0.20
0.2–0.3	0.02–0.03	0.07–0.08	0.12–0.13	0.17–0.18
0.3–0.4	0.01–0.02	0.04–0.05	0.11–0.12	0.12–0.13
>0.4	0–0.01	0.04–0.05	0.08–0.09	0.09–0.11

After CRRI (1976 and 1979), Collected Papers on Landslides, v. 1 and 2, Chinese Railway Research Institute, Lanzhou, China.

$$\tau_r = 0.09 + 0.14 \; \sigma \tag{11.18}$$

Equation (11.18) is based on 200 laboratory tests on plasticity clay. The coefficient of correlation, r^2, is equal to 0.78. Equation (11.19) is the correlation of shear strength along the natural sliding suface $\bar{\tau}$, with laboratory test results of normal stress, σ, and residual strength, τ_r, as follows:

$$\bar{\tau} = \tau_r = 0.06 + 0.15 \; \sigma \tag{11.19}$$

Equation (11.19) is based on 50 field observations. The coefficient of correlation, r^2, is equal to 0.82. It was found that laboratory residual strength, τ_r, is very close to the field shear strength along the natural sliding surface (CRRI, 1979).

(4) Correlation with Residual Strength Coefficient (μ_r)

Voight (1973) summarized various published data from 1967 to 1970 and found a relationship between plasticity index (PI) and the residual strength coefficient, μ_r. The μ_r is defined as:

$$\mu_r = \frac{\tau_f - \bar{\tau}}{\tau_f - \tau_r} \tag{11.20}$$

where μ_r = residual strength coefficient
  τ_f = peak strength (Figure 11.22)
  τ = shear strength at natural sliding surface
  τ_r = residual strength

Results of plasticity index vs. residual strength coefficient are shown in Figure 11.24.

11.9 SOIL DYNAMICS

11.9.1 Dynamic Shear Modulus of Soils

Soil dynamics is part of particle dynamics as discussed in Section 2.9. The dynamic shear modulus of soil is one of the fundamental soil parameters required for the analysis of the response of soil and soil-structure interaction under dynamic loading. Fundamental correlations of shear modulus, damping ratio, shear stress, and shear strain with varying consolidation pressures are determined in the laboratory cyclic shear tests. A comparison is made between

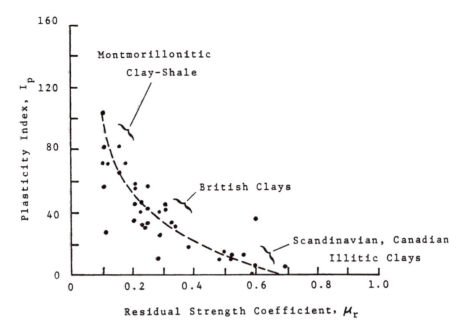

Figure 11.24 Plasticity index vs. residual strength coefficient. (After Voight, B. [1973], *Geotechnique*, v. 23, no. 2, pp. 265–267.)

laboratory test results, *in situ* cross-hole test results, and the building code data (Seed, 1991; Chaney and Pamukcu, 1991). An example of laboratory data on silt with clays and sands is presented and discussed in following section.

For normally consolidated soils such as soft clay, as an example, the use of 'reference strain' approach to normalize stress-strain data provides a curve which is independent of the effective stress path and sample disturbance. Hence, the normalized stress-strain curve produced by this technique should apply to *in situ* conditions. The first procedure was presented by Hardin and Drnevich (1972) where the shear modulus, G, was shown to be equal to the following expression:

$$G = \tau/\gamma = 1/a + b\gamma \tag{11.21}$$

where G = dynamic shear modulus
τ = shear stress
γ = shear strain
a,b = constants

It was also shown that the ratio of shear modulus (G) to its maximum value (G_{max}) evaluated at low shearing strains ($\gamma \leq 10^{-4}\%$) could be given by:

$$G/G_{max} = 1/1 + \gamma/\gamma_r \tag{11.22}$$

where γ_r = reference strain = τ_{max}/G_{max}
τ_{max} = shear stress at failure

Combining Equations (11.21) and (11.22) and solving for constants **a** and **b**, the following relations are obtained:

$$a = 1/G_{max}$$

$$b = 1/\tau_{max} \tag{11.23}$$

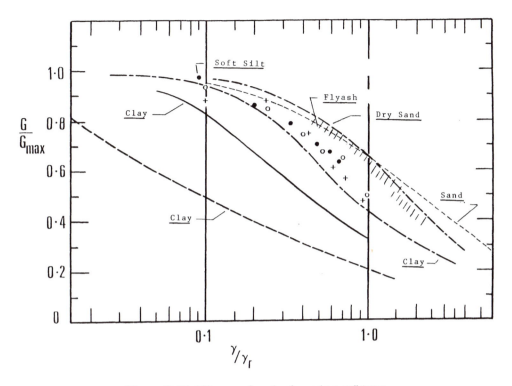

Figure 11.25 G/G_{max} vs. γ/γ_r ratios for various soil types.

Rearranging,

$$\gamma_r = \frac{\tau_{max}}{G_{max}} = \mathbf{a}/\mathbf{b} \qquad (11.24)$$

Constants **a** and **b** are determined by plotting γ/τ vs γ. The G_{max} values obtained by this procedure compared closely against those obtained by using the empirical equation proposed by Hardin and Black (1968) as follows:

$$G_{max} = \frac{1230(2.973 - e)^2}{1 + e} (OCR)^k (\bar{\sigma}_o)^{1/2} \qquad (11.25)$$

where e = void ratio
  OCR = overconsolidation ratio
  k = constant (for low plasticity soil, $k \cong 0$)
  $\bar{\sigma}_o$ = $\bar{\sigma}_v (1 + 2K_0/3)$
  $\bar{\sigma}_v$ = vertical effective stress
  k_0 = coefficient of earth pressure at rest

Using Shanghai soft silt as an example, final results are plotted in Figure 11.25 (Fang et al., 1981) together with published data. A review of Figure 11.25 shows that the cyclic response of Shanghai silt lies in the region between sand and clays.

11.9.2 Dynamic Shear Modulus of Contaminated Soil

In recent years, many hydraulic structures, reservoirs, and river banks have required seismic resistance considerations. However, these considerations are based on soil param-

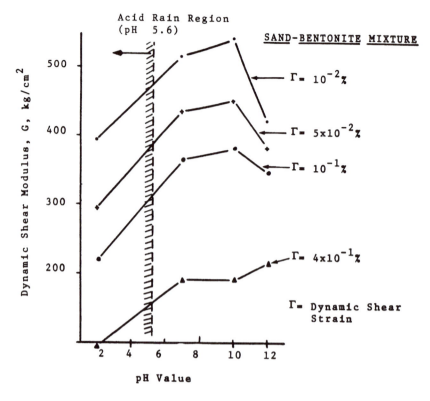

Figure 11.26 Dynamic shear modulus vs. pH value. (After Du et al. [1986].)

eters under normal conditions where the contact pore fluid is not polluted. Unfortunately, in actual cases the contact water, in general, has some degree of pollutants caused by various liquid, solid wastes, and acid rains and acid drainages. Therefore, the dynamic behavior of soil under normal conditions may be somewhat different in comparison to polluted water. Figure 11.26 shows the effect of acid (pH value) on dynamic shear modules of sand-bentonite mixture.

11.10 BLASTING DYNAMICS

11.10.1 General Discussion

Blasting techniques are used in geotechnical engineering in two ways. The first is for quarry blasting which produces the aggregates, stones, and sands to be used as components of construction material such as concrete or asphalt mix. The second use is for removing some undesired objects for roadways, drainage ditches, or for tunneling. In some cases, the blasting technique is used as blasting-expansion pile foundation as a part of the ground improvement system. Regardless of the purpose, the effect on geological formation and ground characteristics relating to the blasting results is important. In this section, we will discuss the environmental geotechnical aspects of quarry and construction blasting dynamics.

11.10.2 Blasting Energy and Safe Limits

Considerable effort has been expended by the U.S. Bureau of Mines on safety limits for blasting that will or will not cause structure damage (Devine, 1966; Wiss and Nicholls, 1974).

Most of their work is in correlating displacement, frequency, peak particle velocities, and safe distance with the blasting energy. A brief review of this work follows.

(1) Safe Distance and Particle Acceleration

On the basis of early instrumentation studies, Rockwell in 1927 stated that structures that are farther than 200 to 300 ft from a blasting would not be damaged. In 1942, the Bureau of Mines reported *particle acceleration* as the best criterion for estimating damage to structures.

(2) Energy Ratio

Crandell, in 1949, presented a criterion based on vibration levels in the ground in the vicinity of the structure. This criterion called for an *energy ratio* which was defined as the following:

$$ER = a^2/f^2 \qquad (11.27)$$

where ER = energy ratio
     a   = acceleration, ft/sec^2
     f   = frequency

(3) Particle Velocity and Scaled Distance

Particle velocity is defined by Devine (1966) as shown in Equation (11.28). The term $(D/W^{1/2})$ in the equation is called *scaled distance*. When particle velocity data from one site are plotted on log-log coordinates as a function of the scaled distance, good grouping of data has generally been obtained. For this reason, scaled distance is a parameter often used to regulate blasting energy.

$$V = H(D/W^{1/2})^{-3} \qquad (11.28)$$

where V = particle velocity, in./sec (ips)
     H = constant for a particular site
     D = distance from blasting to measurement point, ft
     W = maximum charge weight per delay, lb
     β = exponent for a particular site

(4) Fang-Koerner Cracking-Intensity Concept

Fang-Koerner *cracking-intensity concept* (1977) is based on the characteristics of cracking-intensity relationship as shown in Figure 11.27. It presents the type of crack opening response that the previously described strain gauges record. Figure 11.27(a) is a generalized type of response, while Figure 11.27(b) shows the specific signatures of known types of vibrations. The maximum value of crack opening is the critical value to obtain and can be compared to other relevant blasting parameters as they are developed. In addition to measuring the maximum crack opening, a cracking energy can also be obtained by utilizing the cracking duration as shown on Figure 11.27(a). The cracking intensity may be defined as:

$$CI = C\,W_{max}\,t \qquad (11.29)$$

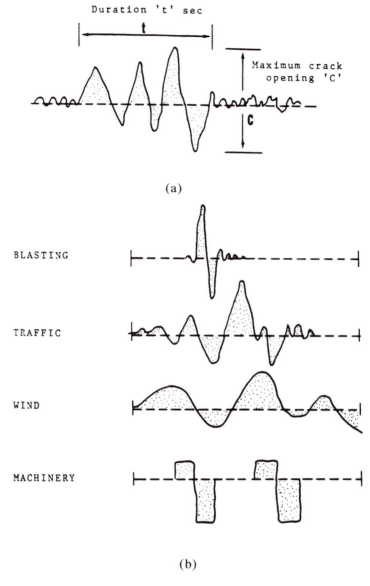

Figure 11.27 Generalized and specific dynamic response curves as monitored using clip-type strain gauges. (a) Generalized dynamic response curve; (b) specific dynamic response curves.

where   CI = cracking intensity, in.-sec
             = constant (depending primarily on material type, it varies from 0 to 1, with brick or concrete block construction being from 0.3 to 0.6)
  W_{max} = maximum crack opening, in.
    t   = duration, sec

The maximum crack opening, W_{max}, and time duration, t, indicated in Figure 11.27(a) and Equation (11.29) must be determined experimentally. The following technique has been used for instrumenting and monitoring structures during nearby blasting operations: two metal ball-points are mounted on the desired location, preferably at a location with initial cracks by use of a fast-setting polymer which can dry in 5 min. A clip-type strain gauge (Figure 11.28) is installed between these two ball-points and is connected to a recording device. A

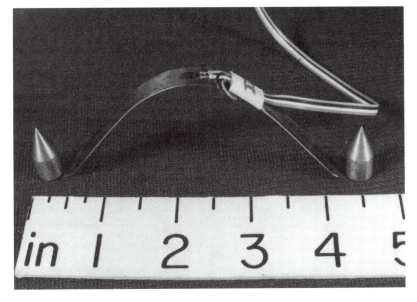

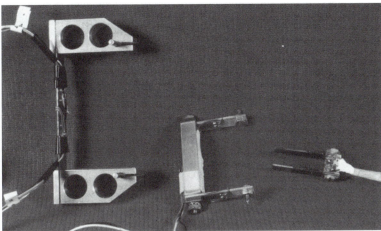

Figure 11.28 Various types of clip gauges for measuring *in situ* structural responses or ground movement due to dynamic effects. (a) Standard gauge for blasting; (b) modifications for other monitoring.

photo shows the various types of clip gauges for measuring the *in situ* structural responses or ground movement due to dynamic effects. Figure 11.28(a) is the standard gauge for blasting, and Figure 11.28(b) is the modifications for other monitoring. Table 11.5 gives a summary of engineering applications, measurable parameters, causes, and locations for use of clip gauges for dynamic vibration monitoring.

(5) Blasting Safe Limits Criteria

Blasting safe limits criteria include: (1) minimum safe distance, (2) particle acceleration, (3) energy ratio, (4) particle displacement, (5) particle velocity, and (6) scale distance summarized in Table 11.6. Human response to blasting vibration is a complicated problem. The human body itself is complex and to study its reaction to vibration, we have to consider not only the pathological and psychological effects, but also the physiological effects.

Table 11.5 Summary of Engineering Applications, Measurable Parameters, Causes, and Locations for Use of Clip Gauges for Dynamic Vibration Monitoring

Problem areas	Causes and measurable parameters	Suggested gauge locations
Existing structures safety	Caused by machine vibration, blasting, moving vehicle, train, wind, construction operation, etc.	Structure walls, floors, ceilings, columns, beams
Highway pavement slabs	Measuring cracks within the pavement slab and between the slabs Caused by moving vehicle and construction operation	At construction or expansion joints or at existing cracks
Airport pavement slabs	Caused by impact-loading during aircraft landing and taxiing	Same as highway pavement slabs
Rock slope stability	Construction operation, subsidence, tectonic movement	Rock joint or fault
Quarry, strip, and deep mines	Effect on nearby structures	Structure walls, floors, ceilings, or any other suspect structural component
Machine vibrations	Effect on soil, rock, and other structures	Foundations
Seismic safety evaluation	Seismic activity	Separate monitoring station
Wave impact	Dynamic forces of waves	Piers, bulkheads, waterfront structures
Environmental effects	Wind; storm, household vibrations, etc.	Suspect structural component

Table 11.6 Summary of Blasting Safety Limits for Structures

Parameters	Safety limits
Charge size and distance	200–300 ft
Particle acceleration	0.1 g
Energy ratio	3
Particle displacement	0.03 in.
	0.9 ips
Particle velocity	2.0 ips
	7–20 ips
Scale distance	50 ft/lb

11.11 SUMMARY

1. Characteristics of unit weight vs. moisture content as interpretated by particle-energy field theory are presented.
2. Laboratory and *in situ* compaction characteristics of soil and factors affecting test result are discussed. Dynamic densification (dynamic consolidation) is also examined.
3. The stress-strain strength of soil and its failure criteria are summarized. Comparisons of shear characteristics of normally consolidated and overconsolidated clays are examined.
4. The concepts of residual strength, test methods, and data interpretation are examined. Correlations of residual strength parameters to other soil constants such as clay contents, liquid limit, plasticity index, and liquidity index are formulated. The comparison of test results from various researchers is presented.
5. A simple method for determination of unconsolidated-undrained (UU) shear parameters, c and ϕ, from unconfined compressive strength, q_u, and tensile strength is discussed and described in Equation (11.11).
6. Dynamic shear modulus of contaminated soil specimens is discussed and comparison made with static load.
7. Basic concept of blasting dynamics is reviewed. A new concept based on cracking intensity criteria is proposed.

PROBLEMS

11.1 From the standpoint of its engineering properties (shear strength, compressibility, and permeability), explain why the specification of dry unit weight alone is not a satisfactory indication of the final performance of an earth dam or a highway embankment.

11.2 Field density and water content determinations on the first few layers of fill averaged as: mass unit weight = 19.3 kN/m; water content = 8.2%. What specific recommendations would you make to the contractor to permit him to meet specifications at a minimum cost to him?

11.3 Most compaction specifications for highway or earth dam construction prohibit the use of any soil which does not have a maximum dry density, as measured in the modified AASHTO compaction test, greater than 100 pcf. Would you ever disregard this specification in the construction of an earth dam? If so, under what circumstances?

11.4 Explain briefly the relationship between settlement under load and the ability to support load. What factors other than consolidation under the load in question can cause settlement? Discuss any one of these factors in detail.

11.5 A monument weighing 250 tons was built some time ago in an area where a 10-ft layer of clay was overlain by a 22-ft layer of sand. The local groundwater level is 11 ft below the ground surface. The present pore water pressure at the center of the clay layer is 1110 psf. Assuming the clay was normally consolidated at the time of construction and that monument load may be treated as a point load, what % consolidation has occurred under the added load of the monument?

11.6 Undisturbed samples are taken from two clay deposits of identical mineralogical composition and particle size distribution. Both soils are normally consolidated and saturated. Clay [A] came from a freshwater lake and clay [B] came from an ocean bottom.
   (a) The two clay samples are consolidated in the laboratory to identical void ratio (or unit weight), then sheared in unconfined compression. Which sample do you expect to have higher strength?
   (b) Both samples are thoroughly remolded and then sheared in unconfined compression. What is the remolded strength in respect to the undisturbed strength? What sample has higher remolded strength?
   (c) The unconfined compression test is run 2 years after the remolding. Which sample is stronger?

11.7 Referring to Problem 11.6, the two clay samples are leached with distilled water in the laboratory until the salt concentration in the pore fluids of the soil is practically the same. Assume that there is no disturbance to the soil structure during the leaching process.
   (a) The two samples are consolidated to the same void ratio and sheared in unconfined compression. Which sample is stronger?
   (b) Which sample has higher strength after thorough remolding?

11.8 Concerning shear strength of a contaminated soil under both static and dynamic loads, why does shear strength decrease when pH value decreases under the dynamic load, and shear strength increase when pH value decreases under static load? Explain.

11.9 What is the genetic approach? Comment on the genetic approach for analysis of the characteristics of shear strength and stability of soil under load.

11.10 Discuss the shear and compressibility characteristics of contaminated unsaturated fine-grained soil.

CHAPTER **12**

Environmental Geotechnical
Engineering Applications

12.1 INTRODUCTION

12.1.1 General Discussion

All civil engineering structures are founded directly or indirectly on ground soil; therefore, the stability of both super- and subsurface structures hinges on the stability of ground soil which is greatly influenced by local environments, as illustrated in Figure 12.1. In general, all structural design is based on loads; however, design criteria based on load factor alone do not give the whole picture and neglect the other important factor which controls the overall stability of all civil engineering structures — the environmental factor design criteria. At present, the load factor design criteria are generally accepted for most civil engineering structural design including ground improvement systems and foundations. Civil engineering design of structures is interdisciplinary in nature and needs the cooperation of all related fields in order to provide a safe and economical structural system. A brief review of this commonly used load factor concept is presented. In addition, the proposed load-environmental factor design criteria are also discussed and compared.

12.1.2 Environmental Geotechnical Applications

As indicated in Figure 12.1, the foundation is a part of an overall ground improvement (or ground modification) system, and it is the heart of environmental geotechnology. Ground improvement is the collective term for any mechanical, hydrological, physicochemical, or biological methods or any combination of such methods employed to improve certain properties of natural or man-made soil deposits. The purposes of this improvement are

1. Strengthen ground soil before failure occurs.
2. Strengthen ground soil during soil's useful life period.
3. Strengthen ground soil after premature or unexpected failure.
4. Temporary ground improvement systems.

Case (1) generally happens where the soil is weak with low-bearing capacity and the groundwater table is high. Case (2) generally refers to proper maintenance or repair of certain potential failure areas for the purpose of prolonging soil's useful life. In case (3), in many cases, ground failure is unexpected. However, it is required to examine the causes of failure before the ground improvement starts. Case (4), a ground improvement system, is used in

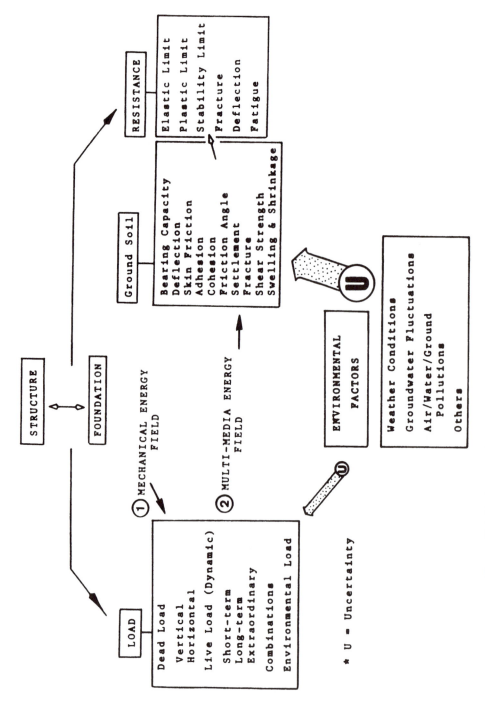

Figure 12.1 Environmental-load factor design criteria for foundations. (After Fang [1987].)

certain conditions and certain locations, such as underwater repair, or where the permanent structure is under construction.

12.1.3 Basic Considerations of Ground Improvement Systems

To make ground improvement systems effective, some basic factors must be considered and the interactions among soil, environment, and improvement systems must be evaluated. Some of these factors are listed as follows:

1. Sensitivity of soil to environment (Section 5.12)
2. Ground soil-water pollution (Sections 1.4 and 5.6)
3. Problematic/adverse ground conditions (Chapter 13)
4. Selection of construction material(s)

In general, ground improvement systems deal with a vast amount of earth materials. Presently, the annual figures of the volume of earth materials used in the construction field are in the billions of tons and the highest among all other construction materials. Other needs to be considered include the recent energy and material shortages. These problems cause burdens on the engineering profession in the search for alternative or low-cost materials to be used in the ground improvement system.

The ground improvement program can affect the social-economic environmental conditions significantly if the program is well planned and technically well performed. It also can receive the maximum possible degree of ground utilization. For example, utilization of sanitary landfill areas can convert waste and deserted lands into useful land which can make significant economic impact. In this chapter, discussions are focused on environmental aspects of ground improvement systems and foundations.

1. Load-environmental factor design criteria: foundation-structure interaction; structure-foundation-soil interaction
2. Environmental loads: lateral pressure; wind; earthquake load
3. Foundations: pile; sheet piling; bulkhead and cellular structures
4. Ground improvement system: reinforced earth; grouting
5. Underwater repair

12.2 LOAD-ENVIRONMENTAL FACTOR DESIGN CRITERIA

12.2.1 Load Factor Design Criteria and Approaches

(1) Load Factor Design Criteria

Load factor design criteria have been used in many structural designs in various engineering projects. From a structural engineering point of view, this approach shows some advantages such as increasing the use of probability in design as a future development, and designs resulting in a more economical structure. Load factor design is a method of proportioning structures for multiples of the service load. The design ultimate load is obtained by applying factors to the different service loads. Load factor design involves:

1. Load function: This involves a consideration of types of loads and the factors to be applied to each. There is dead load, live load, long-term, short-term and extraordinary and the various possible combinations of these.
2. Resistance function: There is a 'resistance function' or limits of structural usefulness. The design process equates the two through analytical techniques.

(2) Load Factor Design Approaches

One of the unique features of load factor design is the use of multiple load factors. Dead loads are subject to less variation and uncertainty than live loads, and on this basis it is not unreasonable to assign a lower load factor to the dead loads than to the live loads. The net result of the application of load factor design is frequently a structure of lighter weight, especially in those cases where live load is a relatively small proportion of the dead load. The resulting economy is the principal reason that one would finally use load factor design (AISC, 1993). The use of load factor design, in the first place, means that:

1. The design condition is the ultimate for a failure state, not the allowable or working condition.
2. There are more design loading conditions to consider.
3. Load factor design opens the way for increased use of probability in design as future development.
4. It can result in a more economical structure.

12.2.2 Environmental-Load Factor Design Criteria

(1) General Discussion

Conventional approach for analysis and design of most foundations or other geostructures is based on allowable or working conditions. Regardless of loading types and environmental conditions, they are using ultimate or failure load divided by a factor of safety. The factor of safety is usually provided by building codes, specifications, standard textbooks, handbooks, or, in some cases, just by individual experience. Unfortunately, ground soil is not like most other construction materials; it is extremely sensitive to the local environments such as temperature, moisture content, and types of pore fluids existing in the soil voids etc., which will significantly change soil behaviors. Numerous unexpected foundation failures frequently occur because in most analysis and design of foundations, these environmental parameters are not included among the factors for consideration.

(2) Environmental Factors

As discussed in a previous section design criteria based on load factor alone do not give the whole picture and neglect the other important factor which controls the overall stability of all civil engineering structures, the environmental factor design criteria. Local environmental factors include weather conditions, groundwater fluctuations, and degree of ground-water-air pollution, as indicated in Figure 12.1. Also, Figure 12.1 illustrates the uncertainty (risk) involved in foundation analysis and design and that the environmental factor is much more critical and significant in comparison with conventional load factor design.

(3) Load/Environmental Factor Design Approaches

As discussed in Sections 3.8 and 5.12, fine-grained soils are more sensitive to environments than large soil particles because the smaller soil particle has a greater surface area. The environmental factor has a more important role than the load factor design in all civil engineering structures founded on ground soil. Civil engineering design of structures is interdisciplinary in nature and needs the cooperation of all related fields in order to provide a safe and economical structure.

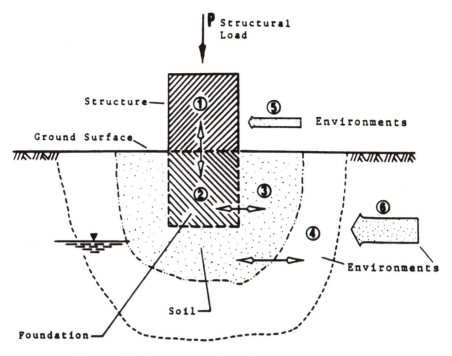

Figure 12.2 Structure-foundation-soil-environment interactions.

12.3 SOIL-STRUCTURE VS. STRUCTURE-SOIL INTERACTIONS

12.3.1 General Discussion

Structure-soil interaction includes wind load and structural load, and soil-structure inter-action includes blasting, machine vibration, moving vehicle, pile driving during construction, and seismic loads which are all related to local environments, as shown in Figure 12.2. Some additional illustrations are added to include lateral earth pressure, pile foundation, and slope stability for further explanation of structure-soil-environmental interactions.

Local environmental factors include weather conditions, groundwater fluctuations, and degree of ground-water-air pollution as indicated in Section 8.2. Also, Figure 12.1 illustrates the uncertainty involved in foundation analysis and design and that the environmental factor is much more critical and significant in comparison with conventional load factor design.

12.3.2 Structure-Soil Interaction

Wind load acting on a building is a typical structure-foundation-soil interaction problem. Figure 12.3 shows the effect of wind loads on a high rise building. The experimental data were obtained from the University of Hong Kong as reported by wind behavior around Hong Kong Island and the South China coastal region (Mackey, 1975). The mechanism of wind load relating to the building-foundation-soil system is still not clearly understood; however, designers believe that a certain amount of wind load has been transfered into the ground soil. DeSimone (1973) pointed out that the passive earth pressure, shear strength, and lateral resistance of a subsurface structure, such as piles or caissons, should be examined, and the behavior of soil affected by wind should also be evaluated.

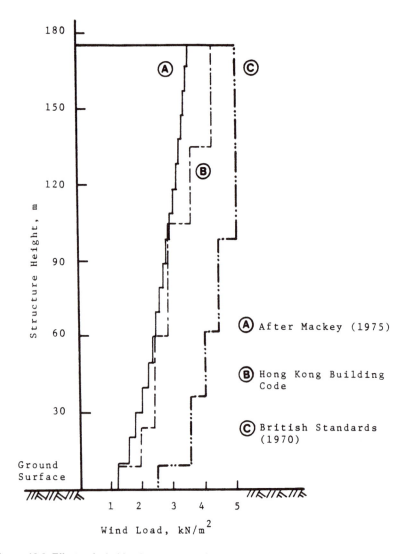

Figure 12.3 Effects of wind load on structure-foundation-soil system (After Mackey [1975].)

12.3.3 Soil-Foundation-Structure Interaction

The soil blasting phenomenon is a typical soil-foundation-structure interaction problem as discussed in Section 11.10. In this section, discussion is focused on the mechanism of this interaction. The phenomenon of this interaction can be visualized in a more simplified form as shown in Figure 12.4. Four distinct actions result from a given blast: (1) the energy transmitted from the blast to the rock or soil adjacent to it; (2) the wave propagation through the rock and soil between it and the structure in question; (3) the energy transmitted from the rock or soil beneath and adjacent to the structural foundation to the foundation itself; and (4) the structural response to the energy it absorbs.

Soil-foundation-structure interaction due to quarry blasting as reflected by wall crack opening is shown in Figure 12.5. It shows the relationship between wall crack opening vs. scale distance for various types of walls and foundations. The experimental data used in the figure are obtained from an eastern Pennsylvania location. The characteristics of blasting are reflected by scaled distance (Equation 11.20). There are significantly different responses for various types of foundations and walls as indicated in Figure 12.5. Figure 12.6 summarizes the soil-structure interaction under dynamic loading and their measurable parameters.

① The energy transmitted from the blasting to the ground

② The energy transmitted through the ground (earth wave)

③ The energy transmitted from the ground to the structure

④ The structure response to the energy it absorbs

⑤ Some of the energy transmitted to some places

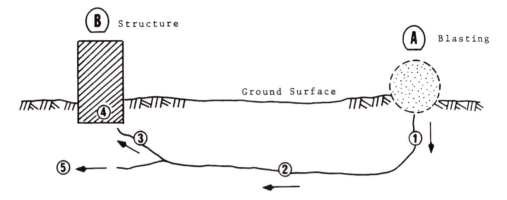

Figure 12.4 Effect of blasting on soil-foundation-structure system.

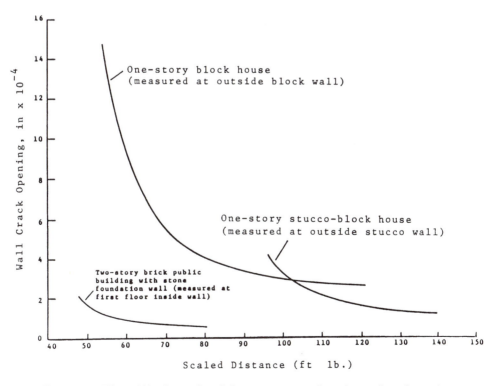

Figure 12.5 Effect of blasting on foundation-structure as reflected on wall-crack opening.

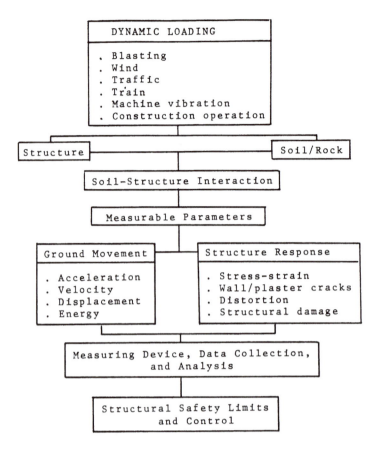

Figure 12.6   Summary of soil-structure interaction under dynamic loading and their measurable parameters.

12.4 LOAD, FACTOR OF SAFETY, AND ALLOWABLE CONDITIONS

12.4.1 Loads and Environmental Loads

(1) Loads

Loads acting on structure include: static, dynamic, and environmental loads. Static load (dead load) includes the weight of the structure and all material permanently attached to it. Permanent and fixed service equipment are usually considered as part of the dead load. Dynamic load includes live load and impact load. Live load includes all loads that are not a permanent part of the structure but are expected to superimpose on the structure during a part or all of its useful life. Vertical loads due to wind or snow are not considered as live load. Human occupancy, partition walls, furniture, warehouse goods, and mechanical equipment are major live loads. The magnitude of live load to be used in the design of various buildings is usually stipulated in local building codes. Railroad and highway bridges subjected to traffic loadings, and reaction from industrial cranes and elevators, sometimes constitute a large portion of the live load.

(2) Environmental Loads

Environmental loads include earthquake, wind, snow, water pressure, and others. Earthquake load may result in lateral force. This lateral force may act on the structure in any horizontal direction. The effect of this lateral force may be evaluated in accordance with the Uniform Building Code. Wind load acts on all exposed surfaces of a structure. Overhanging parts are subjected to uplift pressure. The magnitude of design pressure is usually stipulated in local building codes. Figure 12.3 shows the effect of wind load on structure-foundation-soil system. The design of snow load per square foot of horizontal projection of the roof is given in local building codes. Other loads include water pressure, ice, wave, etc. Structures constructed in a swift river, lakefront, etc. are also subjected to forces due to current flow, ice flow, and wave forces. Total loads acting on the foundation are computed in three conditions: in earthquake zones the force arising from earthquake motion should replace that due to wind. This assumes that the maximum wind pressure and maximum earthquake motion do not occur simultaneously.

12.4.2 Factor of Safety and Allowable Loads

The *factor of safety* or degree of safety is used by engineers to indicate whether or not a foundation soil or earth work will fail under the worst service conditions for which it was designed. The present concept for determining the factor of safety for stability of earth slope is based on Coulomb's law, the principles and application of the methods for computing various factors of safety. They have used probability analysis to calculate the safety of foundations for various loads and strength distributions. Applied load and soil strength are considered to be random variables. The appropriate probability function is determined empirically by fitting to the experimental data. The *allowable loads* can be obtained by computing the ultimate load by a factor of safety.

12.5 BEARING CAPACITY OF GROUND SOIL

12.5.1 Load-Footing Interaction

The load is the mechanical energy; however, the bearing capacity is in the multimedia energy field, because it is influenced greatly by local environmental conditions such as soil properties, groundwater fluctuation, and frost depth. The current approach is that load or pressure develops under a foundation, usually referring to the ultimate capacity at failure of the soil. Allowable or safe, bearing pressure is equal to the ultimate capacity divided by a factor of safety. Furthermore, the failure or movements of foundation may result from foundation failure or collapse from excessive settlement, therefore adequate factor of safety against foundation failure must be provided and adequate margin against excessive settlement must be also considered.

In order to be able to provide an adequate factor of safety against foundation collapse, the ultimate bearing capacity (Q_u) of ground soil must be known. Usually, a factor of safety (F_s) of 2 to 3 is used with the allowable bearing capacity (Q_a); then $Q_a = Q_u/F_s$. The other design parameters include width and depth of the structure and soil properties including cohesion, friction angle, and unit weight. Comparing among these parameters, some are more sensitive to environment than others. If all parameters are equally divided by a same factor of safety, the results are that some are underdesigned and others are overdesigned. Detailed procedures of each method applied to foundation engineering have been discussed by Chen and McCarron (1991).

12.5.2 Bearing Capacity from Building Codes

Many large cities around the world provide their own building code for foundation design. These values are generally based on years of experience in their own particular zone or region.

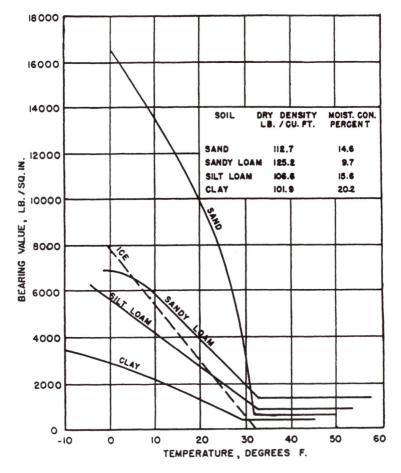

SOIL	DRY DENSITY LB. / CU. FT.	MOIST. CON. PERCENT
SAND	112.7	14.6
SANDY LOAM	125.2	9.7
SILT LOAM	106.6	15.6
CLAY	101.9	20.2

Figure 12.7 Bearing value-temperature curves for various materials. (After Kersten, M.S. and Cox, A.E. [1951].)

These values are simply based on the local experience and often referred to as presumptive pressure. Presumptive pressures are based on a visual soil classification such as soft clay, stiff clay, dense sand, and loose sand. These values obtained from building codes are useful for preliminary analysis and design of foundation systems. However, the bearing capacity based on the building code do not reflect on the groundwater table, depth, and size of footing and other environmental conditions.

Numerous factors affect the bearing capacity of ground soil. Among these factors, the groundwater table and frost depth are most critical. However, both groundwater and frost depth have been discussed in Section 8.3. Bearing capacity of problematic soil deposits and landfill areas will be discussed in Chapters 13 and 17. Bearing capacity of ground soil will be affected significantly by the local environment. In the summer months, in general, the bearing capacity is much higher than in the spring-thawing period. Figure 12.7 shows the temperature effects on bearing capacity as discussed by Kersten and Cox (1951).

12.6 LATERAL EARTH PRESSURES

12.6.1 Active and Passive Earth Pressures

When lateral earth pressure acts on a rigid wall, such as a retaining wall, the pressure distribution along the wall can be divided into three distinct regions, the active, passive, and

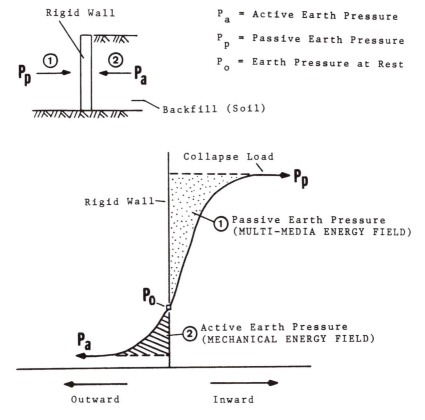

Figure 12.8 Load-displacement relationship of a rigid wall.

at rest as indicated in Figure 12.8. It is apparent that the wall may have two directions of motion: into the bank (soil) or away from the bank. If the wall is initially at rest and held by a force, three pressures acting on the wall are usually defined as the *active* earth pressure (P_a), *passive* pressure (P_p) and pressure *at rest* (P_0). From a mechanical viewpoint, these three pressures can be explained as follows.

As P_p is gradually reduced, the soil undergoes elastic and nonelastic deformation and, finally, will reach failure. This collapse load is generally referred to as passive earth pressure. Let's examine the same problem illustrated in Figure 12.8 by particle-energy-field theory point of view. The active earth pressure is the weight of a soil mass acting as a load against the rigid wall. That weight of soil is treated as mechanical energy. The variation of load representing the weight of soil may be changed from a dry condition to a saturated condition. However, for the passive earth pressure case, when the load (wall itself) acts on the soil (bank), the soil behavior changes when local environmental conditions change. For example, for the same backfill soil, the behavior between springtime and summer months will be significantly different; therefore, in the passive zone, the mechanism must be treated as in the multimedia energy field. When deformation or displacement changes from active to passive, the turning point where there is no pressure in such a case is called earth pressure at rest, P_0. The coefficient of earth pressure is called K_0. Its characteristics and measurements will be discussed in the following section.

12.6.2 Coefficient of Earth Pressure at Rest, K_0

Many naturally occurring sediments as well as manmade fills are deposited and compacted in horizontal layers where no lateral yielding occurs. Under such conditions the ratio of lateral

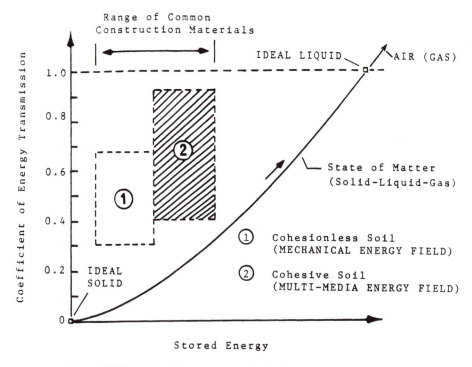

Figure 12.9 Coefficient of energy transmission for various states of matter.

to vertical stresses is known as the coefficient of earth pressure at rest and denoted by the symbol K_0. Al-Hussaini (1981) made a comprehensive study on K_0 and comparison of various methods for determining this parameter. For elastic isotropic material under first loading, the value of K_0 can be expressed directly in terms of Poisson's ratio. Hence, the value of K_0 and stress-strain response of soils under K_0 condition are important for engineering problems involving settlement of fills, lateral pressure, excavations, and other problems where lateral yielding is restricted.

The state of lateral earth pressure, P, acting on the rigid wall is shown in Figure 12.8. The variation of the coefficient of energy transmission, known in geotechnical engineering as coefficient of earth pressure at rest, is shown in Figure 12.9. If a solid body displays a very great cohesion (c $\rightarrow \infty$), K_0 tends to zero. With decreasing binding forces between the elementary particles caused by an increase of temperature (thermal energy) by vibration (mechanical energy), the K_0 value increases to 0.4 to 0.5 (Kézdi, 1964), which is characteristic for grain assemblies. Greater values occur with viscous liquids, in which the internal shearing resistance is much smaller and finally $K_0 = 1$ is reached for ideal liquids with zero internal friction.

In the case of gases K_0 exceeds 1 due to the atomic particle movement. The increase of K_0 can be achieved by transmitting energy to the system either as mechanical energy forcing particles out of the solid, as vibration energy decreasing the number of contacts between the particles, thus reducing the inner resistance in a transient manner, or as thermal energy which enlarges the distances between the particles and increases their speed, thus transforming a solid to a liquid or a liquid to a gas. As indicated in Figure 12.9, the greater the storage energy, the greater the coefficient of energy transmission or coefficient of lateral pressure, K_0.

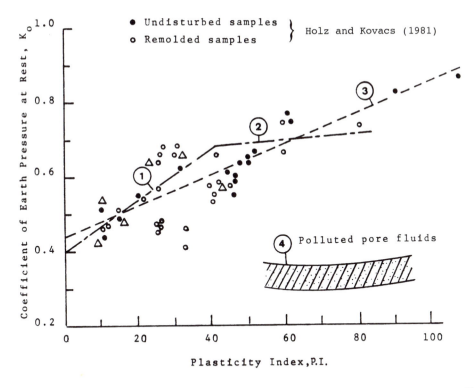

Figure 12.10 Coefficient of earth pressure at rest; K_0 vs. soil types (as reflected by plasticity index, PI).

12.6.3 K_0 for Clay-Like Soil and Sand

(1) Clay-Like Soil

Figure 12.10 shows the relationship of K_0 vs. soil types as reflected by plasticity index. A linear relationship is found in that the increase of plasticity index also increases the K_0. However, there are great variations as indicated in the figure. For a polluted soil-water system, the K_0 value is generally lower because the contaminated soil produced the cracks as discussed in Section 6.6. Figure 12.11 shows the relationship of K_0 and overconsolidation ratio (OCR) for various types of peat soils. The OCR increases as K_0 increases for all types of peat.

(2) K_0 for Sand

Laboratory measurement of K_0 of sand is made by various instruments including linear variable differential transducer (LVDT) and conventional strain gauge as noted by Al-Hussaini (1981). Results between theoretical and experimental results for sand are given in Figure 12.12. Significant differences for these results are obtained.

12.7 FRICTION FORCES AND ANGLES BETWEEN TWO MATERIALS

12.7.1 Friction Angle Between Soil and Soil

As discussed, Section 2.8 indicates the load-deformation process of soil. During the process, there must be a relative motion of soil particles. Force is required to overcome the

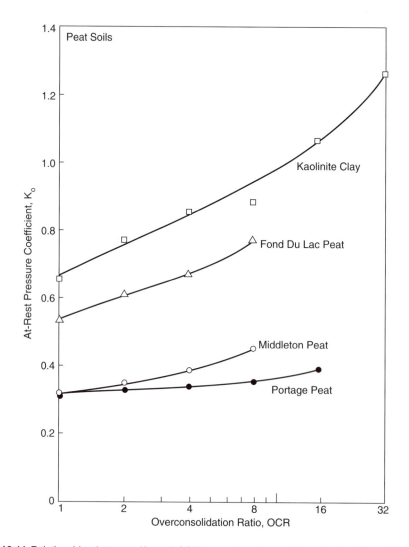

Figure 12.11 Relationships between K_0 and OCR for various types of peat soils (After Edil, T.B. and Dhowian, A.W. [1981], *J. Geotech. Eng. Div. Proc. ASCE,* v. 107, no. GT2, pp. 201–220.)

frictional resistance developed between particles during motion. In geotechnical problems, there are two important cases: (1) the friction force between soil and soil, and (2) the friction force between soil and other than soil. The friction angle between soil and soil generally is referred to as internal friction angle (ϕ), as indicated in Equation (11.1). This angle can be determined by conventional geotechnical procedures such as direct shear test or triaxial shear test as discussed in Section 11.7.

12.7.2 Friction Angle Between Soil and Wall

The friction angle between soil and wall generally is referred to as friction angle (δ). It is the same principle as angle (ϕ). Using the same procedure as described in a previous section except using two different materials, comparison between friction angles of soils (ϕ) and friction angle between soil and wall (δ) is presented in Figure 12.13. In examining Figure 12.13 for granular soil, these two angles are very close; however, for cohesive soil, the (δ) angle is smaller than (ϕ) angle. The main reason is for fine-grained soil where the friction angle between soil and soil is more sensitive to environment than between soil and wall material.

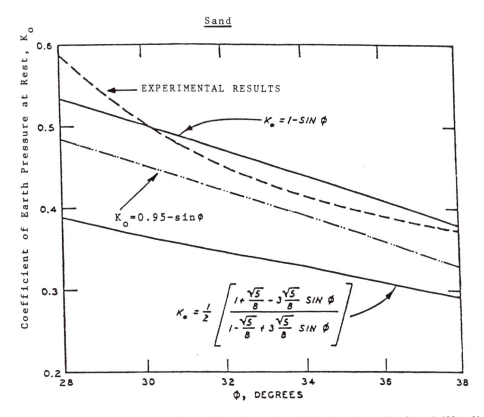

Figure 12.12 Comparisons between theoretical and experimental test results on K_0 of sand. (After Al-Hussaini [1981].)

12.8 PILE FOUNDATIONS

12.8.1 Characteristics of Piles

(1) General Discussion

Pile foundations are one of the major elements in ground improvement or foundation systems. There are numerous books, reports, and published papers that have discussed this subject. However, the pile foundation still is a problem and geotechnical engineers are still looking for an effective solution. The main reason is that the mechanism of soil-pile interaction still is not clearly understood, and it is difficult to predict the performance during pile's useful life. In addition, the skin friction between soil and pile, the penetration, and the end-bearing behaviors are influenced by local environments significantly. The following discussion of piles focuses on environmental aspects of the pile foundations.

Piles are structural members used to transmit structural loads through a material or stratum of poor-bearing-capacity to one of adequate-bearing-capacity material. This load transfer may be by friction, end-bearing, or both, depending on whether the load is resisted by friction along the surface of the pile, or whether the pile end rests on a soil stratum which is strong enough to carry the load. The load carried by friction is called a friction pile, and the load carried by firm soil stratum is called end-bearing pile. The pile, in general, may utilize both friction and end-bearing to vary the imposed structural load. Piles are made from steel, concrete, timber, or composite materials. Piles are made into different shapes such as circular and H-shape.

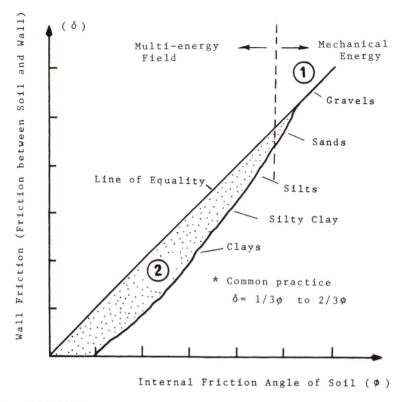

Figure 12.13 Friction angle between soils (φ) vs. friction angle between soil and wall (δ).

The structural load may be static or time-dependent, vertical or lateral, transmitted to the soil stratum from single piles or pile groups. Pile foundations are used (1) for distributing the load; (2) for transfer load to firm soil stratum; (3) to resist the uplift pressure; and (4) to resist inclined or lateral loads. Piles are usually inserted by driving with a steady succession of compaction blows by means of a hammer on the top of the pile. For more efficient and faster means of installing piles, the low and high frequency vibrator methods have been used. The low frequency vibrator operates in the range of 5 to 35 cycles per second and delivers energy by lifting the entire pile, driving it downward on each cycle. The high frequency (resonant) vibratory system utilizes the oscillators having an operating range of 40 to 140 cycles per second. These oscillators make it possible to vibrate a pile at its natural frequency. The basic principal of the vibrator for faster installation of piles is that under vibration the shear modulus or viscosity of a soil is drastically reduced.

(2) Selection of Pile Types

The selection of a pile type and its appurtenances is mainly dependent on environmental conditions. If piles are driven into seawater, the environmental considerations are seawater corrosion, wave action, moving debris, ice, and marine borer attack. If concrete pile is used, strong chemicals in water or in alkali soils could cause serious deterioration. If steel pile is used in an electrolytic environment near electrical currents, they could affect electrolysis deterioration. For some regions, low-cost energy-saving conditions are important; therefore, using sand or stone piles is suitable. Using bamboo-lime composite piles for controlling landslides in earthquake regions is effective has been reported by Fang (1991c). Types of ground soil also affect the selection of pile types. Piles to be driven through obstructions to

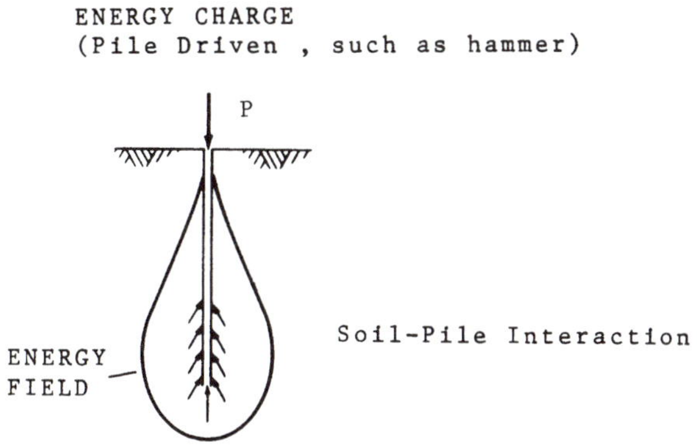

Figure 12.14 Soil-pile interaction under the load and analogies between pile-driven process and energy field. (a) Single pile; (b) group piles.

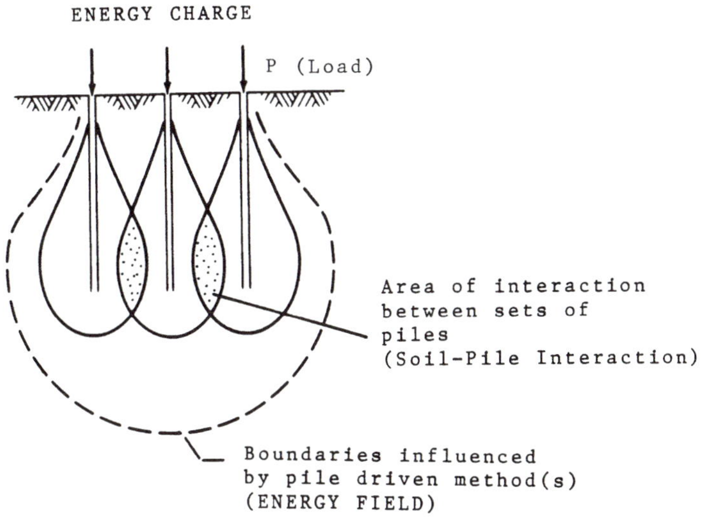

bedrock with the least driving effort and soil displacement would favor a steel H-pile or open-end pipe pile. Further discussion of various types of piles related to the environments are discussed in various publications (Gauffreau [1987]; Dismuke [1991].

(3) Soil-Pile Interaction in the Environment

The mechanism of soil-pile interaction is similar to the particle behavior under load as discussed in the Section 2.8.2. It is a process involving potential energy (energy of position), kinetic energy (energy of motion), and thermal energy. Also, as indicated in Figure 2.2, the relationship between energy charge and energy field is similar to the soil-pile interaction under the load. The analogies between the pile-driving process and energy field is illustrated in Figure 12.14.

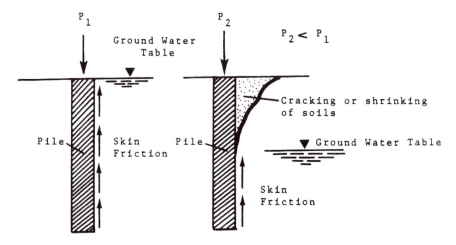

Figure 12.15 Effect of groundwater table on skin friction between pile and soil.

(4) Friction Resistance Between Pile and Soil

There are three basic types of friction resistance which characterize the relative motion between soil particles during loading process. The force required to produce the relative motion includes variables such as: area of the surface of motion; relative velocity of the surface of motion; and distance between the surface of motion. Estimation of friction resistance between soil and pile is important for design of the friction pile. Figure 12.13 may assist with this purpose.

Negative skin friction is a downward drag force on the pile in the soil-pile system. Commonly it occurs in the dewatering process because lowering of the groundwater table will increase the vertical effective stress. It will induce consolidation settlement; or in the case of granular soil placed over a layer of soft clay deposits, it will induce the process of consolidation in the soft clay layer and will produce a downward drag on the soil-pile system.

12.8.2 Environmental Factors Affecting Pile Capacity

There are numerous factors affecting pile capacity including soil types as reflected on cohesion, friction angle, as well as assumed failure mechanisms. However, in field conditions, additional factors are in effect such as groundwater fluctuation, ground temperature, freezing-thawing, pollution intrusion, etc.

(1) Fluctuation of Groundwater Table

The fluctuation of groundwater table relating to the skin friction between soil and pile is illustrated in Figure 12.15. As discussed in Sections 6.2 and 6.3, the groundwater table relates with shrinkage and swelling and consequently affects skin friction. Shrinkage and swelling are not reversible processes, because swelling is under the influence of multi-energies and skrinkage is only controlled by thermal energy.

(2) Ground Temperature and Freezing-Thawing Cycles

Davisson (1970) shows *in situ* tests on how ground freeze affects the ultimate resistance of pile capacity. Also, as indicated in Figure 12.7, bearing capacity of ground soil increases significantly, therefore pile capacity under ground soil condition also increases. Characteristics of frozen ground soil have been discussed in Section 8.7. Pile foundation under freezing-thawing conditions are discussed further in Phukan (1991).

(3) Pollution Intrusion and Corrosion Effects

Ground soil corrosion is one of the most critical environmental geotechnical problems. Frye (1970) and Dismuke (1991) measured the corrosion loss of exposed vertical mild steel in contaminated conditions. Figure 12.16 shows the pile damaged by contaminated soil-water.

(4) Pile Damage during Driving Process

During the driving process it is difficult to visualize what type of pile damage has been done. A case study was done using H-piles driven into a limestone. Six piles were studied: (1) H-pile protected with Pruyn point suffered little or no damage from hard driving into limestone layers; and (2) without protection, H-pile was mangled. Various types of damage and cause are discussed by Fang and Dismuke (1975) as shown in Figure 12.17.

12.8.3 Field Inspection of Pile Foundations

Many pile failures are due to lack of supervision or field inspection. In the field inspection all pertinent information must be observed and recorded. It is recommended that the engineer meet with the field engineers, pile inspectors, and pile-driving contractors prior to the start of driving. It is the duty of the engineer to see that only experienced and competent personnel are employed and that equipment used is adequate for the work at hand. Inspection of piles may be divided into the following steps:

1. Inspection of the pile when ready for driving, which includes the inspection of timber, concrete, and steel piles to meet specifications
2. Handling and driving the pile, which includes inspection of driving equipment, review of the method of driving techniques, recording details such as overdriving of piles, hitting obstructions, driving out-of-plumb, retardation of stroke, variations in cushioning material, and sequence of driving

12.8.4 Special Types of Pile and Their Uses

(1) General Discussion

In many cases, the conventional pile foundation such as steel or concrete may not be suitable due to local environments, or due to availabilty of the material or cost. In such cases, some low-cost and energy-saving piles must be considered. In this aspect, the availability of material and labor should be examined: (a) low-cost/energy-saving pile types, e.g., sand and stone piles; lime pile, mini-concrete pile and composite-bamboo pile; (b) highly acid resistant pile, e.g., polymer concrete pile; and (c) pile used for special applications, e.g., root pile, pin pile, etc.

(2) Highly Acid Resistant Piles

Polymer concrete has shown promise for use as a major construction material. Based on laboratory studies, the strength of polymer concrete is three to four times that of ordinary concrete with very high corrosion resistance and durability. It is particularly suitable for areas

Figure 12.16 Photosgraphs show piles damaged by contaminated soil-water. (a) Concrete pile; (b) steel pile. (Photos courtesy of T.D. Dismuke.)

with high corrosion problems, such as coastal urban areas with problems of saltwater intrusion (Section 14.4).

In the past, piles made of cement were highly susceptible to corrosion in seawater, bending, and shock from impact and were apt to become considerably heavy with great steel reinforcement requirements. Polymer concrete piles were found to be highly suitable for such construc-

Figure 12.17 Pile damage caused by driving process. (a) H-pile with protection; (b) H-pile damaged without protection. (After Fang, H.Y. and Dismuke, T.D. [1975], *JSMFE Div. Proc. ASCE,* v. 101, no. SM6, pp. 594–598.)

tion and have been shown to satisfy all the demands and conditions. Japan is already manufacturing polymer concrete piles for construction of nearshore structures and seagoing barges.

(3) Piles Used in Special Conditions

In recent years, many new types or new forms of piles made of various materials have been on the market for commercial uses. Such piles include root piles, pin pile, mini-concrete piles, lime piles, etc.

Roots piles® are cast-in-place small-diameter reinforced concrete piles used to underpin buildings and to form three-dimensional underground retaining and stabilizing structures. The root pile system is a foundation structure akin to the roots of a tree. The system is formed by the rotary drilling method through the existing structure down to the designed level of depth into the soil below. Each pile will securely anchor the two major components, structure and soil, together. Pin pile is a short smaller-diameter pile. It is used to protect against slope failure in rock slopes.

(4) Piles Made from Waste or By-Products

There are numerous types of piles made from waste or by-products such as solid waste (garbage) or from waste rubber tires. Some of these types of pile made from wastes will be further discussed in Section 17.10.

12.9 UNDERWATER FOUNDATION PROBLEMS

Underwater foundation problems are closely related with the environment such as degree of pollution of underwater conditions, depth of water, current, etc. In this section focus is placed on underwater inspection and repair of bridge substructures and foundations.

12.9.1 Underwater Inspection of Substructures

1. Preparation: The underwater inspection team should be familiar with the construction of facilities and the site conditions before starting the inspection. Proper planning and preparation will save much time in the field and result in a more accurate inspection. The inspection team should consider the inspection schedule, type of inspection, and inspection procedure (Lamberton et al. 1981; Brackett et al. 1982).
2. Cleaning: Cleaning marine growth from the underwater portions of the substructures to facilities inspection and detection of deficiencies is almost always required. Routine inspections, when there are no known defects, generally require only light cleaning in select areas, whereas detailed inspections may require complete cleaning of certain structural elements.
3. Visual examination: Is the primary method used to inspect the underwater portion of substructures. The most obvious limitation to visual inspection is water clarity. For example, for bridges, in many cases, bridges are built in generally turbid water in which the degree of visibility varies from poor to zero. When turbidity is high, underwater visibility can be nonexistent, even if the diver carries artificial lights; direct light is reflected by the turbidity and visibility is reduced. Visibility may be improved if a clear-water mask is attached to the face plate of diving gear. Plastic bags can be used as a clear-water lens to inspect H piles and other irregular shapes. A diver's slate is useful for making notes and sketches under water.
4. Tactile: Planning for tactile inspections requires greater preparation than when working in clear water. Attention should be given to (1) good communications between diver and surface; (2) use of nondestructive testing and other techniques not dependent on sight or touch; (3) increase use of sounding equipment to detect scour, debris buildup; (4) recorder to document voice transmissions; and (5) determination of position and depth of diver.

12.9.2 Maintenance and Repair of Underwater Facilities

After underwater inspection is completed, the maintenance crews and repair teams take over. A prerequisite for the maintenance or repair of an underwater structure is an understanding of what has caused the present condition. Frequently, underwater damage or deterioration is related to local environmental conditions. Construction deficiencies can result in the need for maintenance and repair of underwater structures. Continuing degradation or deterioration may require an investigation process in order to fully understand the type of deterioration that is occurring, why and how it is occurring, and the rate at which it is taking place.

(1) Prevention

Prevention is one of the important factors for controlling the damage of the underwater facilities. Among these factors, the selection of proper materials is one of them. Using polymer concrete as an example, applications of polymer concrete in the geotechnical engineering field have been discussed by many investigators for highway bridge deck, airfield runways, hydraulic structures, etc. In this section, discussion of the uses of polymer concrete system in underwater facilities is presented.

Polymer concrete system has high strength, excellent abrasion resistance, durability to chemical and freeze-thaw attack, impermeability and low ice adhesion, and is considered

Table 12.1 Pile Corrosion Protection and Repair Techniques

Corrosion Protection

Coatings[a]
  Hydrophobic
  Dry film
Barriers[b]
  Jacketing (concrete)
  Plastic (rigid)
  Plastic (flexible)
  Plastic shell and mastic epoxy filler
Sheathing (Monel® etc.)
Cathodic[c]
  Impressed current
  Galvanic anodes

Structural Repair

Section replacement
  Welding
  Bolting (drill and ream or explosive)
Supplementary sections[b]
  Welding
  Bolting (drill and ream)
Structural jacketing[b]
  Dry concrete placement
  Wet concrete placement (tremied)

[a] Requires sand-blasted surface for most coatings.
[b] Usually does not require sand-blasted surface — use of hand scrapers is sufficient.
[c] Does not require surface preparation of piles; for existing piers, ground cables are installed when impressed current systems are used.
Based on Dismuke, T.D. (1991), *Foundation Engineering Handbook,* 2nd ed. Van Nostrand Reinhold, New York, pp. 856–867.

suitable to be the ideal material for seafloor engineering and ocean structures. Polymer concrete barges and multipurpose pontoons, already in production, have also shown high durability, easy handling, and low-cost, while demonstrating their versatility in leisure use, exploitation of natural resources, and construction work.

(2) Corrosion Protection and Repair Techniques

The typical prevention procedures are protective coatings, electrical methods, and modifying the environment. A brief description of each method is presented. A protective coating forms a barrier between the metal and its environment, thereby eliminating the electrolyte from the system. Corrosion cannot proceed with only the presence of cathodes and anodes; it needs an electrolyte. A good protective coating must resist acids, alkalines, salts, moisture, and ultraviolet rays. Organic coatings are probably the most familiar means of controlling corrosion of metals and involve a relatively thin barrier between the substrate material and environment. The most common are paints, tar, pitch, and plastics. The electrical method for corrosion prevention includes cathodic and anodic protection and electrical insulation. A summary of pile corrosion protection and repair techniques is presented in Table 12.1.

Geotextiles are frequently used for waterfront erosion control or for repairs to structural components such as bridge piers and pile foundations. Detailed installation procedures for various applications relating to the nearshore structures are given by Koerner (1994).

Table 12.2 Typical Types of Grouts Used for Various Soil Types

Type of soil	Characteristics of soil	Type of grout	Example
Alluvium and coarse sands	$d_{10} > 0.5$ mm $S < 100$ cm^{-1} $K > 10^{-3}$ m/sec	Binghamian suspensions	Cement grout
Medium and fine sands	d_{10} between 0.02 and 0.5 mm S between 100 and 1000 cm^{-1} K between 10^{-3} and 10^{-5} m/sec	Colloidal solutions	Silica or lignochrome gels By-products of tannins Unsaturated vulcanizable oils Organic or mineral colloides
Silty or clay sands Silts	$d_{10} < 0.02$ mm $S > 1000$ cm^{-1} $K < 10^{-5}$ m/sec	Pure solutions	Phenoplasts Aminoplasts Acrylamide

Data from Caron (1976).

12.10 GROUTING AND INJECTION PROCESSES

12.10.1 General Discussion

Grouting or injection techniques are frequently used to seal cracks, stop leakage, reinforce some types of geostructures such as ground anchors or soil nails, or to repair underground structures due to earthquake damages or other causes. More recently, the application of this technique has been expanded to include hazardous/toxic waste control or storage facilities. Regardless of the purpose for use of this method, the operation process is always in a situation surrounded by difficult environments. For example, sealing off acid drainage from mines or leachate leaking from hazardous/toxic chemical tanks must be faced as a common situation.

The material used for grouting is called *grout*. Grout can be pure solutions, colloidal solutions, or binghamian suspensions. There are two general types of grout, one derived from sodium silicate and the other from aqueous resins. Their main purposes are water-tightness or for both water-tightness and strengthening of soil structure. Grouts are in liquid form so that grout can penetrate into the soil void, but it must transform into a solid after a certain time period. This time period transforming from liquid or semiliquid into a solid is called gelling time. The selection of the appropriate grout is a function of the particle size gradation curve of soil to be treated. Table 12.2 presents the type of soil to be treated and types of grouts to be used.

12.10.2 Grouting Process and Soil-Grout Interaction

(1) Grout and Injection Processes

Grouting is a process between soil and grout. The spacing between two adjacent grout holes is normally about 50 cm. Among the methods used for grouting, the most common method is the Jeziorsky-Joosten technique. This method consists of injecting concentrated sodium silicate (grout) through a first bore hole (soil) and a solution of calcium chlorite through a neighboring hole. The Jeziorsky-Joosten method has two disadvantages: the first requires higher grouting pressure and the second is its difficulty to achieve full effectiveness.

(2) Soil-Grout Interaction

If grout injected into the fine-grained uniform soil is (a) too thick, then the grout may stay on the surface of soil, and (b) too liquid, then the majority of liquid grout will run away by gravity. Some fluid grouts such as silica gels are excellent in fine-grained soil but not suitable for the coarse soil because of the syneresis action (Caron, 1976). In heterogeneous soil, as the injection proceeds, grouts must be more and more penetrating, due to the progressive reduction of permeability of the soil. The logical grouting procedure and soil-grout interaction must consider the rheological behavior in the multimedia energy field.

12.10.3 Grout Used for Controlling Hazardous/Toxic Wastes

(1) General Discussion

In analyzing the grouting process, we always assume that:

1. Grout must be sufficiently fluid to penetrate into both the soil and the voids and fissures of the soil.
2. Grout must stay in the soil without being affected by any of the physicochemical conditions to be applied later to the grouted soil.

In view of these requirements and considerations of the *in situ* conditions, there are significant differences between these two conditions, the assumptions, and *in situ* conditions; therefore, many premature or unexpected failures frequently occur. To understand the grout-environment interactions in the *in situ* condition, it is necessary to examine contaminant-grout interactions.

(2) Grout System for Control of Hazardous/Toxic Wastes

In 1984, a program was begun at Oak Ridge National Laboratories to develop a range of cement-based blended dry soils that, when mixed with specific wastes, produced grouts that when processed would meet regulatory requirements and would have properties meeting long-term performance.

The waste used as grout consists of a concentrated salt solution containing 73% water, 12% NaOH, 6% $NaNO_3$, and smaller amounts of other materials. The water is incorporated into the rigid materials of the solidified grout (Benny, 1990), which forms a physical barrier to the release of the waste constituents into the environment by encapsulating the waste.

The waste consists of multivalent cations which are mostly insoluble hydroxides or carbonates pH above 13. The cementitious material in the grout maintains this high pH which results in a reduced mobility for these species. Some metal ions are also incorprated into the crystal structures of the minerals formed during the curing process. The precipitation and matrix incorporation of heavy metals and other multivalent cations into the cement matrix insures that toxic materials will be retained in the grout product. For further discussions on grout system for control of hazardous/toxic wastes see Bodocsi et al. (1988) and Weaver et al. (1990).

12.11 SUMMARY

1. Load/environmental factor design criteria open the way for increased use of probability in design as a future development, and it can result in more economical and safe foundation structures.
2. The design condition for structure as well as foundations should be the ultimate for failure state, not the allowable or service (working) condition. The factor of safety should consider the local environmental conditions.

3. Mechanisms between soil-structure interaction and structure-soil interaction are discussed and compared.
4. Chemical grouts are being used, only as a remedial means after the traditional methods fail. There is no standard procedure for grouting process; every case must be studied individually under the *in situ* conditions.

PROBLEMS

12.1 Discuss the mechanisms between soil-structure and structure-soil interactions.

12.2 A footing was constructed $10 \times 10 \times 3$ ft with the top at the ground surface of a soft clay deposit with cohesion = 500 psf and a unit weight of 100 pcf. As project engineer you are asked to make a quick, on-the-spot estimate of the maximum allowable bearing capacity for a factor of safety of 3 in order to consider a change in loading for the footing. What is your estimate of the allowable bearing capacity? Explain or justify your choice of procedure.

12.3 What is 'presumptive bearing pressure'? What is a 'factor of safety'?

12.4 Under what conditions are piles justified to reduce consolidation settlement? Explain the statement that misuse of piles may cause more foundation problems than if piles were not used at all.

12.5 Explain why pile-driving formulas are not satisfactory for establishing the load a pile can carry when used as a friction pile in soft to medium stiff clay.

12.6 A group of end-bearing piles were driven until they met refusal on a well-defined rock stratum. The piles were closely spaced. Upon loading, however, many of the piles were observed to settle a substantial amount, the largest settlements occurring under piles driven first. Explain.

12.7 The use of geotextiles is a rather new approach in foundation engineering, and so far only a relatively small amount of research has been done in comparisons with others in this area. However, it is gaining popularity rapidly. (a) Explain why; (b) discuss major applications of geotextiles in foundation engineering.

CHAPTER **13**

Problematic Soils and Rocks

13.1 INTRODUCTION

As discussed in Chapters 1 and 3, some natural soil-rock inherits weaknesses from the parent material. Such deposits include weathering rocks, residual soils, clay shales, dispersive clays, collapsible silts, soluble carbonate rocks, expansive clay, and organic soils. These types of soil have high compressibility, low shear strength, and are very sensitive to local environments, especially to water, and will cause excessive erosion, excessive ground subsidence, slope failure, and landslides. Because these types of soil-rock have special features, they deserve special attention. In this chapter, we shall discuss the engineering properties of these types of soil-rock as related to various analysis and design for geotechnical projects. The discussion includes origin, distribution, and typical geotechnical properties as well as methods for correction of the weaknesses for the following soils and rocks.

- Shales and weathered rocks
- Residual soil
- Expansive clay
- Organic soil (bog, muskeg, and moor soils)
- Karst region (sinkhole)
- Loess (wind-blown silt)
- Dispersive clay
- Laterite soil
- Alluvial sediments and lacustrine deposits
- Saline-alkali soils
- Bentonite and others

13.2 SHALE AND WEATHERED ROCKS

13.2.1 General Discussion

Shales predominate among the sedimentary rocks in the earth's crust. Their properties vary from those of 'solid' rock that must be blasted for excavation to those of soil-like materials that fall within the engineering definition of soil as discussed by Underwood (1967). One differentiates between compaction of soil-like shales in which the bonding material is essentially the water substance even if present only in very small amounts, and cemented rock-like shales in which the particulate components may be cemented by calcareous, siliceous, ferruginous, gypsiferous, phosphatic, or other bonding agents, or may be welded

Table 13.1 Classification of Soil-Like Shale

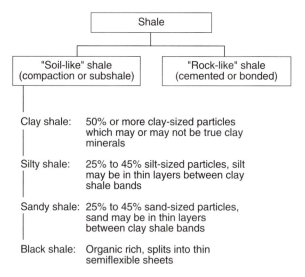

After Mead, W.J. (1936), Trans. 2nd Int. Congress on Large Dams, v. 4, p. 183.

together by recrystallization. The chemical nature of bonding agents can usually be determined by simple field or laboratory tests.

13.2.2 Rock-Like and Soil-Like Shales

Rock-like shales normally preserve their strength and integrity even during repeated exposure to wetting and drying cycles, while *soil-like* shales slake under these conditions. The time required for the slaking of standardized specimens, the size and character of the slaked particles, as well as the general slaking picture represent valuable clues with regard to the engineering behavior of shales, the same as they do for soils.

The effect on the engineering properties of soil-like shales of such physical, physico-chemical, and chemical factors as grain size composition, mineral associations, types and amounts of exchangeable ions, chemical character of electrolytes in the porewater, etc. is of the same nature as their effect on normal soils. The considerations presented here will concentrate on those factors that give soil-like shales their distinguishing characteristics. One of these is the mode of breaking and the degree of fissility. Ingram (1953) recognizes three dominant types of breaking characteristics, viz., massive, flaggy, and flaky. Massive structure and resulting breakage mode can be expected from sediments of predominantly silt, or larger size particles within a limited size range which have not undergone marked shear displacement in their development and consolidation.

Fissility or cleavage characteristics increase with increasing content of plate-shaped minerals (clays and mica) in parallel arrangement because of the manner of sedimentation and overburden pressures. Tectonic disturbances may decrease the parallel order and hence the degree of fissility. The effect of organic matter and of other factors depends on the degree to which they favor or impede the parallel arrangement of the plate-like mineral constituents. The higher the degree of fissility, the greater is the variation as a function of direction of such physical properties as modulus of elasticity, tensile, compressive, and shear strength, and thermal and hydraulic conductivities. Table 13.1 presents the engineering classification of soil-like shale as proposed by Mead (1936).

13.2.3 Engineering Properties of Shales

The porosity, the void ratio, and the degree of packing of a shale depend on its mineral and granulometric composition, its mode of sedimentation, its stress and deformation history, its chemical history, and the duration of exposure to different environmental conditions. Under otherwise comparable conditions, the void ratio decreases with increasing range in particle size. Porosities and corresponding void ratios of shales may range from 3 to 52% and from 0.03 to 1.07, respectively; moisture contents from less than 5% to as high as 35% for some clay shales. Shales are called saturated if their pores are completely filled with water even though they may possess additional water intake and swelling capacity. With respect to hydraulic conductivity, swelling, shrinkage, consistency, and related properties, shales obey the same basic laws as soils.

The strength of shales decreases exponentially with increasing void ratio and water content. With cemented shales it is a function of the strength of the cementing material and the ratio of its volume to the pore volume. Great differences are often observed between the *in situ* strength and elastic constants of shales and the results obtained in laboratory tests. Contributary factors of such differences are

1. Improper or inadequate sampling
2. Disturbance of shale structure
3. Difference in stress conditions
4. Especially the rebound in overconsolidated sediments where the release of strain energy may break the weaker bonds and permit the entrance of air into the expanded pore space

Lowering of strength occurs even when the rebound takes place in water under prevention of air entrance. Vees and Winterkorn (1967) found great differences in the shear resistance of other clays such as kaolinite and attapulgite clays at the same saturated void content and the same normal pressure between virgin specimens and others that had been previously overconsolidated with subsequent pressure release and reconsolidation under water. Residual shear strength of weathered clay shales is determined by repeated-direct shear tests as shown in Figure 13.1. Numbers of cycles during the shear tests significantly affect the residual shear strength parameters (c_r, ϕ_r) as indicated in Figure 13.1.

13.3 RESIDUAL SOILS

13.3.1 Characteristics of Residual Soils

Residual soil is produced by the *in situ* decomposition of the underlying rock and the action of the pertinent soil forming factors such as a microclimate, flora, fauna, and geometric features. Chemical breakdown is particularly active in hot, humid regions with production and decomposition of large amounts of organic materials. The distribution of residual soil and weathered rocks generally occurs in the humid temperatures of the tropical climate.

The physical structure and engineering properties of residual soils are unique. The texture and mineralogy may still reflect the original rock structure with added complication of decreased weathering with increasing depth below the ground surface. The description of the weathering profile is discussed by Deere and Patton (1971). Classification of degrees of weathering is presented in Table 13.2.

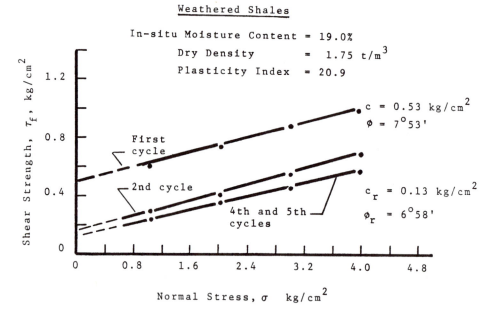

Figure 13.1 Residual shear strength of weathering shales determined by repeated-direct shear tests. (After Tianjin University, [1978].)

Table 13.3 presents the comparison of engineering classification of weathering profile and residual soil for igneous and metamorphic rocks. There are four classifications, namely: Vargas (1953), Sowers (1963), Deere and Patton (1971), and Martin (1977). Each classification has five subgroups as rock types, residual soil horizons, disintegrated rock layer, and unweathering rocks as indicated in the Table 13.2.

13.3.2 Engineering Behavior of Residual Soils

The physical structure and engineering properties of residual soils are unique. The texture and mineralogy may still reflect the original rock structure with added complication of decreased weathering with increasing depth below the ground surface. As reported by Sowers (1963), residual soils exhibit internal friction angle and cohesion, both of which depend on the void ratio and the degree of saturation; they are strongly anisotropic as are other properties. The description of the weathering profile for igneous and metamorphic rocks is summarized and discussed and shows the relative permeability and strength of each profile (Deere and Patton, 1971). Typical values of coefficient of permeability vary from 0.1×10 to 5×10 cm/sec. Description of weathering profile and shear strength parameters of cohesion, and the internal friction angle of various weathering rocks, residual soils, and related materials are summarized in Table 13.4.

13.3.3 Environmental Effects on Residual Soils

Residual soils are sensitive to pore fluids. Figure 13.2 shows the residual soil derived from decomposed granite rock from Hong Kong island vs. degree of saturation indicating significant effects on cohesion when degree of saturation increases for both disturbed and undisturbed soil samples. Figure 13.3 shows the effect of pore fluid and specific gravity on grain size distribution. For the same specific gravity ($G_s = 2.87$), the pH decreases and the D_{60} value increases.

Table 13.2 Classification of Degrees of Weathering

Grade	Degree of decomposition	Field recognition	Engineering properties
VI	Soil	No recognizable rock texture; surface layer contains humus and plant roots	Unsuitable for foundations; unstable when vegetation cover is destroyed
V	Completely weathered	Rock completely decomposed by weathering in place, but texture still recognizable; in types of granitic origin, original feldspars completely decomposed to clay minerals; cannot be recovered as cores by ordinary drilling methods	Can be excavated by hand or ripping without use of explosives; unsuitable for foundations of concrete dams or large structures; may be suitable for foundations of earth dams and for fill; unstable in high cuttings at steep angles; requires erosion protection
IV	Highly weathered	Rock weakened so that fairly large pieces can be broken and crumble in the hands; sometimes recovered as core by careful diamond drilling; stained by limonite	Similar to grade V
III	Moderately weathered	Considerably weathered throughout; possessing some strength — large pieces (e.g., NX drill core) cannot be broken by hand; often limonite-stained	Excavated with difficulty without use of explosives; mostly crushes under bulldozer tracks; suitable for foundations of small concrete structures and rockfill dams; may be suitable for semipervious fill; stability in cuttings depends on structural features, especially joint attitudes
II	Slightly weathered	Distinctly weathered throughout the rock fabric with slight limonsite staining; some decomposed feldspar in granites; strength approaching that of fresh granite	Requires explosives for excavation; suitable for concrete dam foundations; highly permeable through open joints; questionable as concrete aggregate
I	Fresh rock	Immediately beneath weathered rock; fresh rock may have some limonite-stained joints	Staining indicates water percolation along joints; individual pieces may be loose and support may be required in tunnels and shafts

After Little (1967).

Table 13.3 Comparison of Engineering Classification of Weathering Profile and Residual Soil for Igneous and Metamorphic Rocks

Vargas (1953)	Gneiss basalt sandstone	Natural residual soil	Young residual soil		Disintegrated rock layer		Sound rock
Sowers (1963)	Igneous and metamorphic rock	The upper zone	The intermediate zone		The partially weathered zone		Unweathered rock
Deere and Patton (1971)	Igneous and metamorphic rock	IA A-horizon	IB B-horizon	IC C-horizon	Trans. from residual	Part. weathering rock	Unweathering rock
			Residual soil		Weathering rock		Unweathering rock
Martin (1977)	Igneous and metamorphic rock		Residual soil		Disintegrated rock		Rock

After Martin, R.E. (1977), *J. Geotech. Eng. Div. Proc. ASCE*, v. 103, no. GT3, pp. 197–212.

**Table 13.4a Description and Shear Strength Parameters of Residual Soil and Weathered Rocks —
Description**

	Zone	Description	Relative permeability	Relative strength
I Residual soil	IA–A-horizon	Top soil, roots, organic material zone of leaching, and eluviation may be porous	Medium to high	Low to medium
	IB–B-horizon	Characteristically clay enriched; also accumulations of Fe, Al, and Si, hence may be cemented No relict structures present	Low	Commonly low (high if cemented)
	IC–C-horizon (saprolite)	Relict rock structures retained Silty grading to sandy material Less than 10% core stones Often micaceous	Medium	Low to medium (relict structures very significant)
II weathered rock	IIA–transition (from residual soil or saprolite to partly weathered rock)	Highly variable, soil-like to rock-like Fines commonly fine to coarse sand 10 to 95% core stones Spheroidal weathering common	High	Medium to low where weak structures and relict structures are present
	IIB–partly weathered rock	Rock-like, soft to hard rock Joints stained to altered Some alteration of feldspars and micas	Medium to high	Medium to high
III Unweathered rock		No iron stains to trace along joints No weathering of feldspars and micas	Low to medium	Very high

13.4 EXPANSIVE CLAYS

13.4.1 General Discussion

Soils that exhibit the greatest volume change from dry to wet state usually possess a considerable percentage of montmorillonite clay or of related three-layer clay minerals (Table 3.10). Swelling and shrinking soils do not have a continuous granular skeleton with sufficient interstitial porosity to accommodate the volume changes of the silty clay fractions due to increase or decrease in moisture content. In compacted soils the presence or absence of an effective skeleton may be deduced from the size composition of the soil and the Atterberg limits of the #40 or #200 sieve fraction.

There exist in arid regions, however, natural soils of high sand and low silty clay content with such structure that the silt clay fraction forms bridges between loosely packed sand grains. If such soils are exposed to water, the cohesive bridges are weakened and the soil may collapse under its own weight or superimposed loads. This illustrates the importance of

Table 13.4b Shear Strength Parameters

Rock type	Degree of weathering	Strength parameters	
Gneiss (micaceous)	Muram (zone IB)	c = 0.6	ϕ = 23
	Decomposed	c = 0.3	ϕ = 37
Gneiss	Zone IC	—	ϕ = 18.5
	Fault zone	c = 1.5	ϕ = 27
	Heavy decomposition	c = 4.0	ϕ = 29
Schist	Weathered (zone IC)	c = 0.35–0.45	ϕ = 15–18
		$\bar{c}$ = 0.48–0.70	ϕ = 15–21
	Weathered	c = 0	ϕ = 30
Phyllite	Decomposed	c = 0.5–1.2	ϕ = 10–22
Keuper marl	Highly weathered	$\bar{c}$ < 0.1	ϕ = 25–32
		—	ϕ_r = 18–24
Quartz diorite	Decomposed	c = 0.1	ϕ > 30
Quartzite	Decomposed	c = 0.15	ϕ = 35
Granite	Decomposed	c = 0	ϕ = 27–31
	Zone IB		ϕ = 28
	Zone IC		ϕ = 35
	Fine grain		ϕ = 25.5–34
	Coarse grain		ϕ = 36–38
	Decomposed	c = 0.6–1.4	ϕ = 8–18
Granite-gneissic soil	Natural	c = 0.1–0.2	ϕ = 25–35
	Saturated	c = 0–0.2	ϕ = 30–35
Black seams	Zone IC	—	ϕ_r = 10.5–14.5
Diorite	Weathered	c = 0.3	ϕ = 22°
Rhyolite	Decomposed	—	ϕ = 30°

Note: c, ϕ = cohesion and friction angle; c′, ϕ' = effective cohesion and effective friction angle; ϕ_r = residual friction angle; c, c′ = kg/cm²; ϕ, ϕ', ϕ_r = degree.

After Deere, D.U. and Patton, F.D. (1971), Proc. 4th Panamerican Conf. of Soil Mechanics and Foundation Engineering, v. 1, pp. 87–170 and others.

structural features, in addition to granulometric and volumetric relationships in the interaction of soil solids with water.

13.4.2 Factors Effect on Swelling

The total amount of swell of a soil body usually expressed as a percentage of its original volume represents the difference in the sum of the phase volumes of solid, liquid, and gaseous soil constituents in the initial and final state. Assuming the volume of the solid constituents as essentially gaseous constituents, factors that influence the volume changes of swelling clays are

1. The grain size composition and density of the soil
2. The amount, mineral character, and exchange ion of clay fraction
3. Types and concentration of electrolytes in the soil solution and in the source of swelling water
4. Amount and electrolyte content of soil-water system
5. Soil structure as influenced by previous chemical and stress-strain history
6. Environmental temperature and pressure conditions including the effect of surcharges

13.4.3 Identification and Classification of Expansive Clays

(1) Single Index Methods

The USBR uses the colloid content (percent < 0.001 mm), plasticity index (PI), and shrinkage limit (SL) as criteria for identification of expansive clays as shown in Table 13.5.

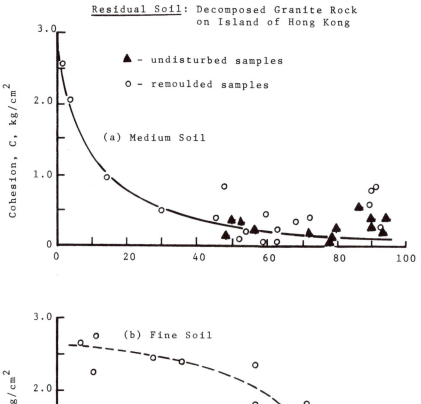

Figure 13.2 Degree of saturation vs. cohesion of decomposed granite rock. (a) Medium soil; (b) fine soil; (c) coarse soil. (After Lumb, P. [1962], *Geotechnique*, v. 12, no. 3, pp. 226–243.)

Since the respective Atterberg limits are routinely determined on soils and the colloid content is also easily determined, this method should be preferred to the single index methods.

(2) Identification Based on Soil Constants

Identification of expansive clays are based on the relative positions of four soil constants, namely, liquid limit (LL), plastic limit (PL), centrifuge moisture equivalent (CME), and field moisture equivalent (FME), in the assessment of soil behavior toward water. An FME > LL indicates the danger of autogenous liquefaction of a soil in the presence of free water. If both FME and CME exceed 30 and if FME > CME, the soil probably expands after release of load and is classified as expansive soil. This procedure is discussed in Section 6.3 and summarized in Table 6.4.

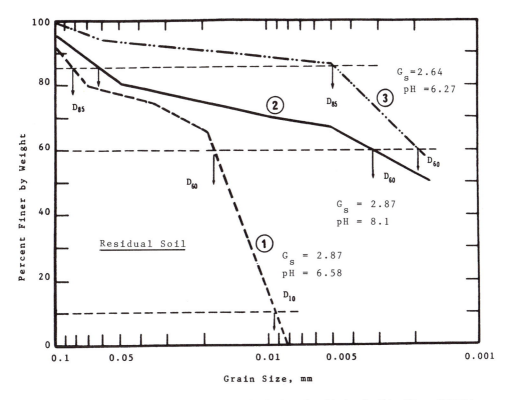

Figure 13.3 Pore fluid effects on grain size distribution of residual soils. (After Wang, [1983].)

Table 13.5 Estimation of Degree of Expansion of Expansive Soils Based on Colloid Content and Atterberg Limits

Data from index tests			Estimation of probable expansion, percent total volume change (dry to saturated condition)	Degree of expansion
Colloid content (percent minus 0.001 min)	Plasticity index	Shrinkage limit (percent)		
>28	>35	<11	>30	Very high
20–31	25–41	7–12	20–30	High
13–23	15–28	10–16	10–20	Medium
<5	<18	>15	<10	Low

After USBR (1973), Earth Manual, U.S. Bureau of Reclamation, Denver, Co.

(3) Dielectric Dispersion Method

The characteristics of dielectric constant and dielectric dispersion have been discussed in Section 9.7. A term called *expansion index* (EI) has been proposed by Fernando et al. (1975) in order to use dielectric dispersion method for identifying expansive characteristics of soil. The procedure for determination of EI is based on one-dimensional volume change developed by Krazynski (1973) and cited by Fernando et al. and refinement is given by Anderson and Lade (1981). The EI is calculated as:

$$EI = 1000 \, (\Delta h)(F) \tag{13.1}$$

where EI = expansion index; Δh = the final dial reading minus the initial dial reading; and F = fraction of sample passing #4 sieve

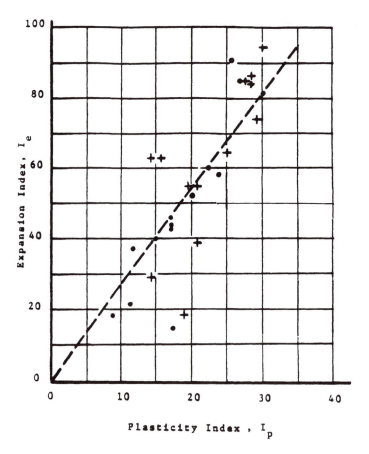

Figure 13.4 Relationship between expansion index (I_e) and plasticity index (I_p). (•, Fernando, Smith, and Arulanandan (1975) and +, Anderson and Lade (1981)).

Table 13.6 Classification of Expansive Soils Based on Magnitude of Dielectric Dispersion

Magnitude of dielectric dispersion	Expansion index	Potential expansion
1–10	1–20	Very low
11–25	21–50	Low
26–45	51–90	Medium
46–65	91–130	High
Above 66	Above 130	Very high

After Fernando, J., Smith, R. and Arulanandan, K. (1975), *J. Geotech. Eng. Div. Proc. ASCE,* v. 101, no. GT9, pp. 1003–1013.

Figure 13.4 presents the relationship between EI vs. soil types as reflected by plasticity index (PI). The classification of expansive soils based on magnitude of dielectric dispersion is presented in Table 13.6. The dielectric dispersion can be determined. Based on data presented in Figure 13.4, a simple equation was developed as follows:

$$EI = 2.72 \ (PI) \tag{13.2}$$

where EI = expansion index and PI = plasticity index.

13.4.4 Properties and Problems of Expansive Clay Deposits

The engineering properties and behavior of expansive clays are influenced significantly by local environmental factors. As discussed in previous sections, there are at least seven factors which may affect swelling. A rational approach to predict swelling soil behavior is proposed by Nagaraj and Murthy (1985). A phenomenological model has been formulated with truncated diffuse-double-layer theory as the scientific base to predict swelling characteristics. Others (Mowafy and Bauer, 1985; Ofer and Blight, 1985) have discussed from various aspects for prediction of swelling pressure and factors affecting the swell behavior of expansive clays in both laboratory and *in situ* conditions. In the following sections some distinct types of expansive clay deposits such as red expansive clays and black cotton clays are used to illustrate their environmental geotechnical problems.

Figure 13.5 shows the shrinkage and swelling behavior of the same red expansive clay. In examining Figure 13.5(a), the rate of volumetric shrinkage increases, and the clay content as reflected on liquid limit also increases. Figure 13.5(b) indicates the volumetric expansion rate increases as the percent of moisture content reduced from its natural state increases. A similar trend is indicated for the swelling pressure as indicated in Figure 13.5(c).

Expansive clays lying above the groundwater table are usually in the unsaturated condition and undergo shrinkage on drying by losing moisture and swell on wetting by absorbing water. When free swelling of these soils is restrained by the presence of a structure on the ground, they exert what is called swelling pressure on the structure, creating a paradox in that soil applies pressure on the structure instead of the latter doing it on the soil. If the structure is not designed to withstand this movement, it ultimately cracks apart. As such, in selecting the most suitable foundation, the degree of expansivity of the soil should be linked to the type of superstructure and the foundation system. The various components and their linkages that are involved in the decision process of design and construction on expansive clays have been developed by Ramana (1996) as illustrated in Figure 13.6.

Foundation analysis and design for building structures on expansive soils, in general, are discussed by many investigators. Such techniques include the use of piers to prevent the uplifting of lightly loaded structures founded on expansive soils. *In situ* study of bearing capacity of driven piles in expansive soil is given in a comprehensive review with design considerations by Chen (1979) and Gromko (1974) and summarized in Table 13.7. Recent developments on various aspects of expansive clay are discussed by Nelson and Miller (1992).

13.4.5 Black Cotton Soils

A large part of central and southern India, Burma, and Kenya are covered with black cotton soils (Wooltorton, 1954), and it is one of the major expansive soils. *Black cotton soil* is black in color, exhibiting excessive swelling and shrinkage and high plasticity characteristics. Due to alternate wetting and drying conditions, deep and wide cracks are observed on ground surface. Such soils are locally called black cotton soils and have been considered as highly problematic for foundations and roadways. Texturally, the soil is classified as clay. Its characteristic features are its high liquid limit, plastic limit, and very low shrinkage limit. Base exchange capacity of this indicates that the clay minerals present in the soil may be montmorillonite or a mixture of montmorillonite and illite.

Ground improvement of this soil type was studied by Katti and Barve (1962) who concluded that upon adding KOH, $Ca(OH)_2$, and NaOH, the soil changes from a plastic clay into a friable soil with increase in SL when the amount of chemical added is beyond 1.5%. NaOH has a pronounced effect on increasing SL. At lower percentages, NaOH actually increases PI. Below 1% both $Ca(OH)_2$ and KOH increase the PI. In general, $CaCl_2$ $MgCl_2$ and $BaCl_2$, as well as a lower LL, increase the PI but have little effect on the SL. KCl beyond

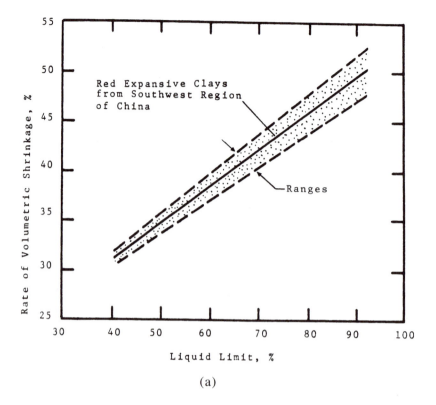

(a)

Figure 13.5 Shrinkage and swelling behavior of red expansive clays from the southwest region of China. (a) Rate of volumetric shrinkage vs. liquid limit; (b) volumetric expansion rate vs. percent of moisture content reduced from its natural state; (c) swelling pressures vs. percent of moisture content reduced from its natural state. (After Tianjin University, 1978.)

2% actually reduces the PI of the soil at all percentages and distinctly increases the SL. $NaCO_3$, $BaCO_3$, $MgCO_3$, and Na_2SiO_3 increase the PI of the soil making the soil highly plastic. Both $FeCl_3$ and $(NH_4)_2HPO_4$ make the soil porous.

13.5 ORGANIC SOILS

13.5.1 Characteristics of Organic Soil

Organic soils are those whose solid constituents consist predominantly of vegetable matter in various stages of decomposition or preservation. They are designated as *bog, muskeg*, and *moor* soils with differentiation between peat and muck soils, on one hand, and coastal marshland soils on the other hand. Muck indicates a higher degree of decomposition of the vegetable matter and of intermixing with mineral soil constituents in contrast to the purely vegetable peats that have well-preserved plant remains. Several types of moors are recognized depending on the source of water supply, such as:

- High moor: water mainly from precipitation
- Low moor: drainage from surrounding areas

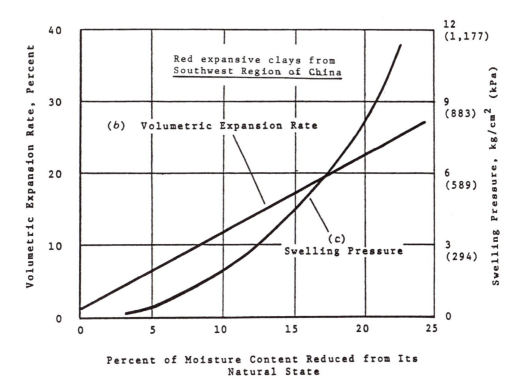

Figure 13.5 b & c (continued)

Pedalogically, organic soils are intrazonal hydromorphous soils which means that they may occur within any macroclimatic zone as long as hydrologic and topographic conditions produce basins of standing water or land areas with a rising water table. Because of the influence of environmental and plant ecologic factors, these soils are best considered as organic terrain or muskeg and classified in accordance with genetic principles as has been done by the Muskeg Subcommittee of the National Research Council of Canada (MacFarlane and Williams, 1974).

The parent material of organic soils is the native vegetation which may cover a wide range of aquatic, marsh, and swamp forest plants. Each visible forest or other vegetation above a land or water surface has a counterpart, a sort of distorted mirror image below this surface. This counterpart is formed by the root systems which are functionally and geometrically related to the normally visible part. Actually, the subterranean or subaqueous part may contribute more to peat and muck formation than the more rapidly decomposed organic matter from the surface vegetation. However, the functional and geometric relationship between the visible and invisible portions of the vegetation proves the importance of the surface cover for purposes of classification of muskeg.

13.5.2 Classification of Organic Soils

Because the parent material of organic soils is the native vegetation which may cover a wide range of aquatic, marsh, and swamp forest plants, the engineering classification of these

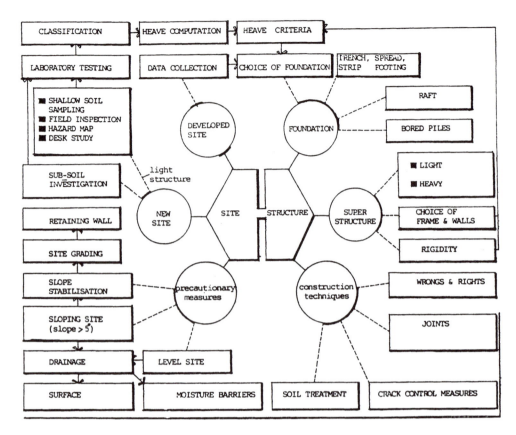

Figure 13.6 Flowchart for decision process of design on expansive clays. (After Ramana, [1996].)

types of soils must be examined for the properties of organic soil itself, including both visible and invisible portions of the vegetation systems. There are several classification systems proposed including the Muskeg Subcommittee of the National Research Council of Canada (MacFarlane, 1969), ASTM (ASTM D2607-69), Arman (1970), and Landva (1983). A brief discussion of each classification system is presented as follows.

(1) NRC Classification (1969)

The National Research Council (NRC) of Canada classification is based on the characteristics of the subsurface material, the organic soil property which is the product of the interaction of topography, vegetation, high water table, and decomposition and preservation processes as shown in Table 13.8. Seventeen types of organic soil are recognized, based on the extent to which the following types of structural components are present:

1. Amorphous and granular
2. Nonwoody and woody fine fibrous
3. Wood particles and coarse woody fibers

(2) ASTM Classification (1972)

ASTM presented a classification system for peats, mosses, humus, and related products (ASTM D2607-72). The term peat refers, according to ASTM, only to organic matter of geological origin, excluding coal, formed from dead plant remains in water and in the absence of air. It occurs in a bog, swampland, or marsh, and it has an ash content not exceeding 25%

Table 13.7 Construction Procedures on Expansive Soils

Total predicted Heave (in.)		Recommended construction	Method	Remarks
L/H[a] = $1\frac{1}{4}$ (1)	L/H[a] = $2\frac{1}{2}$ (2)	(3)	(4)	(5)
$0-\frac{1}{4}$ $\frac{1}{4}-\frac{1}{2}$	$0-\frac{1}{2}$ $\frac{1}{2}-2$	No precautions Rigid building tolerating movement (steel reinforcement as necessary)	Foundations Pad Strip footings Raft (waffle)	Footings should be small and deep, consistent with the soil-bearing capacity; a raft should resist bending
			Floor slabs Waffle Tile	Slabs should be designed to resist bending and should be independent of grade beams
			Walls	Walls on a raft should be as flexible as the raft; no rigid connections vertically; brickwork should be strengthened with tiebars or bands
$\frac{1}{2}-2$	2–4	Building damping movement	Joints Clear Flexible	Contacts between structural units should be avoided or a flexible, waterproof material may be inserted in the joint
			Walls Flexible Unit construction Steel frame	Walls or rectangular building units should heave as a unit
			Foundations Three-point Cellular Jacks	Cellular foundations allow slight soil expansion to reduce the swelling pressure; adjustable jacks can be inconvenient to the owner; three-point loading allows motion without duress
>2	>4	Building independent of movement	Foundation piers Straight shaft Bell bottom	Smallest diameter and widely spaced piers compatible with load should be placed; clearance should be allowed under grade beams
			Suspended floor	Floors should be suspended on grade beams 12 to 18 in. above the soil

[a] L/H is the length-to-height ratio of the wall panel.
After Gromko, G.J. (1974), *J. Geotech. Eng. Proc. ASCE,* v. 100, no. GT6, pp. 667–687.

by dry weight. The classification is based on five major types according to genetic origin of fiber content. Percentages of fiber are based on oven-dried weight at 105°C, not on volume. The five major types are listed as:

1. Sphagnum moss peat (peat moss): The oven-dried peat shall contain a minimum of 66% sphagnum moss fiber by weight.
2. Hypnum moss peat: This peat shall contain a minimum 33.3% content by weight of which hypnum moss fibers shall comprise over 50%.
3. Reed-sedge peat: The oven-dried peat shall contain a minimum 33.3% fiber by weight of which reed-sedge and other nonmoss fibers shall comprise over 50%.

Table 13.8 NRC Classification of Peat Structures

Predominant characteristic	Category	Name
Amorphous-granular	1	Amorphous-granular peat
	2	Nonwoody, fine-fibrous peat
	3	Amorphous-granular peat containing nonwoody fine fibers
	4	Amorphous-granular peat containing woody fine fibers
	5	Peat, predominantly amorphous-granular, containing nonwoody fine fibers, held in a woody, fine-fibrous framework
	6	Peat, predominantly amorphous-granular containing woody fine fibers, held in a woody, coarse-fibrous framework
	7	Alternate layering of nonwoody, fine-fibrous peat and amorphous-granular peat containing nonwoody fine fibers
Fine-fibrous	8	Nonwoody, fine-fibrous peat containing a mound of coarse fibers
	9	Woody, fine-fibrous peat held in a woody, coarse-fibrous framework
	10	Woody particles held in nonwoody, fine-fibrous peat
	11	Woody and nonwoody particles held in fine-fibrous peat
Coarse-fibrous	12	Woody, coarse-fibrous peat
	13	Coarse fibers criss-crossing fine-fibrous peat
	14	Nonwoody and woody fine-fibrous peat held in a coarse-fibrous framework
	15	Woody mesh of fibers and particles enclosing amorphous-granular peat containing fine fibers
	16	Woody, coarse-fibrous peat containing scattered woody chunks
	17	Mesh of closely applied logs and roots enclosing woody coarse-fibrous peat with woody chunks

After MacFarlane (1969).

4. Peat humus: The oven-dried peat shall contain less than 33.3% by weight.
5. Other peat: All forms of peat not classified herein.

Other information relating to peat includes sampling procedures (ASTM D2944-71), techniques for measuring the moisture and ash contents (ASTM D2974-87), volume measurements (ASTM D2978-71), and water-holding capacity (ASTM D2980-71); others are available from ASTM Standards.

(3) Louisiana State Classification System (1970)

For practical applications, an engineering classification system for the organic soil has been proposed by Arman (1970) of Louisiana State University as shown in Table 13.9. The table categorizes organic soils as organic and sandy clay, silty, or clay. It further breaks down major categories into groups according to the amount of organic content. The ranges of the physical characteristics are based on the laboratory studies. For each group, the value of the particular soil is given for use as highway embankment material or for general foundation work.

There are numerous other classification systems for the organic soils; some of most common classification systems are presented in Table 13.10 for comparison.

13.5.3 Engineering Properties of Organic Soils

(1) Water Content and Soil Constants

The engineering properties of organic soils must be determined *in situ*, on location, and in the laboratory, in as close a state of disturbance or nondisturbance as the respective

Table 13.9 Louisiana State Classification System

Major divisions	Group symbol	Organic content (%)	pH range	Specific gravity	Drainage characteristics	Liquid limit (%)	Plasticity index (%)	Value as a base material	Value as an embankment material	Other uses
Sandy clay and organic material	SWO-1E	0–15	>7.0	>2.40	Pervious to fair	26–37	7–8	Excellent to fair	Excellent	Foundation for levees, earth dams, and structures
	SWO-2F	15–20	6.0–7.0	2.18–2.40	Poor	37–48	3–8	Fair to poor	Fair to poor	Foundation for embankments, levees, and light structures[a]
	OS-1F	21–30	6.0–7.0	2.18–2.40	Poor	37–48	3–8	Fair to poor	Fair to poor	Foundation for embankments, levees, and light structures[a]
	OS-2	31–50	5.5–6.0	1.83–2.18	Poor	48–72	0.8–3	None	Poor	Limited use as embankment foundation after densification[a]
	OS-3	>50	<5.5	<1.83	Poor	>72		None	None	None
Silt and organic material	SWO-1G	0–15	>7.0	>2.45	Impervious to very poor	32–41	12–13	Good to fair	Excellent to fair	Foundation for levees, earth dams, and structures
	SWO-2F	15–20	6.0–7.0	2.10–2.45	Very poor to poor	41–56	13–20	Fair	Fair	Foundation for embankments, levees, and light structures[a]
	OS-1F	21–30	6.0–7.0	2.10–2.45	Very poor to poor	41–56	13–20	Fair	Fair	Foundation for embankments, levees, and light structures[a]
	OS-2F	31–50	5.5–6.0	1.75–2.10	Poor	56–72	17–20	Fair to poor	Poor	Limited use as embankment or levee foundation[a]
	OS-3	>50	<5.5	<1.75	Poor	>72		None	None	None

Table 13.9 Louisiana State Classification System (continued)

Major divisions	Group symbol	Organic content (%)	pH range	Specific gravity	Drainage characteristics	Liquid limit (%)	Plasticity index (%)	Value as a base material	Value as an embankment material	Other uses
Clay and organic material	SWO-1	0–15	>7.0	>2.37	Impervious to very poor	65–67	26–40	None	Fair	Levees, earth dams, and structures
	SWO-2	15–20	6.0–7.0	2.15–2.37	Very poor to poor	65–66	18–26	None	Fair to poor	Levees and earth dams
	OS-1	21–30	6.0–7.0	2.15–2.37	Very poor to poor	65–66	18–26	None	None	Embankment foundation after densification[a]
	OS-2	31–50	5.5–6.0	1.81–2.15	Poor	66–76	12–18	None	None	Embankment foundation after densification[a]
	OS-3	>50	<5.6	<81	Poor	>76		None	None	None

[a] Organic deposit must be located under the permanent phreatic surface.

After Arman, A. (1970), Engineering Classification of Organic Soils, HRR no. 310, pp. 75–89.

Table 13.10 Comparisons of Classification Systems of Organic Soil

Ash Content, % of Dry Weight

20	40	60	80	10

Peat — 25 — Muck and other organic rich sediments

Peat — 25

Fuel Peat — 25 — Muck — 25

Organic Soils — 25 — Mineral Soils

Organic Soils — 25 — Soils with Organic Content

Gyttja Soils — 25 25

Soils containing Humus

Peat — 25 — Peaty Muck — 25 — Silty/Clayey Muck — 25 — Silty/Clayey Muck

Soils with Organic Content

Peat — 20 — Peaty Organic Soils — 40 — Organic Soils — 95 99

(Pt)   (PtO)   (O)   (MO,CO)

After Landva et al. (1983).

engineering use may require. Engineering properties of organic soils discussed in this section include: (1) peat and peaty soil; (2) volcanic ash organic soil, and (3) muck. The specific gravity of peats ranges from 1.1 to 2.5, whereby values above 2.0 indicate marked contribution by mineral matter which may be checked by the determination of the ash content. The low specific gravity of both organic matter and water leads to low unit weights for the natural peat. In general, the specific gravity and moisture absorption of soils are related to their organic content; the moisture absorption increases and the specific gravity decreases as the organic content increases.

The organic soil can hold high water content as shown in Figures 13.7 and 13.8. In Figure 13.8, the dry density vs. peaty soil is plotted together with marine clay and wood for comparisons (Ohira, 1977). It can been seen in both figures that organic soil can hold as much as 1500% of water. The aqueous phase of most organic soil is acid, with pH values ranging from 4 to 7, but values as low as 2 and as high as 8 have been encountered.

(2) Permeability, Compressibility, and Others

The permeability of natural peat deposits varies widely despite their high void ratios or porosities. Also, permeability is often much greater in the horizontal than in the vertical directions.

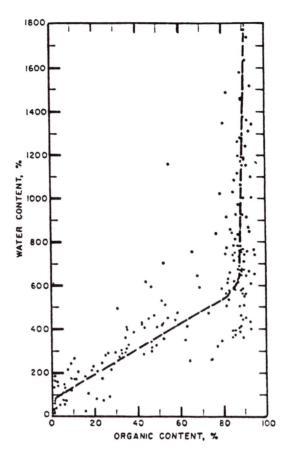

Figure 13.7 Organic content vs. water content (After MacFarlane and Rutka [1962].)

Organic soils are highly compressible; most of the compression takes place in a very short time period. The consolidation characteristics of organic soils containing less than 20% of organic material reasonably conform with Terzaghi's consolidation theory (Arman, 1970). When organic soils are dried, they do not absorb the same quantity of water that they originally contained, and this is a reason that organic soil cannot receive proper compaction.

The relationship between K_0, the coefficient of earth pressure at rest, and overconsolidation ratio, OCR for organic soils is reported by Edil and Dhowian (1981) and indicates that in all cases, the OCR increases as K_0 increases. Other engineering properties and corrosion behavior of organic soils and design and construction procedures are discussed in detail by the Research Committee on Organic Soil, Japanese Society of Soil Mechanics and Foundation Engineering (Yamanouchi, 1977).

13.6 KARST REGION (SINKHOLE)

13.6.1 General Discussion

The sudden appearance of a sinkhole has long plagued mankind in certain geologic areas, particularly in recent times, as heretofore rural land is being used increasingly for residential and industrial development. This major feature in the karst region landscape has been shaped by the removal of calcium and magnesium as bicarbonates from rock deposits ranging in composition from pure limestone through dolomitic limestones to

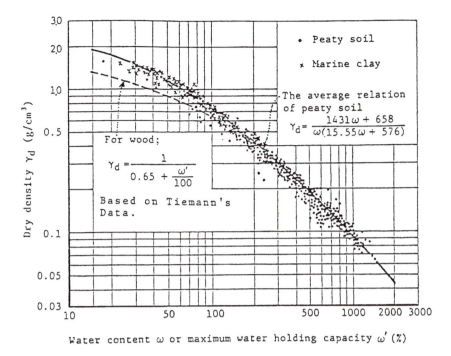

Figure 13.8 Relationship between dry density and water content for peaty soil, marine clay, and wood. (After Ohira, Y. [1977], Engineering Problems of Organic Soils in Japan, Japanese Society of SMFE, pp. 19–34.)

dolomite, the double salt of calcium and magnesium carbonate $(Ca,Mg)(CO_3)_2$, by water charged with CO_2. Solutional features are developed here so extensively and variously as to become geologic models, and the term 'karst topography' is applied to any terrain in which such features occur. Although no other region quite matches this area, some similar solutional forms do appear wherever thick, dense limestone is extensively exposed and well drained. These manifestations also appear where other soluble minerals, particularly rock salt and gypsum, form bedrocks at or near the surface; however, the term karst refers to the geologic consequence of solutions as bicarbonates of pure or mixed carbonates of the alkaline earths, especially of Ca and Mg.

13.6.2 Solution Processes and Mechanisms

Limestones consist mainly of calcium carbonate $(CaCO_3)$, usually in the form of the mineral calcite. Calcite is relatively insoluble in pure water, but its solubility increases rapidly with increasing water acidity. In surface environments, the most abundant source of acidity is carbonic acid derived from dissolved carbon dioxide (CO_2). Carbonic acid's control of groundwater acidity also influences mineral precipitation in caves. There are four basic chemical reactions of limestone solution expressed as follows:

  1. Reaction of calcium carbonate $(CaCO_3)$ in carbon dioxide (CO_2) solution

$$CaCO_3 + CO_2 + H_2O \leftrightarrow Ca^{2+} + 2HCO_3^-$$ (13.3)

  2. Reaction of calcium carbonate in $CaCO_3$

$$CaCO_3 \leftrightarrow Ca^{2+} + CO_3^{2-}$$ (13.4)

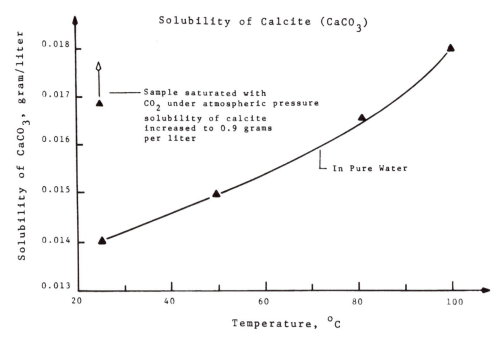

Figure 13.9 Effect of temperature on solubility of $CaCO_3$.

3. Reaction of calcium carbonate in CO_2

$$CO_2 + H_2O \leftrightarrow H_2CO_3 \leftrightarrow H^+ + HCO_3 \tag{13.5}$$

4. Then Equation (13.3) can be rewritten as:

$$Ca^{2+} + CO_3^{2-} + H^+ + HCO_3^- \leftrightarrow Ca^{2+} + 2HCO_3^- \tag{13.6}$$

Simplification of Equation (13.6) then is

$$CO_3^{2-} + H^+ \leftrightarrow HCO_3^- \tag{13.7}$$

Further discussion on the dissolution reaction with carbon dioxide is provided by White et al. (1995). The essentials of the solution processes involved can be exemplified by the chemistry of limestone solution. This rock consists of calcite ($CaCO_3$) in varying percentages. Expressed in grams per liter of pure water, its solubility is shown in Figure 13.9. Already the very small CO_2 content of the normal atmosphere suffices to triple the solubility at 25°C when the water is aerated. Saturation with CO_2 under atmospheric pressure increases the solubility of calcite to 0.9 g/l of water which is about 64 times that in pure water. Flowing and percolating water carries the solute away as it is formed.

Carbon dioxide is absorbed by the water from the atmosphere and from the soil-air where it is concentrated due to the decay of organic matter, or polluted water which may also furnish solution companions that enhance the solubility of calcite over that which is due to the CO_2 content of the water. In some areas carbonate erosion is relatively rapid. For instance, in Kentucky, the rate of limestone removal has been estimated as amounting to about 1 ft of thickness per 2000 years. As the carbonate is removed, poorly soluble impurities such as silica, alumina, iron oxides, and others are left behind, the trivalent iron compounds being responsible for the typical red and yellow colors of the residual soils. Since the prime condition for limestone

Figure 13.10 Photo showing the early stage of sinkhole in a limestone region at Reading, PA.

solution is abundant rainfall to stimulate plant and carbon dioxide production and to provide solvent water, arid regions are not conducive to karst formation. With plenty of water available, karst development is favored by well-jointed strata which allow ready passage of water.

13.6.3 Sinkhole and Its Interaction with Environment

Sinkhole is the most common among the karst features and may occur in such number as to make a countryside appear quite pockmarked. Central Florida is one well-known example; another prominent area is southern Indiana where the sinkhole total is estimated at 300,000. Large numbers of sinkholes also are reported in Pennsylvania and Illinois. Figure 13.10 shows a photo of early stage of sinkhole in a limestone region at Reading, PA.

In limestone country, one can often observe temporary surface streams, whereas only voluminous rivers can maintain a long course above ground on limestone terrain. On well-drained limestone most surface stream channels soon lead to swallow holes which connect with complex subsurface channel systems.

Areas of erosion in limestone are defined by belts of solution-resistant rock. Limestone strata often grade into shale and vice versa. Since shale typically is poorly jointed and its particles are small and well compacted, it tends to block circulation. Therefore, downward percolating water spreads out over the shale stratum or follows the dip until it reaches joints through which downward movement can be resumed. Accumulations of insoluble impurities left over from the solution of limestone also block circulation by filling joints which adds further to the complexity of sinkhole formation.

Impervious or insoluble strata sometime account for disappearing streams. Such streams flow over a resistant stratum until they find a swallow hole; then they follow solution channels along a resistant stratum at a lower level. Sometimes this lower level may lead to the surface where the stream course reappears within a distance of only a few miles. A probability study of sinkhole distribution in the Lehigh Valley, eastern Pennsylvania, was made by Lilly (1976). Perlow et al. (1983) reported polluted water affects the sinkhole growth in the Lehigh Valley, eastern Pennsylvania. Field studies based on boring logs of the limestone region indicate voids in the layers indicating the potential limestone cavity.

13.6.4 Remedial Actions for Sinkholes

All suspected sinkhole areas should be investigated in the field and cores drilled to a depth consistent with the weight of the proposed structure. If sinkholes are discovered, they should be avoided, if possible, since correction is often difficult, costly, and uncertain. Correction procedures include:

1. If a sinkhole cannot be avoided, it must be filled with suitable material or bridged, thereby creating a firm foundation. First, the cavity is cleaned until the solid limestone rock forming its walls is exposed. Water is then poured into the sink, preferably under pressure to determine if any downward breakthrough into underlying solution channels or caves will occur.

2. If the sink is a funnel type or opens into a cavern it may be advisable to build a concrete platform across the hole and fill on top of the platform. The edges of the sinkhole are excavated to a firm foundation and a reinforced concrete slab is placed across the opening. All jointures of the slab with the sinkhole are completely sealed to prevent any entrance of water.

3. Regardless of the corrective methods, surface drainage must be directed away from the site and extreme care must be taken to completely seal the sinkhole and thereby prevent any reoccurrence.

4. Other preventive and corrective measures include foundation alteration, deep foundations, and grouting.

13.7 LOESS (AEOLIAN DEPOSITS)

13.7.1 Origin and Distribution

Loess is an aeolian deposit whose particles are predominantly of silt size but with a certain amount of fine sand and aggregated clay particles present. The valley loesses are typically developed in areas peripheral to those covered by the last ice sheets. Other important loess deposits are in Argentina and China. Typical loess has a calcium carbonate content that acts as a bonding agent which, though weak, allows the loess to form vertical or even overhanging walls on the banks of streams. Slopes at angles less than 90° are easily eroded. Loess distribution and geological aspects in the U.S. are given by Turnbull (1968).

China is well known for having the largest region of loess in the world, with a great thickness and complete sequence of strata. Since 1949, there has been extensive research on the relationship between geological environment and engineering properties of the loess soil in China (CAC, 1961). Loess in China is located to the south of the Mongolia Plateau. As cold airflow from polar region spreads from the north, it carries dust from the desert into the upper air, thus forming dust storms. These often occur in the spring and winter seasons in China. The distribution and geological aspects in China are given by Gao (1983).

13.7.2 Collapsibility Mechanisms and Phenomena

There are various hypotheses and theories concerning the collapsibility of loess soils, e.g., (1) capillary effect, (2) salt dissolution, (3) ion exchange, and (4) a shortage of clay-colloids and wedging effect of water film, as well as undercompaction. These possible collapsibility mechanisms have been explained by means of mechanical, physical, physical-chemical, and geological point of view. Collapse of loess is mainly caused on wetting under pressure, and usually collapsibility is confined to the upper 1 to 15 m. Depending upon the overburden pressure, collapsing loess can be divided into two groups: (1) self-weight collapsing loess and (2) nonself-weight collapsing loess. Some of these mechanisms and phenomena are briefly presented as follows:

1. Coefficient of collapsibility (d_s): an index for specifying collapsibility of loess, is the ratio of the settlement of sample on wetting under a certain external pressure (specified as 0.2 Mpa is commonly used) to the initial thickness of the sample. When $d_s > 0.015$, the loess is specified as collapsing.

2. Coefficient of self-weight collapsibility (d_{zs}): the coefficient of self-weight collapsibility is the ratio of the settlement of the soil sample on wetting under its own weight to the initial thickness of the sample. If $d_{zs} > 0.015$, the loess is specified as self-weight collapsing.

3. Collapsing susceptibility: the collapsing susceptibility refers to the collapsing speed of loess after wetting. The susceptible collapsing loess settles at a rapid speed and can damage structures.

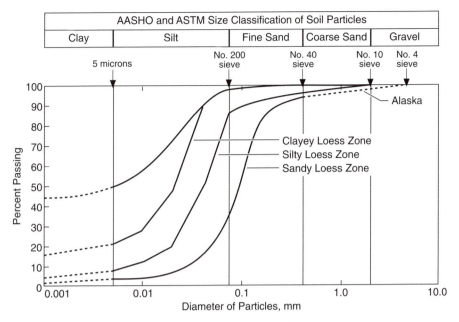

Figure 13.11 Gradation characteristics of loess in the U.S. (After Sheeler, J.B. [1968], Summarization and Comparison of Engineering Properties of Loess in the United States, HRR no. 212, pp. 1–9.)

4. Microstructure examination: to gain an insight into the collapse phenomena, the scanning electron microscope has been used to study the reasons for collapsibility of loess from the clay fabric point of view (Gao, 1983) as:

a. The microstructure of loess shows that the spaced granular structure in loess is the essential reason for collapsibility.

b. The regional variation in microstructure is consistent with the tendency of collapsibility, gradually decreasing due to the direction of airflow.

c. Based on Gao's conclusions of the formation of loess four stages can be formed: (1) the loessization stage, (2) the formation stage, (3) the development stage, and (4) the clayization stage. The four stages of loess regularly exist in different regions of the Loess Plateau and will be influenced by climatic and environmental conditions.

13.7.3 Properties and Foundation Problems

The composition and properties of loess are closely related to the geographic and geological environment. There are three basic types of loess zones in the U.S. The typical gradation curves for each zone are presented in Figure 13.11. Range in values of physical and mechanical properties includes Atterberg limits; compaction and strength tests are summarized in Table 13.11. The physicochemical properties of loess including ion-exchange capacity, percentage of clay content, and pH solution have been discussed in Section 4.8 and Figure 4.11. Coefficient of permeability of loess in natural condition is shown in Figure 13.12. In examining Figure 13.12(a) it is indicated that permeability is influenced by initial moisture content at *in situ* condition. In Figure 13.12(b) the permeability also is affected by pore fluid. Under the same void ratio, kerosene gives larger permeability than water. The relationship between cation-exchange capacity and amorphous material contents of loess is discussed in Section 4.8.

Foundation problems in loess regions include bearing capacity, settlement (subsidence), and erosion. Figure 13.13 presents the allowable bearing capacity of loess vs. liquid limit/void ratio (LL/e) for various moisture contents in natural condition. For a given moisture content,

Table 13.11 Physical and Mechanical Properties of Loess in the U.S.

Property	Location and reference							
	Iowa	Nebraska	Tennessee	Mississippi	Illinois	Alaska	Washington	Colorado
Specific gravity	2.68–2.72	2.57–2.69	2.65–2.70	2.66–2.73		2.67–2.79		
Mechanical analysis								
Sand, %	0–27	0–41	1–12	0–8	1–4	2–21	2–10	30
Silt, %	56–85	30–71	68–94	75–85	48–54	65–93	60–90	50
Clay, %	12–42	11–49	4–30	0–25	35–49	3–20	8–20	20
Atterberg limits								
LL, %	24–53	24–52	27–39	23–43	39–58	22–32	16–30	37
PL, %	17–29	17–28	23–26	17–29	18–22	19–26		20
PI	3–34	1–24	1–15	2–20	17–37	NP–8	<8	17
Classification								
Textural	SL, SCL, SC	SL, SCL, SC	SL, SCL, SC	SL, SCL	SC, C	SL, SCL		SL
AASHO	A-4(8) A-7-6(19)	A-4 A-6	A-4(8) A-6(10)	A-4(8) A-6(9)	A-6(11) A-7-6(20)	A-4(8)		A-6(10)
Unified	ML, CL, CH	ML, CL	ML, CL	ML, CL	CL, CH	ML, CL-ML		CL
Density test data								
Standard OM, %	15–20		14–16			13–18		
Modified OM, %	13–18							
Standard MD, pcf	103–112		100–108			103–112		
Modified MD, pcf	113–119							
In-place MD, pcf	66–99			80–104				76–95
Field moisture, %	4–31		12–25	19–38		11–49		8–10
Shear strength								
Unconfined compression test				2–8				
Unconsolidated undrained triaxial shear								
c, psi		0–67		2–10				
ϕ		31–36°		0–28°				
Consolidated undrained triaxial shear								
c, psi	0–8			3–8				
ϕ	28–31°			28°				
Consolidated drained direct shear								
c, psi	0.3–1.8			0				
ϕ	24–25			32–33°				
CBR				10–13				

Data from Davidson and Sheeler (1952); and others.

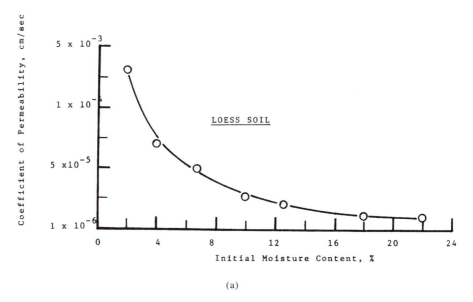

(a)

Loess (Natural Condition)

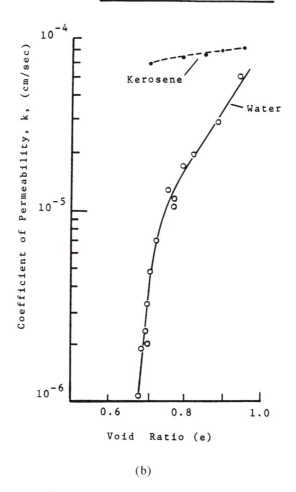

(b)

Figure 13.12 Coefficient of permeability of loess (a) Effect on initial moisture content; (b) Effect on pore fluid. (After CAS, [1961], Research on Loess, Civil Engineering Research Report no. 13, Chinese Academy of Science, Beijing, 157 p.)

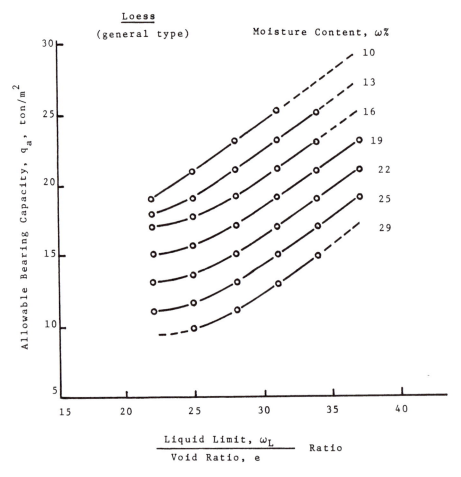

Figure 13.13 Allowable bearing capacity vs. liquid limit/void ratio of loess deposit. (Data from Chinese Building Code for Loess Region [1979].)

as the liquid limit/void ratio increases, the allowable bearing capacity also increases. Subsidence and erosion problems in loess regions are also important and further discussions are presented in Chapters 14 and 15.

13.8 DISPERSIVE CLAYS

13.8.1 General Discussion

Some fine-grain soils, called '*dispersive*' soils, with higher content of dissolved porewater sodium than ordinary soils, rapidly erode forming tunnels and deep gullies by a process in which the individual clay particles go into suspension in slow-moving water (colloidal erosion), damaging earth dams, canals, and other structures. Dispersive soils cannot be differentiated from ordinary soils by conventional geotechnical tests. This clay may have the same visual appearance as the ordinary clays. The term of dispersive clay was introduced by G. M. Volk in 1937. The first clear description of the phenomenon from a geotechnical point of view was provided in the early 1960s by Australian engineers (Aitchison et al. 1963) following investigations of piping failures in many small farm dams. Investigations during 1970–1971 sponsored by the U.S. Soil Conservation Service of piping failures of clay dams in the U.S. confirmed completely the basic Australian conclusions presented by Sherard et al. (1976).

13.8.2 Identification Procedures

Three methods to determine whether a soil is dispersive or nondispersive are available: (1) double hydrometer test, (2) soluble salt in porewater test, and (3) pinhole test. Among these tests, the pinhole method is the simplest method which uses distilled water flowing through a 1.0-mm-diameter hole in a compacted specimen. A typical test setup of Waterway Experiment Station (WES) pinhole erosion apparatus is shown in Figure 13.14(a) and presentation of laboratory test results between dispersive and nondispersive soil is shown in Figures 13.14(b) and 13.14(c). In examining Figure 13.14(b), for dispersive clays, the water becomes colored and the specimen carries a cloudy colored suspension of colloids, as the hole rapidly erodes. For nondispersive clay the water is crystal clear and there is no erosion.

13.8.3 Properties and Foundation Problems

Engineering properties of dispersive clay are controlled mainly by water content and soluble salts in the soil-water system. Figure 13.15 plots the percent of sodium vs. soluble salts. In examining Figure 13.15, it is indicated that higher sodium content is more critical for a soil having dispersive behavior. Figure 13.16(a) shows the effects of pore fluid as reflected by pH on grain-size distribution. Figure 13.16(b) plots the D_{15} vs. pH and larger variations are obtained, especially for the low pH values. Dispersive clays relating to the progressive erosion and landslides will be further discussed in Sections 14.5 and 16.7.

13.9 LATERITE SOILS

13.9.1 Characteristics of Laterite Soil

Laterite soils are extreme types of lastosols which form a suborder of zonal (climatic) soils and include soils formed under forested tropical and humid conditions. Their clay fractions have low silica-sesquoxide ratios, and low activity and base-exchange capacity. Laterites have low contents of soluble constituents of most primary minerals. Their secondary aggregates are usually highly stable and water resistant, and in extreme forms (pisolith) may serve as aggregate in soil stabilization. Lesser degrees of laterization are found in laterite and lateritic soils formed in temperate, warm, and tropical regions and include the yellow podsolic, red podsolic, yellowish-brown lateritic, and lateritic great soil groups.

There are a number of classification systems that have been specifically developed for lateritic materials. Because these systems are generally based on agricultural or mineralogical factors, they are not appropriate for engineering purposes. Vallerga and Van Til (1970) proposed an engineering classification system for laterite materials and information obtained from Southeast Asia and particularly in Thailand.

1. The plasticity characteristics of lateritic soils from Southeast Asia vary over a wide range.
2. A classification system has been devised for lateritic soils by extending the Unified Classification System to include symbols for defining the durability characteristics of gravel and sand aggregate particles and the degree of plasticity of the fine-grained soils in the gravels and sands.

13.9.2 Properties and Foundation Problems

Engineering properties of laterite soils have been discussed and summarized by Little (1967), Vallerga and Van Til (1970) and Gidigasu (1975). Engineering properties reported include physical composition and plasticity characteristics, hardness and durability, strength,

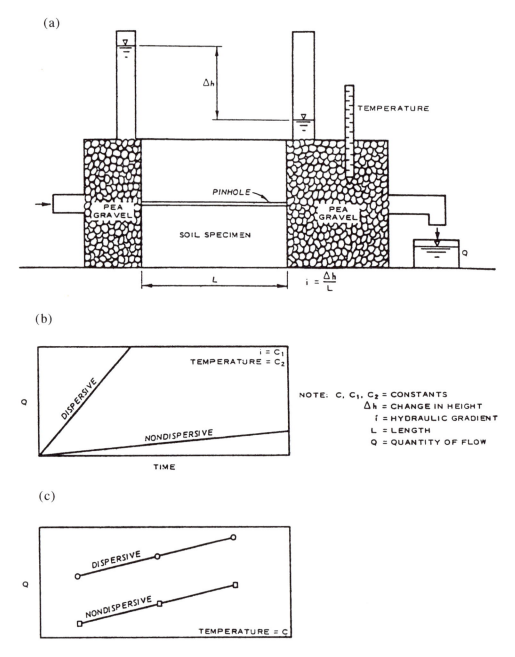

Figure 13.14 Characteristics of dispersive and nondispersive clays. (a) WES pinhole erosioin apparatus; (b) quantity of flow vs. time; (c) quantity of flow vs. hydraulic gradient. (After Perry [1975], Piping in Earth Dams Constructed of Dispersive Clay; Literature Review and Design of Laboratory Tests, Waterway Experiment Station Technical Report S-75-15, 105 p.)

moisture-density, permeability, and dynamic testing. In addition to these basic studies, soil stabilization and pavement design are also evaluated.

Environmental considerations on lateritic soil are also examined. Table 13.12 presents the comparison of strength parameters of laterite soil under acid (H_2SO_4) and nonacid (water) environments. Laboratory and *in situ* shear tests are performed. In all cases, these shear parameters are much less than under normal condition (water). Electrical properties of laterite soil have been fully discussed in Section 9.6 and Figures 9.6 and 9.24. The compressibility

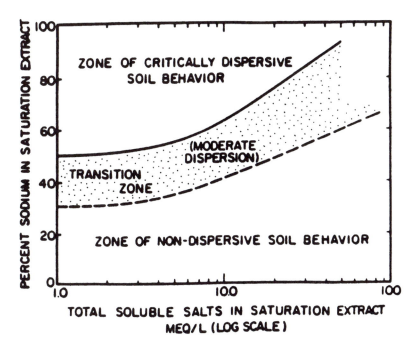

Figure 13.15 Determination of dispersive clay soil behavior. Percent sodium = Na(100)/Ca + Mg + Na + K all measured in MEQ/ℓ of saturation extract. (After Petry and Haliburton [1975].)

of laterite soil is greater under the acid condition in comparision with water as discussed in Section 11.8 and Figure 11.12.

13.10 LACUSTRINE SEDIMENTS AND VARVED CLAYS

13.10.1 General Discussion

The lacustrine deposits are confined to those sediments laid down in lakes and streams associated with the glaciers of the pleistocene, and will include those sediments deposited in lake basins or in valleys occupied by streams which have some characteristics of freshwater lakes, brackish water, and marine sediments, and are thus eliminated by being confined to those sediments commonly called clays but which contain varying amount of silt, sand and stones. Lake sediments consist of marls, tufas, clays, silts, sands, gravels, iron hydroxides, iron carbonate, silicon dioxide, manganese oxide, calcium phosphate, organic matter, and evaporites.

13.10.2 Marls and Tufas

Lake *marl* is a mixture of calcium carbonate with a variable content of impurities. Colors are various, but most marls are gray to white or pale blue, but red and black marls are not uncommon. Some marls have the calcium carbonate in the form of broken shells. Marls have been stated to form only in the epilimnion and this may be the case in some lakes, but bacteria that precipitate calcium carbonate may be present in muds beneath the hydroliminion to depths of water of 24 m, and it is known that extensive deposits of marls have been made to greater depths. Marls of which a large part is composed of shells of gastropods have accumulated over the bottom of green lakes in eastern Wisconsin to depths of greater than 60 m. These deposits are not made in softwater lakes and do not seem to be a common sediment of large lakes.

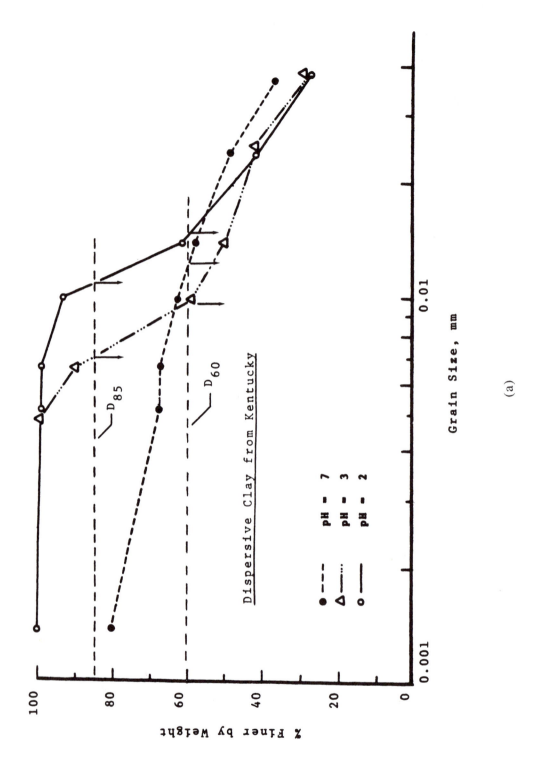

(a)

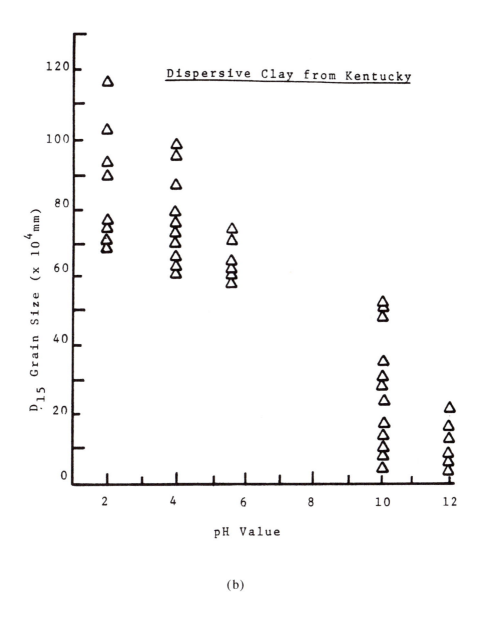

(b)

Figure 13.16 Effect of pore fluid on grain size distribution of dispersive clay. (a, on the previous page) Typical gradation curves; (b) D_{15} vs. pH value

Table 13.12 Comparison of Strength Parameters of Laterite Soil Under Acid and Nonacid Conditions

	Soaking 5-day under	
Shear parameters	**Water (H_2O)**	**Acid (H_2SO_4)**
Laboratory test		
Friction angle (ϕ), degree	22.2	15.2
Coefficient of friction (f)	0.408	0.271
Cohesion (c), kPa	87.5	75.5
In situ measurement		
Static cone penetration, (p_s) kPa	3060	1370

After Sun (1989).

Tufa is a lime carbonate deposit that is more or less porous and banded. The most extensive tufa deposits in the U.S. are those in the remnant lakes of former Luchonstal in Nevada. Hydroxides are common sediments in some lakes. Limonite has been deposited to a thickness of about a half meter in a depth of about 10 m in some Swedish lakes. The colors are shades of black and brown with some vegetable matter and gelatinious silica usually present. Tufas also contain considerable phosphorous in the form of earthy vivianite, and usually some manganese oxide, and considerable organic matter and some silica.

13.10.3 Varved Clays

(1) *Characteristics of Varved Clays*

A paired arrangement of layers in water-deposited materials in a glaciated region reflects seasonal changes during deposition. The fine sand and silt, or rock flour, are deposited in the glacial lake during the summer season, and the finer particles are usually deposited in a thinner layer during the winter (HRB, 1957). Each varve consists of two layers, the lower of which is usually the coarser in grain size and lighter in color. The upper one is finer in grain size and darker in color. These sediments are well stratified. The separation of paired laminae, each of which is nearly identical to those above and below, is characteristic of varves and distinguishes them from other sediments which are stratified but not varved.

In *varved clays* some of the bands may contain little or no clay. In places there is gradation from typical varved clays into varves composed of alternating gravel and sand. Varved clay of glacial origin commonly consists largely of very finely divided quartz, feldspar, and micaceous materials rather than mostly true clay minerals. The fineness of the grain, mineralogic constitution, great variation between horizontal and vertical permeability, and probably other factors related to unstable arrangements of the individual particles sometimes leads to special problems of engineering significance.

(2) *Environmental Factors Affecting Formation of Varved Clays*

Many varved clays are sensitive to environment. The deposits of one lake basin may differ considerably from those of another. Factors which may affect the distribution are

  • Position of lake outlet in relation to the ice front
  • Position of the ice front in relation to the lake
  • Position of the inlet glacial streams
  • Supply of material

The deposits of varved sediments within any one lake basin may also show considerable variation. Factors affecting the deposition in a lake basin are (1) topography of the bed (on an uneven lake bed, varved sediments are normally formed in depressions), and (2) position of glacial stream inlet.

(3) *Engineering Properties and Foundation Problems of Varved Clays*

The thickness of the varves and also the thickness of the individual component layers range from $1/16$ to $1\frac{1}{2}$ in. There will be some variation in layer thickness over one particular site and in many cases the thickness of one layer decreases as the other increases with depth. The clay content ranges from 10% to approximately 70%; in some cases, the field moisture content is lower than liquid limit. In almost all deposits in Ontario, Canada, both on the lake shore and the clay belt areas the field moisture content is near the liquid limit, the liquidity index is about 1.0, and the sensitivity generally is of the order of 4.0. The varved clays that

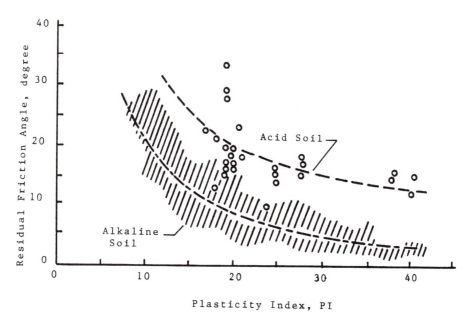

Figure 13.17 Residual friction angle vs. plasticity index for acid and alkaline soils. (After Tianjing University [1978].)

were examined had a maximum liquid limit of about 80 with a minimum of about 30. Plastic limits range from 18 to 28 with varying plasticity indices, reflecting the different origins and geological history of the deposits. It has been generally observed that illite is the predominent clay mineral east of Lake Superior, while montmorillonite is the predominent clay mineral west of that point.

If the preconsolidation loads are in good agreement with the present overburden loads, it indicates the clays are to be normally consolidated or lightly overconsolidated. However, in many cases precompression may have been due to ice loads. The average strength of varved clays ranges from 500 to 1000 psf determined by vane shear tests. Lacustrine sediments and varved clays contain numerous beds of pervious sand and gravel and may also contain considerable quantities of impervious clay or clay silt. Sand and gravel beds are often quite conspicious but also may constitute covered lenslike deposits, either isolated or connected with long stringers of pervious material. Inclusions or pockets of impervious and often highly compressible soft clay may be interlaced with pervious and competent sand or gravel strata. Design of foundations for light structures generally does not present difficulties. Heavy structures being built may be endangered by the possiblity of differential settlement.

13.11 SALINE AND ALKALI SOILS

13.11.1 General Discussion

Saline soil contains an excess of soluble salts, yet which is not excessively alkaline. Saline soils may contain carbonates, sulfates, or chlorides. *Alkali* soil in which sodium occupies 15% or more of the total exchange capacity is usually indicated by a pH value of 8.5 or higher (HRB, 1957). Saline soils can be grouped into two categories:

1. Saline soil deposits existing along the coastal line where the marine deposits are evaporated in saltwater.
2. Saline soil deposits existing naturally in inland regions.

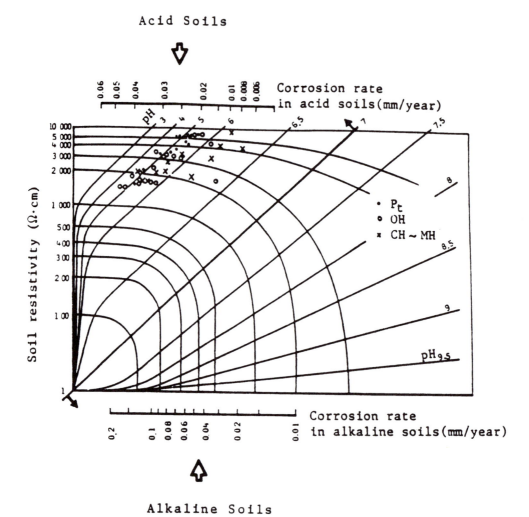

Figure 13.18 Estimation of corrosion rate of steel in acid and alkaline soils by Stratfull's chart. (After Watanable, S. [1977], Engineering Problems of Organic Soils in Japan, Japanese Society of Soil Mechanics and Foundation Engineering, pp. 69–82.)

13.11.2 Properties and Foundation Problems

Engineering properties between saline and alkali soils are significantly different. The residual friction angles vs. plasticity index (PI) is shown in Figure 13.17. Larger variations are found for alkaline soil. Recent research results on salty soil are given by Wang et al. (1996). Foundation problems in saline and alkali soils are discussed by Watanabe (1977). Figure 13.18 presents the corrosion rate of steel in acid and alkaline soils by Stratfull's chart.

13.12 BENTONITE CLAYS

13.12.1 Characteristics of Bentonite Clays

The name *bentonite* was first applied by Knight in 1898 to a highly colloidal, plastic clay found near Fort Benton, WY. It was shown by Hewett (1917) that bentonite clay was formed by the alteration of volcanic ash *in situ*. Later, Ross and Shannon (1926) further studied and

**Table 13.13 Physical Properties of
Commercially Available Bentonite[a]**

Chemical composition	Percentage by weight
SiO_2	56.5
Al_2O_3	20.0
Fe_2O_3	4.8
CaO	1.8
MgO	4.0
KO_2	0.7
Na_2O_3	3.5
Liquid limit	418.
Plastic limit	120.
Plasticity index	298.

[a] Commercially available bentonite in New Jersey.
[b] Chemical composition tested by Department of Geology, Princeton University.

Data from Princeton University.

redefined the term bentonite to limit it to clays produced by the alteration of volcanic ash *in situ*. They pointed out that such clays are largely composed of montmorillonite clay minerals. Until about 1930 bentonite was unknown outside of the Wyoming region.

At the present time, bentonite is a commercial term and defined by the Wyoming Geological Survey as a light gray sedimentary rock that contains at least 75% of the clay minerals beidellite or montmorillonite. Bentonite as a naturally occurring clay is located in Wyoming, South Dakota, Mississippi, and other midwestern states. Most bentonite in the 'Cowboy State' is termed 'Wyoming' or 'Western' bentonite and is different from the other states. The Wyoming bentonite uniquely expands up to 15 times its dry volume upon wetting. Much of the bentonite in this state is interbedded within cretaceous shales.

Bentonite is an important material for the construction of waste control systems such as slurry walls and clay liners (Section 17.8). In general, most bentonite contains montmorillonite minerals, however, some bentonite contains beidellite minerals such as Putnam, Susquehanna, and Lufkin clays (Section 4.10). The engineering properties of these clays will be discussed in a following section. In general, bentonite is the Na-montmorillonite as discussed in Section 4.10.

13.12.2 Properties and Foundation Problems

The natural bentonite clay deposits are located in Wyoming, South Dakota, Mississippi, and other midwestern states. The Wyoming type, only from Wyoming and South Dakota, has a soapy feel. It is a hydrous silicate of alumina containing about 62% of SiO_2 22% of Al_2O_3, 4% of iron oxide, 4% of magnesia and calcia, and 2% of alkali metal oxides (Brady, 1947). The Mississippi type differs from the Wyoming type only in physical properties. The commercially available bentonite from New Jersey is shown in Table 13.13. Mineralogical aspects of bentonite have been discussed in Section 4.10. Bentonite clays are very sensitive to local environments as shown in Figure 13.19.

Due to high water-holding capacity and swelling properties of bentonite clay, it tends to show the greatest amount but slowest rate of compaction. Bentonite soils shrink the greatest amount upon drying. Further, a dried mass of bentonite clay is, in general, harder and tougher than a mass of material composed of the other clay minerals. Bentonite clays can take up the most water, but they may do so at such a slow rate and under such conditions as to be relatively impermeable for all practical purposes. It seems that the presence of bentonite would tend to prevent the development of sheets of water and restrict the movement of groundwater. However, it must be pointed out that bentonite clay is more sensitive to local.

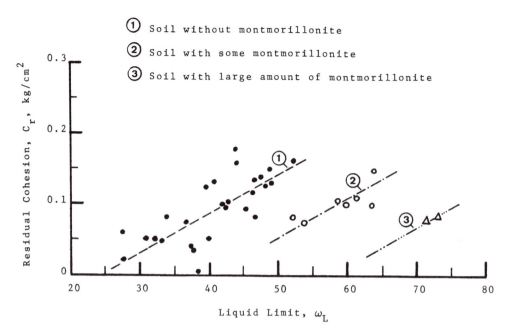

Figure 13.19 Effect of amount of bentonite added on residual shear stress. (Data from Tiangji University [1978].)

environments than others such as illite and kaolinite as discussed in Section 5.12, especially when contamined pore fluids are involved

13.13 TEPHRA, CORAL, AND SEA SHELLS

13.13.1 Tephra (Volcanic Ash and Cinder)

Tephra is a collective term designating all particles ejected from volcanoes, irrespective of size, shape, or composition. Volcanic ash is part of tephra and ranges in size between $1/16$ and 4 mm. Volcanic cinder is an obsolete name for medium-grained tephra. In geotechnical engineering, terms such as volcanic ash and cinder are frequently used in construction. Volcanic cinder and coral has been used as a foundation material, for example in the foundation of a Honolulu municipal office building.

13.13.2 Coral and Sea Shells

Coral is a shiny, hard calcareous material which is a growth composed of the skeletons of certain marine polyps. Their skeleton structures are deposited in extensive masses forming reefs and often resembling leafless trees. In many coastal regions, it is used as a construction material for walls and foundation materials. It also has been used for airport runway foundations. Approximately 30% of coral fill contents can pass through a U.S. sieve #200. Depth of coral deposits vary. Some coral is mixed with sand-gravel.

Clam and oyster shells can be found on banks of coastal regions. Their principal uses are for secondary road surfaces, foundation materials, or protection of embankment slopes. They are excavated generally from the banks and hauled immediately to the designated location without crushing, screening, or other treatment. Crushing and washing processes have been utilized to a limited extent when shells are used as mineral aggregate for concrete and bituminous paving mixtures.

13.14 SUMMARY

1. A soil in its natural location is a dynamic system whose properties depend not only on those of its constituents, but also on the mutual arrangement and interaction of these constituents as functions of the past history of the system and the dynamics of its present environment.

2. The placement of a foundation upon or into a soil system opens up possibilities for favorable or detrimental interactions which should be foreseen and taken into account by the responsible engineer.

3. The severity of unfavorable interaction usually increases with increasing closeness of the new interface to the soil surface where daily and seasonal temperature and moisture changes reach maximum values and where noxious salts may accumulate.

4. While structural sensitivity of cohesive soils and undesirable chemical activity of saline, alkaline, and acid soils are well known and respected, it is less well known that even 'inert' materials such as sands may show greatly varying bearing power depending on their specific location.

5. Also, depending on their modes of deposition, fluvial and marine sands and gravels may be so loose that, while easily supporting static loads, they become macromeritic liquids as a result of earthquake shock.

PROBLEMS

13.1 Define the weathering process. What general changes occur in soil material during soil development?

13.2 What are the differences between rock-like and soil-like shales?

13.3 What are the special features of residual soil?

13.4 What are causes and characteristics of expansive clay?

13.5 A chimney stack, 14 ft in diameter at the base, has a design dead load of 1000 tons (including the estimated weight of the foundation) and an overturning moment of 500 ton-ft due to wind. The stack is to be built in southern Pennsylvania at a location where it is known that the top stratum of soil is very fine sand from 25 to 45 ft thick containing pockets and layers of peat and muck. Beneath the sand is a homogeneous clay stratum some 80 ft thick, which in turn is underlain by hardpan and rock. During the preliminary design stage:
   (a) What kind and how many borings would you specify for this job?
   (b) How deep would you make your explorations?
   (c) What kinds, and how many soil tests would you require? Give reasons for the selections made.
   (d) Describe and sketch the means you would use, to open up and maintain the hole if the footing was 20×20 ft in plan and 10 ft below the ground surface.

13.6 Discuss the differences between loess, aeolian, and wind-blown silt.

13.7 How does one identify dispersive clay in the field?

13.8 Discuss how varved clays developed.

13.9 What are the differences between bentonite and montmorillonite clays? Are Na-montmorillonite and Na-bentonite the same?

13.10 Limestone is a relatively young geological formation which comprises about 10% of exposed rock on the earth's surface and covers a large part of the U.S. Limestone regions are frequently blanketed with a layer of residual soil. Residual soils are often soft and loose; therefore, the designer of a tall building generally requires that foundations be supported directly on the rock surface. However, the properties in a limestone region can vary considerably from poor to good within a short distance. During the preliminary stage, what would you suggest to construct a tall building in such a region?

CHAPTER 14

Wetlands, Coastal Margins, and Soil Erosion Problems

Wetlands, floodplains, coastal margins, dredging, and erosions are part of sensitive eco-logical/geological environmental problems. These areas are exceptionally responsive to envi-ronmental change and especially prone to irreversible ecological upset. In this chapter topics cover wetland, coastal margins, and erosion; in addition, some problems such as saltwater intrusion and greenhouse effects are all included. All explanations are made from the envi-ronmental geotechnical viewpoint.

14.1 WETLANDS

14.1.1 General Discussion

Wetland is a general term which includes marshes, swamps, flood plains, bogs, as well as rice paddies. The formation of wetlands varies greatly in age, especially man-made ones which are relatively recent in origin, while others had their beginning following the retreat of the glaciers. The dynamics of wetland ecosystems are complex. Within a given area, the plants and animals interacting with one another and their environment constitute an ecological system. The economic, ecological, and recreational aspects of these areas have been increas-ingly appreciated in recent years. Wetlands provide a vital habitat for various birds, plants, and fish; storage for water; water purification; and other benefits.

Historical review of wetlands indicates that when European settlers first arrived, wetland acreage in the area that would become the 48 states was more than 200 million acres, according to recent estimates (Brown and Snyder, 1991). By the mid-1970s, however, total wetland acreage was estimated to be about 100 million acres, according to the U.S. Fish and Wildlife Service (FWS). The most recent national survey by the FWS, covering the years 1954–1974, shows a net loss of over 9 million acres or nearly 500,000 acres per year. Agricultural development was responsible for 87% of these losses, with urban development and other development causing the remaining 8 and 5%, respectively.

In wetland areas, most soils belong to organic soils. These soils are solid constituents consisting predominantly of vegetable matter in various stages of decomposition or preser-vation. Wetness in the form of saturated soils or long periods of flooding is another special condition to which wetland plants must be adapted. There are numerous aspects of the characteristics of wetlands. However, in the following sections, only the environmental geotechnical aspects of wetlands are emphasized.

14.1.2 Definitions, Types, and Classifications

(1) Definitions

(a) U.S. Fish and Wildlife Service (FWS) (USDA, 1969)

Land where water is the dominant factor determines the nature of soil development and types of plant and animal communities living at the soil surface. It should be noted that the U.S. FWS has developed a more precise hierarchical system, by means of which each wetland can be mapped or described with reference to its general structure or vegetation, flooding pattern, water chemistry, and soil types (Cowardin et al. 1979).

(b) U.S. Army Corps of Engineers (1975)

Those areas that are inundated or saturated by surface or groundwater at a frequency and duration sufficient to support, under normal circumstances, do support a prevalence of vegetation typically adapted for life in saturated soil conditions. It must be noted that this definition is most commonly accepted by the U.S. Department of Transportation as reported by Thrasher (1983).

(c) Pennsylvania Environmental Research Foundation (ERF, 1982)

Areas where water is at, near, or above the land surface long enough to be capable of supporting aquatic or hydrophytic vegetation and have soils indicative of wet conditions.

(2) Types and Classifications of Wetland Systems

Scientists recognize (Niering, 1985) five major wetland systems: marine, estuarine, lacustrine, riverine, and palustrine. Marine and estuarine habitats include coastal wetlands, such as tidal marshes and mangrove swamps. The other three categories represent freshwater systems. Lacustrine wetlands are associated with lakes; riverine wetlands are found along rivers and streams. The palustrine wetlands include marshes, swamps, and bogs. Table 14.1 summarizes all five types of wetlands commonly used to designate distinct wetland types. Palustrine wetlands listed in Table 14.1 are described in the following section.

(a) Marshes

Marshes are characterized by soft-stemmed herbaceous plants, such as cattails and pickerelweed. Shallow marshes are those with up to 6 in. of water; deep marshes have as much as 2 to 3 ft of water. In the deeper marshes are floating and submerged aquatics. Some marshes begin as shallow lakes and depressions that gradually fill in with decomposed vegetation.

(b) Swamps

Swamps are dominated by woody plants, namely, trees and shrubs. As described by HRB (1957), a swamp is a tract of still water abounding in certain species of trees and coarse grass or boggy protuberances; a tract of wet, spongy land, saturated, but not usually covered with water; a bog, marshland and stream; a slough.

(c) Bogs

Bogs are peatlands, usually lacking an overlying layer of mineral soils. Peat is formed by the building up and gradual decomposition of plant material. According to HRB (1957),

Table 14.1 Classification of Wetland Systems

Saltwater systems
  Marine wetlands (coastal wetlands)
    Tidal marshes
    Mangrove swamps
  Estuarine wetlands: estuary wetland where river meets the sea
Freshwater systems
  Lacustrine wetlands: wetlands around the lakes
  Riverine wetlands
    Wetlands along the rivers
    Wetlands along the streams
  Palustrine wetlands
    Marshes
    Everglades swamps
    Bogs

the bog is a small open marsh which yields under the foot. Some of the interesting wetlands in North America are bogs. They occur primarily in formerly glaciated areas of the northeast, the north-central states, and Canada as recorded. Bogs offer an unusual source of information for biologists as well as environmental geotechnical engineers in the ecological history of the continent. The underlying peat preserves a record in the form of fossil pollens of the kinds of plants that have grown in the area over the last 10,000 to 15,000 years (Niering, 1985).

14.1.3 Natural Characteristics of Wetlands

In general, the hydrologic regime is the controlling feature in wetland's ecology as well as the environment. Deposits in wetland contain the rich organic matters. It is due to the slow rate of organic matter oxidation. This slow-rate oxidation is due to slow O_2 diffusion through stagnant water and low concentrations of mineral nutrients. In warm seasons the high specific heat of water slows the warming rate of the peat. During dry seasons the high thermal resistivity of peat slows warming below the immediate surface. Low O_2 availability is accompanied by high CO_2 partial pressures in most soils and sediments. The carbonic acid-bicarbonate equilibrium then tends to control the soil acidity at about pH equal to 6.0. When microbial activity slows and CO_2 diffusion raises the partial pressure of CO_2, the pH can become more acid if organic acids are being produced, or more alkaline if the soil solution contains high Na^+ concentrations.

Flood plain is the lowland that borders a river, which is usually dry but is subject to flooding when the stream overflows its banks. Flood soils along tropical and subtropical coastlines and river deltas may contain significant quantities of Fe (II) ion. When drained, these sulfides oxidize to H_2SO_4 and the acidic Fe^{3+} ion. Soil acidity can increase to a pH value equal to 2.0 or below. Such conditions are highly phytoxic and can be remedied under aerobic conditions only by extensive leaching and lime applications. If resubmerged, the pH of the acid sulfate soils rises rapidly to near neutrality as the Fe(III) and sulfate are reduced back to Fe (II) sulfides.

14.1.4 Rice Paddy (Man-Made Wetland)

Rice paddy is a man-made wetland used for agricultural purposes. The behavior of rice paddy is similar to flood soil but contents in the flood plain and rice paddy are different. The rice paddy contains large amounts of fertilizers. The behavior of carbon (C), nitrogen (N), sulfur (S), iron (Fe), and manganese (Mg) generally follows that when rice paddies are drained before harvest, redox potentials rise, Fe^{2+} and Mn^{2+} concentrations decrease, and nitrogen, sulfur, and carbon oxidize. When the soils are flooded again, the reactions reverse. Figure 14.1 illustrates the effect of soaking conditions on behavior of a rice paddy. Figure 14.1(a)

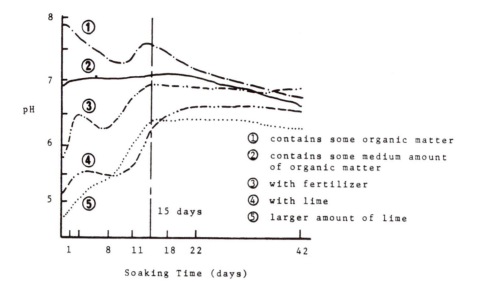

(a)

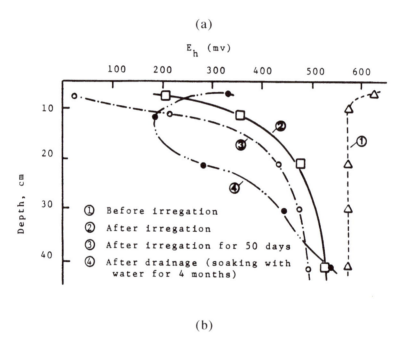

(b)

Figure 14.1 Effect of soaking condition on behavior of rice paddy. (a) Various pH values vs. soaking time; (b) effect of depth on redox reaction. (After Yu, T.Y. and Chiang, H.Y. [1984], *Electro-Chemical Properties of Soils,* Scientific Publishing, Beijing.)

shows the effects of various pH vs. soaking time. In all cases, larger variations of pH values are found during the first 15 days. Figure 14.1(b) shows the drainage conditions' significant effects on redox reaction as reflected by E_h values.

14.1.5 Environmental Geotechnical Problems of Wetlands

Environmental geotechnical problems in wetlands are not only due to a high water table and low bearing capacity, there are numerous detrimental effects in both technical and environmental aspects. The ecological impact assessment process according to Thrasher

(1983) can be divided into three key steps: (1) description of the project, (2) ecological studies, and (3) impact assessments. A description of the project should be provided in detail as much as possible. Ecological studies should include an evaluation of the biotic and abiotic factors. The prediction of impacts is the most important step in the process which is part of the geomorphic processes as discussed in Section 5.10. Impact assessments must also include the consideration of alternatives and means of providing or enhancing positive impacts (FHWA, 1978 and 1980). Procedural guidelines for impact assessment of highways and wetlands are prepared by Erickson and Camougis (1980) and Kent (1994). Some of the problems related to the environmental geotechnology are outlined as follows.

(1) Ecological Effects

Ecological effects include five basic aspects: (1) inhibition of movement of animals, (2) alteration of biological productivity, (3) alteration of nutrient flux, (4) removal of natural filtration systems, and (5) introduction of exotics.

(2) Engineering Difficulties

Engineering difficulties in wetlands are mainly high water content and soft deposits. Dewatering is the first problem dealing with these difficulties: (1) inhibition of storm water and tidal distribution, (2) increased water turbidity, and (3) alteration of water circulation patterns.

Earth fills on wetlands for highway embankments is one of the typical environmental geotechnical engineering problems. Construction on wetland has some special problems which deserve special attention. For example, earth fills on wetland for highway embankments will include the following problems: (1) hydrological effects, (2) physicochemical effects, (3) water quality effects, and (4) erosion and sedimentation effects.

14.2 LAND AND COASTAL MARGINS

14.2.1 General Discussion

Population growth and human activities have brought many changes to various environmental problems; the change in land and coastal environments is one major subject including those as reported by Land Margins Ecosystem Research (LMER) Committee of the American Geophysical Union (EOS, 1992): (1) loss of habitat, (2) interception of water and sediment, (3) increased pollution of near-shore environments, and (4) increasing rate of sea level rise (greenhouse effect). Environmental geotechnical problems in land and coastal margins require information such as characteristics of marine deposits, wave refraction, shore currents, sand transportation, etc. The problems discussed in this chapter include:

- Salt water intrusion
- Estuaries
- Oil spills and other coastal pollution
- Greenhouse effect
- Coastal stability and erosion problems

The rapid changes altering land/coastal environments have made it obvious to researchers and resource managers that there is a pressing need to better understand and predict changes occurring in the entire coastal system. In particular, there is a need to understand how anthropogenic changes will interact with and change ecosystems in estuaries and coastal waters.

14.2.2 Land/Coastal Margin Ecosystem

(1) General Discussion

A land margin ecosystem site study, sponsored by the American Society of Limnology and Oceanography and funded by the National Science Foundation, was held at the Marine Biological Laboratory, Woods Hole, MA (1989). Experts in physical, geological, chemical, and biological oceanography, aquatic and terrestrial ecology, microbiology, and fisheries explored changes in the ecosystems at the land-sea interface. A report of the workshop, titled "At the Land-Sea Interface: A Call for Basic Research", was printed in 1990. It is available from Joint Oceanographic Institutions. Brief discussions on the characteristics of the test sites and environmental geotechnical properties are presented as follows.

(2) Land Margin Ecosystyem Research (LMER) Sites

There are five LMER sites. Columbia River and Tomales Bay located on the west coast, and Waquoit Bay, Chasapeake Bay, and Plum Island located on the east coast. The characteristics of each site are briefly described as follows:

1. Columbia River: The Columbia River is an estuary with large freshwater inflow, short residence times, and a relatively dry watershed. The project studies physical-biological couplings that take place in zones of estuarine turbidity maximum within the estuary's water column.
2. Tomales Bay: Tomales Bay is a medium-sized drowned rift valley in coastal northern California, adjacent to a small watershed mostly covered by grasslands. The Tomales Bay project concerns biogeochemical reactions in estuaries.
3. Waquoit Bay: Waquoit Bay is a small shallow bay in New England and is fed by several forested and urbanized subwatersheds.
4. Chesapeake Bay: Chesapeake Bay is a large Mid-Atlantic estuary, with a very large, complex watershed in a temperate, relatively moist climate. The study focuses on processes of recycling, organic transformations, and exchange between uplands and sea.
5. Plum Island: The Plum Island Comparative Ecosystems Study is located on the coasts of Massachusetts and New Hampshire. This site will explore effects of land use and organic matter-nutrient interactions on estuarine trophic dynamics.

(3) Estuarine Characteristics of LMER Sites

The existing LMER sites provide a wide range of estuarine characteristics summarized in Table 14.2. The sites differ in size, depth, watershed properties, climate, water dynamics, water quality, and biology. In examining Table 14.2 these flow ratios show a large range among the LMER sites, from the rather fast renewal in Columbia River, to the slow rate in Tomales. Sedimentation rates in the LMER systems vary as a function of sediment yield in the drainage basin, land use, the ratio of watershed area to receiving water area, and efficiency of particle trapping. The tidal regime of all these systems is semidiurnal, but the tidal range varies by a factor of 3. Energy associated with freshwater is greatest in Columbia River.

14.2.3 Basic Parameters Needed for Evaluation

(1) General Discussion

Coastal problems are exposed to both man-made and natural adverse environmental conditions such as waves, tides, typhoon, ice, earthquake, marine slides, marine borers, and

Table 14.2 Environmental Geotechnical Characteristics of LMER Sites

	Chesapeake	Tomales	Waquoit	Columbia
Water body area, km²	11,478	28	6.3	420
Water body mean depth, m	9	3.1	0.9	10
Mean tide range, m	0.5	1.0	0.7	1.7
Tidal prism, 10^{-5} m³	1,720	28	4.4	680
Watershed area, km²	164,188	570	46	660,500
Watershed max elevation, m	1,220	794	100	3,500
Runoff, 1000 m³/day	190,000	400	0	600,000
Groundwater flow, 1000 m³/day	10,000	15	115	20,000
Sewage discharge, 1000 m³/day	5,410	0	0	1,000
Precipitation, mm/day	2.7	2.7	3.3	1.3
Evaporation, mm/day	1.3	1.4	1.8	0.7
Freshwater inflow/tidal prism (flow ratio)	0.066	0.008	0.015	0.486
Watershed population	14,500,000	11,000	11,500	4,000,000
Salinity range, %	0–30	0–38	0–32	0–33
Temperature range, °C	−1–30	5–23	1–20	0–23
DIP range, μM	0–4	0.3–7	0–3	0.3–2.2
NO_3 range, μM	0–250	0–90	0–50	0.5–40
NH_4 range, μM	0–50	0–11	0–30	0.3–6.6
Chlorophyll range, μg/l	1–500	0.2–60	0–190	0.5–20
Mean primary production by phytoplankton, g/m²/year	450	330	200	100
Total sediment input, 10^{-6} tons/year	5.5	0.3	0.02	5
Mean sedimentation rate, cm/year	0.2–1.5	0.8	0.2	0.5
Range of total suspended solids, mg/l	5–300	0–200	—	1–1000

Note: LMER = land-margin ecosystems research sites.
After EOS (1992).

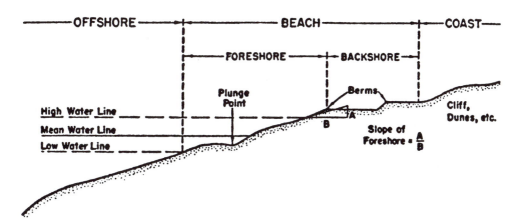

Figure 14.2 Diagrammatic sketch of coastal margin features. (After Beach Erosion Board [1963].)

man-made conditions such as pollution. Zonation or microzonation of geo-environmental and climatological conditions is a major contributing factor relating to the stability of marine structures. It is necessary to examine these conditions before the final design and construction of nearshore/offshore structures can begin. Figure 14.2 presents diagrammatic sketch of shoreline features including beach, offshore, and nearshore. Further explanations of Figure 14.2 will be discussed in Section 14.4

(2) Characteristics of Ocean Waves

Waves are generated by winds. Their characteristics can be determined by the velocity of the wind, its duration, and the fetch length. The *fetch* length is the horizontal length of the generating area (in the direction of the wind) over which the wind blows. When waves approach a straight beach at an angle they tend to swing in parallel to the shore due to the retarding effect of the shallowing water. Such bending of waves is called *reflection* as discussed by Krumbein (1950). In studying coastal margin, weathering around construction sites and nearby areas must be examined. For studying nearshore/offshore, the visibility around the construction areas must also be studied.

(3) Climatological and Geo-Environmental Conditions Along the Coastline

In coastal areas, the lands are generally grouped into three basic types: (1) rock cliff, (2) loose beach sand, and (3) soft clay deposits. They all have difficult engineering problems. In addition, the climatological and geo-environmental conditions are closely related to these problems. Using special features obtained from China coastline studies as an example to illustrate these points, the following aspects are discussed.

Several earthquake and volcanic zones are known to lie along the coast of China such as the volcanic mountain ranges and earthquake zones around Taiwan Island, the Taiwan Straits, and South China Sea regions. These environmental conditions cause problems both during and after earthquakes; the aftershaking commonly causes as much damage as during an earthquake's ground shaking itself. Recent examples of this phenomenon are the Northridge, CA earthquake in 1994 and Kobe earthquake, Japan, in 1995.

Most regions are cooled by trade winds blowing throughout the year, and hurricanes are a feature of this kind of climate. The hurricane in the North Pacific is called a *typhoon* or tropical cyclone. In the southern part of China, the region frequently has periods of unsettled weather, with prolonged rain, muggy spells, gales, and sometimes snowstorms caused by centers of low pressure called depressions. Depressions occur when cold air masses from the polar regions meet warm air masses from the tropics. The typhoon is a violent whirling storm in which winds blow spirally inward toward a center of low pressure. The wind speed ranges from 65 to 174 knots/hr. Hurricanes may be up to 483 km across and can travel at velocities up to approximately 174 knots/hr. They cause tremendous damage to both nearshore and offshore structures and leave a wide trail of destruction. Most typhoons originate from the area surrounding the Philippine Islands most frequently. Wind behavior around Hong Kong Island and the South China coastal region has been discussed in Section 12.3.

Other problems include rainstorms, ocean currents, and broken ice which affects the stability of nearshore/offshore structures. During the construction process, other information such as fog patterns also is needed.

14.2.4 Environmental Geotechnical Problems

In most nearshore locations, there exist large amounts of soft clay deposits, such as New Orleans, Bangkok, Shanghai, and many other regions. Using Shanghai region soft clay deposit as an example, the soft alluvial deposits cover large areas of the populated lower Yangtze Valley. Comparisons of various soft clay deposits along the China coastline and other locations are summarized and discussed in Section 13.6. In addition to the soft clay deposits around the coastline, there are some special types of materials frequently found in these regions such as coral and seashell-soil mixtures. These types of soil have been discussed in Section 13.13.

14.3 BEACHES AND BEACH EROSION

14.3.1 Characteristics of Beaches and Beach Sand

(1) Characteristics of Beaches

A beach is located between the offshore and coastal regions. The Beach Erosion Board has also developed a classification of beach slopes for convenience of description. The following classification system (Table 14.3) is based on a geometric ratio such that each class has twice the slope of the preceding.

Table 14.3 Classification of Beach Slopes

Classification	Slopes
Flat	Less than 1 on 120
Mild	1 on 120 to 1 on 60
Gentle	1 on 60 to 1 on 30
Moderate	1 on 30 to 1 on 15
Steep	Greater than 1 on 15

After Beach Erosion Board (1963).

As illustrated in Figure 14.2, the beach is the zone between the low-water line and the sea cliff which forms the landward limit of the beach. The foreshore is that part of the beach which extends from the low-water line to the crest of the berm if a berm is present. The *berm* is defined as the nearly horizontal surface of the beach which extends inland beyond the normal limit of wave uprush. The berm may actually have a gentle slope toward the landward limit of the beach. On some beaches several berms may be present as the result of different wave conditions. From the crest of the berm to the landward limit of the beach is the backshore, which is subject to wave action only during unusual storm conditions. The slope of the foreshore is measured at the high-water line and is expressed as a 1 in 25, etc. In terms of Figure 14.2, the slope is A/B. Further discussions on beach engineering are given by Krumbein (1950)

(2) Characteristics of Beach Sand

Because of the classifying action of wave attack on the shore material and of the seaward transportation, most marine deposits have a relatively narrow particle size as shown in Table 14.4. These relatively uniform deposits make them susceptible to quake liquefaction (Section 11.9). Also, their void ratio normally exceeds their critical void ratio (CVR) and therefore are in a potential liquid state. They may be changed into actual macromeritic liquids, not only in the specific shear zones but throughout the whole granular system. Relatively small energies, such as machine noises, vibrations, or minor earthquakes acting on the deposits, may trigger the slope failure.

14.3.2 Beach Erosion and Pollution Problems

Shore or beach erosion is a common problem along sandy coastal or lake areas, for example, the north shore of New York's Long Island where homes in certain sections are endangered. A study by the Marine Science Center of the State University of New York at Stony Brook revealed that coastal cliffs, some of them as much as 150 ft high, are eroding at the rate of 1 to 3 ft/year.

Table 14.4 Particle Characteristics of Beach Sands

Location		Particle characteristics: granulometric constants		
		D_{10}	D_{60}	C_u
Alaska	Adak	0.25	0.48	1.9
	Attu	0.40	1.03	2.6
	Kodiak	0.25	0.30	1.2
	Unalaska, Broad Bay	0.64	1.18	1.9
	Wide Bay	0.15	0.33	2.2
Canal Zone	Pima Beach	0.18	0.42	2.3
	Venado Beach	0.38	1.15	3.0
Cuba	Windmill Beach	0.63	1.15	1.8
	Conejo Bay	0.30	0.48	1.6
Florida	Daytona Beach	0.15	0.20	1.3
Guam	Blue Beach	0.48	1.45	3.0
	Purple Beach	0.20	0.68	3.4
	Tarague Beach	0.20	0.65	3.3
	Tumon Beach	0.20	0.55	2.8
Hawaii	Hilo Island	0.07	0.23	3.2
	Oahu, Barbers Point	0.41	0.75	1.8
	Pearl Harbor	0.47	0.90	1.9
Hong Kong		0.42	0.82	2.2
Marshall Islands	Kwajalein	0.60	1.50	2.5
Massachusetts		0.15	0.24	1.6
Midway Islands		0.20	0.44	2.2
New Jersey	Spring Lake	0.30	0.60	2.0
Panama	Taboga Beach	1.43	2.81	2.0
Puerto Rico	Rio Grande	0.25	0.38	1.5
	Navy Beach	0.25	0.43	1.7
	Palo Seco	0.15	0.30	2.0
	Playa, Grande	0.80	1.03	1.3
	Port Maldonado	0.35	0.39	1.4
El Salvador	San Salvador	0.38	0.76	2.4
Taiwan	Taipei	0.46	0.92	2.3
Trinidad	Macqueripe Bay	0.18	0.38	2.1
Virgin Islands	Frenchman's Bay	0.53	0.90	1.7

Note: All samples are taken at low-water mark. D_{10} = effective size; D_{60} = 60% size; C_u = uniformity coefficient = D_{60}/D_{10}, D_{10}, and D_{60} in mm.
After Chaney, R.C. and Fang, H.Y. (1986), Static and Dynamic Properties of Marine Sediments: A State-of-the-Art, ASTM STP 923, pp. 74–111.

More recently, beach pollution is also a problem, based on a newspaper report in 1995 which lists the sources of pollution including 584 closings or advisories prompted by sewer overflows. California led with 910 closures and advisories. New Jersey, which has the most comprehensive beach-testing program, was second with 238. Next came New York with 227, Florida with 215, and Connecticut with 162. The number of closures and advisories in 1994 in other states and territories that did some monitoring were Wisconsin, 148; Ohio, 96; Maryland, 82; Massachusetts, 58; Illinois and Indiana, both 36; Michigan, 26; Hawaii, 16; Maine, 15; and Pennsylvania, 14.

14.4 SALTWATER INTRUSION, ESTUARIES, AND GREENHOUSE EFFECT

14.4.1 General Discussion

Saltwater intrusion or encroachment is shoreward movement of salt water from ocean into coastal aquifers due to the overpumping of groundwater, and it is a dynamic equilibrium

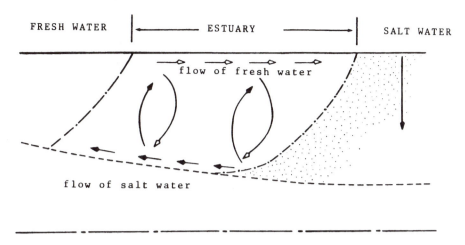

Figure 14.3 Estuaries — where fresh water interacts with saltwater.

of groundwater movement. In coastal regions, successive pumping will also cause saltwater intrusion, consequently leading to the possibility of polluting the groundwater and corroding subsurface structures. Fresh water and salt water are treated as two immiscible fluids; they are separated by an interface with a slope.

14.4.2 Estuaries and Estuarine Systems

The widened channel of the mouth of a river in which influence of tides is felt is known as an *estuary*. In a simple term, the estuary is where river (fresh water) meets the sea (salt water). Estuaries are fragile and easily destroyed and an important part of the ecosystems on earth. They serve as nurseries and spawning and feeding grounds for a large group of marine life and provide shelter and food for birds and wildlife. There are many accepted definitions for estuaries that depend on the point of view of the observer. The most commonly used definition is from Pritchard (1952): "An estuary is a semi-enclosed coastal body of water which has a free connection with the open sea and within which sea water is measurably diluted with fresh water derived from land drainage." Most engineers simply characterize an estuary as a area of interaction between fresh and salt water (Kennish, 1996).

The interaction of fresh and salt water provides a circulation of water and mixing processes that are driven by the differences in density between the two types of water as illustrated in Figure 14.3. The density of salt water depends on salinity and temperature. In estuaries, the salinity range is large and the temperature range is small. Therefore, salinity has the most influence on the behavior of salt water in estuaries.

Estuaries are formed in the narrow boundary zone between the sea and the land. Their form and extent are constantly being altered by erosion and deposition of sediments. Drastic effects are caused by a small raising or lowering of sea level. Although the life of most estuaries is short, they have proved to be an important factor in the development of the world's environment. They have high inflows of nutrients from the land, but tend to have a lesser diversity of life than other aquatic environments. Individual species are numerous, but they are specialized and often tolerant to the large extremes of salinity and temperature that are present.

Estuaries have been the main centers for man's development because of their fertile waters and navigational access to land. The promotion of trade and industry has led to an immeasurable distortion of the natural balance within estuaries. Estuarine topography is directly affected by deforestation. This is because deforestation increases runoff from the land, the flashiness of the discharge, and the sediment load to rivers. It is essential that these effects be studied to prevent further damage to the environment. The main drawback in studying estuaries is that river flow, tidal range, and sediment distribution are continually changing. Therefore, some estuaries may never attain steady-state conditions. Due to the interaction of so many variables, no two estuaries are alike. One never knows whether general principles or unique details are being observed.

14.4.3 Greenhouse Effect on Coastal Environment

Theory says that certain gases in the atmosphere act like greenhouse glass, letting sunlight in but not allowing heat to escape. In such a case, it is called a *greenhouse effect*. Increasing concentrations of atmospheric carbon dioxide (CO_2) and other gases is expected to cause a global warming that could raise the sea several feet in the next century. In 1982, the U.S. EPA organized a project to estimate the magnitude of future sea level rise. From an environmental geotechnical viewpoint, if sea level rises, more wetlands occur promoting problems such as coastal erosion, salinity intrusion, foundation problems, and slope stability in coastal margin. These implications of sealevel rise create various hazardous wastes in coastal flood-plains as reported by Barth and Titus (1984) and Demars and Long (1991).

Loss of coastal wetlands due to sealevel rise is one of the major problems caused by the greenhouse effect. Sea level is projected to rise between 1.4 and 2.2 m by the year 2100 in response to global warming. A 1-m rise in ocean levels world wide has been estimated to result in the creation of 50 million environmental refugees from various countries — more than triple the number in all recognized refugee categories today. The rise in sea level over such a short time span will also result in the loss of coastal wetlands (Chaney and Demars, 1991).

Fluctuations in the position of sea level relative to land can be caused by eustatic, isostatic, and tectonic forces acting independently or in conjuction. Eustatic fluctuations describe sealevel changes as consequence of change in volume of the water in the global ocean, generally associated with the accretion or melting of continental ice sheets. A higher mean sea level would increase the frequency of inundation and exacerbate flood damage. It would inundate fertile deltas, causing loss of productive agricultural land, and increase saline encroachment into aquifers, streams, and estuaries. The increased costs of reconstruction, rehabilitation and strengthening of coastal defense systems could turn out to be crippling for most affected countries.

14.5 MARINE SEDIMENTS AND SEA FLOOR PROBLEMS

14.5.1 Marine Environments

In general, the ocean floor can be divided into two major regions: the continental margins and the ocean basins. The continental margins include the continental shelf and the continental slope and form only a small percentage of the overall oceanic area. The marine environment is predominantly a depositional rather than an erosional environment. As a consequence, marine sediments exhibit more uniformity than normally found on land. Typically, marine sediments are broadly classified on whether the sediments are land derived (terrigenous) or are the result of marine activity (pelagic). The pelagic sediments can be further divided into inorganic or organic materials. Inorganic pelagic materials are typically clay-size material. Deposits with pelagic clay are primarily found off areas of major deserts. The organic materials are primarily the skeletal remains of marine organisms.

There are three general types of deep sea deposits; viz., brown red clays, calcareous ooze, and siliceous ooze. The term ooze denotes that more than 30% of the sediments are of biotic origin. Equatorial and polar regions are especially high in organic productivity. These sediments are calcarous at depths of less than 470 m and siliceous in deeper waters. In areas of lesser biotic activity, calcareous ooze is found in shallow water and brown red clay in deep water. The calcareous ooze is nonplastic, creamy to white in color, with particles of silt size which crush easily. Some oozes contain 50 to 100% water. Siliceous ooze consists mainly of the remains of siliceous plant skeletons which are found in a large belt around the Antarctic. Brown clay of terrestrial origin transported by wind and/or water is found in most of the deeper portions of the ocean.

In coastal margins, some marine deposits contain high percentage of clay (35 to 60%) and silt (40% to 60%) and the sensitivity value ranges from 1.6 to 26.0, therefore indicating very strongly that sediments are in a potential macromeritic liquid state and can become actual liquids after destruction of their interparticle bonds. Any man-made disturbance during the construction period will reduce the strength and bearing capacity significantly, since the liquefaction behavior of the marine deposits is an important factor for analysis and design.

14.5.2 Properties of Seawater

The ocean is actually a vast solution of ions and other substances in which plants and animals exist. Most of the dissolved constituents of the ocean are ions. The range of near-surface temperature of seawater is between 35°C in equatorial latitudes to freezing or –2°C in polar latitudes. The bulk of seawater has a temperature less than 6°C and nearly half is under 2°C. The temperature gradient is the rate of change of temperature with depth as well as one associated with latitude. It is indicated that approximately the upper 200 m of the surface layer is strongly affected with seasonal or environmental variations (Bennett et al. 1991).

The terms salinity and chlorinity are used to characterize the properties of sea water. The *salinity* as defined by oceanographers is the mass in grams of the solids in 1 kg of seawater evaporated to a constant mass at 480°C. Since the relative amounts of dissolved substances are somewhat invariant, the salinity of seawater can be directly related to the chlorinity of the seawater. *Chlorinity* is defined as the number of grams of chloride ion (Cl^-), bromide ion (Br^-), and iodide (I^-) contained in 1 kg of seawater. The experimentally observed relationship between salinity and chlorinity is given by Dickson (1974) as:

$$S = 1.805 \ (C) + 0.03 \tag{14.1}$$

where S = salinity, grams per kilogram of seawater and C = chlorinity, grams per kilogram of seawater.

The relationship between temperature, salinity, and density of seawater is linear. For a given density of seawater, when temperature increases, the percent of salinity also increases. As indicated in Section 5.5, the ocean is actually a vast solution of ions. Therefore, the electrical conductivity of the seawater is important as discussed in Section 9.4.

14.5.3 Engineering Behavior of Marine Sediments

The geotechnical characteristics of marine sediments are governed by their location in the marine environment at the time of deposition. The typical geotechnical properties of marine clays are summarized in Table 14.5. A state-of-the-art review of engineering properties is given by Chaney and Fang (1986). However, some special features on environmental aspects of marine clay are further discussed as follows.

The effect of pore fluid on Canadian sensitive clay as reflected on liquid limit is presented in Figure 14.4. Laboratory experiments were conducted on temperature effects by the consolidation test. The soil specimens were obtained from Mississippi Delta marine clay with results of the tests presented in Figure 14.5. The temperature variations are from 5 to 22°C.

**Table 14.5 Typical Geotechnical Data and
Their Ranges of Marine Deposits**

Parameters	Range
Sizes Composition, %	
Clay 2 μm	35–60
Silt	40–60
Sand	10–
Clay Minerals, %	
Illite	60–75
Kaolinite	10–
Montmorillonite	5–20
Physical Properties	
Moisture content (*in situ*)	60–180
Activity (A)	0.33–1.33
Sensitivity	1.60–26
Liquid limit (LL)	72–121
Plastic limit (PL)	34–51
Field moisture equivalent	65–78
Centrifuge moisture equivalent	55–68
Shrinkage limit (SL)	7–10
Void ratio (e)	0.5–9.0
Compression index (C_c)	2.3–

Data from Winterkorn and Fang (1971); cited
by Chaney, R.C. and Fang, H.Y. (1986), Static
and Dynamic Properties of Marine Sediments:
A State-of-the-Art, ASTM STP 923, pp. 74–111.

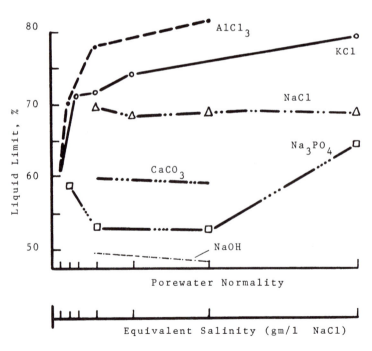

Figure 14.4 Effect of pore fluid on liquid limit of marine clays. (After Torrance, J.K. [1975], *Can. Geotech. J.,* vol. 12, no. 3.)

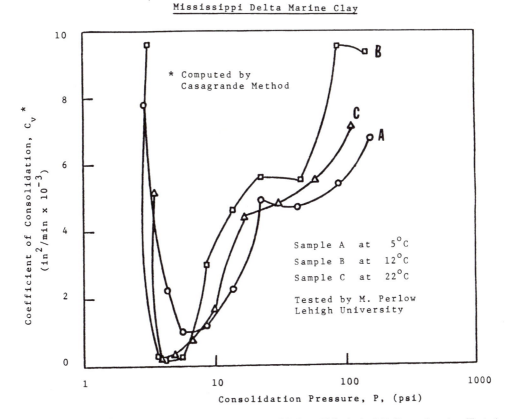

Figure 14.5 Effect of temperature on coefficient of consolidation of Mississippi, Delta marine clay. (Tested by M. Perlow, Lehigh University.)

The coefficient of consolidation is computed by routine Casagrande method. In all cases, both temperature and the consolidation pressure play important roles.

14.5.4 Pollution and Contaminated Marine Sediments

For the last several decades numerous industrial and urban wastes have been dumped into the ground, rivers, and oceans, especially in the areas with highly industrialized harbors where nearshore structures are located. All types of pollution have direct or indirect effects on soil-water systems and superstructures. Corrosion is one of the major problems in marine environments. Acid rain and salty wind affect many exposed marine structures.

Biological attack is the major cause of damage to timber structures in harbor areas and also can occur in any part of nearshore/offshore structures. The principal organisms which attack timber structures are marine borers and fungi. The geographical distribution of marine borers along the U.S. and China coast is presented in Table 14.6.

Corrosion caused by microorganisms in marine environments is a potential hazard to all nearshore/offshore structures. Metabolic processes occurring at metal interfaces have a significant effect on a variety of different corrosion reactions. Some of the basic mechanisms by which microorganisms initiate or accelerate corrosion have been discussed in Section 5.11. The rate of biological attack depends on the level of dissolved oxygen content in the adjacent water and the water temperature; therefore, the South China Sea has more problems than the Yellow Sea or Bohai Gulf as shown in Table 14.6.

Table 14.6 Geographical Distribution of Marine Borers

Marine bore species	Atlantic Coast	Pacific Coast	Gulf Coast	South China Sea
Bankia (various species)	Massachusetts to Canal Zone	Gulf of Alaska to Canal Zone	Florida; Texas	Hainan Dao; Zhongsha Qundao
Limnoria lignorum	Maine to Straits of Florida	Alaska to Canal Zone	[a]	Nansha Qundao; Zhongsha Qundao; Philippines
Limnoria quadripunctata	Bahama Is. Bermuda	California; Gulf of California	Florida; Texas	Hong Kong; Xisha Qundao; Zhongsha Qundao
Limnoria tripunctata	New Hampshire to the Canal Zone	California to Mexico	Florida to Texas	Gulf of Tonkin; Hainan Dao; Xisha Qundao
Martesia striata	North Carolina to Canal Zone	None reported	Florida; Texas	Guam; Hainan Dao; Hong Kong
Teredo (various species)[a]				

[a] Distributed in the Atlantic, Pacific, and Gulf coasts and the South China Sea.

Data from Brackett, Nordell and Rail (1982); Fang and Chaney (1985); and others.

14.5.5 Marine Structures and Slopes

(1) Marine Structures

Nearshore/offshore structures are a major part of marine structural systems. From Figure 14.2, the general offshore area adjacent to the beach is usually defined to extend indefinitely outward from the low-water line, although there is some tendency to define the zone between the low-water line and a depth of 30 ft as the nearshore area, with the offshore area lying at depths greater than 30 ft. The offshore platform depth has been increased from 6 m in 1947 to more than 300 m in 1978. Nearshore/offshore structures are exposed to extremely adverse environmental conditions. These conditions are shown schematically in Figure 14.6 and a summary of the various causes of instability of the structure-foundation system is presented in Table 14.7. A review of Figure 14.6 and Table 14.7 shows that marine structures are constantly under both man-made and natural environmental loads such as waves, currents, typhoons (hurricanes), ice, earthquake, marine slides, and marine borers.

(2) Slopes of Submarine Sediments

Dangerous conditions exist at the foot of slopes of submarine sediments. The particulate components of such sediments are usually of narrow size range as shown in Table 14.4, which makes them especially susceptible to quick liquefaction. Also, their void ratio normally exceeds their critical void ratio. Relatively small energies such as machine noises and vibrations may start the process, and the conversion of the potential energy of the particles from the upper parts of the slope into kinetic energy may do the rest.

14.6 DREDGING AND DREDGING MATERIAL

14.6.1 General Discussion

Excavated river and harbor materials constitute a problem for many coastal states because they are often contaminated with heavy metals and organic compounds. These contaminants

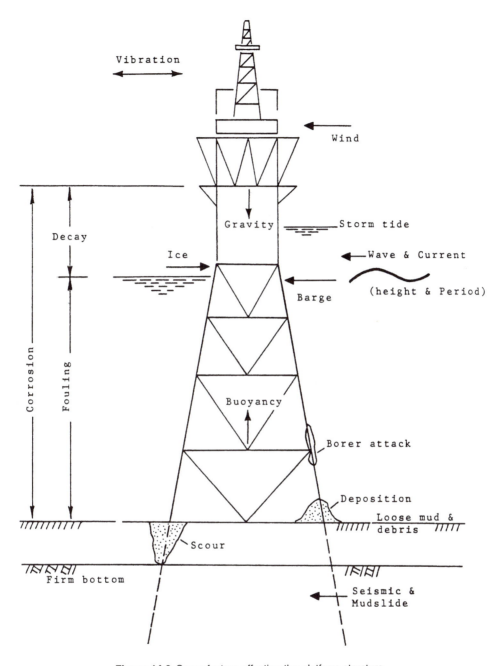

Figure 14.6 Some factors affecting the platform structure.

are a result of industrial discharges and storm runoff into adjacent streams or rivers. The U.S. Army Corps of Engineers currently supports disposal of contaminated dredge materials at a shallow ocean coastal site in a mound that can be capped with clean material. The objective of the cap is to isolate the contaminated sediments and to prevent or control the release of contaminated interstitial waters. According to Chaney and Demars (1991), their, are four factors which are related to geotechnology which must be considered:

Table 14.7 Major Causes of Foundation-Structure Instability of Nearshore/Offshore Structures

Loadings
  Man-made loads
    Equipment vibration
    Barge impact
  Environmental loads
    Volcanic loads
    Earthquake
    Ice
    Wave and tsunamis
    Current
    Rainstorm
    Typhoons (hurricanes)
Freezing-thawing, dry-wet
Marine borers
Marine slides and other mass movements
Pollution (air-water-soil)
Corrosion
Design deficiencies
  Errors in assumed loading
  Design methods
Construction deficiencies
  Weather effect
  Field inspection and supervision
  Construction problems in concrete
    Mixing and placing; hot and cold weather
    Overloading on form; premature form removal
  Construction problems in steel
    Hand and building difficulties for long piling
    Pile and sheet piling connection problems
    Welded or riveted tees or wyes
    Driving conditions
    Splicing sheets from different manufacturers

1. Predicting the initial radial spread of dredged sediment discharged into the coastal zone.
2. Predicting the initial configuration (height, slope, diameter) of sediment mounds. The stable configuration of a sediment mound is highly dependent upon the geotechnical properties of the dredged sediment.
3. Designing mound caps if they are required. Mounds of dredge spoil that are hazardous to the environment must be capped. The geotechnical properties of the cap determine its slope stability, compressibility, and hydraulic conductivity.
4. Predicting erodibility of the dredge spoil. For cohesive dredge spoil the shear strength of the sediment will influence the size of bottom currents that are required to initiate and continue sediment transport.

Dredging practices have been developed to achieve not only the greatest possible economic returns through maximizing production, but also minimizing environmental impacts. Dredges have become more sophisticated and highly instrumented. Methods for evaluating and selecting dredged material disposal alternatives are highly advanced. Advances have also been made in long-term planning techniques for dredging and dredged material disposal. Dredging technology is not a simple matter, as it requires knowledge from other disciplines. International cooperation such as Japan, the Netherlands, U.K. etc. is well established. At present, most research areas focus on: (1) case studies, (2) management techniques, and (3) beneficial uses of dredged material.

14.6.2 Case Studies and Management Techniques

General conclusions based on the results from case studies are (1) the dredged material disposal is the site location suitable for disposal areas. Site location problems are generally related to environmental concerns; (2) confined disposal facility construction programs require extensive coordination and cooperation among regulatory agencies, project sponsors, and local interest groups; (3) decrease in available land for confined disposal facilities requires accurate forecasting techniques to be used in the selection of dredged material disposal sites. Sites should fill long-term needs, not just short-term requirements; (4) erosion protection from wave action both inside and outside of large confined disposal facility is a major design consideration.

According to a U.S. Army Corps study (1984) and Patin (1987) the management techniques include: (1) sediment transport evaluation, (2) transport of high-density slurries, (3) long-term management techniques for disposal areas, (4) management models, (5) benthic resources assessment, and (6) the management of bottom sediments containing toxic substances. The beneficial uses of dredged material are also an important subject to be considered. Some reuse techniques will be further discussed in Section 17.10.

14.7 SOIL EROSION

14.7.1 General Discussion

Soil erosion is caused by the drag action of wind, rainfall, or wave action on the surface of bare or unprotected soil surfaces. It involves a process of both soil particle detachment and transport. The sediment that may be produced by erosion restricts the efficiency of drainage and irrigation ditches, fills reservoirs, pollutes surface water, damages agricultural land, homes, roadways, and bridges, and upsets the natural ecology and environment. Based on U.S. Department of Agriculture estimates, the total cost of sediment damage and dredging resulting from soil erosion is approximately $500 million yearly. It costs approximately $16 million annually just to remove sediment from irrigation ditches. These figures do not take into account the damage to agricultural land, homes, roadways, bridges, and recreational areas which results from the loss of soil. Estimation of erosion rates worldwide as reported by HRB (1973) indicated erosion rates by various causes:

	10^9 metric tons/year
Carried by river	
Dissolved load	2.7–3.9
Solid load	8.3–58.0
Combined both	3.8–17.5
Carried by wind from land	0.06–0.36
Carried by glacier ice estimated	0.1

14.7.2 Erosion Causes and Mechanisms

Soil erosion causes generally fall into two catagories, erosion by water and wind. The basic erosion mechanisms in both cases are the same. The phenomena of erosion occur when water- or wind-induced shearing stresses on a surface reach values great enough to cause particle removal from that surface. The soil-wind interaction and the mechanism of wind erosion will be discussed in Section 18.4. In this section erosion mechanism focuses on water erosion.

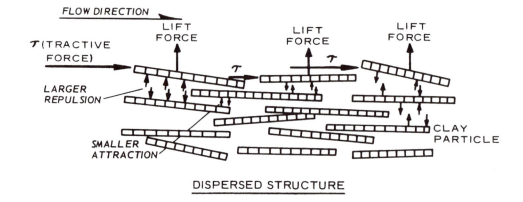

DISPERSED STRUCTURE

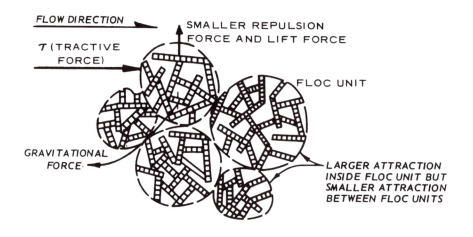

FLOCCULATED STRUCTURE

Figure 14.7 Influence of clay structures on the equilibrium of soil under flowing water. (After Liou, Y.D. [1970], Hydraulic Erodibility of Two Pure Clay Systems, Ph.D. dissertation, Department of Civil Engineering, Colorado State University, Ft. Collins.)

The mechanism of water attack on dry cohesive soil system has been discussed in Section 5.7. Influence of dispersed and flocculated clay structures on the equilibrium of soil under flowing water has been studied by Liou (1970). The forces acting on clay particles for both cases are shown in Figure 14.7. The characteristics of structural units discussed in Figure 14.7 have been presented in Section 4.3. Detailed discussions on causes and mechanism of cohesive soil erosion include:

1. Mode of erosion
2. Open network system formed by floc deposition
3. Surface force distribution
4. Reorientation or relocation of soil particles

Figure 14.8 illustrates these causes and mechanisms. Application of chemical and electrical parameters to predict erodibility was by Arulanandan et al. (1973) and Paaswell (1973). Based on laboratory studies, it was found that both types and concentrations of ions in the

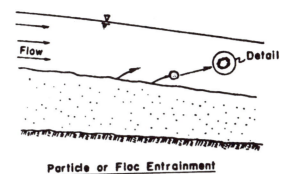

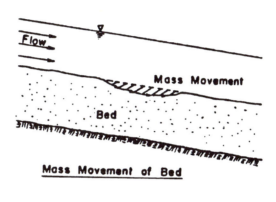

(a)

Figure 14.8 Causes and mechanisms of cohesive soil erosion. (a) Mode of erosion; (b) open network system formed by floc deposition; (c) surface force distribution; (d) reorientation or relocation of particles. (After Paaswell, R.E. [1973], Causes and Mechanisms of Cohesive Soil Erosion: The State of the Art, HRB Special Report 135, pp. 52–74.)

pore and eroding fluids and types of clay minerals have dominant effects on soil erosion. The types and amount of clay minerals and soil structure can be determined by electric dispersion measurements. The type and amount of pore and eroding fluid compositions can be determined by SAR (Section 5.6) and electrical conductivity (Section 9.8).

14.8 WATER EROSION

14.8.1 Soil Erosion by Water

(1) Equation Estimation by Water Erosion

Equation estimation of soil loss by water erosion began in 1940. In 1958 a semiempirical equation was developed and is known as the *Universal Soil Loss Equation* (USLE) by Wischmeier and associates of U.S. Agricultural Research Service. The USLE was developed

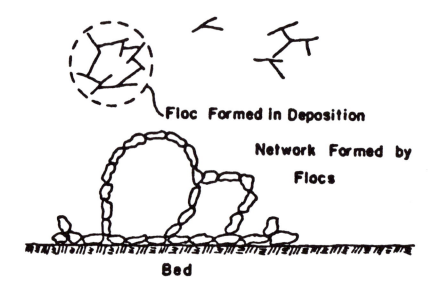

(b)

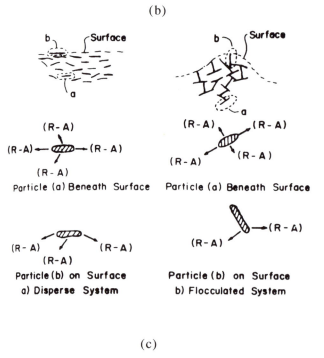

(c)

Figure 14.8 (continued)

for agricultural lands east of the Rocky Mountains. A modified equation, based on the USLE for predicting soil loss due to water erosion on highway construction sites, was proposed by Israelsen et al. (1980a, 1980b). The improvement equation considered the following parameters as:

1. An improved rainfall-erosion index
2. A method of evaluating cropping-management effects on the basis of local climate conditions

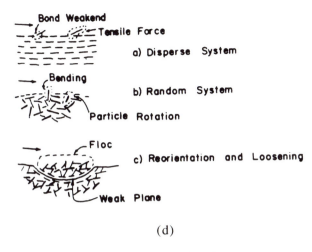

(d)

Figure 14.8 (continued)

3. Quantitative soil-erodibility factors
4. A method of accounting for effects of interrelationships of such variables as productivity level, crop sequence, and residue management

The modified USLE for estimation of soil loss by water erosion is

$$A = R \cdot K \cdot LS \cdot VM \tag{14.2}$$

where A = soil loss per unit area, tons/acre/year, R = rainfall factors, LS = topographic factor (length and steepness of slope), and VM = erosion control factor.

(2) Rainfall Factor (R)

The rain factor (R) in Equation 14.2 is the number of erosion index units in a normal year's rain. The erosion index is a measure of the erosive force of specific rainfall and is defined for a single storm as:

$$R = EI/100 \tag{14.3}$$

where E = total kinetic energy of a given storm, I = the maximum 30-min rainfall intensity.

(3) Soil Erodibility Factor (K)

The soil erodibility factor (K) in Equation 14.2 has a value ranging from 0.1 to 0.7 and is a numeric representation of the ability of the soil to resist the erosive energy of rain. Soils increase in erodibility as the value of K becomes larger.

(4) Topographic Factor (LS)

The topographic factor (LS) in Equation 14.2 is a numerical representation of the length-steepness combination to be used with the rainfall factor (R) and the soil erodibility factor (K) to estimate the erosion rate potential for a particular construction slope. An equation developed by Foster and Wischmeier (1973) and Wischmeier and Smith (1978) is presented as:

$$LS = \left(\frac{\ell}{72.6}\right)^m \left(\frac{65.41\ S^2}{S^2 + 10,000} + \frac{4.6\ S}{\sqrt{S^2 + 10,000}} + 0.065\right) \qquad (14.4)$$

where LS = topographic factor, ℓ = slope length, ft, s = slope steepness, %, m = exponent dependent upon slope steepness, 0.2 for slope < 1%, 0.3 for slopes 1% to 3%, 0.4 for slope 3.5 to 4.5%, and 0.5 for slope > 5%.

(5) Erosion Control Factor (VM)

The erosion control factor (VM) in Equation 14.2 is a single unit and accounts for erosion control measures, whether they are vegetative, mechanical, manipulation of the soil surface, chemical treatments, etc. It does not include structures such as berms and ditches. From research results reported by Israelsen et al. (1980b), the typical factor values are listed as:

Condition	Ranges of VM factor
Compacted fill	1.24–1.71
Asphalt emulsion on bare soil	0.01–0.70
Dust binder	0.29–1.05
Seedings	0.01–0.40
Mulch	0.01

As discussed in a previous section, the phenomena of soil-erosion are complex. All equations such as Equations (14.1) and (14.2) are semiempirical forms. These equations are useful as a guideline for the preliminary studies for soil loss by water erosion.

14.8.2 Factors Affecting Water Erosion

In addition to flow velocity, the water chemistry is the major controlling factor affecting the water erosion. Figure 14.9 presents correlation between soil chemistry tests on soil porewater extract and performance of earth dams. Zones 1 and 2 are classified as highly erodible clays. Zone 3 is the erosion-resistant clays, and zone 4 is the transition zone. Most samples in this zone had low dispersion when tested in the laboratory. The lower boundary of the zone is not well established by the data. Percent sodium in Figure 14.5 is computed as shown in Figure 13.10.

$$\text{Percent sodium} = \frac{Na}{Ca + Mg + Na + K}\ 100 \qquad (14.5)$$

For soils low in potassium, the percent sodium is directly related to the *sodium adsorption ratio* (SAR) and the total soluble salts as shown in Figure 14.10. The term SAR is proposed by Richards (1954) and can be computed from the following equation:

$$SAR = \frac{Na}{0.5\left(Ca + Mg\right)} \qquad (14.6)$$

Figure 14.11, relating the plasticity index of the soil to the product of the depth times the slope of the channel for the design of drainage ditches, was suggested by Smerdon (1967) as cited by Perry (1975). This chart also gives the relative erosiveness of soil as indicated.

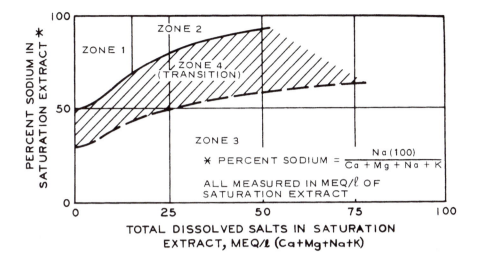

Zone 1 & 2 include nearly all of the clay samples from dams which failed by breaching in Oklahoma and Mississippi. Samples generally have high dispersion when tested in the laboratory, highly erodible clays.

Zone 1 includes all samples from 16 clay dams which were damaged by tunnel erosion from rainfall in Venezuela, Oklahoma, Mississippi, Arkansas. Tennessee and Texas.

Zone 3 includes the test results for most of the "control" samples, probable range of ordinary erosion resistant clays.

Zone 4 is the transition zone, most samples in this zone had low dispersion when tested in the laboratory. The lower boundary of the zone is not well established by the data.

Figure 14.9 Correlation between soil chemistry tests on soil porewater extract and performance of earth dams. (After Sherard [1972].)

14.9 WIND EROSION

14.9.1 Soil Erosion by Wind

(1) Estimation by Wind Erosion

Wind erosion potential may be estimated in a manner similar to that for water by the use of a soil loss equation (Equation 14.2). The wind erosion equation for estimating soil loss due to wind on highway construction sites presented herein is based on Chepil's work (1945) as:

$$E' = I' \cdot C' \cdot K' \cdot V' \cdot L' \qquad (14.7)$$

where E' = soil loss by wind, tons/acre/year, I' = soil wind erodibility factor, C' = local wind erosion climatic factor, K' = soil surface roughness factor, V' = vegetative factor, and L' = length of the unshielded distance parallel to wind in the direction of the wind fetch.

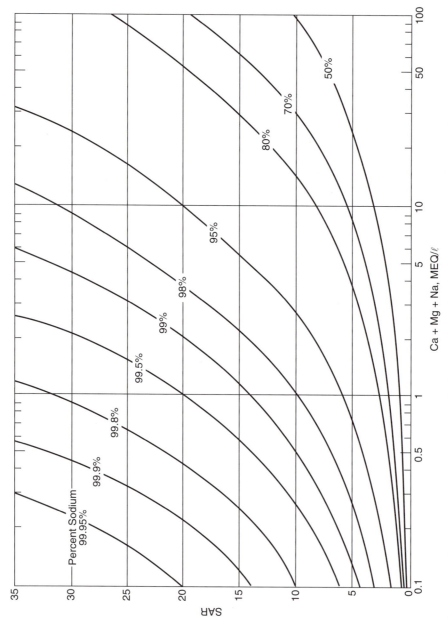

Figure 14.10 Percent sodium as a function of SAR and total soluble salts for low-potassium soil. (After Sherard [1972]; and Perry, E.B. [1975], Piping in Earth Dams Constructed of Dispersive Clay; Literature Review and Design of Laboratory Tests, Waterway Experiment Station Technical Report S-75-15, 105 p.)

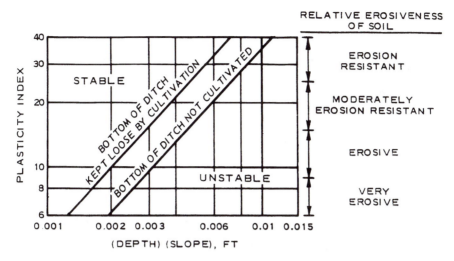

Figure 14.11 Design method for drainage ditches based on plasticity index of soil. (After Smerdon [1966]; and Perry, E.B. [1975], Piping in Earth Dams Constructed of Dispersive Clay; Literature Review and Design of Laboratory Tests, Waterway Experiment Station Technical Report S-75-15, 105 p.)

(2) Soil Wind Erodibility Factor (I′)

The procedures for determination of various parameters used in Equation (14.5) are presented as follows:

$$I' = I \cdot I_s \text{ tons/acre/year} \tag{14.8}$$

where I = soil wind erodibility index and I_s = knoll erodibility factor.

The soil wind erodibity index (I) is the potential soil loss in tons/acre/year from a wide unsheltered, isolated, bare, and smooth noncrusted soil expanse. The I_s value is determined in the *in situ* condition by dry-sieving a soil sample through a U.S. sieve #20 (sieve opening 0.84 mm). Knowing the percentage of particles larger than sieve #20 and if there is no crushing, the tons per acre can be obtained. Whenever the slope is facing the dominant wind direction so that the wind impinges against the slope, erosion is accelerated. This acceleration is known as the windward knoll effect, and the knoll erodibility factor, I_s, is used to correct the soil wind erodibility index, I, for this exposure.

(3) Local Wind Erosion Climatic Factor (C′)

C′ is the cubic of the mean wind velocity for each month divided by the square of the annual precipitation effectiveness index, PE, developed by Thornthwaite (1931) as:

$$C' = 34.483 \text{ V}^3/(PE)_2 \tag{14.9}$$

where C′ = wind erosion factor, V = mean monthly wind velocity at a height of 30 ft for all winds in excess of 12 mph, PE = PE index = Thornthwaite's precipitation effectiveness index = $115(P/T - 10)^{1.111}$, P = mean annual precipitation, and T = mean annual temperature.

(4) Other Factors

Other factors listed in the Equation (14.5) such as K', V', and L' can be determined by graphical procedures developed by Israelsen et al. (1980). Numerical examples on each case are also provided. A brief discussion of characteristics of each parameter is presented as follows.

(a) Soil Surface Roughness Factor (K')

The surface roughness factor, K', is a measure of the natural or artificial roughness of the soil surface in the form of ridges or small undulations. It can be determined by knowing the height of the individual roughness elements.

(b) Vegetative Factor (V')

The V' factor represents equivalent pounds of vegetative matter as a roughness element. The V' value is obtained by wet sieving the air-dried soil to separate the organic matter from the mineral portion. The organic is then dried and weighted. The weight in thousands of pounds per acre is entered by graphical procedure.

(c) The Unshielded Wind Fetch Distance (L')

The unshielded wind fetch distance, L', is defined as the distance parallel to the preponderant wind direction in excess of the shielded distance. In the field, the preponderant direction is laid out with a compass or transit, then the distance across the exposed area in excess of ten times the height of any barriers is recorded in feet as L'.

14.10 EROSION CONTROL IN CONSTRUCTION AREAS

Highway construction causes many problems because of sites close to rivers and streams. Although some methods such as drainage ditches, covers, terracing, contour cultivation, fences, soil stabilization, and straw, hay, or artificial turf have been used during construction for both water and wind erosion control, conventional soil stabilization is generally of a permanent nature. Since construction and maintenance are temporary, conventional stabilization methods may not be suitable. New materials and/or methods are available to stabilize the soil surface on a temporary basis in order to minimize soil erosion from both water and wind during construction or maintenance.

There are a number of structural measures that may have a place either as temporary or as permanent installations. These techniques include: (1) diversions, (2) grassed or paved waterways, (3) buried pipe outlets, (4) bench terraces or berms, (5) retaining structures, (6) soil stabilization, etc. Table 14.8 summarizes some methods for preventing or controlling soil erosion.

14.11 SUMMARY

1. Wetlands, floodplains, coastal margins, dredging, and erosions are part of sensitive ecological/geological environmental problems. Such areas are exceptionally responsive to environmental change and especially prone to irreversible ecological upset.
2. In addition to these sensitive environmental problems as stated in (1), additional topics such as saltwater intrusion, estuaries, and greenhouse effects are also discussed.

**Table 14.8 Methods for Preventing and
Controlling Soil Erosion during
Highway Construction**

Water erosion control
  Surface protection
    Vegetation
    Riprap
    Hay-ball
    Cover
    Protective films
    Protective mulches
    Paving
    Stabilization of soil
  Reducing water flow
    Hays
    Turfs
    Straws
    Filters
    Roughened surfaces
    Contour cultivation
  Interception
    Diversions
    Berms
    Drainage ditches
    Drainage pipes
    Channel changes
    Settling basins
Wind erosion control
  Tillage
  Wind strips
  Dunes
  Fences
  Soil surface stabilizers
  Vegetation

3. Zonation or micro-zonation of geo-environmental and climatological conditions is a major contributing factor relating to the stability of marine structures. It is necessary to examine these conditions before the final design and construction of a structure can begin.
4. Soil erosion is caused by the drag action of wind, rainfall, or wave action on the surface of bare or unprotected soil surfaces. Erosion causes and mechanisms for various actions are examined. Methods for estimation of water and wind erosion are presented. Erosion control technology in construction areas is summarized.

PROBLEMS

14.1 Discuss the ecology/geology sensitive sites. What are those sites?
14.2 Discuss the estuarine phenomenon. Why is it an important part of the earth's ecosystems?
14.3 Define the greenhouse effect and how this effect will affect the environmental geotechnical problems.
14.4 Why does saltwater intrusion cause groundwater pollution?
14.5 Why is most beach sand uniformly distributed?
14.6 An earth dam is to be constructed by the hydraulic-fill method. The mass unit weight of the mixtures of soil and water issuing from the discharge line of the dredge is 84.0 pcf. In its natural state, the soil is found to be only partially saturated with a dry unit weight = 102.0 pcf, and a specific gravity of solids, G_s = 2.72. How many cubic feet of pumping will be required for each cubic foot of excavation?

14.7 A soil is being excavated from a river bottom by means of a dredge. In its natural state the soil is found to have a dry unit weight of 102.0 pcf, a specific gravity of solids of 2.72, and a degree of saturation of 98%. The mass unit weight of the mixture of soil and water issuing from the discharge line of the dredge is 84.0 pcf. If this line is discharging at the rate of 100 gal/min, how long will it take the dredge to excavate 50,000 yd^3 from the river bottom?

14.8 Discuss the beneficial uses of dredged materials and management of bottom sediments containing toxic substances.

14.9 Discuss the mechanism of soil erosion caused by contaminated water.

14.10 Discuss the mechanism of oil-water interface.

14.11 A canal and a river run parallel an average of 200 m apart. A stratum of sand sandwiched between strata of impervious clay intersects both the river and the canal below their water levels. Elevation of water in the canal is 620 m and in the river is 600 m. The sand stratum is 5 m thick and has a permeability (*in situ*) of 0.120 m/min. Compute the seepage loss from the canal in cubic meters per second per kilometer.

CHAPTER 15

Ground Surface Subsidences

15.1 INTRODUCTION

Ground surface subsidence is an old problem. It has existed but went unnoticed until recent years due to the population growth and the rising living standards, as discussed in Section 1.2, as more marginal land is being developed. Subsidence covers large areas subjected to underground mining, pumping of groundwater, oil, and natural gas, as well as floods, underground fire, tree roots, construction operations, and pollution intrusion. The term *subsidence* was first used in geology in 1853 by Charles Lyell. The American Geological Institute (Gary et al., 1972) describes subsidence as: (1) a local mass movement that involves principally the gradual downward settling or sinking of the solid earth's surface with little or no horizontal motion and does not occur along a free surface; and (2) a sinking of a large part of the earth's crust due to tectonic movements.

Encyclopaedia Britannica (1974) defines the subsidence as: 'sinking of the earth's surface in response to geological or man-induced causes.' Benson (1978) proposed a term named *'localized subsidence.'* The localized subsidence will be directed to problems associated with cavities or piping in soluble rocks and unconsolidated materials. In the early 1970s most studies concentrated on geomechanical behavior of land subsidence and focused on hydrology, geotechnology, and geologists. However, more recently, this problem includes geochemistry, physicochemistry, and geothermal as well as bacterial effects on soil-water systems. In addition, subsidences in hazardous/toxic areas are also discussed. A review of subsidence including mechanisms, analysis procedures, and typical case histories is given by Scott (1978). A review of the methods used in investigation of subsidence is given by Saxena (1978).

15.2 CHARACTERISTICS AND CLASSIFICATIONS OF LAND SUBSIDENCE

15.2.1 Characteristics of Land Subsidence

When a load is applied to a soil mass deformation may result from: (1) immediate elastic and inelastic deformation of the soil structure; (2) porewater drained from the soil mass; (3) continuous time-dependent or viscous flow under shear stress resulting in reorientation of the soil particles; and (4) a combination of all the above which in most cases occurs simultaneously. However, it will depend upon soil properties, drainage condition, stress history, and environmental factors. Case (2) is generally referred to as consolidation (Section 11.5),

Table 15.1 Geotechnical Subsidence Classification

Based on Origin

Man-made	Natural
Fluid removal (water, gas, oil)	Earthquakes
Mining	Floods
Deep excavations, trench, quarry, borrow pit	Roots, flora, and fauna
	Surface and subsurface erosion
Construction operations (pile driving, compaction)	Limestone sinkholes
	Tectonic activity
Vibration (blasting, traffic)	Permafrost
Geothermal	

Based on the Load-Deformation Mechanism

Unbalanced stresses	External loading	Fluctuation of pore fluid content or temperature
Mining	Building loading	Fluid removal
Quarry and borrow pits	Earthquakes	Geothermal
Deep excavation, trench	Vibration	Shrinkage and swelling
Tunneling	Blasting	Wet-dry
Limestone sinkholes	Flooding	Freezing-thawing
Tectonic movements		

Based on Appearance of Surface Movement

Slow subsidence occurs	Rapid subsidence occurs
Fluid removal	Abandoned mining
Floods	Deep excavations, trench, quarry, and borrow pits
Tectonic activity	Limestone sinkholes
Geothermal	Tunnels

Based on the Soil Particle Moving Velocity[a]

Subsidence type	Velocity (m/year)
Extremely rapid	0.6–1.0
Very rapid	0.4–0.6
Rapid	0.2–0.4
Moderately	0.08–0.2
Slow	0.006–0.08
Very slow	0.0004–0.006

[a] The above classification is just to give a general idea. If there is a vertical subsidence, it may be referred to as settlement analysis. If particle velocity is larger than 0.06 m/year on a steep slope, or a back-fill slope angle is larger than the internal friction angle, f, it may cause a landslide; if the particle velocity is larger than 3 m/sec it will be classified as an earthquake. If vertical subsidence is not uniform, it may give a differential settlement problem.
After Fang, H.Y. and Cleary, T. [1976], *Analysis and Design of Building Foundations*, Envo Publishing, Bethlehem, PA, pp. 467–518.

case (3) as creep (Section 11.7), and case (4) as subsidence. In the broad sense of the term, subsidence may be defined as the deformation or settlement of a soil mass at any direction caused by various external loading, internal stress, and environmentally unbalanced factors. In a simple, nontechnical term it may be called ground movement. There are several classification systems for ground subsidence. Some of these classifications are presented as follows.

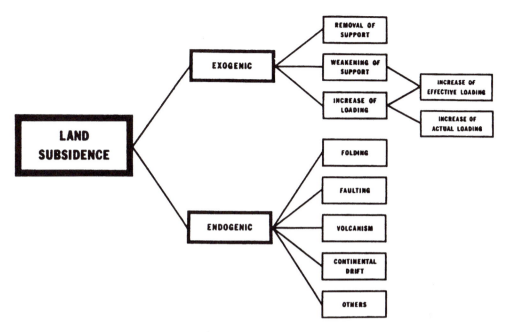

Figure 15.1 Genetic classification of land subsidence. (After Prokopovich, N.P. [1978], Evaluation and Prediction of Subsidence, ASCE, New York, pp. 389–399.)

15.2.2 Classification of Ground Subsidence

(1) Geotechnical Engineering Classification

From the geotechnical engineering point of view, subsidence can be classified based on its origin, load-deformation mechanism, appearance of surface movement, and soil particle-moving velocity. In addition, some subsidences are natural and some are man-made. A geotechnical subsidence classification is proposed as shown in Table 15.1.

(2) Genetic Classification

Prokopovich (1978) proposed genetic approaches for land subsidence as summarized in Figure 15.1. In a broad genetic sense, subsidence can be classified into two major groups — endogenic and exogenic. A brief description of each group is presented as follows.

(a) Endogenic Subsidence

Endogenic subsidence is related to processes that originate essentially within the planet. Such subsidence could be further subdivided based on the causal processes into subsidence related to folding, faulting, volcanism, continental drift, etc.

(b) Exogenic Subsidence

Exogenic subsidence is related to processes that originate near the earth's surface including human activity. The exogenic processes are supported by extraterrestrial energy, particularly solar. Exogenic subsidence is basically a surface expression of compaction of deposits at depth due to natural or man-induced processes. It could be subdivided into:

1. Subsidence related to the removal of solid support such as subsidence due to mining, karst, etc.
2. Subsidence related to an increase of loading on compacting strata because of withdrawal of water, oil, gas, etc. Several typical examples of exo-and endogenic processes are provided.

15.3 MECHANISMS AND PHENOMENA OF GROUND SUBSIDENCE

15.3.1 Creep and Progressive Failures

Some movement in the form of creep takes place in almost all ground subsidence areas. The rate of these movements varies during the year, which are often confined to the surface layer. The rate increases as failure approaches and the actual time of a subsidence can frequently be predicted by monitoring the ground movements.

Seasonal creep or surface subsidence is caused by temperature-moisture variations and vegetation life cycles and insects or worms traveling inside the soil mass. The depth of the surface layer which is affected by seasonal soil particle movement is often equal to or less than the depth of seasonal temperature and moisture variations. In temperate climates seasonal creep is caused primarily by freezing and thawing. High porewater pressures develop in the spring, decreasing the shear strength and causing the soil to move. The subsidence in the early stage may be very small but can lead to failure if the rate increases. Continuous subsidence is caused by gravitational forces and building loads occur below the surface zone of seasonal variations. In shales (Section 13.2) this appears to occur in relatively thick zones.

15.3.2 Phenomena Associated with Subsidence

The subsidence in its early stage may be very small but can lead to failure if the rate increases. There exists a relationship between time to rupture and strain rate. It has been found that the time to rupture is inversely proportional to the strain rate and is independent of the soil type. Phenomena associated with subsidence include surface cracks, progressive erosion, ground settlement, etc. Some of these phenomena are similar to the landslides as discussed in Sections 16.2 and 16.3 and illustrated in Figure 16.2.

15.4 GROUND SUBSIDENCE DUE TO NATURAL CAUSES

15.4.1 Tectonic Movement and Earthquake

Tectonic movements include tectonic creep and tectonic earthquake. Tectonic creep is an apparent continuous displacement along a fault at a slow but varying rate, usually not accompanied by observable earthquakes. Tectonic earthquakes are associated with faulting or other structural processes. Tectonic movements in the earth's crust are a form of subsidence which requires special precaution. Geologically active faults must constantly be monitored to determine the rate of movement in order to compensate for any future displacement in the design of a structure or foundation. One of the most intensively studied faults is the San Andreas Fault of California, which moves at a rate of about 5 cm/year (Legget, 1973). During the 1906 San Francisco earthquake, the ground displacement along the San Andreas Fault was 6 m as reported by Seed (1991).

Subsidence can be initiated by earthquakes and vibrations. If the rate of cyclic loading is greater than the time required for porewater dissipation, a continual buildup of residual porewater pressure after each cycle will result. Subsidence in the form of liquefaction will then occur once this accumulated porewater becomes greater than the total overburden

Figure 15.2 Photo shows the ground floor subgrade soil subsidence due to 1976 Guatemala earthquake.

pressure on the soil. This forces a breakdown of structure in which the soil particles lose contact with one another and the soil behaves essentially as a heavy liquid. There is a definite loss of shear strength which will cause large settlement and bearing capacity problems and which can also initiate landslide failure. Figure 15.2 shows subgrade soil subsidence which separated a concrete floor caused by earthquake shock in downtown Guatemala City in the 1976 Guatemala earthquake.

Subsidence is caused naturally and/or is man-made. In most cases, these are combined together. These causes are geothermal, freezing-thawing, and subsurface piping. The withdrawal of steam for geothermal power has been reported as a cause of subsidence (Poland and Davis, 1969). Effects of temperature on shear strength, volume change, and settlement of surrounding soils are discussed (Section 8.7). Freezing-thawing cycles will cause reduction of soil strength. The subsurface piping or subsurface erosion is the horizontal flow of groundwater into a subsurface crack or cavity which can cause a progressive erosion of fine grained soil particles or piping away from the fissure under seepage flow. In cohesive soils, this erosion can create an extensive tunneling effect in which the cohesive strata support the remainder of the soil mass across the crack opening. Closure of this cavity due to stress increase or progressive strength loss will result in a rapid ground surface subsidence.

Differential subsidence on reclaimed marshland peat in metropolitan New Orleans has been reported by Traughber et al. (1978). Salt motion following nuclear waste disposal has also been examined by Dawson and Tillerson (1978). Thermomechanical subsidence associated with underground coal gasification has been reported by Shoemaker et al. (1978). Many other causes such as tree roots, insects, and animals are further discussed as follows.

15.4.2 Flora and Fauna of the Soil

Soil is generally subjected to the corrosive power of carbon dioxide of respiration and fermentation, to acids produced during the decomposition of successive vegetation, and to enzymes secreted by micro-organisms (Section 5.9). Soil as we have it today is the cumulative result of ages of such attack, combined with the physical effects of wet-dry and freezing-

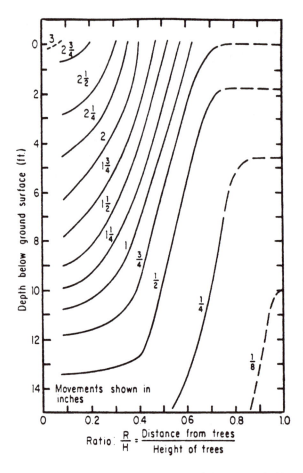

Figure 15.3 Ground surface movement caused by tree roots. (After Bozozuk, M. and Burn, K.N. [1960], *Geotechnique*, v. 10, no. 1, pp. 19–32.)

thawing processes. The organisms vary in size, and they have changed the surface layer of the soil from an aggregate of mineral particles to a mass teeming with organisms and honeycombed by visible channels made by roots of plants or burrows of animals and insects (Thom and Smith, 1938; Stevenson, 1982). Between the soil particles there are also invisible channels and spaces of various sizes, intricate networks of microscopic channels whose surfaces are smeared with collodial slime, and a variable water solution carrying mineral and organic materials.

The process of decomposition is complicated because soil contains various kinds of bacteria and other foreign matter such as azotobacter, actinomycetes, fungi, mycorrhiza, protozoa, myxomycetes, worms, and burrowing animals. These factors are causes of subsidence because they can initiate subsurface cracks which then propagate due to root growth and, most commonly, the infiltration of water. Heavy rains or melting snow can penetrate subsurface fissures causing an increase of pore water pressure, the result being a reduction of shear strength and further crack growth to relieve the excess pore pressures.

Ground surface movement caused by tree roots is a common phenomena. It can be observed around the sidewalk, patio, and playground. Bozozuk and Burn (1960) measured ground surface movement caused by tree roots as shown in Figure 15.3, which indicates that ground surface movement relates with the distance from tree and the height of tree. Tree roots causing ground movement also produce ground surface heave as reported by Samuels and Cheney (1975) and Gray and Leiser (1982). A case study describes a cottage at Windsor founded on London clay. The observations show that the buildings have been subjected to heave of a long-term nature due to gradual swelling of the clay following the removal of trees during site clearance. It is indicated that the buildings have been affected by ground movement and are discussed in terms of angular strain.

I5.4.3 Flood or Rainstorm

Subsidence due to flood is based on the effects of water content changes on the soil-water system and is of concern in both partially saturated and saturated soils. If such a system is subject to flooding, the added weight of flood water will increase the total overburden pressure and cause both consolidation and soil particle reorientation. In partially saturated soils where pockets of air previously prevented a more dense packing, there will be additional subsidence if these air pockets are eliminated during drainage. Similar to dewatering effects, the rate of subsidence is directly related to the rate of drainage which takes place as the system returns to a static equilibrium condition. If flood water is polluted, the effects of pore fluids on soil behavior will significantly change as discussed in Chapters 5 and 6. Consequently, the ground surface subsidence will also significantly increase.

Subsidence can also occur in cases where volume changes do not result from external loads. Shrinkage and swelling are caused by changes in water content and in the internal stresses affected by the water, such as wet-dry, freezing-thawing, or the seasonal effect of groundwater fluctuation, dewatering during construction. Shrinkage results from capillary tension which forms when a saturated soil dries. Swelling can occur in soils when the moisture is allowed to increase and is caused by such factors as the osmotic attraction of the clay minerals for water, and the electrical repulsion of clay particles and their positively charged diffuse-double layers from one another. In cases where overburden has been removed, swelling can occur due to elastic rebound of the soil grains and also through expansion of air trapped in the soil voids. Ground subsidence due to shrinkage can be computed from basic soil constants such as shrinkage limit (SL), natural moisture content, and specific gravity of soil solids. A typical example is presented as follows.

Example 15.1

For a given soil, the natural moisture content is 50%. The laboratory tests indicate that the shrinkage limit (SL) of soil is 35% and specific gravity of solids is 2.70. Determine the amount of subsidence (vertical settlement) or volume change as natural water content is reduced to the shrinkage limit.

Solution

Based on solid-water-air relationship, $S = \omega G_s/e$, where S = degree of saturation (assumed 100% in this problem), ω = natural moisture content = 50%, G_s = 2.70, and e = void ratio in natural condition.

At natural condition: $e = \omega G_s = (0.50)(2.70) = 1.35$
At shrinkage condition: $e = (0.35)(2.70) = 0.945$

The changes of void ratio, e, from natural condition to shrinkage condition is

$$e = 1.35 - 0.945 = 0.405$$

From basic geotechnics, volume change relating to the change of void ratio is

$$\Delta V = V\, e/1 + e = (1)(0.405)/(1 + 1.35)$$

$$= 0.172 \text{ m}^3/\text{m}^3$$

The change of volume per unit volume of soil is 0.172 m^3/m^3, which is the maximum to be expected as a result of shrinkage for element of the given soil initially at a moisture content of 50%.

15.5 DEWATERING

15.5.1 Phenomena and Mechanism of Dewatering

Subsidence due to dewatering has been a common problem in different parts of the world for the last several decades. In areas where the amount of fresh surface water available is limited, or wherever increasing industrial and municipal needs must be met, the only solution left lies in the pumping of water-bearing aquifers. In places where the water table has been lowered in highly compressible soil sediments, shrinkage cracks can develop at the ground surface; due to evaporation through these initial cracks, they are able to extend deeper into the ground. For example, 15 m of subsidence observed in Mexico city has been reported by Zeevaert (1972). Figure 15.4 presents ground surface subsidence profile of the city of Houston due to dewatering as reported by Marshall (1973). In examining Figure 15.4, subsidence in the Houston region is continuing at an ever-increasing rate and has reached 8 ft in the Pasadena-Deer Park area. The location and the amount of subsidence are directly related to declines in groundwater level resulting from withdrawal of water from wells. By the year 2000, total subsidence is expected to reach 19 ft in the Pasadena-Deer Park area.

Shrinkage effects in terms of negative skin friction caused by dewatering are also a problem when considering deep foundations. In the case of friction pile, a shrinkage of the penetrated layer away from the pile reduces the pile-soil adhesion and in turn causes a reduction of pile capacity and possible differential settlements. For end-bearing piles, the confining layer does provide lateral support to the extent that pile buckling and fatigue cracks are prevented. If a shrinkage of this layer away from the pile is to occur, the lack of lateral support may induce the pile to buckle with corresponding damage to the superstructure. Ground surface deformation phenomena during dewatering include:

1. Vertical displacement: Absolute vertical subsidence seldom causes direct damage if all points of a structure move by the same amount in the same direction, since no stresses will be transferred to the structure (as opposed to differential settlements where large stresses are induced into the building frame). Problems arise where head clearance in bridges is reduced, where stairways settle below floor levels, where buildings settle below street level, etc.
2. Horizontal displacement: Similar to vertical displacements, uniform horizontal movement of a structure causes practically no damage. Differential displacement in the form of horizontal curvature is more serious in that a distortion of horizontal alignments can occur.
3. Slopes: The slope of the subsidence trough can change the gradients of roads, railroad tracks, and gas and water mains. It induces tilting which can be serious for tall buildings and where it is necessary to keep machines vertical or horizontal.

15.5.2 Stress Distribution During Dewatering Process

(1) Unconfined Aquifer

An unconfined aquifer exists when the water table serves as the upper boundary of soil saturation in the absence of an overlying impermeable strata. Hence, a piezometric tube penetrating an unconfined aquifer will show a water level corresponding to the water table.

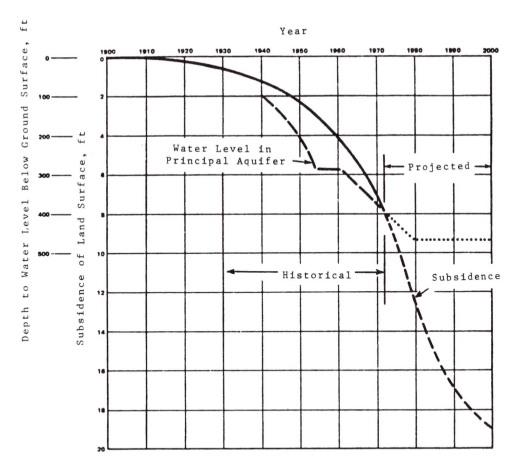

Figure 15.4 Past and estimated future subsidence in Pasadena-Deer Park area, the city of Houston. (After Marshall, A.F. [1973], The Slide Rule, Houston Engineering & Scientific Society, v. 33, no. 2, 6p.)

During or after pumping, the phreatic surface will have dropped with the resulting change being termed the drawdown. The increase of effective stress due to lowering of the water table in an unconfined aquifer is explained with the conventional effective stress concept.

(2) Confined Aquifer

In a confined aquifer, a different situation exists. Here the groundwater is kept under pressure by relatively impermeable overlying and underlying strata, such that water will rise above the aquifer top in a piezometric tube. This rise measured from the piezometer base is termed the piezometric or pressure head and represents the fluid pressure existing at the base of the tube. Once an outlet is provided, water is released under pressure both by compression of the aquifer and adjacent clay beds and by expansion of the confined water. This compression of aquifer and clay beds may be permanent (inelastic) as well as elastic; hence full rebound of compression, through recirculation of water under pressure, may not be possible due to this inelasticity.

Lowering of the piezometric head in an artesian aquifer causes a change in effective stress. It is assumed that the upper confined layer has no change in water table height. Lowering of the artesian head does not change the total pressure. In actuality, some reduction in total head does occur due to expansion of water during removal; this decrease is assumed negligible and is not reflected in the stress distribution. When water is removed from and pressure is decreased in an artesian aquifer, stored water is derived from:

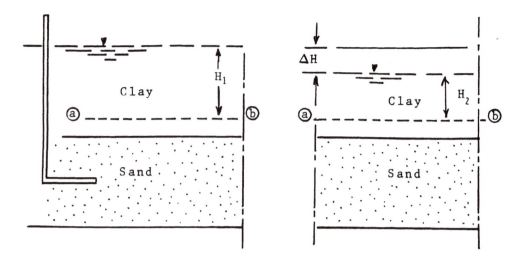

Figure 15.5 Diagrams illustrating cause of subsidence (settlement) of clay layer due to pumping from underlying water bearing sand (left) Piezometric level at ground surface; (right) piezometric level lowered due to pumping.

1. Expansion of confined water
2. Compression of the aquifer
3. Compression of adjacent clay beds

Compression of the aquifer is usually elastic, as in the case of sand or sandstone, while plastic deformation is characteristic of clay beds or lenses which are adjacent to the aquifer.

(3) Effect of Pumping on Clay Strata

If the subsoil contains layers of soft clay, silt, or peat, lowering of the groundwater table may cause large settlement. The physical causes of this phenomenon are illustrated in Figure 15.5, which represents a section through a bed of saturated clay overlying a pervious sand layer. The following explanations are given by Terzaghi and Peck (1967). In Figure 15.5(a) the piezometric level is assumed to be at the ground surface; in Figure 15.5(b) it has been lowered through the distance ΔH by pumping from the layer of sand. Before pumping, the effective pressure on section **ab** is

$$p_{ab} = \gamma' H \tag{15.1}$$

where γ' is the submerged unit weight of the soil. During and after pumping the effective pressure gradually increases to:

$$p_{ab} = \gamma \Delta H + \gamma' H_2 \tag{15.2}$$

where γ is the unit weight of the saturated soil. The change in effective pressure due to the lowering of the piezometric level is

$$\gamma \Delta H + \gamma' H_2 - \gamma' H_1 = \Delta H (\gamma - \gamma') = \gamma_\omega \Delta H \tag{15.3}$$

In examining Equation (15.3), indications are that lowering the water table by a distance ΔH ultimately increases the effective pressure on a horizontal section through the clay layer by an amount equal to the weight of a column of water ΔH in height.

(4) Storage Coefficient

Lohman (1961) derived an equation for determining the amount of elastic compression of artesian aquifers based on decline in artesian pressure and hydrologic properties of the aquifer. This equation is fundamentally based on the *coefficient of storage*, defined as the volume of water an aquifer releases per unit surface area per unit drop of piezometric head normal to that surface. Further modifications on the coefficient of storage are made by Poland (1961). When applied to Lohman's equation, it is assumed that the portion of water derived from storage in the clay is released instantaneously. This is not the case due to the low permeability of clays, in general, and the resulting time lag between pressure drop and the appearance of water released from storage. Realizing this phenomenon, Poland was able to combine short-term field pump test data from Los Banos-Kettleman City, CA in 1956 to determine the contribution of each source of stored water to the total storage coefficient.

From comments based on a field pump test, Poland (1961) concluded that for the area tested the stored water released by plastic compression of clay beds is 50 times greater than the water released by both the elastic-expansion of the water and the elastic-compression of the aquifer. In other words, the coefficient of storage found using only an elastic short-term analysis which considers only compression of the aquifer is negligible in comparison to the long-term storage coefficient (15 to 25 years in this case) which considers compression of the clay beds. On this basis then, it is seen that any realistic calculation of subsidence must consider the compressibility of clay layers over a period of time along with its response to any change in the fluid field operating within it.

15.5.3 Ground Subsidence Prediction

(1) General Discussion

Control methods lie mainly in the prediction of the subsidence trough under various pumping rates with future dewatering then being limited by a rate which would be unacceptable for the given surrounding condition. The first consideration involved in this prediction is the geologic setting under which groundwater removal occurs and the resulting pressure changes which come about from this removal to cause the subsidence.

Groundwater occurs both under unconfined and confined conditions as discussed in a previous section, and hence it is of interest to analyze the stresses involved in lowering the water table for an unconfined aquifer and pressure head in a confined aquifer. There are numerous methods for predicting the subsidence due to dewatering. However, in this section, only two basic methods, the *Terzaghi* (Terzaghi and Peck, 1967) and *Zeevaert* (1972) methods, are presented in detail. Also, brief comparison of mechanisms between these two methods are also made.

(2) Mechanisms of Subsidence Due to Dewatering

(a) Subsidence Due to Primary Consolidation

Terzaghi's theory of consolidation explains the hydrodynamic process whereby excess pore pressure created by an increase of load is transferred to the actual soil structure resulting in a larger effective stress. This added effective stress causes an elastic-plastic deformation and volume change of the soil structure resulting in a ground surface subsidence (settlement).

In materials of low permeability, the transfer of load to the soil structure occurs gradually and is best represented in the form of time vs. consolidation curves. Theoretically, once all the excess pore pressure has been relieved, 100% consolidation has occurred and no further settlement should be observed. The time vs. consolidation curve should then become horizontal; however, instead of approaching this horizontal asymptote, settlement curves for most clays continue to slope showing further settlement. This progressive deformation is secondary consolidation and is due to the gradual adjustment of the soil structure to stress combined with the resistance offered by the viscosity of absorbed water to a slippage between soil particles. This occurrence cannot be explained by means of mechanical energy concepts (such as the piston-spring analogy) as discussed in Section 11.5, which is used to describe primary consolidation by the Terzaghi one-dimensional consolidation theory.

(b) Subsidence Due to Secondary Consolidation

Any clays which show appreciable amounts of secondary consolidation (such as lacustrine and marine silty clays and clay silts) must include an allowance for secondary consolidation if accurate settlement predictions are to be made. Further, because this viscous flow takes place at very low hydraulic gradients in the type of saturated soils mentioned, it cannot be omitted from an analysis of subsidence due to groundwater removal. Two prediction equations based on consolidation theory are presented:

1. Terzaghi method: Based on primary consolidation alone, this method can be applied to areas where this accounts for most of the subsidence.
2. Zeevaert method: This method is an extension of the Terzaghi method and incorporates added subsidence due to secondary consolidation.

15.5.4 Terzaghi and Zeevaert Methods

(1) Terzaghi Method (Terzaghi and Peck, 1967)

(a) Assumptions

Solution by Terzaghi's method gives the land subsidence as a function of drop in groundwater table for various overburden pressures and soil properties. The assumptions are similar to the conventional one-dimensional consolidation theory as:

1. Any elastic subsidence due to expansion of water and compression of the aquifer is negligible.
2. The soil mass is saturated, homogeneous, and isotropic; hence the coefficient of permeability (k) and the coefficient of volume compressiblity (m_v) are the same at every point in the layer and for every stage of consolidation.
3. One-dimensional consolidation theory is considered; hence excess water drains vertically and deformation of the soil is vertical.
4. The change in void ratio or volumetric strain approaches a finite value as the process of primary consolidation ends.

(b) Formulations

Terzaghi's solution gives the settlement or land subsidence as:

$$\Delta H = \frac{c_c}{1 + e_0} H \log \frac{p_1 + \Delta p'}{p_1} \tag{15.4}$$

where ΔH = ground subsidence, c_c = compression index, e_0 = initial void ratio, H = layer thickness, p_1 = overburden pressure, and $\Delta p'$ = change in porewater pressure (Equation 15.3).

The assumption is made that the change in porewater pressure is given in Equation (15.3) as:

$$p' = \gamma_\omega h \tag{15.5}$$

where h = drop in groundwater table and γ_ω = unit weight of water.

Noting that c_c and e_0 in Equation (15.4) are soil properties, we define the soil constant, C_1, as:

$$c_1 = \frac{c_c}{1 + e_0} \tag{15.6}$$

Equation (15.4) can now be written as:

$$\frac{\Delta H}{H} = c_1 \ \log \frac{p_1 + \gamma_\omega h}{p_1} \tag{15.7}$$

Equation (15.7) is solved for various values of (h) and various values of (C_1) and (p_1). The land subsidence can then be estimated for a given soil condition, due to a drop in the groundwater table, h.

(2) Zeevaert Method (1972)

In practice, fine sediments of medium to high and very high compressibility show viscous behavior as well as elastic-plastic behavior that should be considered in the calculation of ground surface subsidence. A complete discussion of the theory involved in elastic-plastic-viscous behavior is given by Zeevaert (1972, 1983) with some additional explanation given by Fang and Cleary (1976).

Comparison between observed and computed ground surface subsidence caused by dewatering is presented in Figures 15.6 and 15.7. The computed data are obtained from the Terzaghi equation. The observed data were directly measured.

15.6 MINING SUBSIDENCE

15.6.1 General Discussion

Mining activity took place in largely agricultural areas, far from centers of population, and as a result, surface subsidence was not of major consequence. However, due to the scarcity of land resulting from increased urban sprawl, these areas now are being considered as potential building sites. Also, the fact that large amounts of coal reserves remain under urban regions coupled with an increased demand for coal as energy and the economic necessity for maximum extraction, consideration of subsidence effects must also extend to damage of existing structures due to active mining.

According to British National Coal Board (BNCB, 1966 and 1975) and Brauner (1973), the ground surface movement due to mining can be categorized resulting from active or abandoned mines as one of two types. Russell et al. (1978) proposed a mining subsidence

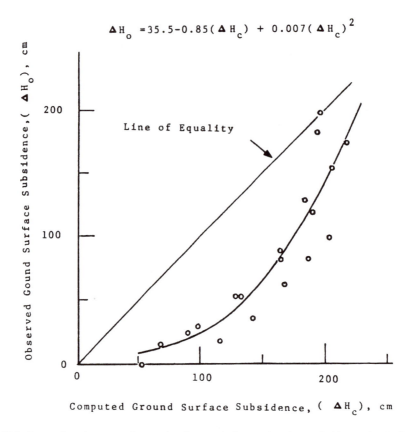

$$\Delta H_o = 35.5 - 0.85(\Delta H_c) + 0.007(\Delta H_c)^2$$

Figure 15.6 Comparison between observed and computed ground surface subsidence due to dewatering process at Taipei Basin, Taiwan. (After Wu, C.M. [1976], Proc. 2nd Int. Symp. on Land Subsidence, Anaheim, CA, pp. 389–398.)

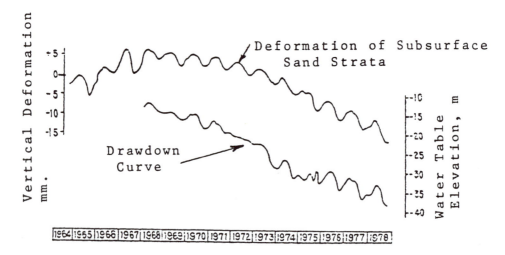

Figure 15.7 Effect of pumping on the deformation characteristics on the subsurface sand strata in Shanghai soft clay region. (After Su, H.Y. [1979], *Chin. J. Geotech. Eng.*, v. 1, no. 1, pp. 24–35.)

classification system with morphological characteristics of each system as shown in Table 15.2. Mining subsidence is grouped into three types as: (1) pothole subsidence, (2) linear subsidence, and (3) regional subsidence. Subsidence patterns, and surficial and geological characteristics of each group are also described.

Table 15.2 Mining Subsidence Classification and Morphological Characteristics

Type	Pattern	Surficial characteristics	Width/ratio	Geological character
Pothole subsidence	Single features	Single circular or rectangular depressions		Thick overburden
	Alignments	Linear series of discrete circular or rectangular depressions	Ranges from 1:1 for new to 5:1 for stabilized potholes	Various bedrock attitudes (0°–90°)
	Lattice	Network of closely spaced potholes		
Linear subsidence	En echelon to curvilinear	"Torn" appearance of bed rock cracks	Much deeper than wide; range of 1:2 to 1:10+	Thin overburden; horizontal to gently dipping bedrock
	Joint-controlled	Relatively smooth bedrock breakage along joint planes		
	Interconnected	Formed by the connection of several potholes in a line, usually along an outcrop		Moderate to steep dips
Regional subsidence	Irregular to circular or rectangular-shaped moist areas or dry depressions	Large (greater than 1 acre) Swampy or water-filled areas differentiated from natural features by age; criteria include tree stump remnants, chaotic vegetation assemblages, and evidence from historical photography	Much wider than deep; large areal subsidence; vertical subsidence ranges from 1 to 10 ft	Thick overburden, common in alluvial valleys; horizontal to gently dipping strata

After Russell et al. (1978).

15.6.2 Ground Surface Movement

(1) Discontinuous Deformation

Underground mining can cause major fractures of the ground surface and the main forms are grouped as: (1) open cracks in the ground surface which can range from inches to feet wide, (2) steps in the ground surface where the ground slides on inclined planes which can be inches to feet deep, (3) cave-in pits which are a local collapse of the ground surface which can be dozens of feet wide and deep. The magnitude of discontinuous deformations cannot be predicted, but conditions for their preferred occurrence are as follows:

1. When mining ore bodies with large vertical dimensions (depth)
2. Due to sudden changes in the type of rock, faults, or minor cracks which favor steplike and cracklike fractures
3. Under conditions where a discontinuity has already formed; further undermining will cause additional discontinuities
4. In sections along the continuous subsidence trough where large tensile strains exist

(2) Continuous Deformation

Continuous deformations are characterized by the formation of a subsidence trough similar to the ground movement associated with fluid removal. There are two types of displacements.

(a) Absolute Displacements

1. Vertical displacement: Vertical displacement can cause foundations to sink below shallow groundwater levels which can reduce bridge head clearance etc.
2. Horizontal displacement
3. Strain: Strain is defined as the derivative of the horizontal displacement and is either tensile (positive) or compressive (negative). It can induce tensile and compressive fractures and buckling of structural elements. The NCB of Great Britain correlates the mining damage to the horizontal ground strain multiplied by the length of the structure with the resulting classification given in Table 15.3.

(b) Differential Movements

1. Slope is the derivative of the vertical displacement with respect to the horizontal variable. It induces tilting of buildings, change in the gradients of roads, railways, utility lines, etc.
2. Curvature is taken as the derviative of the slope. It can cause distortion of framed structures which is defined as angular change in right angles. Also, it can induce bending stresses in the building frame.

15.6.3 Prediction of Possible Surface Damage

(1) General Discussion

Structures located over areas of active mining can be subject to extreme damage due to subsidence effects if proper countermeasures are not taken. Research on the extent and distribution of surface movements resulting from mining and necessary structural precautions to minimize subsidence damage has taken place mainly in Europe. The approach used for control or minimization lies in the prediction of the subsidence trough under complete or partial extraction which can be coupled with underground precautions, such that temporary and final deformations do not exceed allowable values as shown in Table 15.3. Existing approaches to predicting possible surface damage caused by mine subsidence can be divided into two broad categories: empirical and phenomenological methods. The empirical method includes profile function method and influence function method.

(2) Empirical Method

Empirical methods are based largely on experience, degree of underground extraction, and observation of ground movements. Brauner (1973) reduces the fundamentals of empirical precalculation methods into two principles.

(a) Principle of Superposition

This assumes that displacements due to any extraction area are not affected by the existence of displacements due to another extraction area. Hence, the total subsidence which results from more than one area being mined is simply the sum of the displacements that would occur if each area was mined alone. This principle includes the assumption that the maximum possible subsidence (S_{max}), also termed the full subsidence, is dependent on the seam thickness and can be determined from the following linear expression:

$$S_{max} = a \cdot m \tag{15.8}$$

where S_{max} = maximum possible subsidence, a = subsidence factor (Table 15.4), this factor is dependent on the kind of packing and mining method, and m = seam thickness.

Table 15.3 Damage in Relation to Intensity of Ground Strain

Class of damage	Change of length of structure	Description of typical damage
Very slight or negligible	Up to 0.1 ft (3 cm)	Slight cracks showing in walls and ceilings inside buildings, but not visible on outside
Slight	0.1 ft (3 cm) to 0.2 ft (6 cm)	Slight cracks showing inside the building; doors and windows will not close
Appreciable	0.2 ft (6 cm) to 0.4 ft (12 cm)	Slight cracks showing both outside and inside building; doors and windows will not close; drains, sewers, and gas pipes fracture
Severe	0.4 ft (12 cm) to 0.6 ft (18 cm)	Drains, sewers, and gas pipes fracture; open fractures through walls of building; window and door frames distorted, floors noticeably sloping, walls leaning or bulging noticeably; some loss of bearing of beams on walls; porticoes and floors buckle
Very severe	More than 0.6 ft (18 cm)	Worse than above and requiring partial or complete rebuilding; roof and floor beams lose bearing and walls lean badly and need external support; Windows broken and distorted; severe slopes, buckling, and bulging of roofs and walls occur

After BNCB (1975), *Subsidence Engineer's Handbook,* British National Coal Board, London.

(b) Principle of Equivalence

This second principle states that the maximum subsidence (S_{max}) over a subcritical area depends on the size of the mined area in relation to its depth. It means all extraction areas having the same width-to-depth ratios produce the same subsidence, assuming equivalent rock characteristics throughout.

(3) Profile Function Method

Profile function method is based on the construction of the subsidence profile for the extraction area under consideration. Actual profile functions are derived by interpolation between several surface points for which vertical displacements are specified. For critical areas, the central region of the trough has constant subsidence equal to S_{max} (Equation 15.8) at length 'd' from the mine edge (Figure 15.8); the vertical subsidence equals $^1/_2$ S_{max}. The general relations shown in Figure 15.8, such as the inflection point at distance 'd', are in effect derived from results of field observations of similar mined cross sections. Model investigations have also been used in this regard (Lee and Shen, 1969). Actual profile functions are derived and presented by Brauner (1973) and Chen et al. (1974) along with more detailed discussion.

(4) Influence Function Method

In this approach, the extraction area is considered as consisting of an infinite number of infinitesimal elements. The principle of superposition is then used to obtain the contribution of each one of these areas to the displacement at any point along the subsidence trough. This contribution of an extraction element can be expressed as the product of its own extraction area. Influence functions can be derived from observational results or can be based on

Table 15.4 Subsidence Factor (a) and Ratio of Maximum Horizontal Displacement to Maximum Possible Subsidence

Coal field and method of packing	Subsidence factor (a)	Subsidence ratio
British coal fields (NCB, 1966)		
Solid stowing	0.45	0.16
Caving or striping-packing	0.90	0.16
Ruhr coal field, Germany		
Pneumatic stowing	0.45	0.35–0.45
Other solid stowing	0.50	0.35–0.45
Caving	0.90	0.35–0.45
North and Pas de Calais Coal field, France		
Hydraulic stowing	0.25–0.35	0.40
Pneumatic stowing	0.45–0.55	0.40
Caving	0.85–0.90	0.40
Upper Silesia, Poland		
Hydraulic stowing	0.12	—
Caving	0.70	—
U.S.S.R. (Muller et al., 1968)		
Donbass district	0.8	0.3
Lvov-Volyn district	0.8–0.9	0.34
Kizelov district	0.4–0.8	0.3
Donets, Kuznetsk, and Karaganda districts	0.75–0.85	0.3
SubMoscow and Cheliabinsk districts	0.85–0.90	0.35
Pechora	0.65–0.90	0.3–0.5
U.S.		
Pennsylvania (Room and Pillar)		
General	0.5–0.6	—
50% extraction (Zwartendyk, 1971)	0.10	—
Pittsburgh area (White, 1974)	0.15	—
Arizona (copper ore)		
San Manuel Mine (Johnson and Soule, 1963)	0.7–1.0	—

Note: Subsidence ratio = maximum horizontal displacement/maximum possible subsidence.

After Brauner, G. (1973), Information Circular 8571 and 8572, U.S. Bureau of Mines; Chen, Y.C., Chen, Y.N. and Gaffner, D.V. (1974), Architectural Measures to Minimize Subsidence Damage, Report no. ARC-73-111-2551, Michael Baker, Jr., Beaver, PA, 130 p.

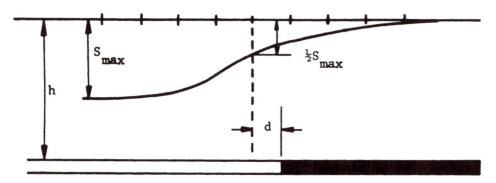

Figure 15.8 Subsidence profile function. (After Brauner, G. [1973], Information Circular 8571 and 8572, U.S. Bureau of Mines.)

independent strata parameters such as the limit angle. Further discussion in this regard and examples of influence functions are given by Brauner (1973).

(5) Phenomological Method

This method involves mathematically replacing the ground above a mine void with an idealized material that deforms in accordance with the principles of continuum mechanics. The theories involved include elasticity, viscoelasticity, and postyield solutions and are summarized by Voight and Pariseau (1970).

15.6.4 Underground Precaution for Active Mining

Surface areas can be protected through two underground measures: safety pillars and controlled undermining. Important in both cases is the concept of critical depth which is defined as the smallest depth of undermining, beyond which surface deformation become less than a given value, i.e., the allowable deformation of a structure. Consequently, through certain underground precautions the critical depth can be reduced.

(1) Safety Pillars

Design of safety pillars has received considerable attention in Europe and is based on the use of the influence function from which it is possible to determine the horizontal distance from a structure at which support must be provided. More common in the U.S. is the angle of protection concept. Gray and Meyer (1970) describe an example of this method (Figure 15.9) in which the extent of mine support is determined by providing a zone 15 ft in width around the periphery of the protection structure. This area is then projected downward and outward at an angle of 15° from the vertical to the level of the mine, with this projected area being the recommended area of support.

(2) Controlled Undermining

This method involves control of undermining such that allowable surface deformations, either temporary or final, must not be exceeded. This can be done by avoiding unfavorable patterns of mining, by stowing, by partial extraction, by rapid undermining, and by using special mining geometries.

(3) Stowing

Stowing effect is reflected in the subsidence factor 'a' from Equation (15.8) and is dependent on the stowed material and the method of stowing. Subsidence factors between 0.3 to 0.7 have been determined for pneumatic, slusher, and hand methods of stowing, while hydraulic stowing induces a subsidence factor ranging from 0.10 to 0.30. To achieve the lower limit of the subsidence factors, it is desirable to fill as early as possible after excavation and to minimize open spaces by tight and careful packing.

(4) Partial Extraction

This involves leaving a regular pattern of opening and pillars and has been referred to as a room and pillar, or panel and pillar system. The effect is such that ground movements

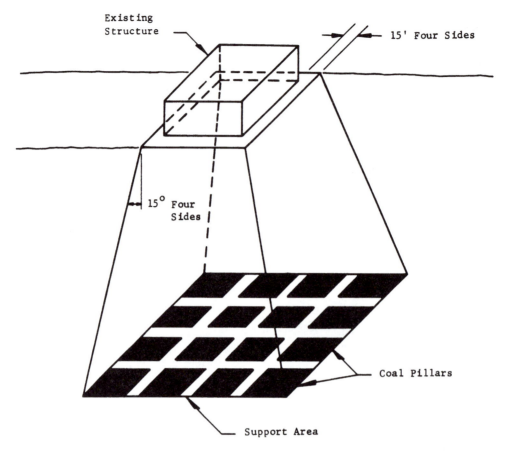

Figure 15.9 Angle of protection method for supporting structure. (After Gray, R.E. and Meyer, J. [1970], *JSMFE Div. Proc. ASCE,* v. 96, no. SM4, pp. 1267–1287.)

outgoing from individual excavations are usually superimposed on a uniform overall subsidence trough (which forms over the entire area covered by pillar and excavation), so that no undulation occurs on the ground surface. This method can be used in conjunction with stowing to reduce the subsidence and improve pillar stability.

(5) Rapid Undermining

The purpose of this method is to extract a supercritical area beneath a structure relatively fast, in order to reduce the time in which unfavorable mining patterns exist. Once the supercritical area is extracted, the structure should undergo full subsidence, but theoretically should not suffer any differential displacements.

(6) Special Mine Geometries

This method involves the phased removal of ore from a critical area, such that the ground surface is lowered relatively smoothly and horizontal strains are reduced or neutralized. Panels are worked simultaneously, but in staggered positions such that traveling strain waves for each panel will tend to cancel each other, i.e., maximum extension due to one working coincides with maximum compression due to another working. This method has been termed harmonic extraction and is discussed further by Voight and Pariseau (1970), Brauner (1973), and Chen et al. (1974).

(7) Structural Precautions

Modifications of existing structures to minimize subsidence damage can take the form of permanent strengthening, temporary supports, or removal of structural elements. In this regard, trenching around the structure, cut slots, and removal of brickwork or glass can reduce compression effects. Tensile cracks which have formed due to position (tension) traveling strain waves should not be filled immediately as compression can follow as the total subsidence trough develops. In pipe lines, flexible and telescopic joints should be incorporated into the pipe system to allow for greater movement without fracture.

15.6.5 Abandoned Mines

(1) Characteristics of Abandoned Mine

The risk of subsidence above abandoned mines is dependent on several factors:

1. The size and distribution of existing coal pillars
2. Condition of the rock above and immediately below the mine
3. The weight and thickness of the overburden
4. Changes in conditions within the mine, i.e., fluctuation in water level

Given these factors, the mechanism of failure involves a slow creep of the mine roof and possibly the mine floor beneath a coal pillar which, coupled with a spalling at the edges of pillars, will cause a reduction in the effective support of the pillar. The failure of one pillar will then cause an increase in loading on surrounding pillars and eventually result in widespread failure and collapse of the mine roof, the effect being felt on the ground surface as a rapid subsidence. Control of this problem is obtained either through specialized support methods or through certain structural precautions which are incorporated into the design of the proposed structure. Tables 15.5 and 15.6 and Figure 15.10 give some information on limit settlement and allowable deformation for various structures.

(2) Support Methods and Structural Precautions

Three basic support methods are commonly used for providing the necessary support for foundation design: (1) grouting the entire mine, (2) installing grout columns to support the main mine roof, and (3) installing cast-in-place drilled caissons or bore piles for foundation support. Prior to the decision making, the economic feasibility of each method must be investigated.

The grout column methods have been used in the state of Pennsylvania for more than 30 years. The main purpose of this method is to reduce the span length of openings within the mine and strengthen broken rock layers above the mine in order to prevent subsidence without filling the entire void and all fractures. Most structural design precautions are of the following forms and can be used either alone or in combination with one another.

1. Strengthening of the foundation or superstructure by rigid design, such that forces transmitted by ground deformations could be withstood
2. Flexible design which would allow a structure to adapt itself to the deformations transmitted without loss of strength
3. Reducing the forces transmitted by the ground deformations by:
   a. Using a small plan area of the structure
   b. Providing gaps between adjacent units of large buildings which would allow independent deformation of the units

Table 15.5 Limiting Settlement for Various Types of Movement

Type of movement	Limiting factor	Maximum settlement[a]
Total settlement	Drainage	6–12 in. (15.2–30.5 cm)
	Access	12–24 in. (30.5–61.0 cm)
	Probability of nonuniform settlement	
	Masonry-walled structures	1–2 in. (2.5–5.1 cm)
	Framed structures	3–4 in. (7.6–10.2 cm)
	Smokestacks, silos, mats	3–12 in. (7.6–30.5 cm)
Tilting	Stability against overturning	Depends on height and width
	Tilting of smokestacks, towers	0.004 L
	Rolling of trucks etc.	0.01 L
	Stacking of goods	0.01 L
	Machine operation — cotton loom	0.003 L
	Machine operation — turbogenerator	0.0002 L
	Crane rails	0.003 L
	Drainage of floors	0.01–0.02 L
Differential movement	High continuous brick walls	0.0005–0.001 L
	One-story brick mill building, wall cracking	0.001–0.002 L
	Plaster cracking (gypsum)	0.001 L
	Reinforced — concrete building frame	0.0025–0.004 L
	Reinforced — concrete building curtain walls	0.003 L
	Steel frame, continuous	0.002 L
	Steel frame, simple	0.005 L

[a] L = distance between adjacent columns that settle different amounts or between any two points that settle differently. Higher values are for regular settlements and more tolerant structures. Lower values are for irregular settlements and critical structures.
Data from Sowers (1970); and Chen, Y.C., Chen, Y.N. and Gaffner, D.V. (1974), Architectural Measures to Minimize Subsidence Damage, Report no. ARC-73-111-2551, Michael Baker, Jr., Beaver, PA, 130 p.

Table 15.6 Allowable Mining Ground Subsidence

Coal field	Strain Compression	Tension	Slope (tilt)	Radius of curvature	Coefficient of safety[a]	Specific application
England[b]	1×10^{-3} for 100 ft (30 m) of structure[c]	—	—	—	—	
France[d]	$1–2 \times 10^{-3}$	0.5×10^{-3}	—	—	—	For pipelines
Germany	0.6×10^{-3}	0.6×10^{-3}	$1–2 \times 10^{-3}$ for house 0.5×10^{-3} for machine foundation	—	—	—
Poland	1.5×10^{-3}	1.5×10^{-3}	2.5×10^{-3}	—	—	—
Soviet Union						
Donets	2×10^{-3}	2×10^{-3}	4×10^{-3}	12 4 mi (20 km)	400–550	—
Karaganda	4×10^{-3}	4×10^{-3}	6×10^{-3}	1.86 m (3 km)	250–300	—
Cheliabinsk	—	—	—	—	300–400	—
Japan[e]	0.5×10^{-3}	0.5×10^{-3}	—	—	—	For concrete foundations
	1×10^{-3}	1×10^{-3}	—	—	—	For wooden buildings
	5×10^{-3}	5×10^{-3}	—	—	—	For soil pond bottoms

[a] Coefficient of safety — safe depth of mining/seam thickness.
[b] From NCB (1974).
[c] Allowable strains depend on the length of structure (see Figure 15.14).
[d] From Grard (1969).
[e] From Voight and Pariseau (1970).
After Brauner, G. (1973), Information Circular 8571 and 8572, U.S. Bureau of Mines; and Chen, Y.C., Chen, Y.N. and Gaffner, V.V. (1974), Architectural Measures to Minimize Subsidence Damage, Report no. ARC-73-111-2551, Michael Baker, Jr., Beaver, PA, 130 p.

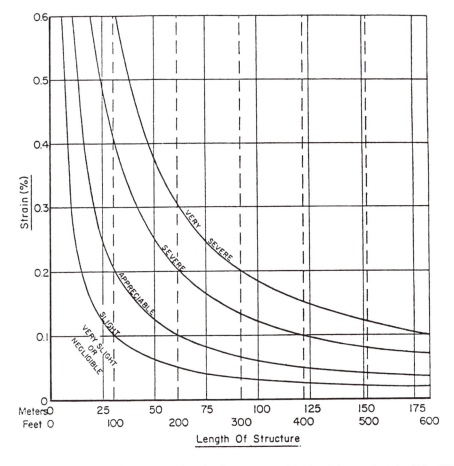

Figure 15.10 Relationship of damage to length of structure and horizontal ground strain. (After BNCB [1975], *Subsidence Engineer's Handbook,* British National Coal Board, London.)

    c. Reducing the bond between the ground and structure

    d. Cutting trenches around buildings in order to absorb compressive ground strains before they can reach the building

4. Orientation of the proposed building with respect to existing mining conditions, so that prospective discontinuities can be avoided (Also, the major lateral axis of a building should be paralleled to lines of equal subsidence in order to reduce strain effects.)

15.7 CONSTRUCTION OPERATIONS

Three basic construction operations which can cause ground surface movement are (1) deep excavation, (2) pile driving operation, and (3) blasting.

15.7.1 Deep Excavation

Subsidence due to deep excavation as trenches, quarries, tunneling, etc. is based on the same principle which was applied to mining subsidence, namely, the existence of unbalanced stresses in the area surrounding the open cut. Again, there is the tendency to reach a state of equilibrium and force a closure with the result being a surface subsidence extending outward from the area of excavation. In addition, consideration must be given to the effect of lateral stresses induced by nearby structures on the stability of the excavation. The rate of subsidence in deep excavations is caused mainly by adjacent existing structures because of the unbalanced stress.

Factors which increase this tendency for closure include seepage forces which act on the sides of the cut during the dewatering phase of construction. Continuous creep under gravity forces can decrease shear strength and induce failure. Lateral loads from soil overburden and surface structural loading must be also considered. The critical height of the excavation and minimum clearance between the excavation and the surface surcharge loading or building loads must be computed. These values can be obtained by conventional limit equilibrium or limit analysis slope stability techniques from standard geotechnical textbooks.

One present method of predicting soil response to excavation is through empirical means, similar to the case of active mining. Butler and Hampton (1975) describe a method and are able to obtain a good approximation of the observed settlement. They point out causes of lost ground entering the excavation due to loss of soil cohesion from induced hydraulic gradients during dewatering and from soil disturbance due to construction practices. They note that an effective means of controlling subsidence effects on surrounding structures is through underpinning methods.

15.7.2 Pile Driving Operation

The effect of pile driving or caisson in terms of surface movements is the occurrence of soil heave. Whenever piles are driven, a certain amount of soil will be displaced both vertically and horizontally, with this amount dependent on how quickly excess pore water pressures which build up during driving can dissipate. For cohesionless soils, the net soil displacement will be small due to quick drainage and resulting compression in and around the pile. A clay, however, is too impermeable to undergo volume change during the short period of driving, the result being that a volume of clay equal to the volume of the pile is displaced laterally and upward. This effect is compounded by a close spacing of piles where zones of displacement can overlap causing a greater surface heave.

15.7.3 Dynamic Compaction and Blasting

Dynamic compaction process is also cause for ground subsidence during construction operation. The principle of the dynamic compaction has been discussed in Section 11.4. During the dynamic process, some settlement or subsidence occurs. Machine vibrations and construction blasting causing local liquefaction are frequently observed.

15.7.4 Subsidence Caused by Local Environmental Conditions

(1) Underground Openings

Underground openings involve many causes such as: deep excavation, sinkholes, pollution intrusion, underground erosion (piping), etc. The mechanism of subsidence due to these underground openings has been discussed by Sowers (1976). Because of the complexity of the phenomena, Sowers pointed out that many subsidences have been mistakenly diagnosed and considerable amounts of money spent for remedial measures that were little or no benefit. The evaluation of the causes of subsidence requires a clear understanding of the geology of the soil and rock formations involved. Subsidence due to solution channels in soluble formations such as sinkholes has been discussed in Section 13.6; pollution intrusion will be discussed in a later section.

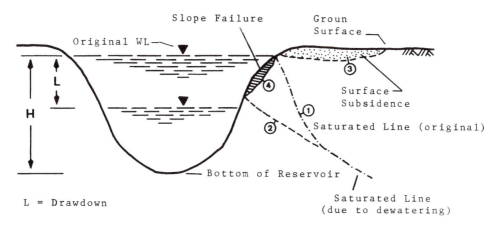

Figure 15.11 Subsidence due to rapid drawdown of an abandoned quarry in congested urban environment.

(2) *Rapid Drawdown*

Rapid drawdown of lake and reservoir causing land subsidence in urban areas is a critical environmental geotechnical problem. Figure 15.11 presents a case study of subsidence in an abandoned quarry due to rapid dewatering occurring in a congested urban environment. In examining Figure 15.11, indications are that points (1) and (2) are the saturated lines before and after the dewatering. Point (3) is the area of ground subsidence, and point (4) is the area of slope failure. Environmental geotechnical problems involved in this situation include dewatering system (Section 7.4), surface erosion (Section 14.8), rock and earth slope stability (Section 16.4), as well as remedial actions (Section 16.14).

(3) *Hazardous/Toxic Wastes Disposal*

There are three major processes for ground subsidence caused by hazardous/toxic wastes: (a) soil decomposition, (b) state of matter changes, and (c) soil-water structure changes. A brief explanation of each case is presented as follows.

(a) *Subsidence Due to Soil Decomposition*

A typical case is a landfill site. Soil decomposition is a part of the geomorphic processes as discussed in Section 5.10 and illustrated in Figure 5.16.

(b) *State of Matter Change*

Behavior of soil elements changes when the state of matter changes. In the thermal energy field, any element can be changed from solid into liquid and gas or visa versa.

(c) *Soil-Water Structure Change*

As local environmental conditions change, the soil-water structure will also change. Consequently, the volume changes, hydraulic conductivity, settlement, and shear strength will all change. The degree of change depends on local environmental conditions.

15.8 SUMMARY

1. Ground surface subsidence is one of the major problems in environmental geotechnology. The causes of subsidence are both natural and man-made. Among these causes, dewatering and mining are most critical.
2. Various proposed subsidence classification systems and subsidence mechanisms together with their failure phenomena are examined and discussed.
3. Land subsidence by both dewatering and mining causes is discussed in detail, including predicting and controlling techniques.
4. Subsidence caused by construction processes and environmental factors such as deep excavation, pile driving operation, blasting, moving vehicles, etc., as well as hazardous/toxic wastes, is also examined.

PROBLEMS

15.1 Distinguish between settlement and subsidence and what are the basic phenomena of subsidence?

15.2 It is predicted from laboratory tests that a building above a 7.6-m layer of compressible soil will change the average void ratio, from the initial value of 1.12 to a final value of 1.03. What will be the subsidence (or settlement) of the building?

15.3 Given data: load, P = 500 kips; mass unit weight of soil = 120 pcf; specific gravity of solid = 2.70; and liquid limit = 35. Footing was placed 8 ft below the ground surface. Compute the subsidence (vertical settlement) of the footing, if groundwater level is dropped 5 ft from ground surface. Assume soil is normally consolidated.

15.4 How and why will underground fire cause ground surface subsidence and local environmental problems?

15.5 How and why will tree roots cause ground surface subsidence? What is the mechanism between tree-root-soil system?

15.6 How and why will moving traffic cause ground surface subsidence? What is rut depth of a pavement? How is the rut depth measured?

15.7 How and why will bacteria and insects cause ground surface subsidence?

15.8 An explosion of a charge of 249 N of dynamite was claimed to have caused cracks in a concrete wall (or subsidence) located 40 m from the charge. The intervening distance was across a river with a limestone and shale bottom. The wall and the charge were both located in shale. You are asked to investigate the claim for the construction company that furnished the dynamite.
    (a) Outline what you would do to perform the investigation.
    (b) Discuss the validity of the claim using computations to illustrate the discussion.

15.9 How and why do hazardous/toxic substances cause ground surface subsidence? What is the mechanism during the process?

CHAPTER 16

Slope Stability of Earth Slopes and Landslides

16.1 INTRODUCTION

Stability of earth slopes and landslides belongs to one system. *Landslide* is the result of slope instability. Landslides occur in many parts of the world, especially in those areas with problematic soils and adverse environmental conditions. In analyzing the landslide problem, the engineers and geologists often look at it from different points of view. The geologist regards landslides as one of many natural processes acting as part of the geological cycle. They are interested in only the ground movement with respect to the geological and hydrological features. On the other hand, the geotechnical engineer is interested in the soil types, their engineering behavior, the maximum height of the slope, and maximum slope angle in terms of a safety factor. In most cases, they do not understand the geological formation and environmental factors causing the landslide. Even within the engineering group, the practitioner often is interested in the measurements of soil-rock properties, ground movements, and environmental conditions; however, the theoretician is interested in the idealized failure surface in order to fit it into a mathematical model.

Landslides occur in almost every conceivable manner, slowly or suddenly, and with or without any apparent provocation. They are usually caused by excavation, undercutting the foot of an existing slope, improper surface and subsurface drainage, tunnel collapse of underground caverns, surface and subsurface erosion, or by a shock caused by earthquake or blasting which liquifies the soil. Since the landslide problem is not a simple matter, it requires knowledge from other disciplines; therefore, a joint effort from geologists, geotechnical engineers, and seismologists is required to tackle this problem. There are numerous state-of-the-art publications concerning slope stability and landslides with these various aspects emphasized (Broms and Wong, 1991; Fang and Mikroudis, 1991; and Morgenstern, 1992). The purpose of this chapter is to present the environmental aspects of stability of earth slopes and landslides.

16.2 CAUSES AND PHENOMENA ASSOCIATED WITH LANDSLIDE

16.2.1 Causes of Ground Movement

The causes of ground movement can be divided into two groups — the external and the internal. The external causes include loadings and environmental conditions. The loading condition includes surcharge, seismic, and blasting vibration, and the environmental factors include rainstorms, dry-wet cycles, soil types, and, in coastal areas, the ocean waves, tsunami, and marine erosion. In most cases, the causes of slope movement involve more than one reason. The internal causes are contributed to from external factors. For example, heavy rainstorms may cause progressive erosion, reduce the shear strength of soils such as residual soils, cause piping failure in dispersive clay, change the porewater pressure, change the temperature and moisture gradients in all types of soil which create massive progressive cracking, and, finally, all these factors will accelerate a landslide condition as illustrated in Figure 16.1.

16.2.2 Prefailure Phenomena Associated with Landslide

Figure 16.2 shows a schematic diagram illustrating some prefailure phenomena associated with landslide. In most cases, the landslide or slope failure will not occur at once. It happens gradually or progressively. The associated phenomena include cracking, shrinking, erosion, and surface creep, which then leads to surface slip and excessive settlement. When the slope soil reaches a certain level, such as from points **a** to **b** in Figure 16.2, the soil's internal resistance is no longer able to hold together due to the external loads, and at that point, the landslide or ground failure begins.

16.3 PROGRESSIVE FAILURES AND SURFACE CREEP

16.3.1 General Discussion

Progressive failures relating to landslides, erosion, and surface slip have been recognized by geologists, and agricultural scientists, since the early days and for geotechnical engineering, considering the design of an earth dam, examined the shear strength of a clay cohesion and progressive failure along a possible sliding surface and pointed out that progressive failure is best evaluated on the basis of empirical relationships between the effective shearing resistance computed from actual failures and the strength as derived from the laboratory tests.

16.3.2 Progressive Failure and Surface Creep

Numerous researchers have examined quantitatively the possible effects of progressive failure. Bjerrum (1967) suggested that the energy released by straining *in situ* might be different from that obtained in laboratory testing and, thereby, cause progressive failure. Since surface movement is always related with surface creep and landslides, many researchers have attempted to measure *in situ* creep rates. The rate of these movements varies during the year and movements are often confined to the surface layer. The rate increases as failure is approached and actual time of a landslide can frequently be predicted by monitoring the movements.

Chowdhury (1978) evaluated the progressive failure for natural slopes and examined the *in situ* stresses at a site along with the high velocities attained by the sliding mass. Numerous case histories on progressive failures related with landslides are also given by Leonards (1979) who examined the progressive failure relating to soft clay including total and effective analysis, strength anisotropy, and strain and strain rate.

16.4 SLOPE STABILITY ANALYSIS PROCEDURES

16.4.1 General Discussion

There are numerous methods currently available for performing the slope stability analysis as discussed in various standard textbooks on geotechnical engineering. The majority of these

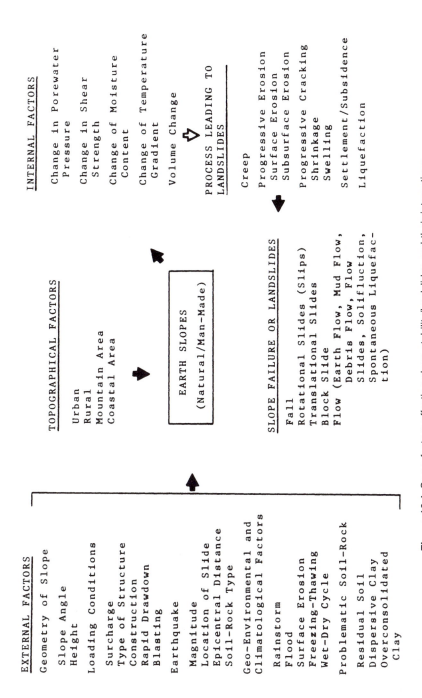

Figure 16.1 Some factors affecting slope stability/landslides and their interactions.

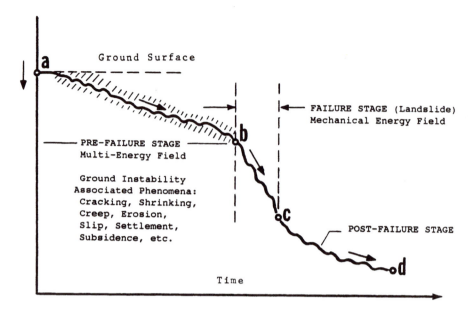

Figure 16.2 Prefailure phenomena associated with landslide.

methods may be categorized as limit equilibrium and limit analysis methods. The limit equilibrium method is widely used at the present time due to its simplicity.

16.4.2 Limit Equilibrium and Limit Analysis Methods

(1) Limit Equilibrium Method

The basic assumption on the *limit equilibrium* approach is that Coulomb's failure criterion is satisfied along the assumed failure surface, which may be a straight line, circular arc, logarithmic spiral, or other irregular surface. A free body is taken from the slope and starting from known or assumed values of the forces acting upon the free body, the shear resistance of the soil necessary for equilibrium is calculated. This calculated shear resistance is then compared to the estimated or available shear strength of the soil to give an indication of the factor of safety.

The limit equilibrium method is widely used at the present time; however, it has inherent weaknesses, among these the most important of which is that it neglects the soil stress-strain relationship. Many improvements have been made on this method including defining more acceptable failure surfaces, handling forces acting on the failure surface, and stress distribution along the failure plane. The mechanism of limit equilibrium concept for slope stability analysis is shown in Figure 16.3. Figure 16.3(a) shows forces acting on earth slope, where **W** is the weight of the soil of the wedge, **S** is the shear strength of soil along the failure surface, and **R** is the result. During the structure's lifetime, **W** may be changed slightly due to fluctuation of groundwater table, dessicating, or wetting. However, the shear strength, **S**, of the soil can change dramatically, as discussed in Section 5.10 and 11.8. The force diagram shown in Figure 16.3(b) for the earth slope does not represent an equilibrium condition at any time, because they are in different energy fields. The weight of soil of the wedge is in the mechanical energy, while the shear strength is in the multimedia energy field.

There are numerous limit equilibrium methods available for evaluation of slope stability, such as the Swedish circle method, Bishop method, Morgenstern method, Huang method, Koppula method, and many others as reviewed by Fang and Mikroudis (1991).

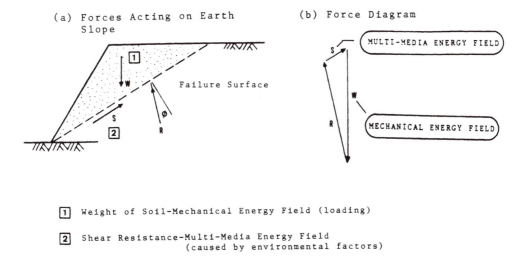

(a) Forces Acting on Earth Slope

(b) Force Diagram

1   Weight of Soil-Mechanical Energy Field (loading)

2   Shear Resistance-Multi-Media Energy Field
        (caused by environmental factors)

Figure 16.3 Limit equilibrium concept for slope stability analysis. (a) Forces acting on earth slope; (b) force diagram.

(2) Limit Analysis Method

The *limit analysis* method uses the concept of a yield criterion and its associated flow rule. The method is based on two theorems for any body or assemblage of bodies of elastic-perfectly plastic material; namely, lower bound theorem and upper bound theorem. According to the upper bound theorem, it is necessary to find a compatible failure mechanism in order to obtain an upper bound solution. A stress field satisfying all conditions of the lower bound theorem is required for a lower bound solution. If the upper bound as provided by the velocity field and stress field coincides, the exact value of the collapse load is determined.

The limit analysis method was introduced to the slope stability problem by Drucker and Prager (1952). Since then this method has been refined and the application has been extended. The major advantage of this method is that in most cases it provides a close-form solution. It also provides engineers with the possibility of a physical picture and helps them to visualize the mechanics of the failure. In some cases, limit analysis can help to solve complicated problems other than limit equilibrium. On the other hand, limit analysis is a relatively new approach and it is not familiar to most practical engineers and has not been widely used as compared with limit equilibrium method.

16.4.3 Selection of Strength Parameters and Factor of Safety

(1) Selection of Strength Parameters

There are various stability analysis requirements for various strength parameters. For example, for short-term slope stability analysis, the total strength is needed. However, for the long-term analysis, the effective strength parameters are required. For overconsolidated clay, the residual strength concept is suggested. Therefore, the selection of strength parameters is a major part of stability analysis as discussed in Sections 11.7 and 11.8.

For correction of the shear parameter for analysis of stability of soft clay deposits, a method has been proposed by Chekiang University (1975) cited by Fang (1980). They claim that shear strength increases with the overburden pressure increase, and the shear strength decreases due to soil creep or other adverse factors. Therefore, correction must be made for stability analysis from conventional shear strength test results by the following procedures:

$$S = (S_0 + S_c)$$
$$= [S_0 + k(- u)]$$
$$= [S_0 + k (1 - u/)] \tag{16.1}$$

$$S = (S_0 + k U) \tag{16.2}$$

where S = corrected shear strength, S_0 = shear test from conventional shear test, S_c = shear strength increase caused by consolidation, = maximum principle stress, u = change in pore water pressure due to increaee in stress, U = degree of consolidation with respect to, = coefficient, mainly caused by creep of soil, this value ranges from 0.75 to 0.90, and k = sin ϕ_{cu} cos $\phi_{cu}/ 1 + \sin_{cu}$ or,

$$S_c = {}_z U \tan \phi_{cu} \tag{16.3}$$

where $_z$ = vertical stress and $_{cu}$ = consolidated undrained friction angle, then

$$S = S_0 + {}_z \tan \phi_{cu} \tag{16.4}$$

(2) Factor of Safety, F_s

As discussed in Section 12.5, the factor of safety or degree of safety is used by engineers to indicate whether or not an earth work or ground soil will fail under the worst service conditions for which it was designed. The present concept for determining the factor of safety for most earth slopes is also based on Coulomb's law. Fellenius (1927) has used the ratio of actual shear strength to critical shear strength, or the ratio of required shear strength (S_R) to the available shear strength (S_A) as:

$$F_s = S_R/S_A \tag{16.5}$$

Theoretically, when F_s is greater than 1.0, the slope is stable; if it is less than 1.0, the slope will fail or/and become a landslide. The concept of safety as applied to geotechnical engineering is discussed by many investigators (Jumikis, 1967). The factor of safety also can be obtained from practical experience, as illustrated in Table 16.1, which provides some guidance in selecting appropriate factor of safety for slope stability analysis.

16.4.4 Uncertainties and Probability of Failure

Since ground soil involves many uncertainties concerning the soil profiles — the states of stress and stress history, the expected changes in loading conditions, the validity of assumptions made in analysis, the variation of soil properties as determined *in situ* or in the laboratory, and the variations during construction — test results and probability of failure may change in computing the factor of safety. Numerous investigators have attempted a rational means of assessing the margin of safety by use of the theory of probability.

Figure 16.4 shows the probability of failure and the factor of safety vs. the slope angle (β). The s_u is the mean value of undrained shear strength. The coefficient of variation, Vs_u is the ratio of standard deviation to the mean; it is generally accepted that $0.2 < Vs_u < 0.5$ for the sedimentary clay deposits as proposed by Lumb (1966) is acceptable. The probability of failure is the ratio of the shaded area to the total area of the frequency-distribution curve in Figure 16.4.

Different shapes of the frequency distribution and various values of Vs can be obtained from experimental design. It can plot the probabilities of failure vs. slope angle (β), for values

Table 16.1 Recommended Factors of Safety for Slope Stability Analysis in Residual Soil Region

		Factor of safety	
Class	Cutting type	(A) Comprehensive site investigation[a]	(B) Cursory site investigation[b]
1	Road cutting or cutting in remote area where probability of life at risk, owing to failure, is small	1.1	1.2
2	Road cutting on main arterial route where main line communications can be cut and risk to life is possible	1.2	1.3
3	Areas adjacent to buildings where failure would affect stability of building, e.g., car park, risk to life significant	1.2	1.4
4	Cuts adjacent to buildings where failure could result in collapse of building; risk to life very great	1.4	Not applicable

[a] Such a site investigation would, in addition to normal boring and drilling, include a program of laboratory testing to determine shear strength parameters for both soils and rock failures. Joint system surveys would be carried out and likely effects of heavy rainfall on the slopes would also be considered. These effects would be included in the soils and rock stability analyses.

[b] Site investigation under such a classification would be limited to determination of the boundaries of the various grades of material, the type of rock, and also predominant joint patterns in the case of rock stability problems. Shear strength parameters would be derived from back-analysis of failures.

After Binnie and Partners (1971), Construction on Slopes Manual, Hong Kong; and Chiang (1979).

of $Vs_u = 0.1$, 0.2, and 0.5, assuming a Gaussian frequency distribution. Leonards concluded that if the variability in strength is relatively large, the probability of failure is also large. The relationship between the factor of safety and the probability of failure is highly nonlinear. The relation between variation in soil properties and probability of failure depends on the soil stratigraphy and on the geometry and nature of the problem. Figure 16.4 can assist in the analysis and design of slope stability structures. Other studies on reliability and probability in stability analysis are given by Christian et al. (1992).

16.5 EFFECT OF RAINFALL ON SLOPE STABILITY

16.5.1 Rainfall and Rainfall Intensity

The soil-water interaction relating to the stability of soil mass has been discussed in Chapters 5 and 6. In tropical regions, the main cause of landslide is rainfall. Barata (1969) described the association of landslides and rainfall in Rio de Janeiro, Brazil. The rain of 1966 and 1967 fell in a period of 2 or 3 days and was followed by another heavy rain of similar duration 3 to 5 weeks later. The rainfall intensity recorded reached 104 mm in 1 hr, 321 mm in 24 hr, and 675 mm in 72 hr. Barata noted that the areas where the rain caused the greatest damage were those involving slides and falls of residual and colluvial soils. The slopes where damage was greatest generally ranged from 15 to 45° (Nunes, 1969).

Deere and Patton (1971) reported that in Bonaventura, Colombia the occurrence of more than 100 slides followed several days of high-intensity rainfall. The slides initiated at the crests of slopes and appeared to involve the upper several meters of the residual soil layer and a few meters of weathered rock over a variety of igneous and metamorphic rocks. The slopes varied from 30 to 45° with most of the slides becoming destructive debris avalanches

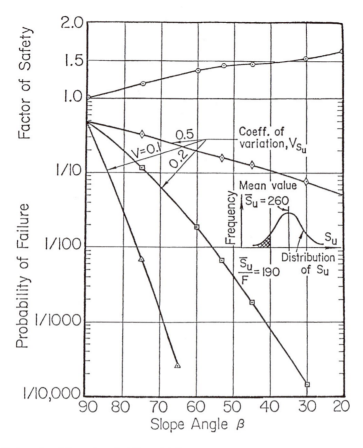

Figure 16.4 Factor of safety and probability of failure vs. slope angles. (After Leonards, G.A. [1979], Proc. 6th Panamerican Conf. SMFE, v. 1, pp. 225–274.)

Table 16.2 Landslide Pattern in Residual Soil Region Caused by Rainfall

Homogeneous slopes	Normally intense rainfalls	Planar slide
		Rotational slide
	Very Intense rainfalls	Avalanches (rapid flow)
Heterogeneous slopes	Normally intense rainfalls	Rolling
	Very intense rainfalls	Complex slides

After Morgenstern, N.R. and de Matos, M.M. (1975), Proc. 5th Panamerican Conf. SMFE, v. 3, pp. 367–383.

which left scars 20 to 30 m wide and hundreds of meters long down the slope. Morgenstern and de Matos (1975) reported that slope failure patterns are related with rainfall intensity and slope types as summarized in Table 16.2.

The effect of rainfall and groundwater flow on slope movements has been explored and it was found that slide movement was regulated by the direction and magnitude of shallow soil hydraulic gradients and wave-like variations of soil pore pressure resulting from heavy intermittent rainfall. The presence of these downward hydraulic gradients instigated slope motions. Debris flows initiated by intense rainfall relating to slope instability have been studied by Brand (1982) and Sitar et al. (1992). Rainfall-induced debris flows may occur as a result of loss of strength of the soil during saturation when rapid pore water pressure rises during periods of intense rainfall. Other factors of stability and performance of slopes caused by rainfall are given by Fukuoka (1992).

16.5.2 Soil-Water Interaction Relating to Slope Stability

There are numerous proposed mechanisms on soil-water interaction relating to slope failure or landslides. Some of these mechanisms interacting with surface creep and progressive failure have been discussed in Section 14.5. In the following section, three possible mechanisms of soil-water interaction related to slope stability are presented.

(1) Mechanism Based on Physicochemical Concept

The mechanism is based on direct interaction when water attacks the dry soil system based on physical and physicochemical concepts. These phenomena have been discussed in Sections 5.6 and 5.7. Winterkorn (1942) suggested using physicochemical concepts to explain the mechanics of dry cohesive clay attacked by water. Terzaghi (1943), considering the slaking process, explained the physical phenomena of shrinking and slaking of unprotected ground soil as the starting point of the causes of progressive failure of ground slopes. The application of electrical concept to the prediction of soil-water behavior, erodibility, and progressive failure related to the ground failure is given by Arulanandan et al. (1973). They indicated that laboratory tests can show that soil erosion behavior of remolded soil requires shear stress to initiate erosion and is affected by types and concentration of ions in the pore fluids and clay minerals. The types and amounts of clay minerals are indicated by dielectric dispersion (Section 9.7).

(2) Mechanism Based on Ion-Exchange Effect

As discussed in Section 4.8, the effects of ion exchange in the soil-water system have two basic elements: first the ion exchange which changes soil properties such as compressibility and shear strength; the second changes in the characteristics of soil's surface electrochemistry as discussed in Section 9.4. The ion-exchange reaction and other chemical effects relating to slope stability and landslides have been brought to our attention by Seifert et al. (1935), Matsuo (1957), Mitchell and Woodward (1973), and others.

A comprehensive study on the effect of cation exchange on the stability of slopes is reported by Matsuo (1957). The effects of sodium chloride, potassium chloride, hydrochloric acid, calcium chloride, and magnesium chloride in solution upon the Atterberg limits of three Japanese soils have been studied. The study included various concentrations of each of these electrolytic solutions and their effect on the liquid limit, plastic limit, and plasticity index. For the silty clays (MH and ML of Unified Soil Classification) no clear change in the plasticity index is observed for various concentrations of the electrolytes utilized. However, for the clay soil (CL), the plasticity index as well as the liquid limit show a marked increase for concentrations of all of the salt from 0.01 to 0.2 N solutions. An increase in the liquid limit indicates an increase in the shear strength.

(3) Mechanism Based on Linear Elastic Fracture Mechanics (LEFM) Concept

Fang (1994a) presented an analogy between progressive failure by use of the linear elastic fracture mechanics (LEFM) concept to explain how the water in the cracks relates to the progressive cracking and subsequently leads to slope failure. For any clay slope which exists, some cracks will fill with water due to nonuniform soil particle size, temperature and moisture gradients in the soil mass, and nonuniform compaction during the construction, seasonal groundwater table fluctuation, rainfall, melting snow, etc. Because the porewater pressure in the cracks changes due to these environmental factors, a capillary tension will be created

a

Figure 16.5 Photo shows the typical slope failure related to the cracking and tensile behavior of soil. (a) Developing cracks; (b) slope failure. (Photo courtesy of CDMB, Colombia, S.A.)

between the soil particles. When a saturated soil dries, a meniscus develops in each void of the soil structure which produces tension in the soil water and a corresponding compression in the soil skeleton. This compressive stress is just as effective in producing soil compression as an external load. Pressures of 5 to 10 kN/m can be produced in fine-grained soils.

These initial cracks and the porewater pressure fluctuation between the soil particles are similar to the notch and cyclic loading in fraction mechanics. Therefore, the crack growth, fracture load, and the Griffith strain-energy release rate can be estimated from laboratory tests (Section 6.8) and the LEFM theory. It is reasonable to say that water pressure fluctuation in the cracks will loosen up the soil structure which creates progressive cracks and eventually lead to slope failure. Numerous earth dam failures relating to these surface cracks have been reported by many researchers and cracks relating to the slope stability analysis have been reported. Figure 16.4 shows the typical failure or landslide related to the cracking and tensile behavior of soil. The ground soil started with small cracks, gradually developing into surface creep, and then surface slips, as shown in Figure 16.5(a). Figure 16.5(b) shows the failure mode caused by progressive cracking due to heavy rainstorms in the residual soil region in the Bucaramanga Mesa area in Colombia. The main reasons for the ground failure were due to the moisture which caused the decrease in the tensile strength (Section 6.7) of the soil.

(4) Mechanism Based on Particle-Energy-Field Theory

Figure 16.6 shows the slope stability caused by rainfall relating to various energy fields. In examining Figure 16.6, it indicates the stage of soil-water interaction along the potential failure surface. In stage [1], as rainwater enters into the slope, it is caused by potential energy;

b

Figure 16.5 (continued)

as water moves into the soil, it is caused by kinetic energy. The stage [2] development of soil-water interaction is discussed in Chapter 5.

16.5.3 Depth of Saturated Zone (Wetting Band)

(1) Depth of Saturated Zone (No Surface Cracks)

In analyzing the slope stability, the effect of the rainfall on slope layers is important. In general, the slide occurrences are limited to the upper zones approximately 3 to 7 m below the ground surface. Rainfall of the order of 35 to 50 mm/day will significantly affect the decomposition of granite rock in Hong Kong as reported by Lumb (1962) and discussed in Section 13.3. The effect of rainfall on stability in thick mantles of residual soil will be affected only when the intensity of rainfall is of the same order of magnitude as the permeability of the soil. The duration of the rainfall must be sufficiently long in order to reach a significant depth. Further study of slope stability relating to the depth of the zone of saturation or wetting band for a particular rainstorm is given by Beattie and Chau (1976). The depth to which the *wetting band* can be developed in a particular slope is dependent on the environmental factors such as:

1. Intensity of rainfall and duration
2. Infiltration capacity of surface soil
3. Slope topography
4. Vegetation covering

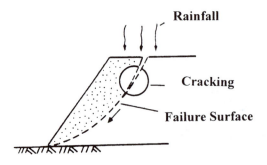

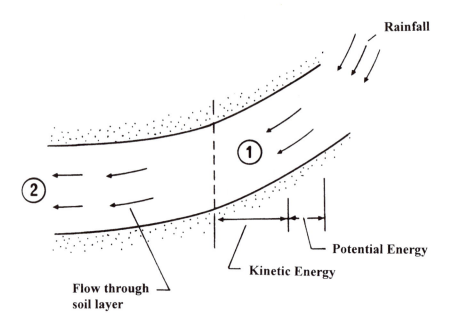

① MECHANICAL ENERGY FIELD

② MULTI-MEDIA ENERGY FIELD (Mass transport phenomena)

Figure 16.6 Slope failure mechanism caused by rainfall relating to various energy fields. (a) General view; (b) close view.

The thickness of the wetting band can be determined from the limiting rate of infiltration. For estimating the wetting band thickness for rainstorms having various return periods, the recommended return period by the government of Hong Kong is 1 in 1000 years. The thickness of wetting band or depth of saturated zone can be estimated by Equation (16.6). The thickness of wetting band vs. permeability of soil is shown in Figure 16.7.

$$h = \frac{kt}{(S_f - S_0)n} \qquad (16.6)$$

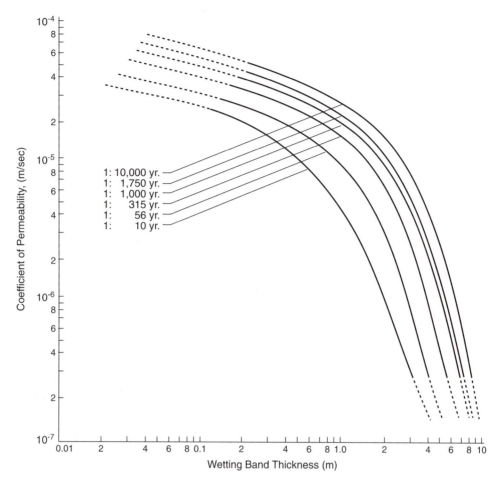

Figure 16.7 The thickness of wetting band vs. permeability of soil. (After Beattie, A.A. and Chau, E.P.Y. [1976], *J. Hong Kong Inst. Eng.*, February. Cited by MAA Group (1980).)

where h = thickness of wetting band (depth of saturation zone), m, k = permeability of ground soil, m/sec, t = duration of rainfall, sec, s_f = final degree of saturation, %, s_0 = initial degree of saturation, %, n = porosity of ground soil.

(2) Corrections of Wetting Band Due to Ground Surface Cracks

Equation (16.6) is designed for the ground surface soil without any cracks. However, in some regions, the ground surface exhibits significant cracks; therefore, Equation (16.6) has been modified to count the surface cracks as follows:

$$h = \lambda_c \frac{kt}{\left(S_f - S_0\right)n} \tag{16.7}$$

where λ_c = correction for surface cracks and this value varies from 1.05 to 1.15.

Types of cracks	Corrections, λ_c
Hair cracks	1.05
Alligator cracks	1.10
Galley cracks	1.15

16.6 LANDSLIDES IN RESIDUAL SOIL AND WEATHERING ROCKS

16.6.1 Characteristics of Residual Soil

Residual soil and weathering rock generally occur in a humid temperature and tropic climate, particularly during periods of intense rainfall as described in a previous section. As discussed in Section 13.3, the residual soil is produced by the *in situ* decomposition of the underlying rock due to the percolation of water through the rock fissures and rock pores, thereby breaking down the unstable rock material. Chemical breakdown is particularly active in the hot, humid regions. Two case studies are presented herein illustrating the landslides in residual and dispersive clays in Bucaramanga, Colombia and residual soil region in Hong Kong. The brief discussions of the problems and findings are presented as follows.

16.6.2 Case Study

(1) Bucaramanga, Colombia

The city of Bucaramanga is located in the northern part of Colombia. It is built on a mesa having its urban development and the living conditions of its inhabitants seriously affected with greater intensity each year by the problems of erosion and landslides which occur especially on the scarps of the mesa. The landslide-erosion problem in the residual soil and dispersive clay area has a landslide or landslides following almost each rainfall. At present, the control methods focus on the surface drainage systems. It has been shown in Figure 16.5 and again in Figure 16.8 that there are critical environmental problems in dispersive clay regions. Progressive failure of slope in residual-dispersive clay region is mainly due to loss of tensile strength of soils.

(2) Hong Kong Landslides, 1972

Because of the hilly topography of Hong Kong, agricultural land is extremely restricted. The upland areas are mostly covered with foliage but in places severely eroded. Population in Hong Kong increased from about 600,000 persons in 1945 to over 4,000,000 at the end of 1971. This latter figure has made Hong Kong one of the most densely populated areas in the world with an average density of about 9800 persons per square mile for the whole colony.

In the Hong Kong landslides, the soil and geological conditions play important roles. Based on Lumb (1975), the residual soils are produced by the *in situ* decomposition of the underlying rock owing to percolation of water through the rock fissures and into the rock pores, thus breaking down the more unstable rocks. In mineral content the granites and volcanics are similar, but in fresh granite the quartz crystals range in size from about 0.2 to 0.5 mm, whereas in the volcanic the size range is from about 0.02 to 0.2 mm. Observations made indicate depths of decomposition ranging from 20 to 200 ft in the granites, compared with 10 to 50 ft in the volcanics.

The nature and variability of the weathered rock have been described. Decomposition starts in the joints and fissures and works inward to produce a matrix of decomposed rock

(a)

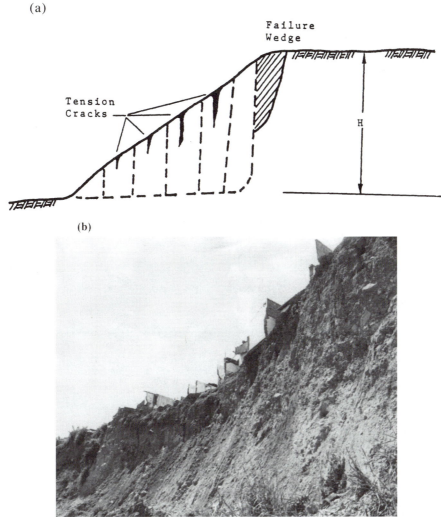

(b)

Figure 16.8 Progressive failure of slope in residual-dispersive clay region due to loss of tensile strength of soils. (a) Sketch shows the tensile failure of earth slopes; (b) photo shows a close view. (Photo courtesy by CDMB, Colombia, S.A.)

around isolated "core-stones" or boulders. Transported soils are produced from the residual soils either as landslip debris, known as *colluvium*, or through erosion of the soil surface by rainstorms and streams as alluvium. Other types of transported soil arise from marine deposits, beach sands, and man-made fill. The main cause of the Hong Kong landslide in 1972 was the heavy rainfall on residual soils. The recorded heavy rainfall occurring in May and June 1972 played the dominant role in causing a landslide that destroyed several tall buildings.

There is no standard method(s) for landslide remedial action. For a particular landslide or potential landslide there is seldom one and only one method of treatment. In general, the most economically effective means of prevention consists of a combination of two or more of the methods as discussed in Section 16.9. For the Hong Kong landslides the surface and subsurface drainage systems are most critical; therefore, the surface drainage system and slope stabilization by use of precast concrete grid are effective as shown in Figure 16.9.

Figure 16.9 Photo shows the precast concrete grid used in Hong Kong for protection of landslides. (a) General view; (b) close view.

16.7 LANDSLIDES IN DISPERSIVE AND EXPANSIVE CLAY REGIONS

16.7.1 Dispersive Clay Regions

As discussed in Section 13.8, indications are that some fine-grained soils, called dispersive soils, possessing higher content of dissolved porewater sodium than ordinary soils, rapidly erode forming tunnels and deep gullies by a process in which the individual clay particle goes into suspension in slow-moving water such as seepage flow. Slope failure and progressive erosion can destroy an entire community located in dispersive clay areas such as in Bucaramanga, Colombia.

16.7.2 Expansive Clay Slopes

Earth slopes which undergo volume changes when subjected to moisture variations due to seasonal climatic changes consequently accelerate slope failures. Landsliding in expansive soils is a significant geomorphic process. The long-term behavior of natural slopes of these soils, in the humid tropical climate, is a complex phenomenon in view of the changes these soils undergo due to environmental causes. Soil weathering coupled with seasonal swelling and shrinkage alters the soil properties and affects shear strength and stress-strain characteristics significantly. Creep deformation influences the landslide mechanism and progressive failure criteria become more appropriate in the slope stability assessment.

The conventional slope stability procedures are inadequate in addressing the long-term environmental effects on shear strength and slope deformation process. Therefore, there is need to understand the natural mechanisms that operate for a realistic assessment of the stability of natural slopes. Some of the factors suggested by Ramana (1996) are crucial for slope stability analysis:

1. Climate, soil weathering, soil pollution, and their effect on soil properties and shear strength
2. Selection of proper shear strength criteria (Section 16.4.3)
3. Perched water table conditions resulting in pore pressure rise, seepage parallel to the slope
4. Incorporation of swell pressures in the stability computations
5. Stratigraphy and thickness of the weathered layer to judge planar or circular slip surface; use of electrical resistivity survey (Section 9.5)
6. Differences between *in situ* behavior and laboratory testing

The above factors must be reviewed in order to get a better insight into expansive clay slopes. Detailed discussions on expansive clay have been presented in Sections 6.3 and 13.4.

16.8 LANDSLIDES IN OVERCONSOLIDATED CLAY AREAS

16.8.1 Failure Mechanism in Overconsolidated Clays

Numerous slope failures occur in overconsolidated soil deposits, especially in slope cutting and excavation projects. These failures frequently fail without warning and, in many cases, the excavation had been completed for a long time, as reported by many investigators. There are various causes for the slope failures in overconsolidated clays. Various researchers explain with their own reasoning on the following three important issues:

1. Natural behavior of overconsolidated clay
2. Delay in failures
3. Occurrence of failures

The majority of investigators believe that for overconsolidated clay deposits the excess negative porewater pressures are developed because of unloading resulting from excavation. Thus net porewater is lowest at the end of excavation and then the increase of porewater pressures due to equilibrium values occurs slowly as a result of the low permeability of cohesive soils. For analysis of shear characteristics of overconsolidated clays, ordinary shear tests are not suitable because they give too high a shear value. Skempton (1964) proposed a test concept showing that the strength remaining in laboratory samples after large shearing displacements corresponded closely with the computed strength from *in situ* landslides as discussed in Section 11.9.

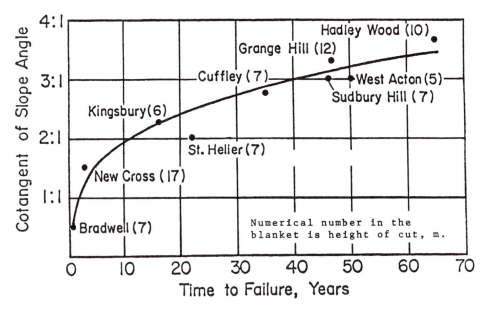

Figure 16.10 Slope angle vs. time to failure of cuttings in brown London clay. (After Leonards, G.A. [1979]. Proc. 6th Panamerican Conf. SMFE, v. 1, pp. 225–274.)

16.8.2 Review of Various Failure Mechanisms

Many overconsolidated clays are highly fissured (cracks) and mass permeability may be quite high; therefore, the equilibrium porewater pressures may be attained very quickly relative to the delays associated with slope failures. Bjerrum (1967) showed many case studies that slides in slopes of overconsolidated clays and clay shales are preceded by the development of a continuous sliding surface of progressive failure. The progressive failure may result from a large content of recoverable strain energy of the clay resulting from its geological history. Leonards (1979) presents slope angle vs. time to failure of cuttings in brown London clay. In examining Figure 16.10, a curvilinear relationship was obtained; however, there is no definite relationship between height of cut, slope angle, and time to failure.

Based on extensive research and *in situ* observations on overconsolidated clay deposits, Chinese Railroad Research Institute (CRRI, 1979) indicates that for overconsolidated clays, when preconsolidation reaches residual strength stage, the particle orientation changes and the electrical potential charge increases; therefore, the moisture content at shear surface increases also. When volume expands during the shear, the space between soil grains increases, which gives the overconsolidated clay a negative porewater pressure which temporarily and consequently reduces the shear resistance.

Chowdhury (1980) summarized some of the causes of failure of excavated overconsolidated London clays and explanation of uncertainty for the first-time failure includes the occurrence of failure and delay in failures. The analysis techniques of slope stability for the overconsolidated clays are mainly based on limit equilibrium concept. The material properties assumed are strain-softening model, brittleness and rheological model, and brittle-softening process.

Designs of cut slopes in overconsolidated clays with case studies have been reported by McGuffey (1982). The design of clay slopes for long-term stability is discussed in relation to mode of failure, soil test methods, method of analysis, selection of strength parameters, and time dependency of stability. Slope failures in overconsolidated clays can be evaluated based on drained shear. Failure of slopes in weathered overconsolidated clay is reported by Laguros et al. (1982). Comments and suggestions for stability of overconsolidated clays are also presented.

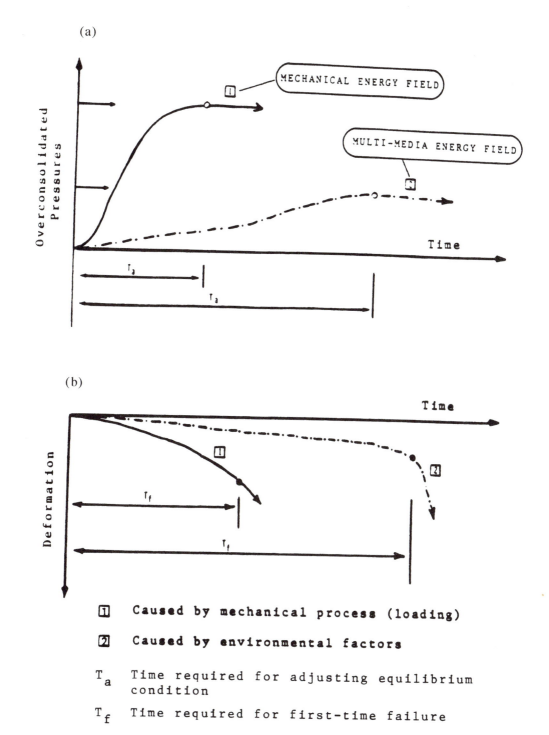

(a)

(b)

①   **Caused by mechanical process (loading)**

②   **Caused by environmental factors**

T_a **Time required for adjusting equilibrium condition**

T_f **Time required for first-time failure**

Figure 16.11 Conceptual model of failure mechanisms of overconsolidated clay. (a) Overconsolidated pressure vs. time; (b) deformation vs. time.

16.8.3 Environmental Geotechnical Approach

The exact causes and mechanisms for overconsolidated clays to fail are not clearly understood. Fang and Mikroudis (1991) proposed a conceptual model as illustrated in

Figure 16.11, pointing out the causes of overconsolidated pressure as discussed in Section 11.7. Further discussions are presented as follows.

In examining Figure 16.11, there are two causes for soil to be overconsolidated in the environment. One is due to loading and the other is due to environmental factors such as acid rain, acid drainage, and hazardous/toxic intrusion which will lead to change in the ion-exchange reaction as discussed in the previous section. In general, the cause by loading is greater than that caused by the environmental factors which will take a longer time to produce such consolidated pressure. On the other hand, the time required for the first-time failure indicates that the environmentally caused case takes a longer time period. For these two causes, the solution for estimation of time for the first-time failure will also be different. In other words, in the loading case, the approach should follow the mechanical energy field as indicated in Table 2.2 and Section 11.7. However, in currently measured methods regardless of the cause, all use a loading approach such as the Casagrande method (Section 11.5). At present, slope failure and landslides in overconsolidated clay regions still cannot be solved effectively.

16.9 EARTHQUAKE TRIGGER-CAUSING LANDSLIDES

16.9.1 Shallow Slope Failure

Causes of landslides involve many factors; however, the earthquake trigger-causing landslides give us more damage than other causes as reported by many published documents (Seed, 1991; Chaney and Pamukcu, 1991; Marcuson et al. 1992). In general, landslides can be divided into two types: the shallow slope failure and deep slope failure. Yen and Wang (1977) reported among various modes of seismically triggered slope failures, the most common type of failure being the shallow hillside slide. In most cases, the shallow depth is less than 2 m in proportion to their length, approximately 10 m or more, with extensive tensile cracks near the top of the slides. Analytical solution for the shallow hillside slope triggered by earthquake developed by Yen and Wang (1977) was based on the one-dimensional model for progressive failure. Also noted is that shallow slope failures strongly reflect the local and geological features (Seed and Schnabel, 1972), soil types, and local environments.

16.9.2 Deep Slope Failure

The deep slope failure is caused mainly by soil liquefaction. A detailed summary of landslides during earthquake due to soil liquefaction has been presented by Seed (1975 and 1991), summarizing all major landslides in the world from 373 B.C. to 1975. Continuing Seed's work, Chaney and Fang (1991) presented additional information on earthquake effects on landslides. The data include earthquake magnitude, location of slide, epicentral distance, type of structure and soil types, together with detailed references with further discussions provided by Chaney and Pamukcu (1991).

16.10 SLOPE STABILITY/LANDSLIDE PROBLEMS IN ADVERSE ENVIRONMENTS

There are numerous types of adverse environments causing the slope failures. Some causes are more critical than others. Regardless of reasons, proper attention must be provided in order to avoid further complications. Some of these causes are frequent and a brief discussion is presented as follows.

16.10.1 Effects of Logging Operations and Tree Roots

Clear-cut logging and associated road construction in forested areas increase local slide activity to approximately five times that of undisturbed areas. Reports on extensive investigations in the Redwood Creek drainage basin find that logging activities correspond to increased slope movements in the area, but that the local geology and soil nature were also important factors in slope mobilization. Large slope movements in the Van Duzen River basin in northern California found significant increases in slope movement occurring as grazing by domestic range animals changed the vegetation patterns on slopes.

Forestry and agricultural practices and the resultant changes in vegetation from these activities may directly affect the stability of slopes on which they occur. Advantages for using tree roots for earth reinforcement are discussed by Gray and Leiser (1982) and Wu (1984). However, there are some disadvantages as some tree roots affect the stability of earth slopes. The presence of vegetation and roots creates many channels for the conduction of free water in a soil mass. Patterns of subsurface flow are disturbed and boundary layers do not form in the soil. The presence of organic debris may increase soil's macropermeability and also may increase free drainage. The saturated micropermeability is increased and some water is removed through plant intake.

Soil-root interaction indicates that the shape of the tree branches is proportional to the shape of the tree roots. This means that tall trees or vegetation have a deeper or larger distribution area of roots. This relationship roughly can be written as proposed by Fang (1991c):

$$(W_T)\,(H_T)\,(\gamma_T)\,(\sigma_T) \approx (W_R)\,(D_R)\,(\gamma_R)\,(\tau_R) \tag{16.8}$$

where W_T, W_R = width of tree branches or roots, H_T = height of tree, D_R = depth of tree roots, γ_T, γ_R = density of tree branches or roots, and σ_T, σ_R = tensile strength of tree branches or roots.

In order for the tree to be stable above ground, the roots will serve just as foundations do for building structures — the taller or bigger the tree, the deeper or wider the root distribution. Bamboo is the one exception, because bamboo itself can grow up to 80 to 100 ft above the ground. However, the depth of the root system only penetrates into the ground from 18 to 24 in. In order for bamboo to stand above the ground, the strength of bamboo roots and roots distribution areas are extremely strong and cover a larger area. The live bamboo roots can penetrate into a 6-in. concrete wall. Many roadways, pavements, and drainage facilities have been damaged by live bamboo. Figure 16.12 shows the tree roots' effects on stability of an earth slope.

In examining Figure 16.12, (1) represents the earth slope, (2) presents the potential failure plane which can be estimated from routine slope stability analysis and can be obtained from standard textbooks, (3) is the potential failure zone, (4) represents the tree and branches above the ground surface, (5) represents the tree roots below the ground surface, (6) represents the wind load, when wind load acts on a tree, and a certain portion of soil layer will be disturbed through the tree roots as shown in (7) and (8). In such cases, the potential slope failure surface of the earth slope changes. In other words, tree roots can stabilize the slope, on one hand, and can disturb the stability of slopes as indicated in the figure.

16.10.2 Effects of Overcutting or Excavation

Overcutting frequently happens in the congested hilly urban area, largely because of the scarcity of flat land. The high cost of site forming results in the overcutting of slopes to make more space for buildings. In overcutting areas, retaining structures are commonly used to

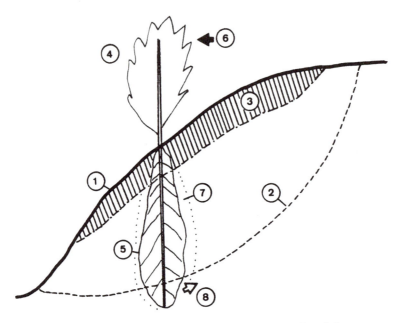

Figure 16.12 Effect of tree roots on stability of slopes.

support the earth pressure surrounding the building site. It is important that the critical location, minimum depths, and minimum strength of the retaining structure be known during the planning stage.

16.10.3 Acid Rain and Acid Drainage Effects on Slope Stability

Acid rain or acid drainage both are natural electrolytes as discussed in Sections 5.4 and 5.5. Acids will change the following soil behaviors:

- Increase ion-exchange reaction activity (Section 4.8)
- Increase redox reactions (Section 5.4)
- Volume changes and cracking patterns (Section 6.8)
- Hydraulic conductivity (Section 7.3)
- Consolidation (settlement) (Section 11.5)
- Shear strength (Section 11.7)

Based on these changes, indications are that acid rain or acid drainage water will increase earth slope surface cracks, progressive erosion, settlement, etc., which will lead to slope failure faster than a slope without acid rain or acid drainage water.

16.10.4 Marine (Underwater) Slope Stability

Marine slopes are commonly fairly gentle, uniform, and homogeneous over considerable horizontal distances. Thus in many cases 'infinite slope' limiting equilibrium methods are applicable. One of the first procedures proposed for evaluating seafloor stability under wave loading was described by Henkel (1970). The analytical model described by Henkel is a total stress analysis (neglecting soil porewater pressures) based on the principle of limiting equilibrium and employing the assumption of a circular failure surface. Finn and Lee (1979) presented results from a more general effective stress method which, in addition to wave loading, also includes earthquake loading and excess pore pressures. Chang et al. (1983) presented the seismic factor of safety of a submarine slope by limit analysis procedure. It must be pointed out that main problems for evaluation of marine slope stability are that field confirmation of slope stability studies are minimal.

The measurement of porewater pressures is now possible, which allows the application of effective stress analysis methods for marine slope stability analysis. Figure 16.13 presents the procedures for determination of the stability of a marine slope proposed by Richards and Chaney (1982). Figure 16.14 presents a process of marine slope evaluation by Chaney (1984).

16.11 LANDSLIDE PREVENTION, CONTROL, AND REMEDIAL ACTIONS

16.11.1 General Discussion

As discussed previously, ground movement or creep is associated with landslides. Therefore, instrumentation and monitoring systems for slope and ground movement are important. Table 16.3 is a summary of instrumentation and monitoring systems which include instrument types and measurable parameters. *In situ* measurement of soil properties at various time periods is also essential for analysis of landslides. Broms and Wong (1991) present slope remediation actions, which cover geometric, hydrological, mechanical, and physicochemical methods. However, landslide control by zonation and microzonation techniques is presented as follows.

16.11.2 Difficult Soil Conditions

Landslide control under difficult soil conditions includes residual soil regions, loess deposits, fissured clays, dispersive clays, and overconsolidated soils as discussed in the following section.

(1) Residual Soils

For the residual soil regions, the following control procedures are proposed by Binnie and Partners (1976) and Chiang (1979).

1. The effective shear strength parameters c and 0 are used in design.
2. Negative porewater pressures are normally assumed to be totally eliminated by infiltration in adverse site conditions with regard to rainfall.
3. In design, the choice of the cohesion intercept c' is complicated by the sensitiveness of the parameter to sampling and handling. A reduction of 40 to 100% of laboratory test results or projected values from back-calculation is usually adopted for analysis and design.
4. The thickness of the wetting band can be estimated from Equation (16.6) for a particular slope site and rainstorm together with records of the groundwater level obtained by monitoring standpipes or piezometers for the estimation of the probable rise in phreatic surface.
5. The depth of tension crack in unsaturated slopes and the development of hydrostatic pressure in cracks are also taken into consideration in stablity analysis.
6. For weathering profiles or relic joints, the methods of analysis assuming circular failure surface are not considered suitable. Alternative methods assuming generalized slip surfaces are more applicable.
7. The factor of safety for slope stability analysis for various slopes and cutting types is indicated in Table 16.3.

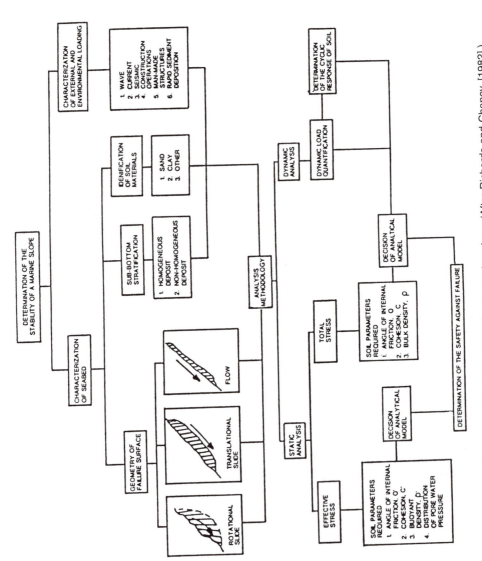

Figure 16.13 Determination of the stability of a marine slope. (After Richards and Chaney, [1982].)

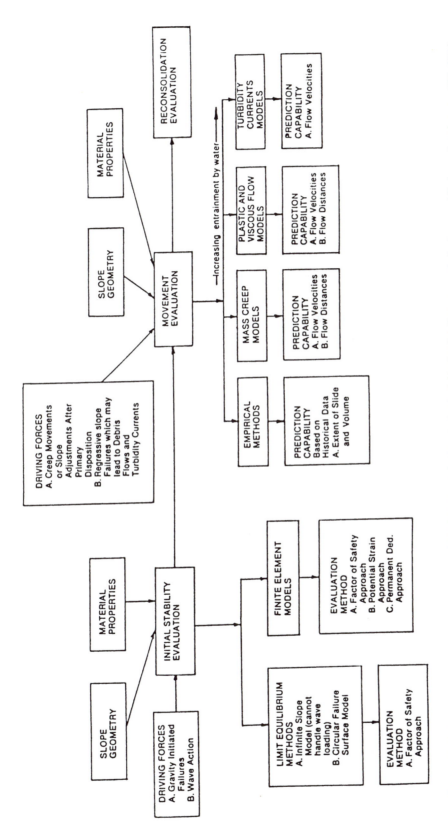

Figure 16.14 Process of marine slope evaluation. (After Chaney, R.C. [1984], *Seabed Mechanics*, Denness, ed. Graham and Trotman, London, pp. 159–167.)

Table 16.3 Instrumentation for Monitoring Slope and Ground Movements

Instrument	Measurable parameters
Short stake	Surface movement
Tiltmeter	Tilts
Inclinometer	Horizontal movement of soil mass
Borehole extensometer	Movement within a soil mass
Shear-strip indicator	Measures the slip-surface
Piezometer	Porewater pressure
Slide warning system	
Cracking-intensity indicator	Measures crack opening and duration
Acoustic emission	Measuring stress, strain deformation by sound waves
Seismic survey	Measures discontinuity of subsurface layers

(2) Loess and Silt

Loess deposits have high permeability in the vertical direction because they contain networks of interconnected channels that were formed from decayed roots. Protecting such slopes from infiltration and erosion prevents vertical percolation and water from destroying the particle bonds and causing slope failure.

(3) Fissured Clays and Shales

Slope failures in stiff-fissured clays and shales may occur progressively over a long period of time as they are subjected to very large shearing displacements. When these displacements exceed the shear displacement corresponding to peak strength, the shearing resistance is reduced to the residual value, although, in some cases, these soils may show stable slopes at steeper angles.

16.11.3 Tropical-Earthquake-Urban Environment

Table 16.3 summarizes some instruments to measure slope movements that also can be used as a warning system to reduce landslide damage. Some recommendations for control and correction of landslides in tropical-earthquake regions with urban environments follow:

1. Prepare landslide-seismic zonation and microzonation maps (Section 16.14) for a given area, especially in urban environments.
2. Enforce the building code and safety regulations by city government agencies.
3. Minimize the cutting of hillsides in order to reduce the risk of creating an unstable slope.
4. Both surface and subsurface drainage systems should be properly installed diverting all surface water away from potential failure areas. Inspect the drainage system regularly.
5. Landslide warning systems or instrumentation should be installed.

16.12 GENETIC DIAGNOSIS OF LANDSLIDE AND INFRASTRUCTURAL ACTIONS

16.12.1 General Discussion

In order to control or correct the landslide, several methods such as geometric, hydrological, physicochemical, and mechanical are available as shown in Table 16.4. For a particular landslide or potential landslide there is seldom one and only one method of treatment. In general, the most economical effective means of prevention consists of a combination of two or more of the previously mentioned methods as listed in Table 16.4. Two approaches for

Table 16.4 Some Methods Commonly Used for Controlling Landslides

Methods	Procedures
Geometric method	Flatten slopes
	Add pressure berms
Hydrological method	Surface drainage
	Sand drains
	PVC drains
	Horizontal drains
Mechanical method	Mechanical compaction
	Precast concrete grid
	Rock bolts and anchor
	Toe walls
	Reinforced earth
	Lime column
	Root piles
Physicochemical method	Chemical injection or grouting
	Lime stabilization
	Cement stabilization
	Asphalt stabilization

controlling the landslide problems are proposed as: (1) genetic approach for diagnosis of the causes of landslides, and (2) infrastructural remedial actions.

16.12.2 Genetic Diagnosis Approach

In many cases, due to complex geological formation and highly nonhomogeneous soil layers, it is extremely difficult to analyze the slope stability by conventional approach such as limit equilibrium and limit analysis as discussed in Section 16.4. In such cases, the genetic approach may be helpful. The genetic approach is a system based on the characteristics of geological formation, stress history, and clay composition, rather than by assumption that soil is homogenous with uniform porewater pressure distribution in the entire soil layer. An example of genetic diagnosis approach is illustrated in Section 17.7.

16.12.3 Infrastructural Actions

Strong public and institutional interest in various infrastructure has grown with the recognition. Intrastructual actions include infrastructural systems and components. There are numerous systems and components available; however, in this section, two types of infra-structural landslide-controlling systems are discussed, namely: root pile® and bamboo-lime composite piles. A brief discussion on root pile and bamboo-lime composite pile is presented as follows.

(1) Root Pile®

There are various types of pile foundations which can be used for ground improvement and landslides; the root pile is a very versatile construction tool which lends itself to individual use in many different situations as well as in clusters or in lattice-type network structures known under the trade name of Fondedile Reticulated Root Piles® retaining structures. Two of the important applications of the reticulated structures are for (1) landslide stabilization and (2) stabilizing ground beside large excavations. Several landslide stabilizations with Reticulated Root Piles® have been reported. Figure 16.15 shows the Monessen landslides in Pennsylvania (Aurilio, 1983); others include Jackson Bridge Interstate 55 in Mississippi and Mendocino landslides on Highway 7 in California.

Figure 16.15 Root pile for controlling landslide, Monessen, PA. (Top) Front view of root piles; (bottom) closed view of root piles. (Photos courtesy of Dr. U. Dash, Pennsyvania Department of Transportation.)

(2) Bamboo-Lime Composite Pile

Bamboo-lime composite pile is a low-cost landslide control system useful as a vertical element (pile) for preventing ground movement or for supporting structures and controlling landslides. All bamboo used for engineering purposes should be seasoned. Bamboo is available in various sizes and lengths and strength; therefore, the length and size relating to the ultimate loads should be predetermined prior to start of a project.

For controlling landslides, the length of the bamboo pile should be longer than the theoretical failure plane. This theoretical failure plane can be estimated by conventional slope stability analysis. The spacing of the bamboo pile is approximately eight to ten times the diameter of bamboo-lime composite element. For reinforcement of new embankment or small earth dams, the horizontal types of bamboo mats or strips are effective. Bamboo-lime system can also be used as anchor, tension pile, or root pile as discussed in a previous section. The principal of the bamboo anchor is the same as a bamboo pile with reversed-load application. For detailed discussion and *in situ* installation procedures see Fang (1991c).

Hashim and Ang (1996) conducting a multidiscipline research program involving hydrology, geology, soil mechanics, climatology, botany, and highway pavement subgroups, describe a technique developed to quantitatively evaluate the safety of slope systems incorporating the various factors that influence slope safety. The primary application is for slope along

Table 16.5 Description of Map Units for Stability Map of Naturally Occurring Slopes

Map unit	Slope stability	Landslide potential	Solifluction potential	Slopes (%)	Geology
1	High	Very low	Very low	<45	Undifferentiated
2	Moderately high	Generally low	Generally low	45–100	Generally stable bedrock
3	Moderate	Moderately low to moderately high	Moderately low	45–100	Medium- to coarse-grained unconsolidated deposits
4	Moderately low	Moderately low to moderately high	High	45–100	Soft, frost-susceptible bedrock, chiefly argillite
5	Low	Moderately high	High (chiefly rockfalls and snow avalanches)	Commonly >100	Bedrock, fractured and in part faulted
6	Very low	High	High (chiefly slumps and earthflows)	15–>100	Unconsolidated deposits underlain by sensitive clay and silt, or incorporating ice

After Dobrovolny, E. and Schmoll, H.R. (1976), *Focus on Environmental Geology,* Oxford University Press, New York, pp. 11–27.

highways, however, the technique can be used elsewhere if proper local environmental factors are modified.

16.13 LANDSLIDE ZONATION AND MICROZONATION

16.13.1 Zonation and Microzonation

Zoning for landslide effects has been used for planning and design of various buildings, earth-retaining structures, dams, and highway embankments (Dobrovolny and Schmoll, 1976). The idea for zoning a landslide is similar to the seismic zonation and microzonation serving as a guide for safer land use and construction. The landslide-seismic zonation and microzonation should be prepared simultaneously. It also should be prepared in conjunction with geological and material maps. The information in the landslide zonation map includes:

1. Soil-geological data
2. Seismicity characteristics
3. Erosion features
4. Drainage systems
5. Solifluction potential
6. Landslide potential
7. Degree of slope
8. Percent of slope etc.

The primary criterion for determining instability is the degree of slope, areas of steep and very steep slopes which will be the chief sites of instability. Also, the degree of instability depends on the geologic or construction materials underlying the slope. Environmental factors such as rainfall, seasonal changes, and hot-humid areas should also be considered in preparation of landslide zonation.

16.13.2 Landslide Zonation and Microzonation Map

Based on information obtained as indicated in Table 16.5, a zonation map for slope stability has been developed as shown in Figure 16.16. Similar techniques to map potential

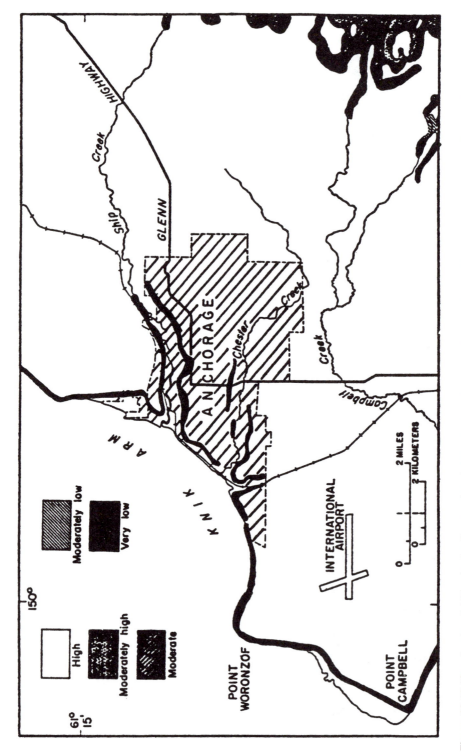

Figure 16.16 Map showing stability of naturally occurring slopes, city of Anchorage and vicinity, Alaska. Map units are described in Table 16.7. (After Dobrovolny, E. and Schmoll, H.R. [1976], *Focus on Environmental Geology*, Oxford University Press, New York, pp. 11–27.)

landslides in urban and forest areas have been developed. These types of map zonation are designed primarily to delineate those areas in which particular problems can be expected and to determine the pertinent site characteristics and their incorporation into the design of various projects to minimize damage to acceptable levels in the event of a destructive disaster.

16.14 SUMMARY

1. Slope stability and landslide belong to one system. Landslide is the result of slope instability. Landslides or slope failure will not occur all at once. It happens gradually or progressively. The associated phenomena before the landslide include ground surface cracking, shrinkage, erosion, and creep which lead to surface slip, excessive subsidences, and landslide.
2. The rainstorm is the major contribution to landslides in tropical regions. The mechanics of soil-water interaction relating to the progressive failure of slope are essential.
3. Cracking-fracture-tensile behavior of soil is directly related with slope failure. The moisture content effect shown in laboratory studies indicates that tensile strength is more sensitive than conventional shear parameters such as cohesion, friction angle and unconfined compressive strength. Using tensile strength as a criterion for measuring slope failure is suggested.
4. The earthquake trigger causing landslides gives us more damage than other causes; therefore, landslide-seismic zonation and microzonation are a useful tool to serve as a guide for safer land use and construction. However, the value to landslide damage potential and seismic damage potential or the combination of these two items should be further evaluated for practical engineering uses.
5. In analysis of slope instability, the procedures or methods for selecting the proper strength parameters are important.
6. For the control and correction of landslides, larger amounts of construction materials are required; utilization of locally available material and economical analysis should be thoroughly examined.

PROBLEMS

16.1 What is the relationship between slope stability and landslide?
16.2 Describe some phenomena at the prefailure stage during landslides.
16.3 You have been retained to act as a consultant to the Department of Transportation of a mountainous state in a temperate climate area. Your first assignment is to investigate two slope failures and suggest corrective measures. One of the failures occurred in a hillside fill on a steep slope, while the second was a nearby cut made at a steep slope angle through clay material.
   (a) Outline in detail the steps you would take during the investigation of each slide with the justification for each step.
   (b) Suggest at least two possible corrective measures for each case that may be applicable depending upon the results of your investigation. Explain each corrective action that you suggest.
16.4 A cut 50 ft deep is to be constructed in a soil of infinite depth and having a shear strength of 700 psf. The bottom 30 ft of the cut lies below water. The slope of the cut is 3:1. A warehouse that will exert a load of 6 tons per linear foot at a distance of 25 ft from the crest of the cut is to be built soon after the cut is made. Assume the soil is dry above the water table and saturated below it. The specific gravity of the soil solids is 2.67 and the soil has a void ratio of 0.56. If the soil has an angle of friction of 15° as well as a cohesion equal to 700 psf, what will be the factor of safety of the cut?
16.5 Why does acid rain and/or acid drainage cause a landslide more than just rainfall?
16.6 Why and how will ion exchange in the soil cause slope instability?
16.7 What is the so-called first-time slope failure? Why does slope failure occur without warning in overconsolidated soil deposits?

16.8 What are the major differences between limit equilibrium and limit analysis for analysis of slope stability?

16.9 When treating slope stability problems in purely cohesive soils, a tension crack usually develops at some distance from the top of the slope. How would you determine the depth of such a crack, and how does it influence the analysis of the stability of the slope?

16.10 A cut is to be made in a homogeneous clay deposits. The height of cut is 20 ft, the slope angle is 30°, the effective cohesion is equal to 0.15 tsf, and effective friction angle is 15°. Typical environmental geotechnical data for dry summer months and wet spring months are presented as: for the summer conditions, *in situ* density is 115 pcf, saturation of soil is low, so that the porewater pressure in the soil can be considered zero. For the spring condition, the *in situ* density is 122.0 pcf, and groundwater table is at ground surface.

(a) What is the factor of safety against a slope failure in the summer months?

(b) In which case (summer or spring conditions) is the value of factor of safety larger? Why?

CHAPTER 17

Solid/Liquid Waste Control and Utilization of Wastes

17.1 INTRODUCTION

17.1.1 General Discussion

Two basic problems relating to solid/liquid waste are discussed in this chapter: first, how to control the solid and liquid wastes; second, how to recover and reuse these wastes as conventional construction materials if possible. All wastes discussed herein are urban refuse generally known as garbage, which covers five aspects: (1) basic characteristics of urban refuse (garbage) including composition, classification, and engineering behavior such as volume change, hydraulic conductivity, bearing capacity, and stress-strain-strength of compacted garbage; (2) decomposition processes of garbage in a landfill including causes, mechanisms and settlement prediction; (3) compaction control during disposal process and the slope stability of the landfill in both newly established and abandoned landfill sites; (4) design and remedies of waste control facilities; and (5) utilization of waste for conventional construction materials.

There are several ways to dispose of garbage such as its use as substitute construction material, building blocks, incineration, or simply just dumping into the ground as part of landfill. Dumping into landfill is the most simple way to get rid of unwanted garbage. Garbage in landfill can be grouped into three major categories: (1) fast degradable material, (2) slow degradable material, and (3) relatively nondegradable material. Recently, due to maximum utilization of land, some of the existing abandoned landfill site(s) are being used as part of highway routes, recreation parks, as well as shopping malls and industrial sites. Using such sites for conventional construction, proper precautions must be taken.

17.1.2 Stability of Urban Refuse (Garbage) in the Landfill

(1) Stability of Fresh Garbage

Material in the landfill has three stages: (1) how long the fresh garbage will stay in the landfill in a stable condition; (2) as time goes on, some degradable landfill material gradually decomposes, and if so, what is the mechanism of the decomposition processes; and (3) estimate what to expect in the final stable condition of the decomposed landfill material.

(2) Stability of Abandoned Landfill Site

An *abandoned landfill* site as defined by the Pennsylvania Environmental Research Foundation (PERF, 1982) is one of inactive hazardous waste disposal or storage facility which cannot be easily traced to a specific owner, or whose owner has gone bankrupt and subsequently cannot afford the cost of cleanup, or a location where illegal dumping has taken place. The abandoned landfill site is a wasteland. In the past, it was not suggested for any use, but now is being used. Therefore, additional problems such as bearing capacity, settlement, slope stability, etc. must be carefully reexamined (Dunn, 1995 and Jefferis, 1995).

17.1.3 Relationship Between Solid and Liquid Wastes

All wastes can be grouped into solid and liquid forms. The majority of the solid waste is urban refuse (garbage). This refuse will decompose through the years and will become a liquid form generally referred to as leachate. According to the U.S. EPA indications are that about 90% of the wastes which includes industrial hazardous wastes is in liquid form. Therefore, controlling solid and liquid wastes means controlling liquid wastes only (Fang and Evans, 1994). The main purpose of all types of waste control systems is to control or prevent the liquid waste from seeping into nonpolluted areas. In order to accomplish such a task, three basic questions must be examined:

1. Why and how does a solid waste become a liquid waste?
2. What is the behavior of liquid waste?
3. What are the interactions between waste and waste control systems (mainly clays)?

The understanding of the interaction between pore fluids (liquid waste) and clay behavior (waste control system) requires the knowledge of the environmental aspects of geotechnical engineering. It is essential to the engineering utilization of naturally occurring materials for the containment of hazardous and toxic wastes. Without this knowledge of clay behavior in response to hazardous wastes, any engineering systems design, such as remedial action programs, can have no sound basis on which to project the long-term behavior of that system.

Solid and liquid waste control facilities are also a complex system; they require interdisciplinary knowledge from geotechnical, hydrogeological, and environmental fields to analyze, design, and construct waste containment systems. Conventional passive hydraulic barriers are frequently adapted as barriers for waste containment; however, special considerations are needed in dealing with these barriers. In this chapter, discussion will focus on the following points:

1. Characteristics of urban refuse including decomposition processes and pollution process mechanism
2. Basic considerations for analysis and design of controlling systems
3. Techniques of controlling waste
4. Precautions and protection for these controlling facilities

17.2 CHARACTERISTICS OF URBAN REFUSE (GARBAGE)

17.2.1 Degradable Garbage

Waste disposal material (garbage) consists of anything that cannot be further used or recycled economically; thus its composition varies from country to country, community to community, as well as from season to season. The density varies from 50 to 400 pcf depending on the amount of metal and debris. Table 17.1 gives the average composition of solid waste

Table 17.1 Average Compositon of Solid Waste

Types	New York City (1968)	U.S. National (1973)	Kawasaki, Japan (1979)	Osaka, Japan (1980)
Paper	58.8	51.6	35.3	37.1
Food wastes	9.2	19.3	13.5	13.2
Metal	7.6	10.2	7.5	5.5
Glass	8.6	9.9	19.6	12.3
Wood	2.5	3.0	3.6	2.5
Textiles	0.8	2.7	3.0	4.0
Rubber, leather	0.8	1.9	0.5	0.3
Plastic	0.8	1.4	9.7	15.2
Garden wastes	10.1	—	2.9	2.5

Data from Remson, I., Fungaroli, A.A. and Lawrence, A. (1968), *J. Sanitary Eng. Div. Proc. ASCE*, v. 94, no. SA2; Sowers, G.F. (1973), Proc. 9th ICSMFE, v. 4, pp. 297–310; and Yamamura, K. (1983), *Waste Manage. Res.*, v. 1, pp. 1–15.

based on U.S. national survey results and the City of New York as well as Osaka, Japan (Yamamura, 1983). Eventually these wastes will be dumped into a landfill. Needless to say, some industrial wastes are highly hazardous and toxic.

17.2.2 Engineering Classification of Garbage

In general, there are two groups of garbage, the degradable and relatively nondegradable materials. Table 17.2 presents a *classification of garbage*. Among degradable refuse, some are hazardous and toxic and some are not. There are various states of matter — solid, liquid, and gaseous forms — as illustrated in Figure 2.8, depending on environmental conditions in the landfill.

Many wastes, when mixed with other wastes or materials at a hazardous waste facility, can produce effects that are harmful to human health and environment, such as heat or pressure, fire or explosion, violent reaction, toxic dust, mist, gases, or flammable fumes or gases. A detailed list of these items has been prepared by the U.S. EPA (1990). Characteristics of hazardous wastes abstracted from 1990 EPA List is presented in Table 17.3. All generators of waste materials are required to determine whether the waste is a hazardous waste in one of two ways:

1. They are either waste and spent materials that are hazardous by definition and contained in specific lists (such as U.S. EPA Lists).
2. They exhibit one of four hazardous characteristics such as: ignitability, reactivity, corrosivity, and toxicity.
3. Radioactive wastes are not covered by Resource Conservation and Recovery Act (RCRA) or Act 97. They are specifically controlled by the U.S. Atomic Energy Act and P.Ls 807 and 1380 as discussed in Section 10.8.

17.2.3 Recycling and Energy Recovery from Wastes

Recycling is a process that uses discarded materials and objects in original or changed form rather than disposing of them by sending the material back into the process by which it was first formed. Recently, in the U.S. a successful program of recycling and energy recovery from waste has been developed. Aluminum cans, plastic containers, and used papers have been recycled and used again. Heavy metals such as lead (Pb), mercury (Hg), and zinc (Zn) trapped in the ground soil or in the landfill areas have been recovered using electrokinetic and electromagnetic processes (Section 9.10) as part of soil decontamination and energy recovery programs. Therefore, the composition of landfill material has changed since this program began.

Table 17.2 Engineering Classification of Garbage

Categories	Types	Descriptions	Engineering behavior
Fresh garbage			
Fast degradable			
A-1	Kitchen	Vegetable food wastes; meat scraps; fried fats; citrus peelings	Unit weight varies
A-II	Garden trash	Wood chips and bark; tree leaves and roots; flowers and garden plants; lawn grass; evergreens and shrub trimmings	Bad odors
A-III	Dead animals and manure	Dead animals, birds; manure; oils and paints	
A-IV	Paper and paper products	Cardboard; newspaper; mail; magazines; paper food cartons; tissue paper; wax paper, plastic-coated paper; brown bags	Combustion potentials
Slow degradable			
B-I	Textiles, toys, rugs	Textiles; rugs; clothes; shoes	
B-II	Glass, ceramics	Soft drink and beer bottles; broken glass and dishes	
B-III	Plastic, rubbers, leathers	Plastic bags; rubber tires; rugs; leather products	
Relative nondegradable			
C-I	Metals, appliances	Various metals; electrical appliances; abandoned vehicle parts	Unit wt = 200–400 pcf
C-II	Demolition and construction materials	Broken stones, bricks, walls, concrete and asphalt pavements; sand-gravels; mortar	Unit wt = 110–150 pcf
C-III	Soils	Soils; dust, ash, vacuum cleaner wastes	Unit wt = 100–135 pcf
Decomp. garbage			
D-I	Partial decomposed garbage	Organic soil containing some glass or metal articles	Unit wt = 85–120 pcf
D-II	Decomposed garbage	Organic soils contain some sands	Unit wt = 100–135 pcf

Based on Fang (1995a).

17.2.4 Engineering Problems of Landfill Sites

There are two basic problems of landfill design which need engineering knowledge:

1. During waste disposal period: Problems involved include compaction, drainage systems, stability of landfill slopes, and liners or hydraulic barriers.
2. Utilization of wastes or landfill sites: There are two types of utilization. The first uses waste as conventional construction material, such as scrap waste tires used as lightweight concrete aggregates. The second, utilizes old abandoned landfill sites as housing projects. In such cases, the characteristics of engineering behavior of wastes must be known. Some major elements or factors to be considered for analysis and design of waste control systems are summarized in Figure 17.1.

Among these factors relating to the stability and safety measures of landfill, the compaction control during the garbage disposal process and the slope stability of landfill during and

Table 17.3 Characteristics of Hazardous Wastes

Ignitability

A solid waste exhibits the characteristics of ignitability if the waste exists in any of the following forms

A liquid, other than an aqueous solution containing less than 24% alcohol by volume with a flash point below 60°C

A nonliquid, which under standard conditions is capable of causing fire through friction, absorption of moisture, or spontaneous chemical changes, and when ignited burns in a manner that created a hazard

An ignitable compressed gas

An oxidizer, such as a permanganate, inorganic peroxide, or nitrate

Reactivity

A solid waste exhibits the characteristics of reactivity if the waste

Is normally unstable and readily undergoes violent change without detonation

Reacts violently with water; or forms potentially explosive mixtures with water

Generates, when mixed with water, toxic gases, vapors, or fumes

Is a cyanide or sulfide-bearing waste that generates toxic gases, vapors, or fumes

Corrosivity

A solid waste exhibits the characteristic of corrosivity if the waste

Is aqueous and has a pH less than or equal to 2, or greater than or equal to 12.5

Is a liquid and corrodes steel at a rate greater than 6.35 mm/year at a test temperature of 55°C

Toxicity

A solid waste exhibits the characteristic of toxicity such as arsenic, cadmium, carbon tetrachloride, lead, mercury, selenium, silver, etc.

Abstracted from 1990 U.S. EPA list.

after construction are the most critical from an engineering viewpoint. Further discussions on these aspects are presented as follows.

17.3 COMPACTION OF LANDFILL

17.3.1 General Discussion

When garbage delivery is made daily by truck and dumped into the landfill sites, this is the so-called old fashioned approach for waste disposal. In some cases, this garbage is spread into a thinner layer, mixed with some earth material, and compacted by conventional compaction equipment for the purpose of covering up an unattractive landfill site, minimizing odor, or preventing animal and bird vandalism. There is no standard rule or regulation how garbage should be dumped or compacted at the present time. In this section some compaction control in landfill areas is suggested and past experiences are reviewed.

17.3.2 Review of Sanitary Landfill Test Sections

Conventional compaction or densification process has been discussed in Section 11.2. In landfill area, when proper compaction is applied, it also can reduce potential fire hazards in the landfill. The less air trapped in the landfill the less potential for fire hazards. Several test sections on landfill have been carried out by the states of New York (Moore and McGrath, 1970) and California in early 1970 (Chang and Hannon, 1976). The test sections performed in California are the most comprehensive. A brief discussion is presented as follows.

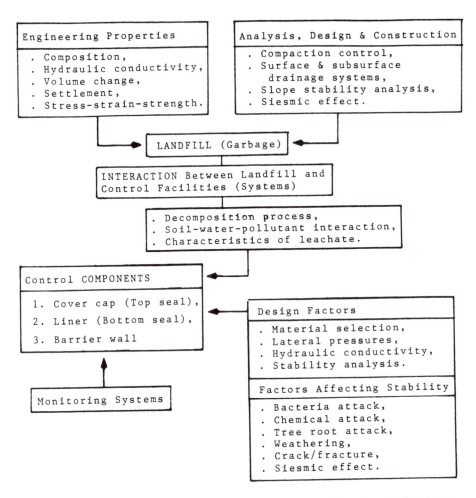

Figure 17.1 Some major elements or factors to be considered in waste control systems.

The test site was located on the proposed Route 52 in San Diego County, California. This is within the city of San Diego's sanitary fill accessible from Mercury Street north of Clairemont Mesa Boulevard. Refuse materials, including glass, wood, kitchen garbage, paper, rags, metals, etc., were dumped in this area between 1964 and 1967 and covered with a thin layer of earth. The area receives only about 10 in. rainfall yearly and little decomposition has taken place, except for food or vegetation waste. Based on this experience the following benefits were obtained:

1. The 50-ton roller can be used to compress a foundation material consisting of dry or unsaturated sanitary landfill to reduce total primary settlement. Maximum prerolling settlement was achieved after 25 passes of the 50-ton roller, 85% of the maximum prerolling settlement developed under 10 passes of the 50-ton roller.
2. Surcharge loading is of major benefit in accelerating consolidation of sanitary landfill. The settlement rate accelerated substantially when the maximum surcharge height of 10 ft was reached; 50% of the average total surcharge settlement occurred prior to completion of the surcharge embankment and 80% had occurred 30 days after completion of the surface embankment.

17.3.3 Surface Compaction Process

The process of surface compaction plays an important role for stabilizing the landfill. However, it requires planning during the waste disposal process period. Following are some suggestions graphically (Figure 17.2) illustrated and discussed by Fang (1995a):

1. Garbage comes in all types and it cannot be uniformly distributed in the landfill. However, within the limit, it should be distributed uniformly within the layer (Figure 17.2a).
2. If not, heavier items should be dumped closer to the center of landfill for the purpose of controlling the stability of the fill (Figure 17.2b) and avoid being dumped around the edges of the landfill (Figure 17.2c).
3. Spread the newly dumped garbage as thin as possible. Since garbage is unstable material, it requires stabilizing in order to reduce excessive settlement/subsidence. Mix locally available soil with the garbage or add fly ash, lime or other as a stabilizing material.
4. Heavy rolling such as the 50-ton roller can be used. The weight of the roller is related to the thickness of landfill layers. Theoretically, the heavier the roller the better the landfill stability. In some cases, use of dynamic consolidation to control the compaction in landfill areas recently has come into practice.

17.3.4 Deep Compaction on Landfill Sites

Conventional compaction techniques used in landfill areas sometimes give ineffective results for compacting highly nonuniform materials. Recently, dynamic consolidation has been used for this situation with good results (Inyang, 1992a). The effective depth or the depth of influence is important for landfill compaction. The method for estimation of effective depth in landfill area by dynamic consolidation technique is shown in Figure 11.3. Most of the design factors that need to be considered in heavy tamping projects are presented in Table 17.4. Numerical ranges for relevant design parameters are also suggested for use in projects that involve contaminant migration rate control. The most important design aspect is the selection of magnitudes of impact value parameters to match site conditions and compact the soil effectively. This selection process requires the use of a numerical relationship that contains parameters such as material (soil) properties, tamper weight, tamper size, drop height, and depth of influence of each impact as discussed in Section 11.4.

For deep compaction of landfill site, dynamic consolidation techniques generally produce effective results. However, careful observation must be made of some side effects such as: (1) vibration produced from dynamic consolidation may affect the liquefaction phenomenon on surrounding structures; and (2) the friction force produced from vibration may create a fire hazard in the landfill. There are several case histories on combustion fire in the landfill due to dynamic consolidation when landfill refuse is in a dry condition, because of the large amounts of gas (Figure 17.3) produced during the garbage decomposition process.

17.3.5 Compaction Control

In general, compaction control in landfill areas is not really controlled because the major intention is to dispose of the wastes with a simple attitude of "who cares". When this wasteland is converted into useful land at a latter date, frequently the regret is expressed that "we should have done that at an earlier date because we did not do it properly for current use." Therefore, during waste disposal process, advanced planning must be made to avoid complications at a later date. Two basic approaches are proposed as: (1) application of ground improvement (stabilization) techniques and (2) compaction control is suggested, with proper records on composition of landfill to be maintained.

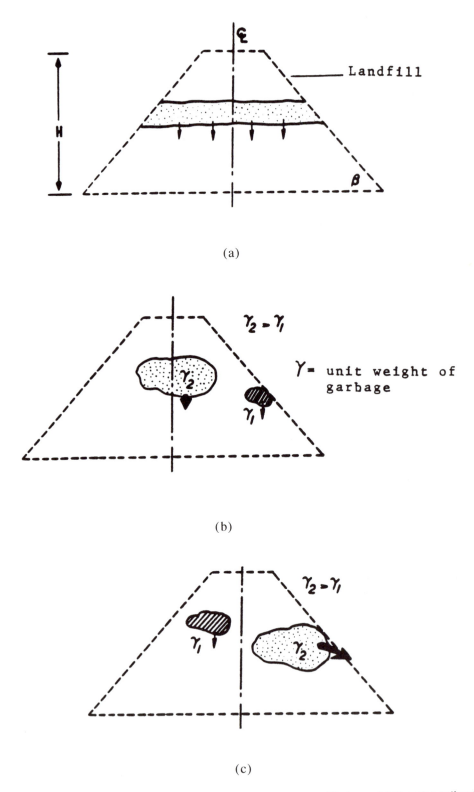

Figure 17.2 Effect of garbage dumping procedure on stability of landfill slopes. (a) Dumping uniformly: equal settlement — stable slope. (b) Dumping irregularly: (Case I); differential settlement — relatively stable slope. (c) Dumping irregularly: (Case II); differential settlement — unstable slope.

Table 17.4 Numerical Ranges of Heavy Tamping Design Factors for Contaminant Migration Rate Control

Factor	Numerical range	Comments
Required depth of compaction (m)	<12	Depends on the thickness of the loose layer
Applied impact energy (t-m/m²)	200–300	Applied impact energy depends on specified improvement in relative density of soil
Tamper diameter (m)	1.0–3.0	
Drop height (m)	10–30	
Drop weight (t)	5–40	
Repetitions (impacts per point)	2–8	Depends on specified improvement in relative density
Spacing of impact points (m)	1.0–4.0	Depends on the desired overlap of the tributary areas
Minimum depth to water table (m)	2.0–3.0	Temporary dewatering may be required to attain this condition
Required improvement in relative density (%)	>40	Depends on desired decrease in soil permeability
Peak particle velocity at nearby structures (cm/sec)	<4.0	Should be evaluated prior to full-scale project implementation

After Inyang (1992a).

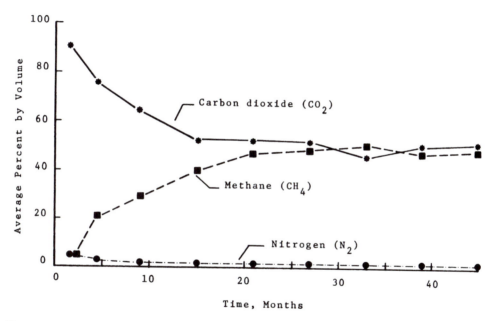

Figure 17.3 Gas distribution near the landfill. (After NRC, [1977], Methane Generation from Human, Animal and Agricultural Wastes, NRC, Washington, D.C.)

Application of the random inspection and sampling plan described by Hilf (1991) commonly used in geotechnical engineering should also be used in landfill areas to reduce the frequency of sampling and permit the estimation of the in-place performance of a landfill site.

17.4 SLOPE STABILITY OF LANDFILL

17.4.1 Slope Characteristic in Landfill Site

Slope stability in the landfill has two cases: one is the fill condition and the other is a cut condition. During the waste disposal process, the condition of the slope of the landfill site is

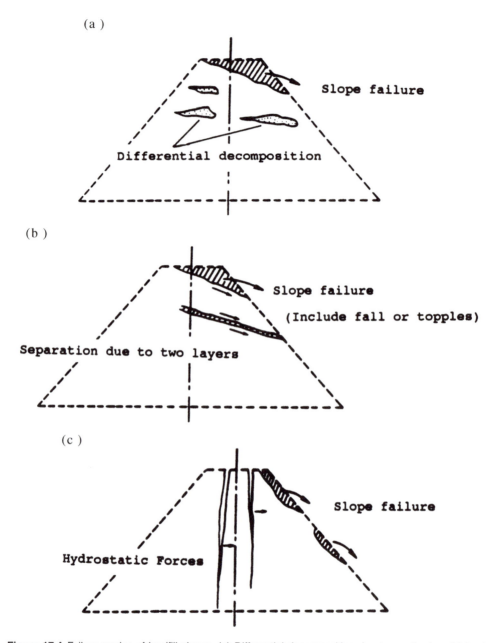

Figure 17.4 Failure modes of landfill slopes. (a) Differential decomposition due to weathering. (b) Lack of cohesion between two or more refuse materials. (c) Hydrostatic or environmental forces acting on decomposed or loosened refuse pieces. (d) Differential decomposition and differential settlement causing slope cracks and fall. (e) Chemical corrosion erodes the support and results in slope failure. Supported refuse material erodes the support. (After Fang [1995a]).

in the fill stage. How high can garbage be piled to stand up without slope failure? The critical height (H) of the slope is important (Figure 17.2a). When an abandoned landfill site is to be used for other construction sites that may require excavation, then the slope angle (β) of the cut is important. In general, landfill slope failure modes are similar to the slope failure in earth slopes. The types of failure also include falls, topples, slips, and slides. Falls and topples are due to lack of cohesion between two loose refuse pieces.

(d)

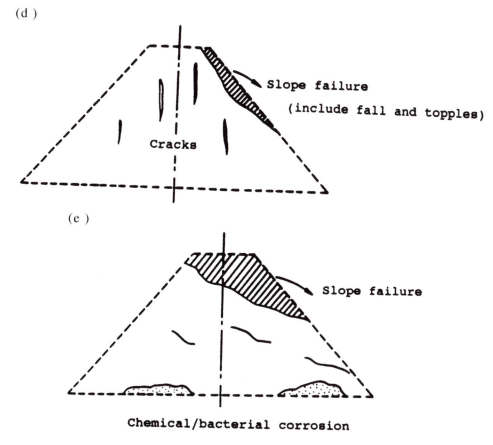

(e)

Figure 17.4 (continued)

17.4.2 Slope Failure Modes

Slope failure potential is directly related to the compaction control during the waste disposal process. The better control of compaction, the less risk for slope failures. Figure 17.4 illustrates some typical landfill slope failure modes.

1. Differential decomposition due to weathering (Figure 17.4a)
2. Lack of cohesion between two or more refuse materials (Figure 17.4b)
3. Hydrostatic or environmental force acting on decomposed or loosened refuse pieces (Figure 17.4c)
4. Differential decomposition and differential settlement causing slope cracks and fall (Figure 17.4d)
5. Chemical/bacteria corrosion of some supported refuse material erodes the support and results in failure (Figure 17.4e)

17.4.3 Slope Stability Analysis

(1) Current Practice

The stability analysis of slopes in refuse dams or hazardous waste landfills may require the use of a large computer program. Chen (1986) reports an analysis of a waste disposal site for the Taipei metropolitan area by the computer program STABLE II. A stability failure

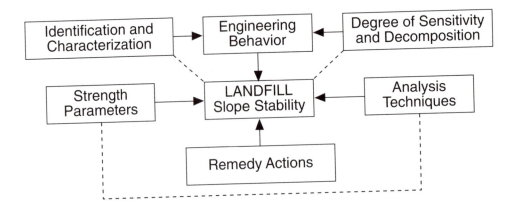

Figure 17.5 Stability of slope of landfill — genetic diagnosis approach.

on a sanitary landfill was analyzed by Dvirnoff and Munion (1986) using the computer program SLOPE.

Recently, various approaches and more refined computer programs have been developed. Current practice for seismic analysis and design of waste fills is given by Seed and Bonaparte (1992), Inyang (1992b), and Ling and Leshchinsky (1996).

(2) *Genetic Diagnosis Approach*

Refuse (garbage) in the landfill site is complex as discussed in previous sections. It cannot be generalized as a conventional geotechnical problem; therefore, a *genetic diagnosis approach* is proposed (Fang, 1995a) which may assist in understanding or solving the landfill slope stability problems. An overall system is presented in Figure 17.5. The general proposed procedures are explained as follows:

1. Identify and characterize garbage
2. Evaluate nature of garbage and engineering properties.
3. Examine the degree of sensitivity toward environment and potential decomposition.
4. If the landfill site is newly established, proper records must be kept for the purpose of future land use. Records should include sources, types, amounts, compaction characteristics, etc.
5. If it is an old/abandoned landfill site and the necessary records are no longer available, in such cases, taking boring logs is recommended.
6. Other information includes: the age of of the refuse deposit, thickness (or depth), local weather conditions, groundwater table, etc.

17.5 LABORATORY TESTS ON COMPACTED GARBAGE

17.5.1 Stress-Strain-Strength Characteristics

Figure 17.6 shows the photos of laboratory-prepared waste disposal material for soil stabilization (from left to right: 100% soil, 83% soil, and 75% soil by weight). The objective of these experiments was to find at which percentage the waste disposal material and soil should be mixed to achieve the optimum condition. The waste disposal material was composed of those components found by a national survey to be in the proportion (by weight) as shown in Table 17.1 (with the food wastes removed). In this case, only tensile strength (Section 6.7)

Figure 17.6 Laboratory prepared waste disposal material for soil stabilization (from left to right: 100, 83, and 75% of soils by weight).

was used. The test samples were molded and compacted in a standard Proctor mold. The graph of Figure 17.7 shows the load stress-strain behavior for specimens of 100, 94, 84, and 25% soil by weight. It was found that the tensile strength was improved for percentages of solid waste up to 25%. For higher percentages of solid waste the tensile strength was decreased.

17.5.2 Strength Parameters and Bearing Capacity

To use the traditional geotechnical engineering approach for estimation or prediction of the engineering behavior of material such as bearing capacity or stability analysis, it is necessary to have strength parameters c and ϕ of the material. Unfortunately, it is difficult to determine c and ϕ for garbage by a routine laboratory approach. However, a short-cut method has been developed for estimation of these parameters for compacted garbage (Fang et al., 1977 and Fang, 1995a). A brief description is presented as follows.

The method proposed was based on the Chen-Drucker modified Mohr-Coulomb failure envelope as discussed in Section 11.7. The test specimens used for this study are compacted natural waste disposal blocks (bale). Each bale $4 \times 4 \times 5$ ft in size was formed with three-dimensional baling forces varying from 1300 to 0 psi. Laboratory compression and tension tests on compacted natural waste disposal bales were tested at Lehigh University. Test specimens used for these studies were provided by the American Hoist and Derrick Co. Figure 17.8 shows the typical setup for compression test. Figure 17.9 presents load-compression and load-penetration curves. Then, the cohesion and friction angle of these materials can be computed from Equation (11.7) as discussed in Section 11.7 and bearing capacity of compacted garbage block can be estimated by the conventional bearing capacity equation.

17.5.3 Hydraulic Conductivity

Laboratory permeability on compacted garbage block (bale) specimens were also submitted to the constant head permeability test. The results of the unit weight of bale vs. coefficient of permeability are presented in Figure 17.10, and a linear relationship was found.

17.6 STABILITY OF GARBAGE DURING DECOMPOSED STAGE

17.6.1 Basic Considerations

From a geotechnical engineering viewpoint, the following basic factors must be examined for analysis and design of landfill controlling facilities:

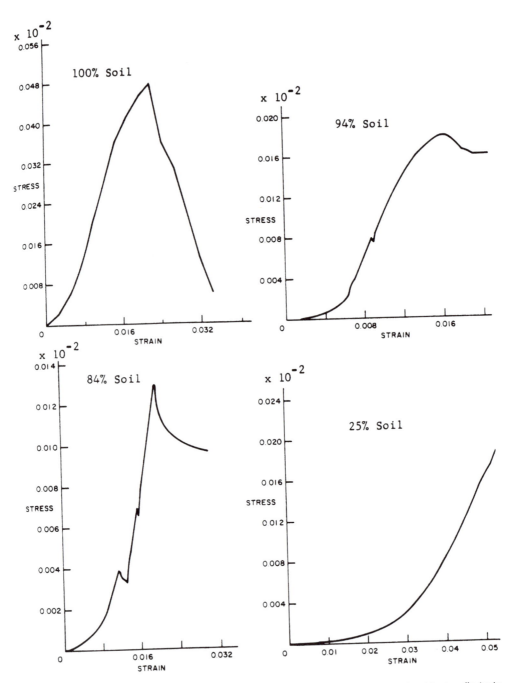

Figure 17.7 Stress-strain characteristics of compacted waste-soil mixtures determined by tensile tests. (After Fang, H.Y., Slutter, R.C. and Steubben, G.A. [1976], *New Horizons in Construction Materials*, v. 1, Envo Publishing, Bethlehem, PA, pp. 127–138.)

1. Technology for prevention or control of any polluted water (leachate) in the landfill seeping into the groundwater aquifer
2. Methods or procedures for covering the landfill, so that no additional water (i.e., rainwater, surface water) and animals disturb the landfill materials
3. Estimation of environmental pressures (or loads) produced from landfill
4. Hydraulic conductivity and mass transport phenomena around and in the landfill site

Figure 17.8 Typical laboratory setup for compression test of compacted garbage block (bale).

17.6.2 Decomposition Processes

Decomposition is a natural process. Any material in a given location, due to various environmental factors after a certain time, will gradually change its form and properties as discussed in Section 5.10. This process's stages between fresh and aging garbage are illustrated in Figure 17.11. In examining Figure 17.11, there are basically four stages of decomposition. The mechanical alteration is considered a short-term process, while the others are long-term processes. The chemical processes are dissolving or deteriorating processes of minerals. Others include physicochemical and geomicrobiological processes. Figure 17.12 illustrates the geomorphic process on a 10-year garbage specimen as reflected on a grain-size distribution curve.

(a)

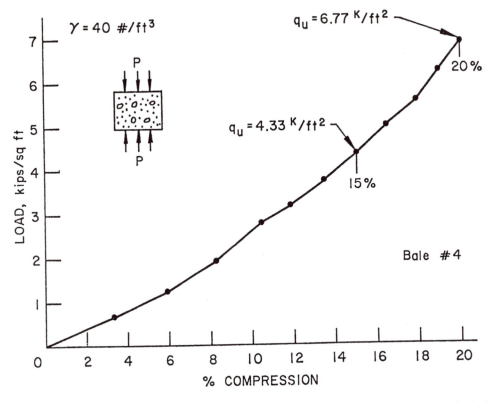

Figure 17.9 Load-deflection curves of compacted garbage block (bale). (a) Unconfined compression test; (b) tensile test. (After Fang, H.Y., Slutter, R.G. and Koerner, R.M. [1977], Proc. Specialty Session of Geotechnical Engineering and Environmental Control, Moh, Z.C., ed. 9th ICSMFE, v. 1, pp. 265–278.

The large amount of settlement in landfill area is contributed to mechanical loads such as surcharge loads, snow, rainwater, etc. Since the refuse is nonhomogeneous and contains organic matter, the settlement in the landfill areas is also nonuniformly distributed. Because of this phenomenon, the top seals (caps) of the landfill covers frequently break down or have excessive cracks.

Leaching and ion-exchange reaction affect soil properties significantly. Soil property changes occur when rainwater or drainage processes remove some soluble clay minerals in the clay liners or around the landfill sites by leaching. During the decomposition process in the landfill areas, the ground temperature increases. In sanitary landfill areas, the refuse is nonhomogeneous and contains larger amounts of organic matter; biological changes caused by decomposition constantly occur. Figure 17.13 shows the schematic diagram illustrating the cumulative settlement vs. time due to decomposition of organic matter. Basically, there are five stages with all stages related to the bacterial activities in the refuse material.

17.6.3 Settlement Prediction

Sowers (1975) suggested that continuing settlement of sanitary landfill is analogous to secondary compression of soil and settlement with respect to time and depth of fill as:

(b)

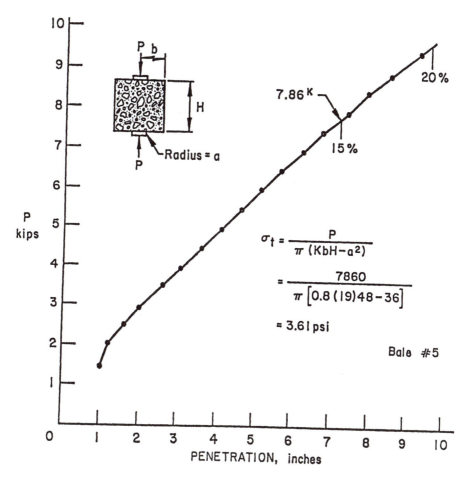

Figure 17.9 (continued)

$$H = \alpha \frac{H}{1+e} \log \frac{t_2}{t_1} \qquad (17.1)$$

where ΔH = total settlement, α = coefficient which depends on field condition, $\alpha = 0.09e$ (for condition favorable to decomposition), $\alpha = 0.03e$ (for unfavorable conditions), H = fill depth, t_1, t_2 = time, and e = initial void ratio.

Yen and Scanlon (1975) suggested that settlement rate can be computed by the following equation:

$$m = \frac{\alpha}{1+e} \frac{H}{t-1} \log(t) \qquad (17.2)$$

where m = settlement rate, ft/month and t = time elapsed, in months.

Other notations in Equation (17.2) are the same as defined in Equation (17.1).

Due to the nature of the problem, the settlement analysis of landfill cannot be solved by mathematical equation(s). Semi-empirical methods such as Equations (17.1) and (17.2) may

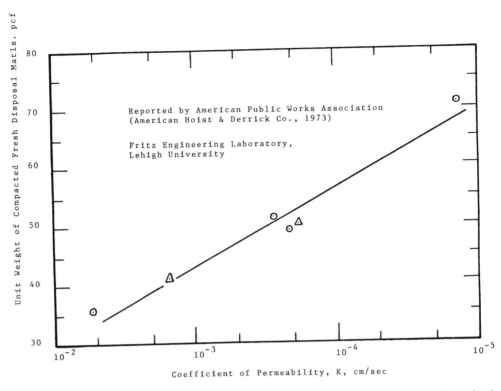

Figure 17.10 Unit weight of compacted garbage block (bale) vs. coefficient of permeability determined by constant head permeability test.

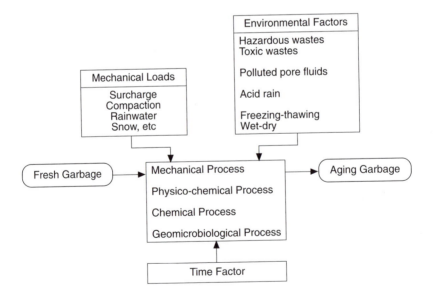

Figure 17.11 Aging process (geomorphic process) stages between fresh and aging garbages.

be the best approach at the present time. Yen and Scanlon suggested that precompaction of landfill can eliminate some of the initial settlement. The rate of settlement in landfill site is an important factor on which numerous investigators have reported including modeling,

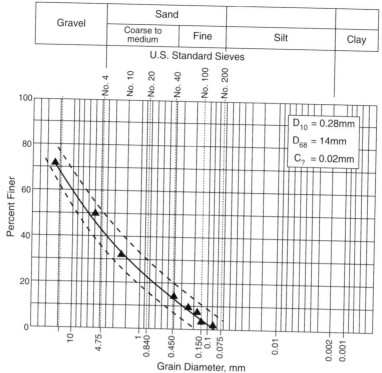

Figure 17.12 Grain size distribution for a 10-year-old garbage specimen. (After Witmer, K. [1987], Proc. 1st Int. Symp. on Environmental Geotechnology, v. 2, p. 389.)

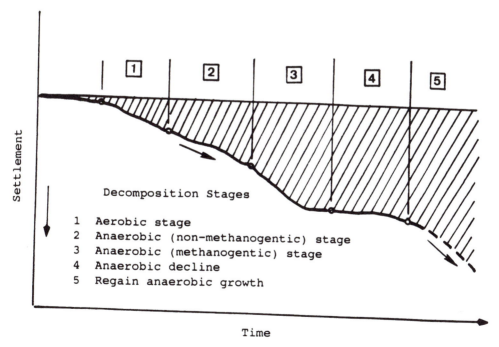

Decomposition Stages

1 Aerobic stage
2 Anaerobic (non-methanogentic) stage
3 Anaerobic (methanogentic) stage
4 Anaerobic decline
5 Regain anaerobic growth

Figure 17.13 Schematic diagram illustrating the cumulative settlement in landfill vs. time.

Table 17.5 Range of Composition of Leachate from Landfil

Component	Range
pH	3.7–8.5
Hardness, $CaCO_3$	200–7,600
Alkalinity, $CaCO_3$	720–9,500
Ca	240–2,400
Mg	64–410
Na	85–3,800
K	28–1,700
Fe	0.15–1,640
Ferrous ion	8–9
Chloride	50–2,400
Sulfate	20–750
Phosphate	0.5–130
Organic-N	3.0–490
Nh_4-N	0.3–480
BOD	22,000–30,000
COD	800–50,000
Zn	0.02–130
Ni	0.15–0.9
Suspended solids	13–27,000

Data from Evans (1990) and Fang (1995a).

laboratory, field observations, and case studies. A recent study of settlement analysis of waste landfill is given by Chen and Lee (1996).

17.6.4 Types of Polluted Water and Pressures Produced

(1) Types of Polluted Water Produced in the Landfill Areas

Polluted water produced in a landfill site generally is referred to as leachate. The properties of leachate also change from community to community as well as season to season. Table 17.5 shows the national survey results and its ranges. Leachate can be grouped into inorganic and organic pore fluids:

1. Aqueous inorganic: Aqueous inorganic are those in which water is the solvent and the solute is mostly inorganic.
2. Aqueous organic: Aqueous organic are those in which water is the solvent and solutes are predominantly organic. The organic part covers organic wastes and organic fluids.
3. Organic wastes: Organic wastes are those in which an organic fluid is the solvent and the solutes are other organic chemicals dissolved in the organic solvent.
4. Organic fluids: Organic fluids can be classified as organic acids and organic bases. Organic acids include those organic fluids which react with bases and include proton donors; and organic bases include any organic fluid capable of accepting a proton to become an ionized cation.

(2) Acidity Exists in Clay-Liner Interface

Most pore fluids produced from landfill are acidic as discussed in Section 5.3. These acids seep through from landfill site into liner, regardless of the types of liners used; the polluted fluid must interact with soil at various stages. All soils contain H ions in their aqueous phases since water itself is dissociated into H and OH ions. Higher H concentrations are due to exchangeable H ions on the soil particles and/or the presence of mineral or organic acids,

Table 17.6 Waste Disposal Facility Classification

Facility category	Facility description	Generic names	Control wastes disposed
I	Past disposal site	Abandoned Inactive Retired Midnight dump Uncontrolled site Orphaned	Little to none
II	Active disposal	Secure landfill Sanitary landfill	Some
III	Future disposal	Waste treatment complex Recycling facility	Well-controlled

After Evans, J.C. and Fang, H.Y. [1982], Proc. National Conf. on Management of Uncontrolled Hazardous Waste Sites, Washington, D.C., pp. 175–1982.

e.g., H_2SO_4 from oxidation of pyrites and other organic acids from the decomposition of vegetation. Soil acidity varies with the season.

(3) Environmental Pressures Produced from Landfill

The environmental pressures or stresses developing with the soil/water system are caused mainly by change of local environmental conditions. There are at least four types of pressure existing (Table 5.5), in addition to the external loads such as surcharge weight from landfill, lateral earth pressures, or loads due to construction activities around the landfill area (Sections 5.7 and 5.8).

(4) Conductivity Phenomena Between Landfill and Liner

Flow through fine-grained liner material has low velocity, is unsteady, nonuniform, or sometimes discontinuous, and contains toxic or hazardous substances in solid or gas forms. There are at least six possible causes which may cause the flow movement from one place to another, as summarized in Table 7.2. In all cases, the pressure or flow and causes for the movement are of small magnitude; however, the accumulations of these small amounts can be significant enough to cause groundwater pollution, thereby affecting the stability of all waste control systems. The bacterial and chemical corrosion or erosion may cause barrier leakage. Many case studies indicate that liners or barrier walls leak due to these unnoticed or commonly considered unimportant factors. For example, the velocity head (Section 7.3) in soil commonly is negligible; however, for prediction of pollution migration purposes, this small item must also be considered.

17.7 BASIC CONSIDERATIONS FOR DESIGN OF WASTE CONTROL SYSTEM

17.7.1 General Discussion

Waste disposal facility classification can be grouped into three categories as shown in Table 17.6. Facility description can be divided into *past* disposal site, *active* disposal, and *future* disposal. A systematic approach to the control of wastes requires the engineer to fully assess both the site and subsurface conditions and evaluate the applicability of containment alternatives. The basic requirements for controlling the wastes are given by Sarsby (1995) and Daniel and Koerner (1995).

To review existing information including historical site data, geological data, and groundwater data, for past disposal sites, it is necessary to obtain as much information as possible on the types of waste disposed, the timetable of waste disposal, and the previous disposal practices, i.e., drum, solid waste, lagoons. Aerial photos can be used. Information regarding the subsurface conditions can be obtained from past records and borings, from the site construction history, and from geological information.

17.7.2 Assess the Existing Site and Quantify Site Conditions

Assess in detail the existing site conditions including geological conditions, groundwater conditions, and contamination distribution. *In situ* investigation will be required at this stage. The use of geophysical instruments and other nondestructive testing (NDT) techniques prior to test boring or monitoring well installations can provide valuable insight into the subsurface conditions.

Quantify site conditions including the direction, volume, and velocity of groundwater flow, the interaction of groundwater with surface water, the distribution of contamination in the groundwater system, and the contaminant loading. The degree of sophistication of this quantification phase may vary from a simple conceptual model to a complex computer model. For example, the GEOTOX knowlege-based expert system can be used for this purpose (Mikroudis and Fang, 1987).

17.7.3 Develop the Containment/Treatment Program

Containment treatment program is closely related to environmental geotechnology. In many cases, subsurface investigations are primarily geohydrologic investigations. Geotechnical properties of soil may not normally be part of the routine investigation. It is desirable to have geotechnical engineering input during the site investigation phase to avoid future data gaps. Thus, testing should include tests for both physicochemical and engineering properties (Collins and Johnson, 1988 and Acar and Daniel, 1995).

17.8 WASTE CONTROL SYSTEM COMPONENTS

Solid and liquid waste control systems can consist of a wide range of components. These components can be classified in two general categories: *active* and *passive* components. The waste control system components are shown in Figure 17.14.

17.8.1 Active and Passive Components

(1) Active Components

Active components of a containment system listed in Figure 17.14 are those which require ongoing energy input. Examples of active components include disposal wells, pumping wells, and treatment plants. The disposal wells include injection wells, the pumping wells include pumping through a pumping ridge, and treatment processes include activated carbon specialty processes.

(2) Passive Components

Passive components of a containment system listed in Figure 17.14 are those which do not require ongoing energy input. Typical examples of passive components include drain tile

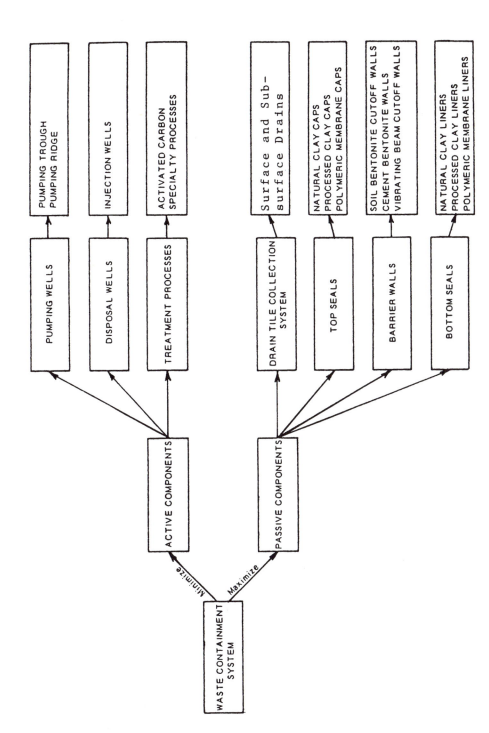

Figure 17.14 Solid and liquid waste control systems. (After Evans, J.C. and Fang, H.Y. [1982].)

collection systems, barrier walls, liners, and covers. While an active system requires ongoing energy input, the passive components typically require maintenance. For example, it is necessary to keep vegetation taproots from penetrating a clay cap. Further discussion on tree and vegetation effects on liner stability will be presented in Section 17.9. Passive components can be further divided into subsystems. Detailed discussions of each case will be discussed in the following sections

> Barriers walls
>> Soil-bentonite cutoff walls
>> Cement-bentonite walls
>> Composite system vertical cutoff barriers
>> Vibrating-beam cutoff walls
> Top seals (caps)
>> Natural clay caps
>> Bentonite caps
>> Polymeric membrane caps
> Bottom seals (liners)
>> Natural clay liners
>> Bentonite clay liners
>> Polymeric membrane liners

17.8.2 Structure of Control System Components

(1) Barrier Walls

The containment of contaminant migration from existing disposal sites or impoundments may necessitate a subsurface barrier to horizontal groundwater flow. Barriers are commonly constructed as soil-bentonite slurry trench cutoff walls or cement-bentonite slurry trench cutoff walls. Vibrating beam cutoff walls, grout curtains, or sheet piling are also used. For barrier walls to be effective, they generally key into an impermeable stratum of natural materials beneath the site.

(2) Soil-Bentonite Slurry Wall

The procedure for design and construction for soil-bentonite slurry trench cutoff walls is well documented. As shown in Figure 17.15(a), a trench is excavated below the ground surface and trench stability is maintained utilizing a slurry of bentonite and water. This slurry maintains trench stability in much the same way as drilling fluid maintains borehole stability. Figure 17.15(b) is a photograph of an actual installation. The bentonite water-slurry is designed by the geotechnical engineer to have proper density, viscosity, and filtrate loss properties which allow for the formation of a filter cake along the walls of the trench resulting in a computed factor of safety for trench stability greater than one. Trench depths are generally limited to about 10 m using conventional backhoe diggers. In order to achieve greater depths, a modified dipper stick is required which can be provided by specialty slurry wall contractors. To go deeper than 17 m usually requires the utilization of a clam shell, and an extended backhoe capable of excavating 22 m has been developed (Evans, 1991 and Bagchi, 1990).

A typical cross section of a slurry wall during excavation and after backfilling is shown in Figure 17.16. After excavation and addition of the slurry, the trench is backfilled with a matrix of soil materials. The mixture of soil and bentonite water slurry typically has a consistency similar to that of high slump concrete. In order to design a soil-bentonite slurry wall for waste containment, studies are required beyond those typically required for other applications Chemical analysis of samples of onsite materials considered as potential backfill materials may be required.

(a)

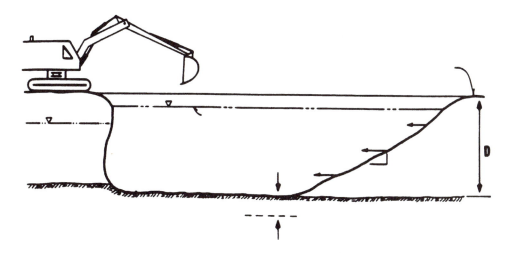

Figure 17.15 Excavation of slurry trench. (a) Schematic section of slurry trench excavation and backfill; (b) photo shows the excavation of slurry trench.

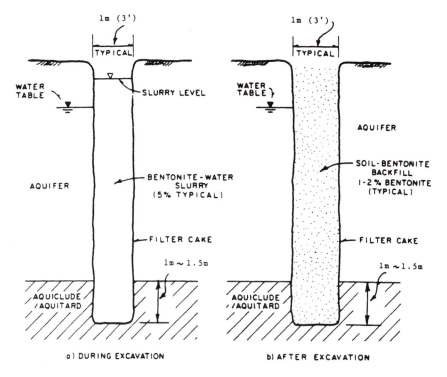

Figure 17.16 Typical section of soil-bentonite slurry trench cutoff wall.

(3) Cement-Bentonite Slurry Trench Cutoff Walls

As an alternative to soil-bentonite cutoff walls, cement-bentonite cutoff walls can be utilized. The trenches are excavated in a manner similar to soil-bentonite walls as described in case (1) utilizing a slurry to maintain trench stability. However, cement rather than soil is used in the slurry. The slurry is left in the trench and allowed to harden. A strength equivalent to stiff clay can be obtained after a period of about a month. Design considerations include the cement and bentonite content and type, and their relationship to the strength and permeability of the backfill. Leachate compatibility tests must be conducted utilizing the site contaminant permeant. The overall permeability of a cement-bentonite cutoff wall is generally higher than for soil-bentonite walls.

(4) Composite System Vertical Cutoff Barrier

A concept for constructing a vertical impermeable barrier to prevent the migration of contaminated groundwater or leachate from a hazardous waste site or waste disposal area is proposed by Druback and Arlotta (1985). The composite system is a hybrid cutoff wall constructed with high density polyethylene (HDPE) and sand backfill and is installed using the slurry trench construction method. When installed, a very low permeability, composite, vertical barrier is established with unique engineering properties, including improved chemical resistance, leak detection, and groundwater migration control. A full-scale construction test project of the system was performed at an existing sanitary landfill in New Jersey to demonstrate the overall fabrication and construction procedures. Detailed procedure on this system includes design construction and performance as provided by Druback and Arlotta in their paper.

(5) Vibrating-Beam Walls

Barriers to horizontal groundwater flow have been designed and constructed using the vibrating-beam injection method. This technique utilizes a vibratory-type pile driver to cause the penetration of a beam of specified dimensions to the design depth. Slurry is added through injection nozzles as the beam penetrates the subsurface soils and as the beam is withdrawn.

The slurry utilized with the vibrating-beam technique is generally either of two types, cement-bentonite or bituminous grout. Mix design considerations for cement-bentonite were discussed in case (2). Bituminous grouts are prepared as a homogeneous blend of asphalt emulsions, sand, portland cement, and water. Fly ash may also be included. It is reported that this bituminous grout can resist strong acids and high saline content wastes. Using thin slurry cutoff walls installed by the vibrated-beam method was discussed by Leonards et al. (1985).

17.8.3 Top Seals (Caps)

The function of *top* seals or *surface* seals (caps) as shown in Figure 17.17 is to control surface water so as to minimize infiltration, or to prevent animals digging into the landfill areas, thereby reducing subsequent leachate production and/or contaminant transport potential. There are several types of material that can be used for top seals such as natural clay, bentonite clay, and synthetic membrane. A brief discussion on each of these is presented as follows.

(1) Natural Clay Caps

The most cost-effective top seal usually is native clay material from locally available sources. If suitable native clays are not available, imported materials or synthetic membranes must be used. In general, clays will last longer than synthetic materials, and clay caps rather than synthetic caps are usually chosen. However, to avoid the bathtub effect (more water entering the facility than can drain out), the use of a synthetic membrane cap whenever the bottom liner is also a synthetic membrane is required (EPA, 1982).

Selection of clay materials for clay caps is based on the hydraulic conductivity of locally available material as discussed by Sarsby (1995). As shown in Figure 17.17, the thickness of the clay cap is typically from 0.6 to 0.8 m. Some factors to be considered in design are

1. The top several inches of clay cannot be as well compacted as the remainder of the thickness. Further, it may be difficult, in the long term, to maintain the clay density in the top few inches due to potential desiccation cracking and frost action.
2. The bottom of the clay cap may become somewhat intermixed with subgrade material during construction. Therefore, the effective thickness of the nominal 0.6-m cap may be less than that.
3. The compaction procedures for the clay seals are similar to the standard compacted fills (Section 11.2).
4. Other protection from degradation due to erosion by rainwater, cracking due to drying, rutting from moving vehicles, as well as tree root attacks must be considered.

(2) Bentonite Clay Caps

Where local natural clay is not available in sufficient quantity, quality, or price, processed clay is a commonly utilized alternative. Processed clay is typically sodium montmorillonite deposits. Bentonite is a hydrophilic colloidal clay which swells in water. The construction

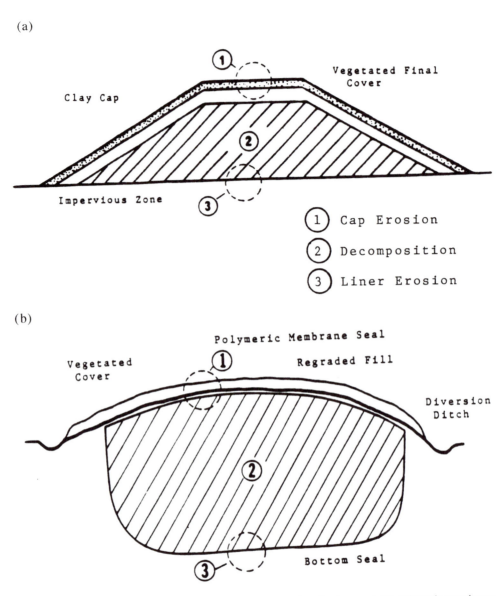

Figure 17.17 Typical section of waste control system. (a) Native clay top seal; (b) polymeric membrane seal.

of a processed clay cap requires the application of the bentonite at a controlled rate (e.g., about 2 psf), adequate mixing with the in-place soil to a predetermined loose thickness, and compaction. The main advantage of this method is low cost with relatively low hydraulic conductivity. The disadvantages are (1) it is difficult to obtain uniform application at the blending rate and (2) bentonite clay is very sensitive to pore fluid. If polluted water is introduced, significant changes of clay behavior are expected (Anderson et al., 1985).

(3) Synthetic Membrane Caps and Others

Synthetic membranes (Figure 17.17b) can be utilized as top caps for waste containment. Membranes are available in a wide range of materials from numerous manufacturers. Recent developments in both materials and/or technology for caps or covers have been reported.

Kargbo et al. (1993) discussed the environmental significance of acid sulfate 'clays' as waste covers. It is suggested that clay cap designers test the acid-generating capabilities of potential clay cap materials taken at the sources, by incubating a sample of the candidate cover soil (pH > 3.5) under moist aerobic conditions at room temperature. The soil will be said to contain sulfidic materials if it shows a drop in pH of 0.5 or more units to a pH value of 4.0 or less within 8 weeks. Decisions should then be made as to whether the clay soil should be abandoned or used with amendments to overcome the acidity.

Zimmie et al. (1995) used waste paper sludge for landfill cover material. Paper mill sludges have a high water content ranging from 150 to 250% and have a high degree of compressibility. Composed of 50% kaolinite and 50% organics, the sludge behaves like a highly organic soil. Based on laboratory standard tests on *in situ* specimens indications are that hydraulic conductivity met the $1 \times 10-7$-cm/sec requirement for landfill covers. Recent studies on covers are given by Lee (1996) and Ling and Leshchinsky (1996).

(4) *Other Surface Water Controls*

In addition to surface sealing, surface water diversion and collection systems can provide short- and long-term measures to isolate waste disposal sites from surface water inputs. Techniques used to control flooding and offsite erosion transport of cover and surface seal materials include dikes and berms, interceptor ditches, diversion dikes and berms, terraces and benches, sheets and downpipes, levees, seepage ditches, and sedimentation basins and ponds.

Surface seals provide multiple functions in the overall liquid and solid waste containment control system. The main function has been discussed as the control of infiltration by minimizing water infiltration and/or maximizing surface runoff away from the side as well as the minimization of erosion.

17.8.4 Bottom Seal (Liners)

In new containment facilities, it is frequently necessary to provide a liner system. The major function of a liner is to prevent leachate or waste from entering the groundwater regime; liners, as with covers, can consist of native clays, processed clays, or synthetic membrane liners. It is important to note that under the 'U.S. EPA's 1982 Interim Regulations', a synthetic membrane liner is considered best to 'prevent' migration of wastes, whereas a clay liner will 'minimize' migration of wastes.

(1) *Native Clay Liners*

The compatibility between the natural clays and the waste is an important design consideration for the use of natural clays as liners. It is important to ascertain the volume change and permeability characteristics of the proposed clay liner material. The bulk transport of liquid waste through cracks as discussed in Section 17.6 must be precluded. Bulk transport of liquid through clay liners could occur due to differential settlement of the foundation base materials.

(2) *Bentonite Clay Liners*

Design and construction considerations for the use of processed clay for liners must include waste-liner compatibility as well as considerations previously discussed. The volume change characteristics of the processed clay are especially important. Generally, the processed clay is mixed with the subgrade material to form the impermeable liner. The impedance to groundwater flow is primarily due to the processed clays, especially when the matrix soil is

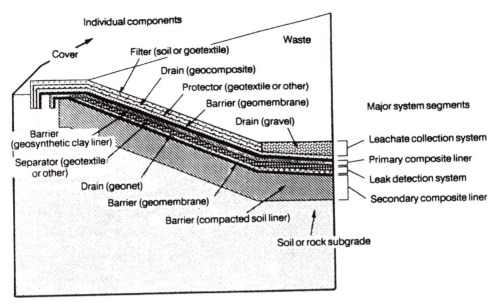

Figure 17.18 A double-composite landfill liner system. (After Daniel, D.E. and Koerner, R.M. [1991], Civil Engineering, ASCE, New York, pp. 46–49.)

relatively free of natural fines. Hence, if the processed clay shrinks upon exposure to the waste, large increases in permeability can occur. The hydration of a processed clay liner with uncontaminated water prior to waste disposal is recommended. The triaxial permeability tests (Section 7.3) with the actual proposed clay subgrade material, groundwater, and leachate contaminants discussed previously should be conducted in this situation.

(3) Geosynthetic Membrane Liners

As with other liner types, waste compatibility is a major design consideration. However, the permeability of a polymeric liner can also be increased due to liner stretching. Thus, total and differential foundation settlement can impact the liner design. Close construction control is essential to the overall system performance. The 'permeability' of an installed membrane liner system is generally a function of bulk transport through seam, joints, tears, holes, and pinholes. The long-term durability of geomembranes is discussed by Koerner (1994). Studies included ultraviolet radiation, chemical degrations, as well as swelling, oxidation, and temperature. In addition, the predictive methods for evaluation of long-term durability are also proposed and discussed.

17.8.5 Double-Composite Liner System

A *double-composite liner* system proposed by Daniel and Koerner (1991) is based on the combination of clay and geotextile structures as shown in Figure 17.18. This system is the minimum standard of care to which future landfills that can produce significant quantities of threatening leachate should be held. These systems require careful design, testing, and construction. The system consists of four major parts:

- Leachate collection systems (layers 1, 2, and 3)
- Primary liner (layers 4, 5, and 6)
- Leak detection system (layer 7)
- Secondary line (layers 8 and 9)

These major parts are made up of nine layers. A tenth layer, the soil or rock subgrade, completes the double-composite liner system.

(1) Leachate Collection Systems

The leachate collection system is the uppermost segment as shown in Figure 17.18. The system is composed of a highly permeable granular material on the base of the disposal unit and a high-transmissivity geocomposite material for side slopes provides timely cover with adequate protection.

(a) Layer 1: Filters (Soil or Geotextile)

This layer separating the lowest portion of select waste or initial operations layer of soil from the leachate collection and drainage medium is essential. If suspended particles from the leachate travel into the drainage system, it may clog. To minimize this problem, it is suggested that filters be used only when they are truly needed and high-permeability filters be employed. A geotextile filter is part of the geocomposite drainage layer along side slopes. Geotextiles are sensitive to ultraviolet-light degradation if left exposed.

(b) Layer 2: Drain (Gravel for Base, Geocomposite for Side Slopes)

The leachate collection layer requires high in-plane transmissivity and pores sized to resist plugging.

(c) Layer 3: Protector (Geotextile or Other)

This layer prevents materials in the drainage layer from puncturing the primary geomembrane liner. Protectors are usually thick, needle-punched, nonwoven geotextiles.

(2) Primary Composite Liner (Figure 17.18)

(a) Layer 4: Barrier (Geomembrane)

This component of the primary liner can be made from polymeric materials, including polyvinyl chloride (PVC), chlorinated polyethylene (CPE), chlorosulfonated polyethylene (CSPE), ethylene interpolymer alloy (EIA), high-density polyethylene (HDPE), and very low-density polyethylene (VLDPE).

(b) Layer 5: Barrier (Geosynthetic Clay Liner)

For the soil component of the primary liner, the geosynthetic clay liner (GCL) is recommended. GCL, previously called prefabricated clay blankets and other terms, are factory-manufactured dry bentonite clay layers sandwiched between geotextiles or attached to a geomembrane.

(c) Layer 6: Separator (Geotextile or Other)

To avoid migration of clay particles from GCL into the underlying geonet, adequate separation is needed. Best results in experimental tests have been obtained with (1) nonwoven, needle-punched geotextile with small apparent openings; (2) nonwoven, heat-bonded geotextile; and (3) geomembranes.

(3) Leak-Detection System (Figure 17.18)

This leak-detection system is the third major part of the system. It is also known as the secondary leachate collection system. It identifies leakage from the primary lining system and enables it to be collected and removed.

(a) Layer 7: Drain (Geonet)

A geonet is preferable to granular materials for the leak-detection layer because it is much easier to place on side slopes and can be placed with lightweight equipment, and because granular materials can puncture a geomembrane. Geonets also offer faster detection of leaks than most granular materials. Hazardous waste facilities typically must detect leaks within 24 hr, and often a geonet is the only material that can do this.

(4) Secondary Composite Liner (Figure 17.18)

(a) Layer 8: Barrier (Geomembrane)

Technical requirements for the secondary geomembrane liner are generally the same as for the primary layer (layer 5), so the same type and thickness of material are usually used.

(b) Layer 9: Barrier (Compacted Soil Liner)

A compacted soil liner can be constructed without risk of damaging any underlying liner system components.

(5) Layer 10: Subgrade (Soil or Rock)

For large-scale facilities, a complete subsurface soil investigation is necessary. Once foundation stability is assured, the site should be proofrolled and graded.

17.9 FACTORS AFFECTING THE STABILITY OF WASTE CONTROL SYSTEMS

17.9.1 Chemical and Tree Roots Attack on Liners

Control and prevention in the stability of the waste control system are important from an economic viewpoint, because repairs in any facilities in hazardous/toxic sites are very costly. There are numerous factors affecting the stability of the system; however, the chemical, bacteria, and tree-root effects are most significant. When a liquid interacts with a solid, it only happens on the solid's surface. For a clay mineral, the surface is electrically charged. The liquid phase, on the other hand, is more complex than the solid phases. Therefore, the discussion concentrates on liner-polluted liquid interaction.

Landfill areas, in general, are planted with tree/vegetation for the purpose of beautification of unwanted landfill sites, but planting trees or vegetation also has a detrimental side effect. Two important phenomena on how tree roots attack liners are

1. Tree roots looking for nutrition (food), the landfill area is an ideal place; roots are capable of and will penetrate liners, and
2. Pore fluid (leachate) in landfill will seep out through root hairs' suction force.

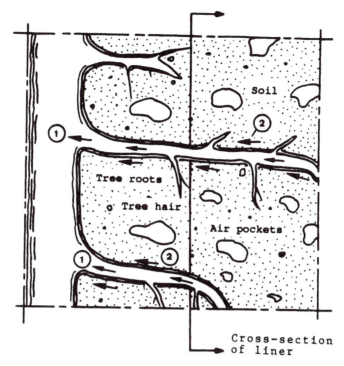

Figure 17.19 Mechanisms of soil-root-pore fluid (leachate) interaction at landfill area. (1) Leachate in landfill seeps out through root hairs by suction force; (2) leachate in landfill seeps out between root and soil by capillary action.

All higher plants require immense quantities of water, practically all of which is absorbed from the soil through their root systems. Special organs are provided for absorption which include the tip of roots and innumerable root hairs. The root tips are always seeking new 'feeding grounds'. Garbage dump site is one such rich food storage area.

How pore fluid (leachate) in landfill seeps out through root hairs is illustrated schematically in Figure 17.19. This phenomenon of moisture migration within the root hair and between root hair and soil is due to suction force and capillary action. Plant species differ not only in their water requirements, but also in their adaptation to water conditions. Some plants are so constituted that they can tolerate an abundance of soil water, whereas others can endure limited moisture. Therefore, tree and vegetation planted around the landfill area must be examined.

17.9.2 Leaking from Impervious Ground Soil Layers

Leakage through the landfill bottom seal will end up at barrier wall and the impervious ground soil layer as indicated in Figure 17.20. At present, only the safety of the barrier wall (Figure 17.16) is the major target for our concern. Designers do not suspect that leaks can pass through the so-called impervious soil layer. There is little information available about failure in this assumed 'safe zone.'

From an environmental geotechnology point of view, there is no such thing called 'impervious soil layer'. As discussed in Sections 3.2 and 3.3, it is indicated that:

1. Soils are three-dimensional systems; they have a two dimensional areal extent and a third depth dimension.

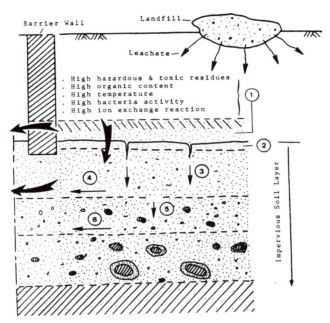

Figure 17.20 Leachate leaks from impervious soil layer.

2. They are geological deposits or formed on site by the interaction of geologic parent material, climatic factors, topography, and living organisms. Soils *in situ* share many essential properties with 'living' systems such as microflora and microfauna (Section 5.11) dispersed in soils.

3. Natural soil is not a mixture of their components. Horizontal as well as vertical transition into another soil type may be gradual or abrupt depending on geologic and soil-forming factors (Section 5.10).

4. Vertical differentiation as caused by soil-forming factors is called horizons, as illustrated in Figure 3.3, and voids exist between horizons as indicated in Figure 17.20.

When leachate leaks through a bottle seal of landfill it will end up at zone [1]. It is in zone [1] where the leachate has:

- High hazardous and toxic residues
- High organic content
- High temperature
- High bacteria activity
- High ion-exchange reaction

As discussed in Chapter 5, the soil-water interaction in the multimedia energy fields and the sensitivity of soil to environment are critical. The mass transport phenomena in zone [2] are significant and some leachate is seeping through this so-called 'impervious soil layer' as indicated in by points [3], [4], [5], and [6]. In addition, there is horizontal capillarity action (Section 7.2) as indicated by points [4] and [6] of Figure 17.20. It is recommended that for analysis and design of waste control facilities, the conditions of impervious soil layer must be carefully examined.

17.9.3 Dynamic Load Effects

Earthquake or dynamic load affects contaminated clay liner material, so for analysis and design of hydraulic barriers or any other waste control facilities two important aspects must be considered: (1) the behavior of soil-pollutant interaction under dynamic load and (2) what the mechanism is between load and clay liner.

As discussed in Section 11.11, with limited experimental data shown in Figures 11.16 and 11.26, some comments and suggestions may be made. The result from static load test of clay liner material is shown in Figure 11.16. In this experiment, temperature variations are introduced. As temperature increases, the shear strength decreases as expected, but when acid concentration increases, as reflected in the pH value, the shear strength also increases, because the catalytic effect on soil-acid reactions will increase the bonding action; consequently, the shear strength increases. The result from dynamic test at room temperature is shown in Figure 11.26 and is a contradiction as indicated by the circle. Under dynamic loading, if the dynamic excitation force is greater than bonding force created by the catalytic effect, higher acid concentration will not benefit the catalytic effect. As indicated by Fang (1994b), dynamic loading may cause more damage than static loading in contaminated ground areas.

17.9.4 Effects of Nearby Sinkholes and Others

If sinkhole(s) are developing nearby the waste control facilities, there is a good possibility instability will be caused in the control facilities. As discussed in Section 13.7, the contaminated pore fluid will accelerate the development of the sinkhole if a waste control system is located in a karst region. Gabr and Hunter (1992) used plain-strain finite element analyses to estimate the stresses and deformation induced in landfill liners constructed on existing or potential sinkholes. The analyses demonstrated the significance of using structural geogrids to reduce the magnitude of deformation and possibly the potential for collapse of landfill liners. In variations of analysis, key parameters include depth of overburden (D) and width of the sinkhole (B). For a D/B ratio of 2 and a (D) of 40 ft, the tensile strain is reduced from 18%; for the case of no reinforcement, to 5% when reinforcement is included.

If landfill site is located near mudflats and marsh deposits (Mitchell et al. 1992) and other problematic soil/rock deposits, proper precautions must be taken in order to avoid further complications as discussed in Chapter 13. Factors affecting long-term properties of clay liners and contaminated clays have been discussed by Mitchell and Jaber (1990), Meegoda and Rajapakse (1993), and others.

17.10 UTILIZATION OF WASTES FOR ENGINEERING PURPOSES

17.10.1 General Discussion

New construction, expansion, rehabilitation, and maintenance are all dependent on the supply of large amounts of material. Since economic consideration is one of the major decision factors, it is necessary to examine low cost, energy saving, and utilization of wastes for these activities. In order to select the existing material or to develop new substitute material, knowledge of the present status of production, consumption of materials, and the economic energy and environmental aspects for development of the material needs must be examined (Inyang and Bergeson, 1992).

Based on recent study, Ahmed and Lovell (1992) conducted a synthesis study to identify those waste materials which have demonstrated technical, economic, and environmental feasibility for use in highway construction. Current uses of waste products in the U.S. highway industry are given in Table 17.7. A total of 27 waste products have been reported. Of the 27 waste products, only 11 are presently used by three or more respondents which are listed in descending order of number of reported users in Table 17.7. Using garbage as part of highway construction material has been discussed in Section 17.3.

Table 17.7 Current Uses of Waste Products in U.S. Highway Industry

Waste material	No. of states using material	Material is used as additive to/material is use as[a]			
		(a)[b]	(b)[b]	(c)[b]	(d)[b]
Reclaimed paving materials	43	23/8	26/16	14/8	6/5
Coal fly ash	33	22/1	6/2	7/—	5/2
Rubber tires	30	22/—	6/1	1/—	3/3
Blast furnace slag	17	5/4	3/5	—/3	1/2
Steel slag	9	4/1	2/2	1/—	—/2
Coal bottom ash	7	2/—	2/1	1/1	1/1
Boiler slag	7	4/—	1/1	—/1	—/1
Used motor oil	7	Used as: fuel, 1; recycle, 3; embankment, 1			
Waste paper	7	Used as: recycle, 2; mulch for landscaping, 5			
Mine tailings	5	—	—	—	—/5
Sewage sludge	3	Wearing course, 1; landscaping, 3			

[a] Figures in each column indicate the number of states using the waste product either as additive to/or as replacement for conventional materials.
[b] Key to pavement components: (a) wearing course; (b) base; (c) subbase; and (d) subgrade/embankment.
After Ahmed, I. and Lovell, C.W. (1992), Proc. Mediterranean Conf. on Environmental Geotechnology, pp. 409–418.

17.10.2 Technology on Recovery and Reuse of Waste Materials

The subject of resource recovery and re-use of waste materials has gained much attention within the past decade, principally due to the increased number of enviromental statutes and regulations that necessitate minimizing of waste disposal (Collins and Ciesielski, 1994 and Sharma and Lewis, 1994). The benefit of re-use of stabilized and/or solidified waste materials should be twofold:

1. Reduce environmental hazards: Compliance with regulations will thus help reduce environmental hazard.
2. Added economy: Effective utilization of these materials as an inexpensive alternative for conventional materials can produce much economy provided that: (1) there are no adverse effects to the environment, and (2) the created material performs similarly or equally well as the one it replaces.

The technology for chemical fixation and solidification of liquid and sludge wastes increased vastly (EPA, 1985). The use of CFS for waste materials dates back to the 1970s (Conner, 1990). However, the development and use of the technology in ground improvement methods (e.g., soil stabilization and grouting) are relatively new.

17.10.3 Scrap Waste Rubber Tire

(1) General Discussion

The U.S. Environmental Protection Agency indicates that production of rubber tires for automobiles increased 42% from 1960 to 1980. Currently, an estimated 280 million waste tires are discarded annually in the U.S. alone. Among this amount, only a small amount of tires are recovered for recycling purposes (5.6% was recovered in 1988). Most waste tires are dumped into landfill space. The scrap tire piles are growing each year and pose two major threats to the public:

1. Fire hazard: Once set ablaze, it is almost impossible to extinguish such a fire, causing serious air pollution problems.
2. Health hazard: The debris/water trapped in the empty spaces of the tires provides an ideal breeding ground for insects, small animals, as well as bacteria.

(2) Current Practice

Current methods for reuse of scrap waste tires can be grouped into three general categories:

1. Use as a fuel source or as raw material in production of other polymeric materials
2. Use as additive or ground into small particles as part of asphalt-rubber mixtures for use as substitute in asphalt pavement material mixtures
3. Use as lightweight aggregates for concrete mixtures or as soil reinforcement element, either cut into small pieces, shredded, or whole tires

If the waste tires are used as described in categories (1) and (2), additional refining processes are needed, consequently requiring additional investment and creating additional environmental problems. Also, results obtained from these categories for large scale field applications are still in the trial and error stage with no definite benefits reported.

(3) Uses of Scrap Rubber Tire

(a) Construction Fills

Scrap rubber tire used as aggregate in construction fill has been studied. There are two types of construction fills with the first being compacted fill generally used as highway embankment. The second type is low-strength fill used to fill up empty space around bridge abutments, retaining walls, and building foundations.

(b) Repair for Sinkhole

Scrap rubber tire added into soil-cement used to fill-up sinkhole has been studied (Morrison et al. 1993). This low-strength mixture with higher slumps could be used in filling sinkholes, since the material must be flowable enough to move through the sinkhole and plug up the subterranean cavities that may be associated with the sinkhole.

(c) Scrap Rubber Tire-Clay Liners

Scrap rubber tire-clay liners against petroleum-based contaminants have been evaluated by Baykal et al. (1992). Good interaction between tire and clay or fly ash has been achieved. When gasoline was used as the permeant, the hydraulic conductivities of both clay and fly ash increased. A significant decrease in hydraulic conductivity values was observed for mixtures with rubber. The rubber particles are lathlike and with varying lengths. A typical grain-size distribution curve of rubber particles is obtained. Swollen rubber fibers upon contact with gasoline decreased the pore space in the mixtures and reduced the hydraulic conductivity.

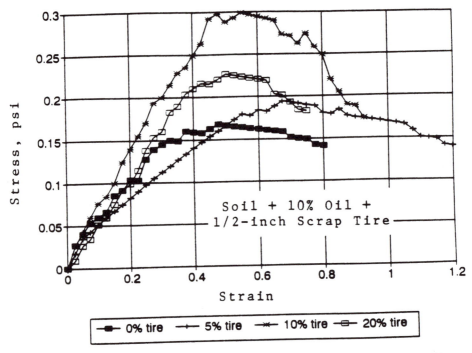

Figure 17.21 Stress-strain relationship of soil-tire-oil mixture. (After Fang et al. [1992].)

(d) Other Uses of Rubber Tire

The whole or cut-half rubber tires are frequenty used for river banks, harbor, or earth slope protection. Scrap rubber tires have been used as light-weight aggregates for base materials in low-volume roadways (Morrison et al., 1993 and Frascoia and Cauley, 1995), parking lots, and secondary earthquake-resistant structures.

17.10.4 Other Waste Uses

Solid, semisolid, or liquid waste generated from municipal, commercial, or industrial waste treatment facilities, wastewater treatment plants, waste supply treatment plants, or air pollution control facilities exclusive of treated effluent from a wastewater treatment plant show promise. Dredged materials can be reused as conventional construction material as reported.

The problems of oil-contaminated soils around the coastal margin have been discussed in Section 14.4. Utilization of these oil-contaminated soils to be used in conventional construction have been studied by numerous investigators. Most of their work concentrates on how this contaminated soil can be reused. One typical example is illustrated in Figure 17.21. In examining Figure 17.21, 10% of oil and 10% of 0.5 in.-diameter scrap rubber tire give better results, as indicated.

17.11 SUMMARY

1. Almost 90% of all hazardous wastes are in liquid form. An understanding of the pore fluids on liner behavior is essential to the design of various components of waste control facilities. Without a good understanding of the liner-pore fluid interaction, there is no sound basis on which to project the long-term behavior of these systems.

2. Passive techniques presented in this chapter can and have been used to mitigate contaminant migration. Consideration must be given to the identification of all contaminant pathways and selection of the most appropriate control technique. Each technique then must be evaluated as to its effectiveness and the design and construction must incorporate all site specifics and technical considerations.

3. Garbage itself is an unstable material. Its strength is low and its stress-strain relationship is irregular. Without proper stabilization or ground improvement techniques being applied, garbage itself cannot take large loads both vertically or horizontally.

4. Surface compaction is an important factor for controlling the stability of landfill slopes. Proper compaction control is needed during the waste disposal period. Fresh garbage dumped into the landfill site should be well spread within the layer, if not, the heavier items (Figure 17.2a) should be dumped closer to the center of the landfill and not around the edges of the landfill.

5. Garbage in the landfill area is complex. It cannot be generalized as a conventional geotechnical problem; therefore, a genetic diagnosis of landfill refuse may enhance our understanding. A detailed approach and procedure are outlined.

6. Computer-aided design or a knowledge-based expert system can be of assistance in solving slope stability analysis in landfill areas. However, human experience based on genetic diagnosis is more effective.

7. Waste recovery from landfill such as scrap rubber tire can be used effectively for conventional construction materials.

PROBLEMS

17.1 Define the following terms: domestic waste; residual waste; urban waste; solid waste; liquid waste.

17.2 What is the relationship between solid and liquid wastes?

17.3 Comment on prediction of settlement in landfill.

17.4 Designers do not suspect that leaks can pass through the so-called impervious soil layer. What are your comments about this statement? From the environmental geotechnology viewpoint, is there such a term called an 'impervious soil layer' existing in the field? How would you define the 'impervious soil layer'? If there is a leak, discuss the leakage route(s) and how it (leachate) can pass through the 'impervious soil layer'.

17.5 Discuss the conventional slope stability analysis approach such as limit equilibrium or/and limit analysis approach for highly nonhomogenous material such as garbage? Do you have comments on the genetic approach for analysis of the stability of landfill slopes? Is there any merit for such an approach?

17.6 A new industrial plant is to be constructed at a selected site. When completed the plant will pump waste water which contains a small to modest amount of sodium chloride into the local river. Downstream from the plant site, the river is used for irrigation of the highly calcareous ($CaCO_3$) soil of the region. During your soil investigation of the plant site, a group of farmers file a suit against your client claiming that the salts from the waste water will make the river water unsuitable for irrigation. Is the farmers' claim valid? As a soil engineer, how are you going to argue in court?

17.7 What is a construction material and what are the basic requirements for such a material (or a waste material) to be suitable as conventional construction material? List the economic-technological factors affecting ground improvement planning if waste material is considered for use.

17.8 What type of interaction mechanism exists between rubber tire and soil? Using scrap rubber tire in conventional construction, what are three basic problems?

17.9 Discuss the optimum use of materials (utilization of local waste material) and energy in transportation construction.

17.10 Discuss the site investigator, designer, and contractor's viewpoints of a ground improvement program if waste material is considered for use.

CHAPTER 18

Arid Lands, Desert, and
Antidesertification Measures

18.1 INTRODUCTION

The world's arid lands appear to be enlarging, and droughts are contributing to the economic devastation of the whole world. Both industrialized and developing countries suffer from this crisis. As pointed out in Section 1.7.2, the deterioration of productive land into desert-like conditions continues without letup at the rate of 14.8 million acres a year. In the U.S. it is reported that desertification faces falling water tables and increased groundwater salinity in the southwestern states. Arid lands have been exploited for their agricultural potential. Water practices developed for temperate climates may not be suitable in arid lands. For environmental and economic reasons, it is necessary to reevaluate the problems existing in the arid lands as well as in desert areas. The contents of this chapter include:

1. Evaluation of arid land and desert-like environments for sand-heat interaction, sand-wind interaction, and sand-water interaction
2. Development of techniques for increasing the supply of usable water
3. Reducing the demand for water supply, as well as delivery, and proposing an integral system for the purpose of antidesertification measures

18.2 CHARACTERISTICS OF DESERT AND DESERTIFICATION

18.2.1 Definition

A *desert* is defined loosely as a deserted, unoccupied or uncultivated area. In recent years, the term '*desertification*,' a form of the word 'desert,' has been put into use throughout the world. Desertification can be defined as the diminution or destruction of the biological potential of land that can lead ultimately to desert-like conditions; grazing lands cease to produce, dry land agriculture fails, and irrigated fields are abandoned owing to salinization, water logging, or some other form of soil deterioration. Desertification is the classic example of the process described by Hardin (1968) in his paper "Tragedy of the Commons". As illustrated in Figure 18.1, desertification is a self-accelerating process, "feeding on itself," and as it advances, rehabilitation costs rise exponentially (Eckholm and Brown, 1977 and Biswas and Biswas, 1980).

Desert defined by a pedologic engineer (HRB, 1957) is a region lacking sufficient moisture to support vegetation of the character commonly found in other areas of approximately the

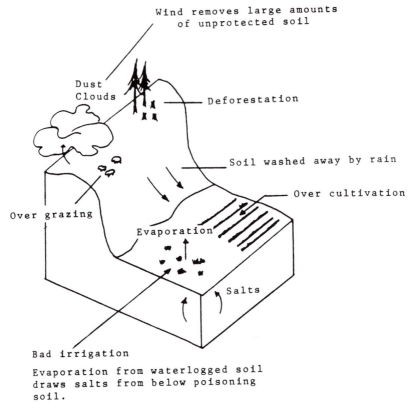

Figure 18.1 Schematic illustration of the many faces of desertification.

same latitude and necessary to sustain human habitation, frequently covered with considerable growths of cacti, coarse bunch grass, mesquite, and other shrubs. A desert is not necessarily a plain, but may be and often is broken by the sharp escarpments, buttes, and sand dunes which are common to the arid regions, or by features of volcanic origin. A desert may include a great variety of topographic features, often of considerable difference in elevation.

18.2.2 Desert and Civilization

A short review of history shows us many examples of the effect of desertification on civilizations. Mayan civilization over a period of decades collapsed from a peak of 5 million with a density comparable to that of the most agriculturally intensive societies of today, to 1/10 of that value. The reason for this collapse is believed to be due to soil erosion draining the productivity of the cropland. Further examples are given by the Tigris-Euphrates area where irrigation systems had no drainage components, the underground water table gradually rose, and waterlogging and salting ensued. Similarly, North Africa was once the granary of the Roman Empire, but Libya and Algeria now import half of their grain from North America. More recently, the Turpan Depression in China has been damaged by windblown sand: winds greater than 36 m/sec carried sands that totally buried all the houses in the area.

Most people think that a desert is caused by lack of rainfall; unfortunately, rainfall is only a part of the picture. There are four major factors which lead to the development of a desert-like environment:

- Low precipitation
- High temperature

- High proportion of evaporation to precipitation
- Poor land management

18.2.3 Social and Economic Impacts

As discussed in Section 1.7, desertification currently affects approximately 100 nations. The spread of deserts threatens 35% of the earth's land surface. There is a great social and economic impact on our society. Desertification hazards also have an indirect influence on the entire world through:

1. Unplanned mass-migration with socio-economic and socio-political repercussion on adjoining countries
2. Dust storms that bring to the global circulation of air masses added loads of particulate matter whose impact on the global climate has not yet been assessed
3. Loss of genetic resources of plants and animals

18.2.4 Desert Distribution

According to Walker (1982), indications are that approximately one third of the land surface of this planet is considered to be desert, barren land with meager rainfall that can support only sparse vegetation and a limited population. The majority of desert environments are located around the equator in areas where the temperature is high and there is a lack of rainfall, as shown in Figures 18.2(a) and 18.2(b). Death Valley in southeastern California is the most intensely desert area in the U.S. Also, a majority of people believe that the desert is a geological cycle which has a natural cause. However, this concept is not the whole picture, because deserts are found in many semiarid regions. These deserts are caused by human behavior due to poor land management and a lack of soil-water conservation systems. Deserts give us the false impression that there is 'nothing' there. This impression is not true because deserts contain many natural resources such as coal, petroleum, and rare minerals in addition to the recovery of useful land and farmlands. Table 18.1 shows interrelationship among antidesertification measures, natural resources recovery potentials, and required knowledge from other disciplines in order to evaluate desert-environment interaction.

18.3 DESERT ENVIRONMENTS

18.3.1 Climates

There are three major features controlling the desert climate: the high temperature, low precipitation, and high evaporation rates. The general classification of tropical and desert regions based on rainfall is shown in Figure 18.3. A review of Figure 18.3 shows that the annual rainfall in desert areas is less than 6 in. (15 cm). In general, desert regions can be divided into three categories according to the amount of rainfall. Extremely arid lands have at least 12 consecutive months without rainfall and have no regular seasonal cycle of rainfall. Arid lands have less than 250 mm of rainfall annually, whereas semiarid lands have a mean annual precipitation of between 250 and 500 mm.

The desert receives 90% of incident solar radiation which heats the ground and lower air layer. Only 10% is deflected by dust particles and clouds. The highest temperature ever recorded on earth, 58°C in the shade, was recorded at Azizia, in the Libyan area of the Sahara Desert. Highest temperature officially recorded at Death Valley was 42.5°C. In addition to low rainfall and high temperature, most deserts have high winds. The wind exists as a result of convection. During daytime hours, air over especially hot places is heated and rises, and

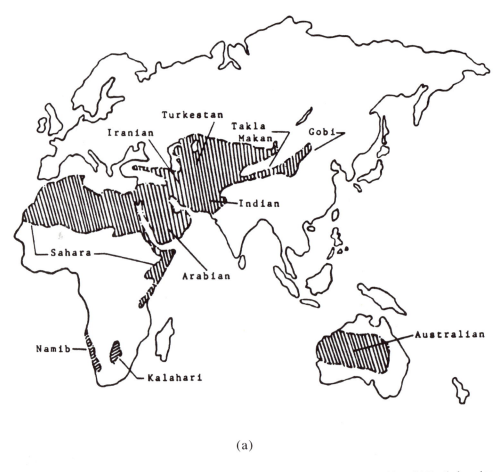

(a)

Figure 18.2 Location of deserts, a worldwide distribution. (a) Asia, Australia and Africa; (b) North America
and South America.

this allows cooler surface air from other locations to move in and take its place. Low rainfall
is a major feature in the desert; however, in some places there is also heavy rain. Convection
causes much of this precipitation on deserts. As the columns of air rising over hot places are
cooled, the moisture they contain can condense and fall localized as heavy rains, widely
known as cloudbursts.

18.3.2 Whirling Dust and Long-Distance Dust

Whirling dust frequently occurs in the desert area and is called "*dust devils*" or '*djinns*',
the term in Arabic, or '*willy-willies*' in Australia. This phenomenon is caused by high-
velocity wind vortices with pressures ranging from 20 m/sec at their center to 100 m/sec
high. In some exceptional cases, it can reach as high as 1.0 km. They are short-lived
phenomena and wander randomly over the ground surface in desert regions. They are
most frequently seen in lowland valleys and locations where fine sand dune or dust
accumulations occur.

Dust in the desert can be transported over thousands of miles by air currents and is referred
to as 'long-distance dust'. There are numerous recorded (by rainfall) examples of long-
distance dust: the most well-known dust routes such as from the Sahara to the British Isles
"red rain"; from West Africa to Miami, Mongolia to Hawaii, and Ukraine to Sweden as
reported by Allan and Warren (1993).

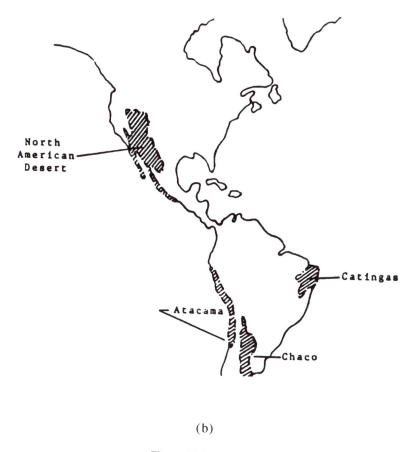

(b)

Figure 18.2 (continued)

18.3.3 Trees, Vegetation, and Animals

Vegetation in the deserts is a direct reflection of the dry climate. Intense solar light can damage pigments and high temperatures can disrupt the biochemistry of cells. In general, desert plants have a lower density of stomata in their leaves and will hem to conserve water. Most vegetations are not continuous but tend to be thin and grow only in patches. The plants consist of low bushes, growing rather far apart, with bare areas between them. This pattern of vegetation promotes active movement of sediment by both wind and moving water. Some desert plants are active only for limited periods. These are known as the annuals and ephemerals, and included among them are rockroses, poppies, and chenopods. The root systems of desert plants are often extremely deep, tapping supplies of water beneath the surface to replenish their own lost water. Further discussion on sand dune stabilization by vegetation or trees will be discussed in Section 18.6.

Animal life on the desert is completely dependent upon plant life for sustenance. Only a green plant can manufacture organic food from nonorganic material (carbon dioxide, water, and solar energy). The world of desert animals is a complex society; some eat plant food directly, some eat each other. In the desert, there are two phases of growth: the lush period after a rainfall and the long drought, when the only available foods are plant stems, roots, drought-resistant leaves, and seeds dropped to the ground. Details of desert flora and fauna behaviors are given by Leopold (1980), Wichens (1985), and Allan and Warren (1993). The geological and geotechnical aspects of the desert environment are given by Longwell et al. (1969), and Doehring (1981).

Table 18.1 Interrelationship among Antidesertification Measures, Natural Resources Recovery Potentials, and Others

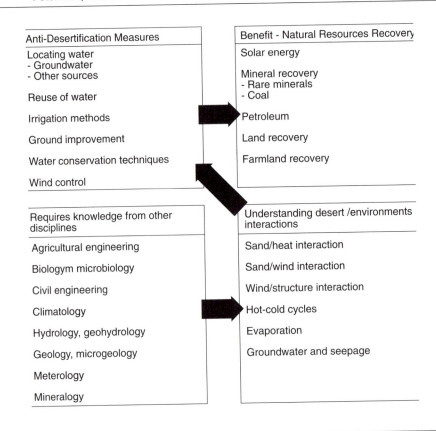

Anti-Desertification Measures	Benefit - Natural Resources Recovery
Locating water - Groundwater - Other sources	Solar energy
	Mineral recovery - Rare minerals - Coal
Reuse of water	
Irrigation methods	Petroleum
Ground improvement	Land recovery
Water conservation techniques	Farmland recovery
Wind control	

Requires knowledge from other disciplines	Understanding desert /environments interactions
Agricultural engineering	Sand/heat interaction
Biologym microbiology	Sand/wind interaction
Civil engineering	Wind/structure interaction
Climatology	Hot-cold cycles
Hydrology, geohydrology	Evaporation
Geology, microgeology	Groundwater and seepage
Meterology	
Mineralogy	

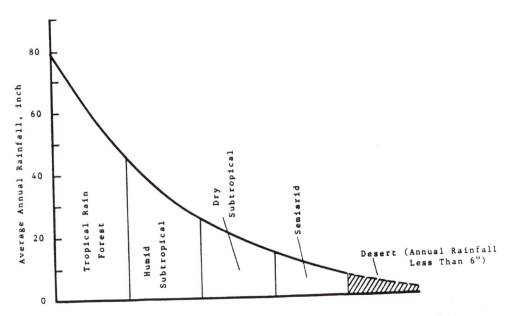

Figure 18.3 General classification of tropical and desert regions based on annual rainfall.

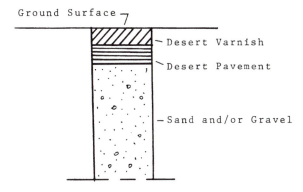

Figure 18.4 Typical desert soil profile.

18.4 CHARACTERISTICS OF DESERT SOILS

18.4.1 Profile of Desert Soils

A desert soil profile consists of three basic layers: (1) the surface layer covers desert varnish and desert pavement; (2) the main part of surface layer is sand or sand-gravel; and (3) subsurface layer. A typical soil profile is shown in Figure 18.4.

18.4.2 Desert Varnish and Desert Pavement

Surface soils in the desert region have four distinguishing characteristics: (1) this surface dust contains various constituents from the rock debris with which it mingles; (2) because desert areas are less moist, desert soils are less chemically altered than soil in humid regions; (3) because of lack of water and lack of the leaching (Section 5.6) process in desert area, soils are generally saltier than humid soil, and (4) because the chemical effect in desert soils is less, they retain many more features from the past.

(1) Desert Varnish

The uppermost layer on many desert rock surfaces is a thin, dark surface patina known as *desert varnish*. Most desert varnishes are very old, but some were formed recently. The causes and formation of desert varnish are still questionable. Early geomorphologists thought that desert vanish is caused by the evaporation process which will carry iron and manganese from underlying rock layers. Some scientists have found evidence that in some desert varnishes, elements are scavenged and fixed by lichens and bacteria. Desert varnish has been found on rock in the Sinai desert. The coating of varnish takes thousands of years to form. Desert varnish has been found on ancient rock carvings, for example, in the Australian desert (Allan and Warren, 1993).

(2) Desert Pavement

The topmost layer of most desert soils consists of angular stones, known as the *desert pavement*. Desert pavement is a dense layer; most vehicles can drive on this layer without developing rut depth. Desert pavement is found in the Algerian Sahara. The upright stones have been edged up as the soil expands and contracts when it is subjected to the wetting-drying and heat-cool cycles. Differences between desert varnish and desert pavement are shown in Figure 18.4.

18.4.3 Subsurface Soil Layer

The soil beneath the desert pavement in most cases has soluble salts. In humid areas, these salts are washed out entirely. However, in desert areas, there is a sequence of these accumulations. There are three distinct groups:

1. Very dry desert soil; for example, the Sahara Desert has a layer of salt (sodium chloride, NaCl).
2. In slightly moist area, such as southern Tunisia and part of New Mexico, the common salt is washed out and gypsum (calcium sulfate dihydrate, $CaSO_4 \cdot 2H_2O$) is washed down to form a soil horizon.
3. On the wetter edges of the desert, as most areas bordering the Mediterranean, large swathes of Australia, and southwestern U.S., the subsurface soil layer is formed by calcretes (calcium carbonate, $CaSO_4$), common salt and gypsum having been leached away.

Some desert soils have other hard subsurface layers or horizons which are due to a wetter past. These layers include laterites (high iron content) and silcretes (high silica content).

18.4.4 Valley Soils in Desert Region

In the lower parts of desert landscapes, there are two major types of soil existing in these areas:

1. In some semiarid lands, where there is not too much salt, silica washes down in drainage water mixes and combines to form black-color organic types of clay known as vertisols, which also form through the breakdown of basalt rock in semiarid conditions (HRB, 1957). They are very fertile when irrigated and, for example, support cotton growing in the Deccan Plateau of India and in central Sudan.
2. The second type of soil is far from fertile. In the silt deposited by rivers or streams, salt may be brought to the surface by capillary action from a water table a few meters below the ground surface. When surface water evaporates, the residual salt crystals remain on the surface. This kind of salinization (Section 13.11) occurs at alarming rates in many irrigation schemes, where the water table has been brought too close to the surface by heavy uses of water. Today, salinization is active in many countries and is a danger in any irrigation scheme in the dry parts of the world.

In general, many desert soils have a pebbly surface, created partly when the wind removes fine grains, partly when occasional storms wash grains away, and partly when pebbles rise to the surface as the soil is intensely heated and cooled. Field observations for many years made by the author indicate the following: the Egyptian Sahara Desert is formed mostly of fine sand and the surface layers are very hard. According to geological reports, indications are that layers of salt (sodium chloride) are in the surface layers. However, in the Sinai Desert are mixed sand and gravel. In most areas of the Gobi Desert are gravels, boulders, and pebbles. As reported by Allan and Warren (1993), indications are that subsurface layers in Tunisia and part of New Mexico are relatively softer, because the common salt is washed out and gypsum is washed down to form a soil horizon. From a geotechnical engineering viewpoint, the desert soil covering is mainly sand-gravel, silt, and clay as discussed in Chapter 3. However, sand is the major material in a desert area.

18.4.5 Floods

Rain patterns in the arid regions are unpredictable. The reason for the unpredictability is caused by local environments and can be explained based on a hydrological viewpoint. It is

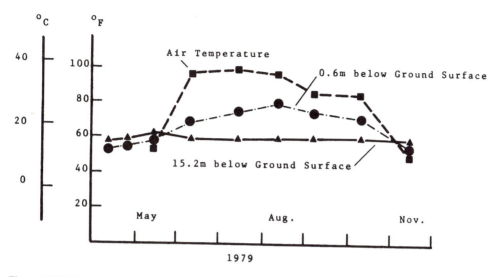

Figure 18.5 Seasonal air and ground temperature variations on the Nevada desert. (After Chaney, R.C., Ramavjaneya, G., Hencey, G., Kanchanastit, P. and Fang, H.Y. [1983], *ASTM Geotech. Testing J.,* v. 6, no. 4, pp. 220–225.)

indicated that most deserts are located beneath semipermanent anticyclones into which moist air, rain-bearing frontal systems, or tropical cyclones can only occasionally penetrate. When it happens, rapidly rising hot desert air, which cools and condenses as it rises, causes highly localized rainfall.

In areas with a fair amount of vegetation and thick layer of soil, most water generated by rainstorm can be absorbed. In arid lands, the surface layers become saturated much more quickly. As discussed in Section 7.4, if the intensity of the rain is greater than infiltration and percolation, surface runoff rapidly courses down hills to cause flash floods. Flash floods carry great loads of sediment and it can be a problem in many rivers.

18.5 DESERT SAND-HEAT INTERACTION

18.5.1 General Discussion

The ground surface in a desert receives 90% of the incident solar radiation which heats the ground and lower air layer. In contrast, humid lands absorb 40% of the incident solar radiation. The remaining solar radiation in humid land is dispersed in a variety of ways: 30% is deflected by water surface and land cover, 10% by dust particles, and 20% by clouds. At night, the desert turns cold because 90% of the heat generated by the solar radiation escapes back into the atmosphere (Anthes et al., 1978). Temperatures in desert areas have been shown to vary in a somewhat regular pattern reflecting both the annual and diurnal cycles of solar radiation. Superimposed on these regular cycles are fluctuations of variable durations and amplitude created by changing climatic conditions. The seasonal air and ground temperature variations in the Nevada Desert are shown in Figure 18.5. In examining Figure 18.5, the ground temperature is shown to remain constant at the depth below 15.2 m from the ground surface (Chaney et al., 1983).

18.5.2 Thermal Resistivity of Desert Sand

The thermal resistivity of a soil is influenced by its saturation, porosity, and to a lesser extent its mineralogy. The effect of saturation on the thermal resistivity of a soil from the Nevada Desert is shown in Figure 18.6. A view of Figure 18.6 shows that for both sand and clay, an increasing level of saturation results in decreasing thermal resistivity. The variation of thermal resistivity shown in Figure 18.6 shows that, in general, the higher the porosity the greater the thermal resistivity. The relationship between thermal resistivity and porosity of sand and clay in laboratory condition is presented in Figure 18.7. The range of *in situ* condition of desert sand is also shown in the figure for comparison. Explanations of behavior of sand-heat interaction in the thermal energy field has been discussed in Chapter 8.

18.6 DESERT SAND-WIND INTERACTION

18.6.1 Causes and Types of Wind

The predominant erosive factor in the desert environment is the combination of sand and wind. Therefore, the sand-wind interaction is critical. In desert region, wind dominates the desert sand. The characteristics of sand-wind interaction are a critical feature and must be carefully examined.

18.6.2 Mechanisms of Sand-Wind Interaction

In general, there are three possible mechanisms as shown in Figure 18.8:

1. Loose particles of sand and silt picked up by the wind and carried from one place to another. This mechanism is shown in Figure 18.8(a) where two sand particles strike against each other in the air.
2. Wind-driven sand particles strike against pebbles or boulders on the ground with the result that additional particles hit other sand particles on the ground, making the sand particles on the ground jump up as illustrated in Figure 18.8(b).
3. Similar to case (2), but the sand particle on the ground is smaller. The height and moving distance between cases (2) and (3) will be different with case (2) moving farther and higher.

18.7 CHARACTERISTICS OF SAND DUNES

18.7.1 General Discussion

The direction of wind and its speed are also important for the antidesertification measures, because the direction and speed of the wind will control the patterns of sand dunes. There are four basic patterns of sand dunes caused by wind as reported by Longwell et al. (1969), namely: transverse, longitudinal, star, and barchan patterns. Breed and Grow (1979) have developed a system of classifying aeolian sand features based on a global inventory of deserts seen on Landsat images. They categorize dunes according to their external shapes and the relative positions of their slip faces, with steep slopes generally found on one or more sides of a dune. Sand dunes are classed as:

1. Linear (straight, elongate, longitudinally symmetrical sand ridges)
2. Crescentic (crescent-shaped mounds or ridges)

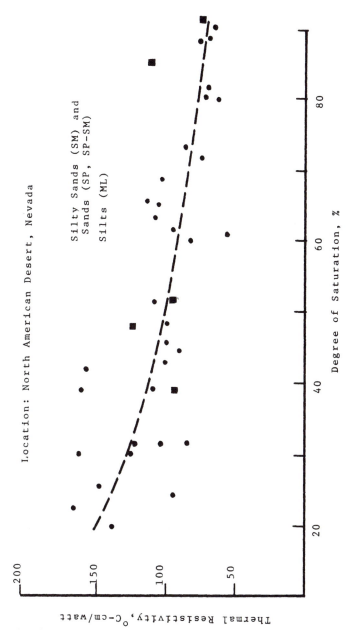

Figure 18.6 Variation of thermal resistivity with level of saturation. (After Chaney, R.C., Ramavjaneya, G., Hencey, G., Kanchanastit, P. and Fang, H.Y. [1983]. *ASTM Geotech. Testing J.*, v. 6, no. 4, pp. 220–225.)

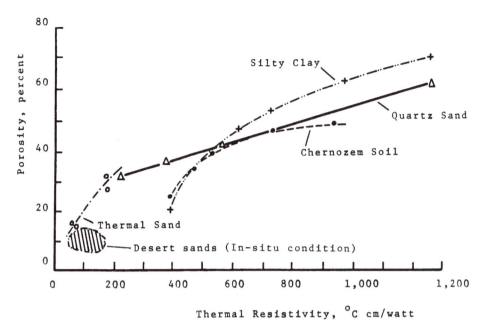

Figure 18.7 Relationship between thermal resistivity and porosity of sand and clay.

3. Dome-shaped (semicircular mounds with no slip faces)
4. Star, or parabolic
5. Irregular patches of aeolian sand without slip faces and classified as sand sheets
6. Sharply defined elongated sand patches referred to as streaks

Walker (1982) claimed that the Breed and Grow classification system which has global applicability transcends regional language barriers.

18.7.2 Types and Causes of Sand Dunes

Types of sand dune patterns as discussed in the previous section and some causes of these patterns can be explained as follows:

1. Transverse dunes: These dunes are the product of moderate, one-directional winds which move only light or loose sand. Tumbling air eddies swirl heavier grains to the side which tends to make ridges.
2. Longitudinal dunes: These dunes occur when stronger one-directional winds move both fine and coarse sand particles cutting long troughs parallel with the path of the wind.
3. Star dunes: These dunes form in areas where the wind blows from all directions. Star dunes remain stationary.
4. Barchan dunes: They form where sand is relatively scarce. The crescent shape results from a one-directional wind blowing the sand more readily over the dune's low tips than its center.

In addition to these typical sand dune patterns, there are many other shapes of sand dunes existing in various places and discussed in many references. A comprehensive discussion of the deserts of North America from Oregon to Mexico is given by Findley (1972) and MacMahon (1985); the Sahara Desert by Gautier (1953); Taklamakan and Gobi Deserts by Zhu et al. (1980 and 1987); Walker (1982), and others (Allan and Warren, 1993).

From an engineering point of view, when we build houses, highways, or railroads, dig a ditch, or plant a tree in desert areas, we must know the prevailing direction of the wind in a

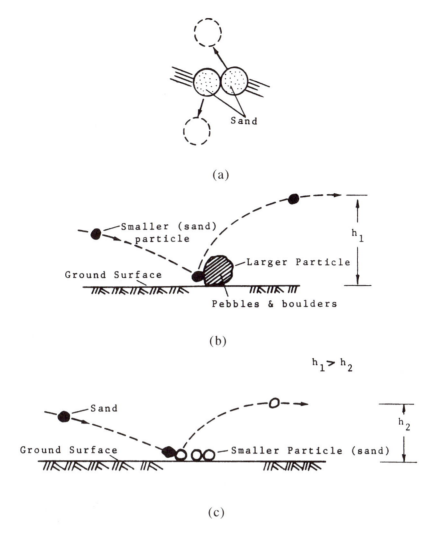

Figure 18.8 Sand-wind interaction related to the particle size in a desert environment. (a) Sand-wind interaction in the air; (b) smaller particle interaction with larger particle; (c) larger particle interacts with smaller particle. (Based on Longwell et al. [1969] and Fang and Chaney [1987].)

particular desert; otherwise, the sand dune will move in and destroy everything if the proper protection is not made. To prevent this occurrence, we must understand the interactions of sand and wind.

18.7.3 Characteristics of Sand Dune Movement

There are very little experimental data available for sand-wind interaction at the *in situ* conditions. Figure 18.9 is obtained from published data on the Taklamakan and Gobi Deserts in China. Sand dune movement in the southwest area of Pishan in the Taklamakan Desert is shown in Figure 18.9(a). It indicates that between May 1961 and July 1962 the sand dunes moved more than 60 m. Figure 18.9(b) presents the sand dune movement routes. Figure 18.9(c) is the cross section of sand dune movement. The sand particle movement as well as sand dune movement route is not only due to wind speed and direction, but also to particle size, density, degree of saturation, as well as local environmental conditions such as vegetations, hills, and surrounding structures.

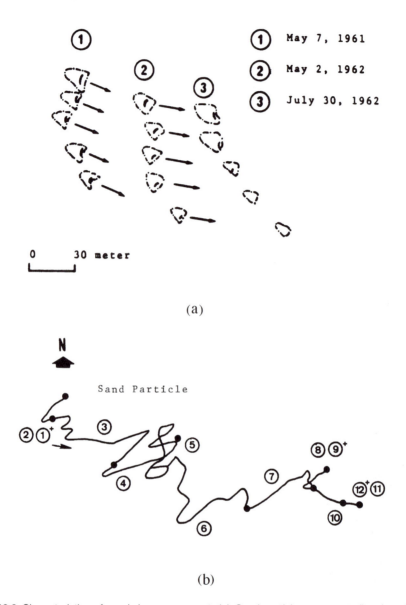

Figure 18.9 Characteristics of sand dune movement. (a) Sand particle movement direction; (b) sand dune movement route; (c) cross section of sand dune movement. (After Zhu et al. [1980].)

The characteristics of sand movement can be described as either creep or jumping movement. Sand particle movement near the ground surface caused by wind is called creep. The jumping movement (Figure 18.8c) is a dynamic action when moving sand hits other sand on the ground surface. In contrast, creeping movement is the horizontal movement of sand. The initiation of either creeping or jumping movement as a function of sand content and wind velocity is shown in Figure 18.10(a). It shows that for a constant wind velocity the initiation of creeping movement begins at a lower sand content than jumping movement. The rate of sand dune movement as a function of wind velocity is shown in Figure 18.10(b). It shows that the rate of movement of sand dunes increases with the increasing wind velocity. In addition, the unit change of rate of movement is less below a wind velocity of 7 m/sec

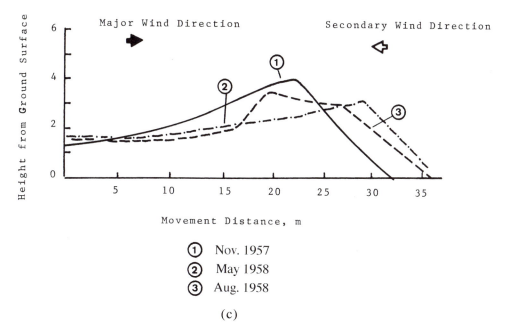

(c)

Figure 18.9 (continued)

than above. The wind velocity in Figures 18.10(a) and 18.10(b) was measured 2 m above ground surface.

The height from the ground surface as a function of sand content and wind velocity is shown in Figure 18.11(a). It shows that a linear relationship exists for all three cases. In addition at 3 cm above the ground surface, the same sand content is shown to exist for all wind velocities studied (point A in Figure 18.11a). Sand dune height vs. movement in meter per year as a function of location is shown in Figure 18.11(b), which shows that the greater the height of a sand dune the greater the movement. It is also shown in Figure 18.11(b) that the rate of movement to dune height is a function of location.

Sand dune movement distance can also be estimated from the following equation as (Liu, 1960; cited by Zhu et al., 1980):

$$D = Q/\gamma H \qquad (18.1)$$

where D = sand dune movement distance at unit time, Q = amount of sand passing though a unit width at unit time, γ = unit weight of sand, and H = height of sand dune.

Figure 18.12 also shows the height of sand vs. wind velocity. Numerical numbers indicated in the figure are the equal sand content in percentage. For a given wind velocity, the sand content varied with the height.

18.8 DESERT SAND-WATER INTERACTION

18.8.1 Characteristics of Water in Desert Region

Without water there is no life; therefore, water in the desert is of great importance. The acquisition of water and, once obtained, how to save it is the topic in the following section. In a desert region, the annual rainfall is less than 15 cm (Figure 18.3). Rainfall in Death Valley is about 1 to 2 in. annually, and it has experienced years with no rain at all. Reports from the Atacama Desert in northern Chile indicate periods of more than 10 consecutive years without rain.

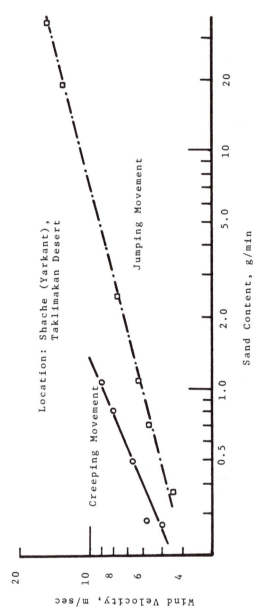

Figure 18.10 Sand-wind interaction — wind velocity vs. (a) sand content in air; (b) rate of movement of sand dune. (After Wu et al. [1965]; cited by Zhu et al. [1980].)

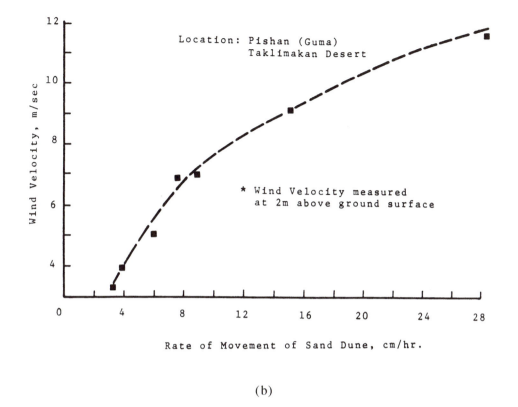

(b)

Figure 18.10 (continued)

Water in desert regions occurs in various forms. Periodically, there are heavy rains widely known as cloudbursts. The surface water resulting from this kind of rain disappears by a combination of evaporation, infiltration (Section 7.4), and surface runoff. In a desert the regolith is generally loose and dry and where bare, it is easily eroded. Basins formed by faulting and other movement of the crust play a larger part in determining the general sculpture of the land in an arid region than a moist one, because only rarely is water abundant enough to fill the basins. In the western U.S. such a basin is an ephemeral shallow lake called a playa lake and is frequently seen in dry regions. One of the most famous playa lakes is named '*Lop Nor*' located in the Taklamakan Desert, China.

18.8.2 Surface Water — Lakes and Streams

Many playa lakes contain high concentrations of salts in their waters and are called salt lakes. The chemical character of the water in the lake and the type of precipitates that result depend on the types of rock formation underlying the lake basin. In general, lakes are called various names depending on their chemical makeup. For example, lakes rich in sodium chloride (NaCl) are called *salt* lakes, those rich in sodium (Na) and potassium carbonate (K_2CO_3) are known as *alkali* lakes, and lakes rich in sodium sulfate (Na_2SO_4) are called *bitter* lakes.

Based on field observation by the author in 1983 and 1985, around the Urumchi and Turfan areas, there are numerous saltwater and freshwater ponds or lakes. The source of the fresh water is the melted snow from Heavenly Mountain (Himalayas).

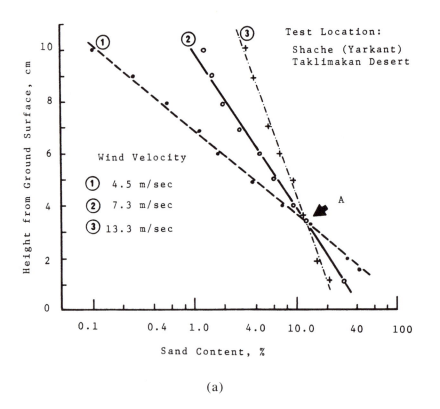

(a)

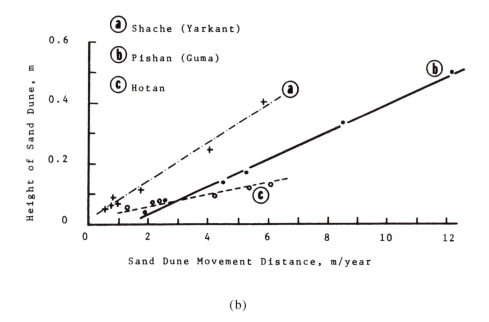

(b)

Figure 18.11 Effects of height of sand dune. (a) Sand content in air; (b) sand dune movement distance. (After Wu, [1965]; cited by Zhu et al. [1980].)

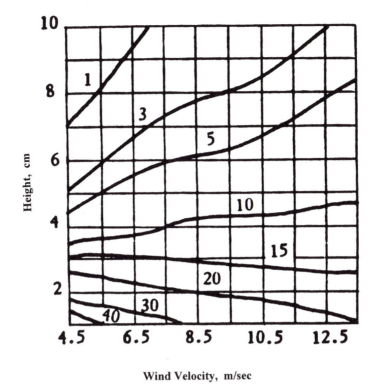

Figure 18.12 Height vs. wind velocity with equal sand content in air. (After Zhu et al. [1980].)

18.8.3 Underground Water Resources

In the deserts, although there is less surface water, there are many underground rivers. Certain regions have rich groundwater reservoirs. In the Gobi and Taklamakan Deserts, large amounts of water are obtained from melted snow runoff from the Heavenly Mountain. To develop these sources of water, new techniques are needed. Using solar energy to melt these snows is one such technique that has been proposed. In addition, methodologies for transporting these waters from one location to another and its reuse will also have to be developed.

The scarcer the water, the greater the need for effective utilization of advanced technology for the acquisition and development of water supplies along with a realistic resource management program. Some important items relating to these management techniques are discussed in the following sections. Where groundwater is available, surface and groundwater supply and delivery systems should be considered in combination or in conjuction for the optimum use of the total water source. Reducing evaporation of water from the ground surface is also an important consideration. Because evaporation is invisible, it is seldom regarded as a serious drain on stored water, but annual evaporation losses particularly in desert areas are very great.

18.8.4 Water Resources and Conservation Techniques

(1) General Discussion

Where groundwater is available, surface and groundwater supply and delivery systems should be considered in combination for optimal use of the total resources. Usually, farmers

Table 18.2 Water Management in Arid and Desert Regions

Water sources	Water conservation techniques
Rainfall	Reducing waste water
Utilization of groundwater	Rainwater collected from hillslopes
	Runoff agriculture
Utilization of dew	Irrigation of saline water
Exploration of underground water	Reducing evaporation from water surface
	Reducing seepage losses
Reuse of water	Reducing evaporation from ground surface
Utilization of saline water	Reducing transpiration
Utilization of melting snow/ice from high mountains	Utilization of solar energy from desert to melt snow/ice, then to desert in irrigation

tend to overirrigate when water is available. This can lead to problems of waterlogging and salinity and leaching of fertility. On the other hand, over irrigation may be needed to remove accumulating salts. This section discusses the technology for enhancing water supplies. Table 18.2 is a summary of commonly used water management techniques for arid and desert regions. Other sources of water include solar distillation, the use of satellite and aircraft for detecting water in arid lands, and rainfall augmentation.

(2) *Horizontal Wells and Horizontal Drains*

In temperate regions, some small springs are often neglected. However, in arid or desert regions, every drop of water is needed and worth saving. Engineers must carefully investigate the sources of such water supply. Horizontal wells (NAS, 1974) are a typical example. In many remote and arid mountain regions, springs are the safest, most dependable sources of water. The horizontal well system is installed in certain geological formations to collect water trapped in the soil or rock joints for irrigation or drinking uses. A typical condition is illustrated in Figure 18.13.

The principle of the horizontal well is the same as horizontal drains. Since the principle of horizontal drains is well developed in geotechnical engineering, some knowledge on horizontal drains is presented herein. A horizontal drain is simply a small-diameter well that is drilled nearly horizontally into a hillside. In geotechnical applications, these horizontal drains are used to remove groundwater. Many of the early installations were made by water-driven drills known as "Hydrauger". Also, horizontal drains often are installed as part of a planned stabilization of cut slopes in wet, unstable ground, or used to correct the stability of earth slopes. Frequently horizontal drains are installed at several levels, and many effective installation techniques have been developed for highway construction and ground improvement system. These useful techniques can also be applied to arid areas and desert regions in search for a water supply. Of course some additional modifications must be made in order to satisfy the different environments.

(3) *Underground Rivers or Streams*

Underground rivers and streams are one of the major water resources in desert region. Geological data indicate there are numerous hidden underground rivers or streams existing in desert regions. Underground water transportation system is also an important item to be considered. Underground water resources exist in desert regions, but transport of this water from one place to another is also important. To eliminate the surface evaporation in desert region, an underground water transportation system must be evaluated. One such system sucessfully built in Taklimakan Desert has been observed.

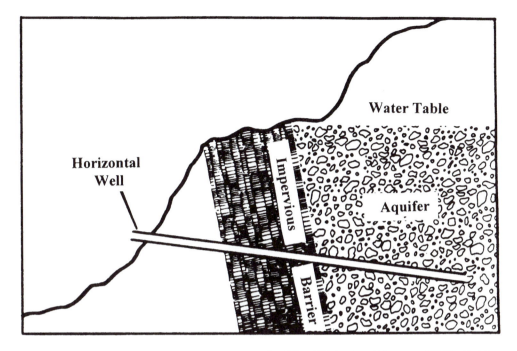

Figure 18.13 Typical water supply from horizontal wells. (After Welchert and Freeman [1973]; cited by NAS [1974], *More Water for Arid Lands,* NAS Press, Washington, D.C., 153p.)

18.9 LOCATING WATER SUPPLY SOURCE BASED ON PREFERRED PLANES CONCEPT

18.9.1 General Discussion

A method based on *two preferred planes* for locating water supply source has been successfully developed by Luo et al. in 1982, and it is particularly valuable for locating groundwater supply sources in arid regions. Based on field observations indications are that the most likely groundwater supply source is where the rock mass will be deformed or have a failure location. It is also suggested that the deformation and failure location(s) usually occur along the discontinuities of the rock mass. The following procedures are suggested by Luo et al. (1982, 1984) to evaluate the occurrence, characteristics, and the interactions of these discontinuities.

18.9.2 The Concept of Preferred Planes

According to the concept of two preferred planes and based on long-term research and accumulated data, there have been established 6 types and 14 evolutionary models of the deformation and failure (geomechanical models) of rock slopes in igneous regions.

1. Simple model: It is assumed that a failure plane is a simple form of discontinuities and structural weakness planes. Basically, it is based on limit equilibrium techniques and can be used for preliminary investigations for various rock stability problems.
2. Hoek's model: This model for analysis of rock mass stability is a commonly used method developed by Hoek in 1974 (Hoek and Bray, 1981). The method is based on discontinuities through the center of pole concentration to a preferred plane. There is simple classification on the failure types of rock slopes, namely, circular, planar, wedge, and toppling failures.

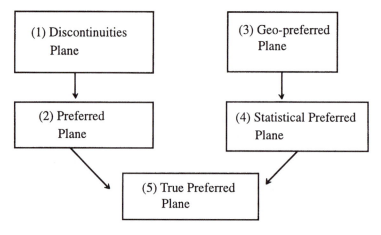

Figure 18.14 Luo's preferred plane concept (1984).

3. Luo's model: Luo's model is relatively new, the theory has been confirmed by numerous case studies. A brief description of the method and fundamental concept is presented as follows.

18.9.3 Procedures for Locating the Preferred Planes

(1) Determination of Failure Types and Models

The deformation and failure of rock mass usually occur along the discontinuities that exist in the rocks. The basic idea is that the discontinuity is to determine the true preferred plane. The Lou's preferred plane concept (1984) is presented in Figure 18.14. There are five planes: the discontinuities, the preferred, geo-preferred, statistical preferred and true preferred, planes.

It must be noted that Hoek's preferred plane is somewhat equal to or similar to the statistical preferred plane of Lou's concept as indicated in Figure 18.14[4], yet Hoek's plane is a numerical approach and not a statistical approach. In examining Figure 18.14[3], the geo-preferred plane is a characteristic plane of rock mass. If we find hundreds of joints or cracks with small size and various directions in the rock mass and a fault dipping out of the rock slope, then in this case the fault is a decisive factor promoting the movement of the slope.

Step [5] in Figure 18.14 is the summation of steps [1] to [4]. After obtaining the true preferred plane [5], the potential failure types or models (geomechanical models) of rock mass can be obtained and a proper mathematical or mechanical model can be formulated. The two-preferred plane concept has setup following procedures for evaluation of true preferred planes as follows:

1. Set up the program or procedure for engineering geology investigations of preferred planes.
2. Set up an analytical procedure based on probability methods.
3. Determine 6 failure types and 15 failure modes of rock mass.
4. Examine local environmental conditions such as regional rock mass stability and water existing in the fissure zone.
5. Locate the preferred planes and their superior index (SI).

(2) Superior Index (SI)

Superior index (SI) is used to identify the stability of rock mass. There are four types of SI existing as proposed by Lou et al. (1982; 1984). A brief discussion of the index is presented. According to the concept of preferred planes, the structure factor is a most unstable factor and the failure types of slopes are controlled by preferred planes. Since there are various types of structural planes in natural rock mass, one can use geologic age as the major factor for determination of SI.

In a natural condition, the structural planes existing in rock mass have different degrees of cementation when the rock mass is excavated. In such cases, the geologic age is the decisive factor. In general, there are two types of faults in the rock mass:

1. Aged fault: A fault which was formed in an old geological tectonic movement. The fractural zone has been cemented or filled by rigid substances such as quartz or calcareous cement.
2. Neotectonic fault: Newly formed fault caused by neotectonic movement, the fractural zone always has loosened and/or been filled by muddy cement. There are three types of neotectonic fault.
   a. Old neotectonic fault: The old fault was formed in Jurassic-Cretaceous period.
   b. New neotectonic fault: The new fault formed in Tertiary-Quaternary period.
   c. Active neotectonic fault: Active fault is in Holocene (Q_4) period.

The new and active faults owing to their young age of geologic period have weak cementation in the fault openings. This type of fault is an important environmental geotechnical problem.

18.10 ANTIDESERTIFICATION MEASURES

18.10.1 General Discussion

Antidesertification measures have been employed by China with moderate success in combating desertification. These programs have ranged from stabilizing the sand dunes of the Gobi and Taklimakan Deserts by planting trees and shrubs and digging artesian wells, to installing sprinklers and irrigation pipes in the Sinai Desert. Table 18.3 presents geotechnical aspects of control in antidesertification measures. Basically there are two general approaches:

1. Reducing wind velocity and decreasing sand content in air
2. Consolidating sand surface and resistance in air-sand interaction

18.10.2 Prevention of Sand Movement

Movement of sand by wind occurs within a few inches off the ground surface. In the extreme, the sand collects in massive dunes which can then migrate onto arable land (Breed and Grow, 1979). In the following sections strategies to deal with both the stabilization of sand dunes as well as minimizing the effect of sand migration across lifelines will be discussed. Stabilization of desert sand can be done by chemical or bioengineering techniques. Desert sand stabilization by various stabilizing agents such as lime, lime/slag, cement, and gypsum is discussed as follows.

The design of a low cost pavement structure for low volume roads in arid areas of southern Africa is described by Strauss and Hugo (1979). Two available sources of material exist, namely, calcrete and windblown sand, both of which vary greatly in quality. Calcrete, a pedogenic material, is formed by the cementation or replacement of existing soils through the deposition of calcium carbonate from soil water. The material can be identified as strongly calcified sands, or hardpans. The windblown sands are the single-sized sand virtually all passing the 2-mm sieve and are nonplastic with a A-2-4 or A-3 AASHTO soil classification. No more than 12% passes the sieve #200. Figure 18.15 gives an indication of the increase in strength through different stabilizing agents and stabilizer contents for windblown sands.

Table 18.3 Geotechnical Aspects of Control in Antidesertification Measurements

Reducing wind velocity and decrease sand content in air	Biological control methods	Stop sand movement by sodding
		Stop sand movement by planting trees
		Plant trees between sand dunes
		Surface sodding
	Mechanical control methods	Tall sand barriers
		Semivisible covered-up-types of barriers
		Using gravel or clay covered on sand layer
Consolidating sand surface to resist the air-sand interaction	Chemical control methods	Using various chemical sprays
		Moving sand control chemical techniques

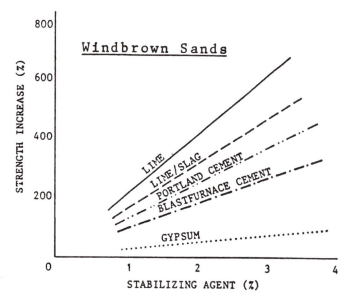

Figure 18.15 Change in strength of windblown sand with type and quantity of stabilizing agent. (After Strauss and Hugo [1979].)

18.10.3 Bioengineering Stabilization of Sand Dunes

Stabilization of a sand dune against the action of wind is shown in Figure 18.16. The process begins by first planting trees along the base of the dune to be stabilized toward the prevailing wind as shown in Figure 18.16(a). As these trees mature, new trees are planted in a new line closer to the top of the dune. When the tree line has reached approximately 3/4 of the height of the dune, the top of the dune is then leveled off as shown in Figures 18.16(c) and 18.16(d). The dune is then completely planted with trees as shown in Figure 18.16(e).

In an arid environment the ability of a plant to survive is directly related to its ability to utilize water sources. For this reason many arid plant species utilize both lateral root systems and tap roots. The lateral root systems play an important part in reducing the dry density of the upper soil layer by preventing the soil from clumping together to such a degree. The tap roots provide the vegetation with vital sources of water from underground sources during the dry seasons.

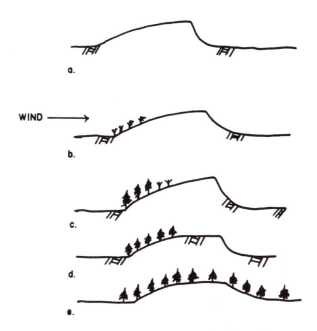

Figure 18.16 Bioengineering stabilization of sand dunes.

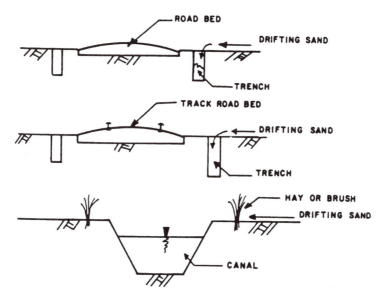

Figure 18.17 Protection of lifelines against drifting sand.

18.10.4 Protection of Lifelines Against Drifting Sand

The technique involves the construction of shallow trenches 1 to 2 ft deep (0.3 to 0.7 m) or hay or brush barriers parallel to the lifeline to intercept the sand transport. Figure 18.17 shows the protection of lifelines against drifting sand. Figure 18.18 shows straw that is used along the open drains to protect the drainage lines against the drifting sand. The photo was taken in 1985 at nearby Dunhuang City, a large oasis along the old Silk Road in China.

In desert regions, there is plenty of sand or gravel, but a lack of other construction materials such as cement, water, and timber; therefore, a movable type of retaining wall is proposed.

Figure 18.18 Straw used to protect the open drains against the drifting sand.

The movable retaining wall is a type of gravity wall. A shell is developed to form a container type of wall. The container wall or shell is made of plastic, or timbers, then it is filled up with locally available materials such as sand or gravel. This type of structure is a temporary, movable, and low-cost element.

Installation of vertical types of barriers will reduce the sand content in the air as illustrated in Figure 18.19. For foundation problems, due to deep sand layer, the conventional compaction techniques may not be effective. Therefore, dynamic compaction (dynamic consolidation) techniques as discussed in Section 11.4 may be used.

18.11 SUMMARY

1. There are four major factors which lead to the development of a desert-like environment: low precipitation, high temperature, high proportion of evaporation to precipitation, and poor land management.
2. The predominant erosive factor in the desert environment is the combination of sand and wind. The rate of movement of sand dunes increases with increasing wind velocity and height of dune. Experimental data are presented for sand-wind interactions.
3. Water management in arid and desert regions is summarized, including water sources and water conservation techniques.
4. Geotechnical aspects of antidesertification measures are summarized, including biological, chemical, and mechanical control methods.

PROBLEMS

18.1 List the most important factors affecting desertification.
18.2 Most people believe that in a desert there is nothing. Would you comment?

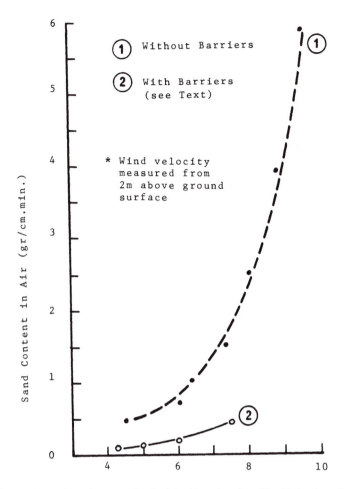

Figure 18.19 Comparison of sand content in air: (1) without barriers (2) with barriers. (After Chu et al. [1980].)

18.3 Deserts continue to expand affecting more than a hundred nations. Why are very few people concerned with this problem?

18.4 Prepare a typical soil profile and horizon for desert soils.

18.5 Define the following terms: desert varnish; desert pavement; gobi; whirling dust; red-rain; djinns; willy-willies.

18.6 What are the major controlling factors for evaluation of sand-wind interaction?

18.7 How do animals affect water supply?

18.8 What is the mechanism of an aerial well? Is this a feasible concept or is it fiction? If not, is it worth further study for possible practical application?

18.9 Define the two preferred planes and how this approach can locate underground water sources.

18.10 Discuss the basic principles of antidesertification measures. Why are very few geotechnical engineers interested in this type of problem?

18.11 How would you propose a water management plan in the arid and desert regions?

Comments and Conclusions

A soil in its natural location is a dynamic system whose properties depend not only on those of its constituents, but also on the mutual arrangement and interaction of these constituents as functions of the past history of the system and the dynamics of its present environment. The placement of a foundation structure or any other disturbance upon or into a soil system opens up possibilities for favorable or detrimental interactions which should be foreseen and taken into account by the responsible engineer. The severity of unfavorable interaction usually increases with increasing closeness of the new interface to the soil surface where daily and seasonal temperature and moisture changes reach maximum values and where noxious salts may accumulate. While structural sensitivity of cohesive soils and undesirable chemical activity of pollutants and problematic soils are well known and respected, it is less well known that even "inert" materials such as sands may show greatly varying bearing power depending on their specific location, e.g., beach sand and the windward or leeward side of wandering dunes in desert regions. Also, depending on their modes of deposition, fluvial and marine deposits and gravels may be so loose that, while easily supporting static loads, they become macromeritic liquids as a result of earthquake shock.

This textbook contains 18 chapters. It attempts to present a synoptic picture of environmental geotechnology, its interactions, and applications. It is hoped, however, that the information contained in this textbook is informative enough to convey to geotechnical engineers the idea of the importance of environmental aspects of geotechnology. The examples given in this text are of sufficient variety to illustrate the many different areas in which environmental geotechnology knowledge and methodology are extremely helpful tools as a new dimension to understand fundamental soil behavior.

However, there is another, perhaps even more important aspect, which is the influence of the type of thinking on the mind of the geotechnical engineer which is basic to:

- Soil-environment interaction
- Soil-water interaction in the environment
- Mass transport phenomena in contaminated unsaturated fine-grained soil
- Geomorphic processes or aging effect of soil and rock properties

It is hoped this additional information and illustration may assist in further understanding the fundamental behavior of the soil, especially when soil interacts with complex local environments.

At the present time, the subject of geotechnology stands at a crossroads. One road still dogmatically follows the classical concept developed by K. Terzaghi in 1925, and the other, the environmental situations that have challenged these concepts and methods for analyzing soil behavior under varied environmental conditions, because they cannot effectively explain all the soil-water-environment phenomena and soil-structure interaction the modern world presents. The situation cannot be changed or corrected dramatically overnight, but it must change eventually without a large gap between the two directions. To cope with this issue, a combined approach is presented which is why the particle-energy-field theory proposed by the author is introduced in this textbook. In other words, this textbook is presented within an old framework with new information blended into it as necessary.

Some of the remarks in this text may be considered controversial to classical soil mechanics; however, research findings indicate that environmental geotechnology has the potential of greatly assisting problematic engineering projects. The recognition of the environmental factor effect on geotechnology is the main theme of this textbook.

APPENDIX 1A — ABBREVIATIONS AND ACRONYMS

AASHO	American Association of State Highway Officials (current name AASHTO)
AASHTO	American Association of State Highway and Transportation Officials
ACS	American Chemical Society
AEE	Association of Energy Engineers
AGU	American Geophysical Union
AID	Agency for International Development
AIEE	American Institute of Electrical Engineering
APCA	Air Pollution Control Association
ASAE	American Society of Agricultural Engineers
ASCE	American Society of Civil Engineers
ASTM	American Society of Testing and Materials
BPR	Bureau of Public Roads (current name FHA)
BNCB	British National Coal Board
CAA	Clean Air Act
CERCLA	Comprehensive Environmental Response, Compensation and Liability Act
CWA	Clean Water Act
DER	Department of Energy Resources
DOE	Department of Energy
EPA	Environmental Protection Agency
FAA	Federal Aviation Agency
FHA	Federal Highway Administration
HRB	Highway Research Board (current name TRB)
HRR	Highway Research Record (current name TRR)
ICSMFE	International Conference of Soil Mechanics and Foundation Engineering
JSMFE	Journal of the Soil Mechanics and Foundations
LMER	Land-Margin Ecosystems Research Sites
NCHRP	National Cooperative Research Project
NAS	National Academy of Science
NRC	National Research Council
NSF	National Science Foundation
OECD	Organization for Economic Cooperation and Development
OSM	Office of Surface Mining
PERF	Pennsylvania Environmental Research Foundation
RCRA	Resource Conservation and Recovery Act
SSSA	Society of Soil Science of America
TRB	Transportation Research Board
TRR	Transportation Research Record
USBM	United States Bureau of Mines
USBR	United States Bureau of Reclamation
USDA	United States Department of Agriculture
USGS	United States Geological Survey

APPENDIX 1B NOTATIONS — SYMBOLS

A, a	Constant; area; crack length (Figure 6.3)
	Subsidence factor (Equation 15.7)
A	Activity of clay; A-horizon (soil profile)
	Atomic weight; mass number of isotope (Equation 10.1)
	Attractive force; ampere; Avogadro number

597

	Soil loss per unit area (Equation 14.1)
Å	Angstrom unit
Atm	Atmosphere pressure
amu	Atomic mass unit
a_v	Coefficient of compressibility (Equation 11.8)
B, b	Constant; width
B	B-horizon (soil profile)
	Boltzmann constant (Equation 2.1)
	Bulk modulus
BRIE	Biogeochemical reactions in estuaries
C, c	Constant; cohesion; unit cohesion (Equation 11.10)
	C-horizon (soil profile); capacitance (Table 9.1)
	Chlorinity (Equation 14.1)
	Velocity of light (Equation 10.2)
	Centigrade scale (Equation 8.2)
c_r	Residual cohesion
c'	Effective cohesion
C'	Wind erosion (Equation 14.7)
C_m	Mass heat capcity (Table 8.1)
C_s	Heat capacity (Equation 8.1)
C_v	Volumetric heat capacity (Table 8.1)
C_{sp}	Specific heat
C_{mv}	Mass heat capacity
CD	Consolidated drained shear test
CE	Compaction effort (Equation 18.1)
CU	Consolidated undrained shear test
CI	Cracking intensity (Equation 11.22)
	Consistency index (Equation 6.2)
CT	Computed tomography
CBR	California bearing ratio
C_i	Curie
C_s	Spacing of transverse cracks (Equation 7.24)
C_v	Mole heat at constant volume (Equation 5.4)
c_c	Compression index
c_s	Swelling index
c_u	Uniformity coefficient
c_v	Coefficient of consolidation (Equation 11.7)
D, d	Diameter; distance; depth; thickness; constant
	Distance between the ions and surface (Equation 4.8)
d	Displacement distance (Equation 9.10)
D	Diffusivity (Equation 7.25)
	D-horizon (soil profile)
D_e	Effective depth (Equation 11.1)
D_r	Rut depth (Figure 15.17)
D_{10}	Diameter at which 10% of the soil is finer (effective size)
D_{60}	Diameter at which 60% of the soil is finer
DM	Dipole moment
DTA	Differential thermal analysis
E	Young's modulus; energy (Equation 10.2); compaction energy Kinetic energy (Equation 11.5); efficiency (Equation 12.2)
	Total kinetic energy of a given storm (Equation 14.2)
	Emittance; electromotive force (Equation 9.11)
E'	Soil loss by wind (Equation 14.5)
E_a	Attractive energy (Equation 4.3)
E_K	Kinetic energy per mole or per particle (Equation 2.1)
E_R	Repulsive energy (Equation 4.7)
E_T	Total thermal energy content per mole (Equation 5.4)
ETM	Estuarine turbidity maximum
e	Void ratio; electronic charge (Equation 4.3)
e_{min}	Minimum void ratio
e_0	Initial void ratio
e_f	Final void ratio

e^-	Electronic charge
emf	Electromotive force
esu	Electrostatic unit
eV	Electron volt
F	Force; freezing index; farad (Equation 5.3)
	Fahrenheit scale (Equation 8.2)
	Percent passing the #200 sieve (Equation 3.2)
F_a	Air freezing index
F_m	Mean freezing index
F_s	Surface freezing index; factor of safety
FS	Factor of safety
f	Frequency; coefficient of friction
f_{vp}	Consolidation function
g	Gram; gravity
	Conductance (Table 9.1)
G	Griffith strain-energy release rate (Equation 6.4)
	Dynamic shear modulus (Figure 11.9)
GI	Group index (Equation 3.2)
G_c	Critical strain-energy release rate (Equation 6.4)
G_{max}	Maximum dynamic shear modulus
G_s	Specific gravity of solids
H, h	Head; height; thickness
h_e	Elevation head
h_p	Pressure head
h_v	Velocity head
h_c	Capillary head
h_x	Height of free drop (Equation 11.3)
H_e	Hydraulic head (Equation 9.16)
HLRW	High-level nuclear wastes
I	Wind erodibility index (Equation 14.6)
	Electric current intensity; amperage (Equation 9.12)
	Rainfall intensity (Equation 14.2)
I′	Soil wind erodibility factor (Equation 14.5)
i	Angle; gradient
i_c	Critical gradient
i_T	Thermal gradient
i_w	Water-content gradient (Equation 7.28)
i_e	Electric gradient
I_c	Consistency index (Equation 6.2)
	Crack infiltration rate (Equation 7,24)
I_f	Flow index
I_L	Liquidity index (Equation 6.1)
I_p	Plasticity index
I_s	Knoll erodibility factor (Equation 14.6)
I_t	Toughness index
I	Energy gradient (Equation 7.26)
J	Joule
K, k	Coefficient of permeability
k	Boltzmann constant (Equation 2.1)
	Thermal conductivity (Equation 8.5)
k′	Physical permeability
k*	Permeability and apparent permeability (Equation 9.17)
k_p	Rate of infiltration (Equation 7.24)
	Pore shape factor (Equation 7.12)
K	Soil erodibility factor (Equation 14.1)
	Stress intensity factor (Equation 6.5)
	Kelvin (absolute) scale (Equation 8.4)
K′	Soil surface roughness factor (Equation 14.5)
	Physical permeability (Equation 7.18)
K_a	Acid ionization constant (Table 5.3)
K_c	Fracture toughness (Equation 6.3)
K_e	Coefficient of permeability of saturated soil (Equation 9.16)

K_h	Permeability at horizontal direction
K_v	Permeability at vertical direction
KE	Kinetic energy (Equation 2.1)
K_0	Coefficient of earth pressure at rest
K	Coefficient of energy conductivity (Equation 7.26)
KMT	Kinetic molecular theory
KBES	Knowledge-based expert system
L	Length; wave length
	Length of the conductor; drawdown (Figure 15.11)
L'	Length of the unshielded distance (Equation 14.5)
LI	Liquidity index (Equation 6.1)
LL	Liquid limit
LEFM	Linear elastic fracture mechanics
LLRW	Low-level nuclear wastes
LMER	Land-margin ecosystems research
LRFD	Load resistance factor design
LS	Topographic factor (Equation 14.1)
M, m	Mass; moment; meter; mass of the particle (Equation 10.2)
	Seam thickness (Equation 15.7)
m	Exponent (Equation 14.3)
m_n	Neutron mass
m_p	Proton mass
m_v	Coefficient of volume compressibility (Equation 11.8)
M	Molecular weight of water (Equation 4.1)
	Molarity
MES	Modular expert system
MDES	Multidomain expert system
meq	Milliequivalents
n	Neutron; porosity; number of piles
	Electrolyte concentration (Equation 9.11)
n_c	Porosity due to capillarity (Equation 7.13)
n_r	Pore space (Equation 7.15)
N	Newton; Avogadro number (Equation 2.1)
	Number of blows; number of traffic lanes (Equation 7.22)
N_c	Bearing capacity parameter (Equation 12.3)
	Number of longitudinal joints (Equation 7.24)
NAPLs	Nonaqueous phase liquids
NSF	Negative skin friction (pile foundation)
OCR	Overconsolidation ratio
OLC	Optimum liquid content
OMC	Optimum moisture content
p	Proton
P	Force; load; pressure; power
	Mean annual precipitation
PC	Personal computer
PI	Plasticity index
PE	Precipitation effectiveness index (Equation 14.7)
PL	Plastic limit
P_u	Pile capacity
PCC	Portland cement concrete
PSI	Pollution sensitivity index (Figure 3.8 and Table 3.18)
p_c	Preconsolidation pressure (Equation 11.9)
p_0	Overburden pressure (Equation 11.9)
pCi/l	Picocurie per liter
pK_a	Negative logarithms of K_a (Table 5.3)
ppm	Part per million
psi	Pound per square inch
Q, q	Load; discharge; amount of infiltration (Equation 7.22)
	Rate of flow; quantity of heat (Equation 8.5)
Q_e	Volume of expelled liquid from soil (Equation 9.11)
Q_{ult}	Ultimate load
q	Water flux (Equation 7.24)

q_a	Allowable bearing capacity
q_e	Unit flow (Equation 9.11)
q_i	Design infiltration rate (Equation 7.24)
q_u	Unconfined compression strength
R, r	Radius
r^2	Coefficient of correlation
r_r	Portion of the radius (Equation 7.15)
R	Repulsion; thermal resistivity (Equation 8.6)
R_s	Shrinkage ratio (Figure 6.3)
RBE	Relative biological effectiveness
RPISU	Radon progeny integrating sampling unit
sec	Second; undrained shear strength (Equation 11.4)
	Soil-water suction; length of the dipoles (Equation 4.8)
	PCC slab length (Equation 7.22)
s	Slope steepness (Equation 14.3)
s_u	Mean undrained shear strength (Figure 16.4)
S	Shear strength; salinity (Equation 14.1)
	Specific surface area (Equation 4.1); pile spacing (Equation 12.2)
	Specific surface per unit volume (Equation 7.12)
	Slope-gradient factor; degree of saturation
	Storage coefficient
	Sensitivity (Equation 11.12)
S′	Sensitivity (Equation 11.13)
S_c	Shear strength increase caused by consolidation (Equation 16.1)
S_N	Suitability number
S_r	Residual shear strength (Equation 11.14)
SI	Superior index (Section 18.9)
SL	Shrinkage limit
SR	Shrinkage ratio
SP	Specific energy (Equation 11.5)
SAR	Sodium adsorption ratio
SEM	Scanning electron microscope
SPT	Standard penetration test
SSR	Silica/sesquioxides ratio (Equation 3.4)
T, t	Time; tortuosity (Equation 7.12); surface tension (Equation 7.1)
t	Double-layer thickness
T	Absolute temperature; mean annual temperature
	Tesla; thawing index; transmissibility
TS	Tensile strength
TSC	Thermal storage capacity (Figure 8.4)
T_a	Air thawing index
	Time required for equilibrium (Figure 16.11)
T_f	Time required for first-time failure (Figure 16.11)
T_m	Mean thawing index
T_s	Surface tension
T_s	Tension at solid-liquid interface (Figure 4.15)
T	Tension at liquid-vapor interface (Figure 4.15)
T_s	Tension at solid-vapor interface (Figure 4.15)
u	Porewater pressure; atomic mass unit (amu)
U	Total heat capacity (Equation 8.1)
	Degree of consolidation (Equation 16.2)
UP	Unconfined penetration test (Equation 6.3)
V, v	Velocity; volume; voids
V	Mean monthly wind velocity (Equation 14.7)
V′	Vegetative factor (Equation 14.5)
VM	Erosion control factor (Equation 14.1)
Vs_u	Coefficient of variation (Figure 16.4)
v_s	Volume of solid
v_v	Volume of void
W, w	Weight
W	Width of crack opening (Equation 11.22)
	Pavement width (Equation 7.22)

Width of aggregate base subjected to infiltration (Equation 7.24)
WL   Working level (Table 10.9)
W_c Length of transverse cracks (Equation 7.24)
W_x Weight of pounder (Equation 11.3)
X, x   Distance
        Chemical symbol of element (Equation 10.1)
XRD   X-ray diffraction
Y, y   Distance
Z, z   Distance; elevation head
Z   Atomic number (Equation 10.1)
Alpha (α)
    Angle; coefficient
    Angle of wetting (Equation 7.1)
    Thermal diffusivity (Table 8.1)
Beta (β)
    Angle
    Beta decay
Gamma (γ)
γ Unit weight of soil
γ_o Dry unit weight of soil
γ_m Mass unit weight of soil
γ_T Total unit weight of soil
γ_ω Unit weight of water
    Submerged unit weight of soil
    Gamma decay
    Shear strain (Equation 11.21)
γ_r Reference strain (Equation 11.22)
    Dynamic shear strain (Figure 11.26)
Delta (δ)
    Angle
    Angle between soil and wall
Epsilon (ε)
    Dielectric constant
ε_ω Dielectric constant for water (Equation 4.6)
ε_m Dielectric constant for montmorillonite (Equation 4.6)
    Energy
    Electromotive force
Zeta (ζ)
    Zeta potential
    Coefficient correction (Equation 16.1)
Eta (η)
    Coefficient of viscosity of liquid (Equation 7.12)
Theta (θ)
    Angle; constant
    Liquid-solid contact angle (Figure 4.15)
Kappa (κ)
$1/\kappa$ Distance from particle surface (Figure 4.13)
    Double-layer thickness (Figure 4.13)
Lambda (λ)
    Specific conductance (Equation 9.17)
    Characteristic length (Section 7.4)
λ_c Surface cracking correction factor (Equation 16.7)
Mu (μ)
    Friction angle
    Dipole moment (Equation 4.8)
    Absolute viscosity; dynamic viscosity
μ_r Residual strength coefficient (Equation 11.20)
Nu (ν)
    Ionic valence (Equation 4.3)
    Kinematic viscosity
    Poisson's ratio (Equation 6.5)
Xi (ξ)

Tension/cohesion ratio of soil (Equation 11.11)

Rho (ρ)

    Density (Table 8.1)

    Electric resistivity

    Specific resistance (Equation 9.13)

Sigma (σ)

    Normal stress on shear plane (Equation 11.10)

    Effective pressure on shear plane (Equation 11.14)

σ_v Vertical stress

$\Delta\sigma$ Deviator stress

σ_t Tensile stress

σ'_v Vertical effective stress

    Surface charge density

Tau (τ)

    Shear stress

$\bar{\tau}$ Shear stress at natural sliding surface (Equation 11.20)

τ_r Residual stress (Equation 11.20)

τ_f Peak strength (Equation 11.20)

Phi (ϕ)

    Angle; friction angle of soil

ϕ_r Residual friction angle

ϕ' Effective friction angle

Psi (ψ)

    Coefficient of effective depth (Equation 11.6)

    Flow velocity in energy fields (Equation 7.26)

    Surface potential (Figure 4.13)

    Stern potential (Figure 4.13)

Omega (ω)

w Moisture content

w_0 Initial moisture content

w_p Plastic limit

w_L Liquid limit

w_s Shrinkage limit

APPENDIX 2A CONVERSION TABLE

Length
  1 in. = 25.4 mm = 2.54 cm = 0.0254 m
  1 ft = 0.305 m
  1 mm = 0.0394 in.
  1 cm = 10 mm = 0.394 in.
  1 km = 1000 m = 3280 ft
Area
  1 in.^{2} = 6.45 cm^2
  1 ft^2 = 0.0929 m^2
  1 cm^2 = 0.155 in.^{2}
  1 m^2 = 10.8 ft^2
Volume
  1 in.^{3} = 16.4 cm^3
  1 ft^3 = 0.0283 m^3
  1 cm^3 = 0.061 in.^{3}
  1 m^3 = 35.3 ft^3
Capacity
  1 ft^3 = 7.48 U.S. gal = 28.3 l
  1 cm^3 = 0.001 l
  1 l = 1000 cm^3 = 61.02 in.^{3} = 0.264 U.S. gal
  1 U.S. gal = 3785 cm^3 = 231 in.^{3} = 3.78 l = 0.00385 m^3
Force
  1 lb = 4.45 N = 453.6 g = 5 × 10^{-4} ton
  1 kip = 1000 lb = 4450 N = 4.45 kN
  1 ton = 8.89 × 10^3 N = 8.89 kN = 9.07 × 10^5 g = 2000 lb
  1 dyn = 2.248 × 10^{-6} lb
  1 N = 102.0 g = 0.225 lb = 1.124 × 10^{-4} ton = 100,000 dyn
  1 kN = 0.2248 kips
  1 g = 9.81 × 10^{-3} N = 2.20 × 10^{-3} lb = 1.102 × 10^{-6} ton = 980.7 dyn
Stress
  1 psi = 6.89 kN/m^2 = 6.89 kPa = 7.03 × 10^{-2} kg/cm^2 = 144 psf
  1 psf = 47.9 N/m^2 = 47.9 Pa = 4.88 × 10^{-4} kg/cm^2 = 6.94 × 10^{-3} psi = 5 × 10^{-4} ton/ft^2
  1 ton/ft^2 = 95.8 kN/m^2 = 95.8 kPa = 9.76 × 10^{-1} kg/cm^2 = 13.9 psi = 2000 psf
  1 ksi = 6.89 N/mm^2 = 6.89 MPa
  1 N/m^2 = 1 Pa (pascal) = 1.02 × 10^{-5} kg/cm^2 = 2.08 × 10^{-2} ton/ft^2
  1 kPa = 1 kN/m^2 = 0.1 N/cm^2 = 0.145 psi
  1 kg/cm^2 = 9.81 × 10^4 N/m^2 = 98.1 kN/m^2 = 98.1 kPa = 14.2 psi = 2.05 × 10^3 psf = 1.02 ton/ft^2
Unit Weight
  1 pcf = 157 N/m^3 = 0.157 kN/m^3 = 1.6 × 10^{-2} g/cm^3
  1 g/cm^3 = 9.81 × 10^3 N/m^3 = 9.81 kN/m^3 = 62.4 pcf
  1 N/m^3 = 1.02 × 10^{-4} g/cm^3 = 6.37 × 10^{-3} pcf
Temperature
  1°F = 0.555°C = 0.555 K
  1°C = 1 K = 1.8°F
  0 K = −273°C = −460°F

APPENDIX 2B

eV = electron volts
    $= 1.602 \times 10^{-12}$ erg
    $= 0.386 \times 10^{-19}$ cal
1 atm = 14.7 psi $= 1.0 \times 10^6$ dyn/cm^2
      = 29.9 in. Hg (mercury)
      = 1,033 cm H_2O
      = 760 torr
1 torr = 1 mm Hg
T K = T°C + 273.2°
1 bar $= 1.0 \times 10^6$ dyn/cm^2 = 14.7 psi
1 Å = Angstrom unit $= 10^{-7}$ mm $= 10^{-8}$ cm
    $= 10^{-4} \mu = 0.1$ μm
    = 0.0001 cm $= 10^{-4}$ cm = 0.001 mm $= 10^{-3}$ mm
1 μm $= 10^{-3} \mu$; 1 μ = 0.0001 cm $= 10^{-4}$ cm $= 10^{-3}$ mm
1 cm Hg (at 20°C) = 5.34 in., water (at 20°C)
                  = 13.57 cm water (at 20°C)
                  = 27.75 lb/ft^2
1 psi $= 6.895 \times 10^4$ dyn/cm^2
                  = 5.17 cm Hg (at 0°C) = 5.19 cm Hg (at 20°C)
                  = 2.03 in. Hg (at 0°C) = 2.04 in. Hg (at 20°C)
                  = 70.29 cm water (at 4° C)
                  = 70.43 cm water (at 20°C)

APPENDIX 2C CONVERSION FACTORS FOR THERMAL CONDUCTIVITY

Units	a	b	c	d	e	f	g
kcal/m-hr-°C	1	$2.778 \cdot 10^{-3}$	$11.63 \cdot 10^{-3}$	1.163	$1.163 \cdot 10^{-3}$	$672 \cdot 10^{-3}$	$56 \cdot 10^{-3}$
cal/cm-sec-°C	360	1	4.187	418.68	$418.7 \cdot 10^{-3}$	242	20.2
watt/cm-°C	86.0	0.239	1	100	0.1	57.8	4.815
watt/m-°C	0.860	$2.39 \cdot 10^{-3}$	0.01	1	0.001	0.578	$48.15 \cdot 10^{-3}$
kw/m-°C	860	2.39	10	1000	1	578	48.15
BTU/ft-hr-°F	1.488	$4.134 \cdot 10^{-3}$	$17.3 \cdot 10^{-3}$	1.73	$1.73 \cdot 10^{-3}$	1	$83.3 \cdot 10^{-3}$
BTU/in.-hr-°F	17.86	$49.6 \cdot 10^{-3}$	0.208	20.8	$20.8 \cdot 10^{-3}$	12	1

APPENDIX 2D ATOMIC NUMBERS AND WEIGHTS

Element	Symbol	Atomic number	Atomic weight	Element	Symbol	Atomic number	Atomic weight
Actinium	Ac	89	227	Mercury	Hg	80	200.61
Aluminum	Al	13	26.98	Molybdenum	Mo	42	95.95
Americium	Am	95	[243]	Neodymium	Nd	60	144.27
Antimony	Sb	51	121.76	Neon	Ne	10	20.183
Argon	Ar	18	39.944	Neptunium	Np	93	[237]
Arsenic	As	33	74.91	Nickel	Ni	28	58.71
Astatine	At	85	[210]	Niobium	Nb	41	92.91
Barium	Ba	56	137.36	Nitrogen	N	7	14.008
Berkelium	Bk	97	[249]	Osmium	Os	76	190.2
Berylium	Be	4	9.013	Oxygen	O	8	16
Bismuth	Bi	83	209.00	Palladium	Pd	46	106.4
Boron	B	5	10.82	Phosphorus	P	15	30.975
Bromine	Br	35	79.916	Platinum	Pt	78	195.09
Cadmium	Cd	48	112.41	Plutonium	Pu	94	[242]
Calcium	Ca	20	40.08	Polonium	Po	84	210
Californium	Cf	98	[249]	Potassium	K	19	39.100
Carbon	C	6	12.011	Praseodymium	Pr	59	140.92
Cerium	Ce	58	140.13	Promethium	Pm	61	[145]
Cesium	Cs	55	132.91	Protactinium	Pa	91	231
Chlorine	Cl	17	35.457	Radium	Ra	88	226.05
Chromium	Cr	24	52.01	Radon	Rn	86	222
Cobalt	Cp	27	58.94	Rhenium	Re	75	186.22
Columbium (see Niobium)				Rhodium	Rh	45	102.91
Copper	Cu	29	63.54	Rubidium	Rb	37	85.48
Curium	Cm	96	[245]	Ruthenium	Ru	44	101.1
Dysprosium	Dy	66	162.46	Samarium	Sm	62	150.35
Erbium	Er	68	167.2	Scandium	Sc	21	44.96
Europium	Eu	63	152.0	Selenium	Se	34	78.96
Fluorine	F	9	19.00	Silicon	Si	14	28.09
Francium	Fr	87	[223]	Silver	Ag	47	107.880
Gadolinium	Gd	64	157.26	Sodium	Na	11	22.991
Gallium	Ga	31	69.72	Strontium	Sr	38	87.63
Germanium	Ge	32	72.60	Sulfur	S	16	32.066
Gold	Au	79	197.0	Tantalum	Ta	73	180.95
Hafnium	Hf	72	178.50	Technetium	Tc	43	[99]
Helium	He	2	4.003	Tellurium	Te	52	127.61
Holmium	Ho	67	164.94	Terbium	Tb	65	158.93
Hydrogen	H	1	1.0080	Thallium	Tl	81	204.39
Indium	In	49	114.82	Thorium	Th	90	232.05
Iodine	I	53	126.91	Thulium	Tm	69	168.94
Iridium	Ir	77	192.2	Tin	Sn	50	118.70
Iron	Fe	26	55.85	Titanium	Ti	22	47.90
Krypton	Kr	36	83.80	Tungsten	W	74	183.86
Lanthanum	La	57	138.92	Uranium	U	92	238.07
Lead	Pb	82	207.21	Vanadium	V	23	50.95
Lithium	Li	3	6.940	Xenon	Xe	54	131.30
Lutetium	Lu	71	174.99	Ytterbium	Yb	70	173.04
Magnesium	Mg	12	24.32	Yttrium	Y	39	88.92
Manganese	Mn	25	54.94	Zinc	Zn	30	65.38
Mendelevium	Mv	101	[256]	Zirconium	Zr	40	91.22

Note: A value given in brackets denotes the mass number of the most stable known isotope.

Source: Lide, D.R., Ed., *The CRC Handbook of Chemistry and Physics,* 77th ed., CRC Press LLC, Boca Raton, FL, 1996.

APPENDIX 2E COMMON OXIDES, CHLORIDES, HYDROXIDES, CARBONATES, SULFATES, NITRATES, AND PHOSPHATES

Oxides
CaO	Calcium oxide
MgO	Magnesium oxide
Al_2O_3	Aluminum oxide
MnO_3	Manganese dioxide (pyrolusite)
FeO	Iron(II) oxide or ferrous oxide
Fe_2O_3	Iron(III) oxide or ferric oxide (hematite)
Fe_3O_4	Iron(II, III) oxide (magnetite)

Chlorides
$NaCl$	Sodium chloride (halite)
KCl	Potassium chloride (sylvite)
$CaCl_2$	Calcium chloride
$MgCl_2$	Magnesium chloride
HCl	Hydrogen chloride
$AgCl$	Silver chloride (horn silver)

Hydroxides
$NaOH$	Sodium hydroxide
KOH	Potassium hydroxide
$Mg(OH)_2$	Magnesium hydroxide
$Ca(OH)_2$	Calcium hydroxide
$Al(OH)_3$	Aluminum hydroxide
$Fe(OH)_3$	Iron(III) hydroxide or ferric hydroxide

Carbonates
Na_2CO_3	Sodium carbonate
$CaCO_3$	Calcium carbonate (calcite, marble)
$MgCO_3$	Magnesium carbonate (magnesite)
$FeCO_3$	Iron(II) carbonate or ferrous carbonate (siderite)
$ZnCO_3$	Zinc carbonate (smithsonite)

Sulfates
Na_2SO_4	Sodium sulfate (thenardite)
$CaSO_4$	Calcium sulfate (anhydrite)
$BaSO_4$	Barium sulfate (barite)
$Al_2(SO_4)_3$	Aluminum sulfate

Nitrates
$NaNO_3$	Sodium nitrate (soda niter, chile saltpeter)
KNO_3	Potassium nitrate (niter, saltpeter)
$Ca(NO_3)_2$	Calcium nitrate

Phosphates
Na_3PO_4	Sodium phosphate or trisodium phosphate
$Ca_3(PO_4)_2$	Calcium phosphate (whitlockite, phosphate rock)

APPENDIX 2F COMMON AND SYSTEMATIC NAMES FOR SOME COMMON COMPOUNDS

Formula	Common name	Systematic name
Al_2O_3	Alumina	Aluminum oxide
NH_3	Ammonia	
$NaHCO_3$	Baking soda or bicarbonate of soda	Sodium hydrogen carbonate
$Na_2B_4O_7 \cdot 10H_2O$	Borax	Sodium tetraborate decahydrate
$CaCO_3$	Calcite or marble	Calcium carbonate
$KHC_4H_4O_6$	Cream of tartar	Potassium hydrogen tartrate
$MgSO_4 \cdot 7H_2O$	Epsom salt	Magnesium sulfate heptahydrate
$CaSO_4 \cdot 2H_2O$	Gypsum	Calcium sulfate dihydrate
C_2H_5OH	Grain alcohol or alcohol	Ethyl alcohol or ethanol
$Na_2S_2O_3$	Hypo	Sodium thiosulfate
N_2O	Laughing gas or nitrous oxide	Dinitrogen oxide
PbO	Litharge	Lead(II) oxide or plumbous oxide
CaO	Lime	Calcium oxide
$NaOH$	Lye	Sodium hydroxide
CH_4	Methane	
O_3	Ozone	
$2CaSO_4 \cdot H_2O$	Plaster of paris	2-Calcium sulfate 1-water
K_2CO_3	Potash	Potassium carbonate
NH_4Cl	Sal ammoniac	Ammonium chloride
$NaNO_3$	Saltpeter	Sodium nitrate
$Ca(OH)_2$	Slaked lime	Calcium hydroxide
$C_{12}H_{22}O_{11}$	Sugar	Sucrose
$NaCl$	Salt	Sodium chloride
$Na_2CO_3 \cdot 10H_2O$	Washing soda	Sodium carbonate decahydrate
CH_3OH	Wood alcohol	Methyl alcohol or methanol
HF	Hydrogen fluoride	
HCl	Hydrogen chloride	
NO	Nitrogen oxide (commonly called nitric oxide)	
N_2O	Dinitrogen oxide (commonly called nitrous oxide)	
SO_3	Sulfur trioxide	
H_2S	Dihydrogen sulfide (commonly called hydrogen sulfide)	
P_4S_3	Tetraphosphorus trisulfide (used in "strike anywhere" matches)	
Sb_2S_3	Diantimony trisulfide (used in safety matches)	
$Na_2B_4O_7 \cdot 10H_2O$	Sodium tetraborate decahydrate (borax)	
$MgSO_4 \cdot 7H_2O$	Magnesium sulfate heptahydrate (epsomite)	
$CuSO_4 \cdot 5H_2O$	Copper sulfate pentahydrate (blue vitriol, cahalcanthite)	
$CaCl_2 \cdot 2H_2O$	Calcium chloride dihydrate	
$Al_2O_3 \cdot 3H_2O$	Aluminum oxide trihydrate	
$Na_2SO_4 \cdot 10H_2O$	Sodium sulfate decahydrate (glauber salt)	
$Na_2CO_3 \cdot 10H_2O$	Sodium carbonate decahydrate (washing soda)	

References

Acar, Y. B. and Olivieri, I. (1989), A study on the interactions of pore fluid on the fabric and hydraulic conductivity of laboratory compacted clay, HRB Annual Meeting, 24p.

Acar, Y. B. and Haider, L. (1990), Transport of low-concentration contaminants in saturated earthen barriers, *J. Geotech. Eng. Div. Proc. ASCE*, v. 116, no. 7, pp. 1031-1052.

Acar, Y. B. and Daniel, D. E. eds. (1995), *Geoenvironmental 2000: Characterization, Containment, Remediation & Performance in Environmental Geotechnics*, ASCE, NY.

Adachi, T. and Takenaka, H. (1973), Engineering properties of cohesive soil from the viewpoint of the strain-rate effect and moisture content, (see Takenaka and Yasuhara, 1977).

AGU (1992), Understanding changes in coastal environments, The LMER program, EOS, Trans. AGU, v. 73, no. 45, Nov. pp. 481-484.

Ahmed, I. and Lovell, C. W. (1992), Use of waste products in highway construction, Proc. Mediterranean Conf. on Environmental Geotechnology, M. A. Usmen and Y. B. Acar, eds., A.A. Balkema, Rotterdam, pp. 409-418.

AISC, (1993), *Manual of Steel Construction-Load and Resistance Factor Design*, 2nd. ed., American Institute of Steel Construction, Inc., Chicago, Sept.

Aitchison, G. D., Ingles, O. G. and Wood, C. C. (1963), Post-construction deflocculation as a contributory factor in the failure of earth dams, *Proc. 4th Austirlian-New Zealand Conf. SMFE*, pp. 275-279 and 339-340.

Akagi, H. (1994), A physico-chemical approach to the consolidation mechanism of soft clays, *Soils and Foundations*, v. 34, no. 4, Dec. pp. 43-50.

Aldrich, H. P., Jr. (1956), Frost penetration below highway and airfield pavements, HRB Bulletin 135, pp. 124-149.

Al-Hussaini, M. (1982), Comparison of various methods for determining k_o, ASTM STP 740, pp. 78-93.

Allan, T. and Warren, A. (1993), *Deserts: The Encroaching Wilderness*, Oxford Univ. Press, NY.

Allinger, N. L., Bigelow, N. J. and McAllister, H. C. (1976), *An Introduction to General, Organic and Biological Chemistry*, Wadsworth Publ., Belmont, CA.

Alther, G., Evans, J. C., Fang, H. Y. and Witmer, K. (1985), Influence of inorganic permeants upon the permeability of bentonite, ASTM, STP 874, pp. 64-73.

Andersland, O. B. and Anderson, D. M., eds. (1978), *Geotechnical Engineering for Cold Regions*, McGraw-Hill Book Co., NY.

Anderson, M. S. (1929), The influence of substituted cations on the properties of soil colloids, *J. Agric. Res.*, v. 38, pp. 565-584.

Anderson, D. and Brown, K. W. (1981), Organic leachate effects on the permeability of clay liners, U.S. EPA Report 600/9-81-002b, pp. 119-130.

Anderson, D. M. and Morgenstern, N. R. (1973), Physics, chemistry and mechanics of frozen ground, North American Contribution, *Proc. 2nd Int. Conf. on Permafrost*, NAS, pp. 257-288.

Anderson, J. N. and Lade, P. V. (1981), The expansion index test, *ASTM Geotech. Testing J.*, v. 4, no. 2, pp. 58-67.

Anderson, D. C., Crawley, W. and Zabcik, J. D. (1985), Effects of various liquids on clay soil: bentonite slurry mixtures, ASTM STP 874, pp. 93-101.

Anderson, D. G., Hushmand, B. and Martin, G. R. (1992), Seismic response of landfill slopes, ASCE Geotechnical Special Publ. 31., pp. 973-989.

Andrews, R. E., Gawarkiewicz, J. J. and Winterkorn, H. F. (1967), Comparison of the interaction of three clay minerals with water, dimethyl sulfoxide, and dimethyl formamide, HRR no. 209, pp. 66-78.

Andrews, J. N. and Wood, D. F. (1972), Mechanisms of radon release in rock materials and entry into groundwater, Institution of Mining Metallurgy, London, *Trans*. Sec. B81 (Applied Earth Science), pp. 197-209.

Anthes, R. A., Panofsky, H. A., Cahir, J. J. and Rango, A. (1978), *The Atmosphere*, 2nd ed., Charles E. Merrill Publ. Co., Columbus, OH.

Arman, A. (1970), Engineering classification of organic soils, HRR no. 310, pp. 75-89.

Armistead, R. A. and Stanley, J. H. (1989), Computed tomography: quantitative 3-D inspection, *Proc. 12th World Conf. on Non-Destructive Testing*, pp. 68-73.

Arulanandan, K. and Smith, S. S. (1973), Electrical dispersion in relation to soil structure, JSMFE Div. *Proc. ASCE*, v. 99, no. SM12, pp. 1113-1134.

Arulanandan, K., Sargunam, A., Loganathan, P. and Krone, R. B. (1973), Application of chemical and electrical parameters to preduction of erodibility, HRB Special Report 135, pp. 42-51.

ASTM (1997), *ASTM Standards*.

Aurilio, G. (1983), Root piles, Recent Advances in Geotechnical Engineering, ASCE-Penn DOT, Harrisburg, PA, pp. 1-36.

Avrorin, V. V. (1981), Production of higher fluorides and oxides of radon, *Radiokhim*, v. 23, pp. 879-883.

Bagchi, A. (1990), *Design, Construction and Monitoring of Sanitary Landfill*, John Wiley & Sons, NY.

Barata, F. E. (1969), Landslides in the tropical region of Rio de Janeiro, *Proc. 7th ICSMFE*, v. 2, pp. 507-516.

Barden, L. and Sides, G. (1971), Sample disturbance in the investigation of clay structure, *Geotechnique*, v. 21, no. 3, pp. 211-222.

Barth, M. C. and Titus, J. G., eds. (1984), *Greenhouse Effect and Sea Level Rise*, Van Nostrand Reinhold Co. NY.

Bartlett, N. (1962), Xenon hexafluoroplatinate (V), Xe^+PtF_6, *Proc. Am. Chem. Soc.*, v. 218.

Basu, R. (1972), Identification and Prediction of Swell of Expansive Earth Materials, MSCE Thesis, Univ. of California at Davis, Davis, CA.

Batschinski, (1913), Untersuchungen uber die innere Reibung von Flussigkeite n., *Z. Phys. Chem.*, v. 84, pp. 643-706.

Baver, L. D. and Winterkorn, H. F. (1935), Sorption of liquids by soil colloids II: surface behavior in the hydration of clays, *Soil Sci.*, v. 40, no. 5, pp. 403-419.

Baver, L. D. and Hall, N. S. (1937), Colloidal properties of soil organic matter, Missouri Agric. Exp. Stn. Res. Bull. 267.

Baykal, G., Yesiller, N. and Koprulu, K. (1992), Rubber-clay liner against petroleum based contaminants, *Proc. Mediterranean Conf. on Environmental Geotechnology*, M. A. Usmen and Y. B. Acar, eds., A.A. Balkema, Rotterdam, pp. 477-481.

Beach Erosion Board (1963), The Annual Bulletin of the Beach Erosion Board (BEB), Department of the Army, Corps of Engineers, v. 17, July, 99p.

Beattie, A. A. and Chau, E. P. Y. (1976), The assessment of landslides potential with recommendations for future research, *J. Hong Kong Inst. Eng.*, Hong Kong, Feb.

Beavers, A. H. and Marshall, C. E. (1950), The cataphoresis of minerals and factors affecting their separation, *Proc. SSSA*, v. 15, pp. 142-145.

Beiser, A. (1988), *Applied Physics*, McGraw-Hill Book Co. NY, pp. 238 & 280.

Benkelman, A. C., Kingham, R. I., and Fang, H. Y. (1962), Special deflection studies on flexible pavements, HRB Special Report 73, pp. 102-125.

Bennett, R. H., Bryant, W. R. and Hulbert, M. H., eds. (1991), *Microstructure of Fine-Grained Sediments: From Mud to Shale*, Springer-Verlag, NY.

Benson, R. C. (1978), Assessment of localized subsidence (before the fact), *Evaluation and Prediction of Subsidence*, ASCE, NY. pp. 47-57.

Benson, C. H. and Charbeneau, R. J. (1991), Use of diffusion theory to predict first-passage times for earthen landfill liners, Dept. of Civil Engineering, Univ. of Wiscosin, Madison, WI, 19p.

Berelson, W. M., Hammond, D. E. and Eaton, A. D. (1987), A technique for the rapid extrtaction of radon-222 from water samples and a case study, *Radon in Groundwater*, B. Graves, ed, Lewis Publ., Chelsea, MI, pp. 271-282.

Berggren, W. P. (1943), Prediction of temperature distribution in frozen soils, *Trans. AGU*, part III.

Bergna, H. E. (1950), Electrokinetic behavior of clay minerals, *Trans. 4th Int. Congr. Soil Soc.*, v. 3, pp. 75-80.

Bernal, J. D. and Fowler, R. H. (1933), *J. Chem. Phys.*, v. VI, no. 8, p. 535.

Beskow, G. (1935), Soil Freezing and Frost Heaving with Special Application to Roads and Railroads, The Swedish Geological Society Series C, no. 375.

Binnie & Partners (1977), *Construction on Slopes Manual*, Hong Kong.

Bisbing, R. E. (1989), Clues in the dust, *American Laboratory*, Nov. pp. 9-14.

Bishop, A. W. (1955), The use of the slip circle in the stability analysis of slopes, *Geotechnique*, v. 5, no. 1, pp. 7-17.

Bishop, A. W., Green, G. E., Gerga, V. K., Anresen, A. and Brown, J. D. (1971), A new ring apparatus and its application to measurement of residual strength, *Geotechnique*, v. 21, no. 4, pp. 273-328.

Biswas, M. R. and Biswas, A. K. (1980), *Desertification*, Pergamon Press, NY.

Bjerrum, L. (1954), Geotechnical properties of Norwegian marine clays, *Geotechnique*, v. 4, no. 2, p. 49.

Bjerrum, L. (1967), Progressive failure in slopes of overconsolidated plastic clay and clay shales, JSMFE Div., *Proc. ASCE*, v. 93, no. SM5, pp. 3-49.

Blackwelder, E. (1954), Geomorphic processes in the desert, California Div. of Mines and Geology, Bulletin 170, Ch. 5, pp. 11-20.

Blanck, E. ed. (1930, 1939), *Handbuck der Bodenlehre*, v. 6, Julius Springer, Berlin.

BNCB, (1966, 1975), *Subsidence Engineer's Handbook*, (1st. ed. 1966), British National Coal Board, London.

Bodocsi, A., Bowers, M. T. and Sherer, R. (1988), Reactivity of Various Grouts to Hazardous Wastes and Leachates, U.S. EPA Report no. EPA/600/S2-88/021.

Bohn, H. L., McNeal, B. L. and O'Connor, G. A. (1985), *Soil Chemistry*, 2nd ed., John Wiley & Sons, NY.

Bolt, G. H. (1955), Analysis of the validity of the Gouy-Chapman theory of the electric-double-layer, *J. Colloid Sci.*, v. 10, pp. 206-218.

Born, M. (1946), *Atomic Physics*, Hafner Publ. Co., NY.

Boudet, G. (1976), Mali, Ecological Bulletin, v. 24, Stockholm.

Bozozuk, M. and Burn, K. N. (1960), Vertical ground movement near elm trees, *Geotechnique*, v. 10, no. 1, pp. 19-32.

Brackett, R. L., Nordell, W. J., and Rail, R. D. (1982), Underwater Inspection of Waterfront Facilities: Inspection Requirements Analysis and Nondestructive Testing Technique Assessment, Naval Civil Engineering Laboratory, CA, Report no. TN-1624, 148p.

Brady, G. S. (1947), *Material Handbook*, McGraw-Hill Book Co., NY.

Brand, E. W. (1982), Analysis and design in residual soils, Engineering and Construction in Tropical and Residual Soils, *Proc. ASCE Specialty Conf. Honolulu*, pp. 89-143.

Brauner, G. (1973), Sunsidence due to underground mining, part 1, theory and practices in predicting surface deformation, part 2 ground movements and mining damage, U.S. Bureau of Mines, Information Circular 8571 and 8572.

Breed, C. S. and Grow, T. (1979), Morphology and distribution of dunes in sand seas observed by remote sensing. A Study of Global Sand Seas. USGS Prof. Paper 1052, pp. 253-302.

Broms, B. B. and Wong, K. S. (1991), Landslides, Ch. 11, *Foundation Engineering Handbook*, 2nd. ed. Van Nostrand Reinhold Co., NY, pp. 410-446.

Brown, R. L., Colbeck, S. C. and Yong, R. N. eds. (1981), Proc. of a Workshop on the Properties of Snow, U.S. Army Cold Regions Research and Engineering Laboratory, Special Report 82-18, 135p.

Brown, T. W. and Thomas, J. C. (1987), Mechanism by which organic liquids increase the hydraulic conductivity of compacted clay materials, *Soil Sci.*, v. 51, pp. 1451-1459.

Brown, M. and Synder, D. W. (1991), Origins and development of environmental mandates/laws/regulations, wetlands protection and property use issues in Pennsylvania, *Proc. Symp. Evaluation and Control of Environmental Hazards*, ASCE Lehigh Valley Section and Lehigh Univ. pp. 9-1 to 9-31.

Brumund, W. F., Jones, E., and Ladd, C. C. (1976), Estimating in-situ maximum past (preconsolidation) pressure of saturated clays, from results of laboratory consolidometer tests, TRB Special Report 163, pp. 4-12.

Brutsaert, W. F. (1981), Geologic and hydrologic factors controlling radon-222 in groundwater in Maine, *Ground Water*, v. 19, no. 4, p. 407.

Buckley, J. D. (1975), Reverse osmosis: moving from theory to practice, *Consulting Engineer*, v. 45, no. 5, pp. 55-61.

Burmister, D. M. (1951), The application of controlled test methods in consolidation testing, ASTM STP 126.

Burmister, D. M. (1964), Environmental factors in soil compaction, ASTM STP 337.

Butler, R. A. and Hampton, D. (1975), Subsidence over soft ground tunnel, *J. Geotech. Eng. Div. Proc. ASCE*, v. 100, no. GT1, paper 11074, pp. 35-49.

Byers, H. G., Kellogg, C. E., Anderson, M. S. and Thorp, J. (1938a), Formation of soil, *Soils and Men*, Yearbook of Agriculture, USDA, pp. 948-978.

Byers, H. G., Anderson, M. S. and Bradfield, R. (1938b), General chemistry of the soil, *Soils and Men*, Yearbook of Agriculture, USDA, pp. 911-928.

Camerson. C. P. (1987), A Review of Radon Emanation and Mobilization in Minerals and Rocks, Waterways Experiment Station, Report no. MSGL-87-27, 42p.

Campanella, R.G. and Mitchell, J. K. (1968), Influence of temperature variations on soil behavior, JSMFE Div., *Proc. ASCE*, v. 94, no. SM3, pp. 709-734.

Carey, W. N., Jr. and Andersland, O. B. (1957), Frost depth determination by electrical resistance measurements, Highway Research Abstracts, v. 27, no. 4.

Caron, C. (1976), Chemical grouting-European techniques, *New Horizons in Construction Materials*, Envo Publ. Co., Bethlehem, PA, pp. 237-247.

CAS (1961), Research on Loess, Civil Engineering Research Report no. 13, Chinese Academy of Science, Beijing, 157p. (In Chinese).

CAS (1969), *Soil-Water Conservation Handbook*, Beijing. (In Chinese).

Casagrande, A. (1932), The structure of clay and its importance in foundation engineering, *Contributions to Soil Mechanics, 1925-1940*, Boston Society of Civil Engineers, pp. 72-125.

Casagrande, A. (1948), Classification and identification of soils, *Trans. ASCE*, pp. 901-992.

Casagrande, L. (1952), Electro-osmotic stabilization of soils, *J. Boston Soc. Civil Eng.*, v. 39, no. 1, pp. 51-83.

Casagrande, L. (1983), Stabilization of soils by means of electro-osmosis, state-of-the-art, *J. Boston Soc. Civil Eng.*, v. 69, no. 2, pp. 225-302.

Chandrasekharan, E. C., Boominathan, S., Sadayan, E., and Setty, K. R. N. (1969), Influence of heat treatment on the pulverization and stabilization characteristics of typical tropical soils, HRB Special Report 103, pp. 161-172.

Chaney, R. C. (1984), Methods of predicting the deformation of the seabed due to cyclic loading, *Seabed Mechanics,* Denness, ed., Graham and Trotman, London, pp. 159-167.

Chaney, R. C. Dangler, L. A. and Fang, H. Y. (1983), Discussion on pulverized coal ash as structural fill, JSMFE Div., *Proc. ASCE*, v. 109, no. GT10, pp. 1356-1359.

Chaney, R. C., Ramavjaneya, G., Hencey, G., Kanchanastit, P. and Fang, H. Y. (1983), Suggested test method for determination of thermal conductivity of soil by thermal needle procedure, *ASTM Geotechn. Testing J.*, v. 6, no, 4, pp. 220-225.

Chaney, R. C. and Fang, H. Y. (1986), Static and dynamic properties of marine sediments: a state-of-the-art, ASTM STP 923, pp. 74-111.

Chaney, R. C., Richards, A. F. and Murray, C. N. (1986), Abyssal plains: potential sites for nuclear waste disposal, *Proc. 1st. Int. Symp. Environ. Geotechnol.*, v. 1, pp. 276-284.

Chaney, R. C. and Demars, K. R. (1991), Geotechnical engineering aspects of the marine coastal environment, *Proc. 1st. Int. Symp. Environ. Geotechnol.*, v. 2, pp. 41-65.

Chaney, R. C. and Fang, H. Y. (1991), Liquefaction in the coastal environment: an analysis of case histories, *Marine Geotechnol.*, v. 10, pp. 343-370.

Chaney, R. C. and Pamukcu, S. (1991), Earthquake effects on soil-foundation systems, part II, from 1975 to 1991, Ch. 16, *Foundation Engineering Handbook*, 2nd ed., Van Nostrand Reinhold, NY, pp. 623-672.

Chang, J. C. and Hannon, J. B. (1976), Settlement performance of two test highway embankments on sanitary landfill, *New Horizons in Construction Materials*, v. 1, Envo Publ. Co., Bethlehem, PA, pp. 139-158.

Chang, C. J., Yao, J. R. P. and Chen, W. F. (1983), Evaluation of seismic factor of safety of a submarine slope by limit analysis, *Proc. Shanghai Symp. on Marine Geotechnology and Nearshore/Offshore Structures*, Tongji Univ. Press/Envo Publ. Co., Bethlehem, PA, pp. 262-295.

Charles, J. A., Burford, D. and Watts, K. S. (1981), Field studies of the effectiveness of dynamic consolidation, *Proc. 10th ICSMFE*, v.1, pp. 617-622.

Chapman, D. L. (1913), A contribution to the theory of electro-capillarity, *Philos. Mag. J. Sci.*, v. 25, no. 6, p. 475.

Chen, F. H. (1979), *Foundations on Expansive Soils*, Elsevier Scientific Publ., NY.

Chen, R. H. (1986), Slope stability analysis of a waste landfill, *Proc. 1st. Int. Symp. Environ. Geotechnol.*, v. 1, Envo Publ. Co., Bethlehem, PA, pp. 37-42.

Chen, R. H. and Lee, Y. S. (1996), Settlement analysis of a waste landfill, *Proc. 3rd. Int. Symp. Environ. Geotechnol.*, v. 1, Technomic Publ., Lancaster, PA, pp. 533-539.

Chen, W. F. and Drucker, D. C. (1969), Bearing capacity of concrete blocks and rock, *J. Eng. Mech. Div. Proc. ASCE*, v. 95, no. EM4, pp. 955-978.

Chen, W. F. and McCarron, W. O. (1991), Bearing capacity of shallow foundations, Ch. 4, *Foundation Engineering Handbook*, 2nd. ed, Van Nostrand Reinhold, NY, pp. 144-165.

Chen, Y. C., Chen, Y. N., and Gaffnev, D. V. (1974), Architectural Measures to Minimize Subsidence Damage, Report no. ARC-73-111-2551, Michael Baker, Jr. Inc., Beaver, PA. 130p.

Chepil, W. S. (1969), Dynamic of wind erosion: I. nature of movement of soil by wind, *Geomorphology*, J. G. Nelson and M. J. Chambers, eds., Methuen Publ., Canada, pp. 307-324.

Chiang, Y. C. (1979), Design and construction practice of slopes in Hong Kong, *Proc. Seminar on Slope Stability and Landslides*, Chinese Institute of Engineers, Taipei, pp. 55-82.

Childs, E. C. and Collis-George, N. (1950), The permeability of porous materials, *Proc. Royal Society, London*, v. 201A, pp. 392-405.

Chowdhury, R. N. (1980), Landslides as natural hazardous-mechanisms and uncertainties, *Geotech. Eng.*, v. 11, no. 2, pp. 135-180.

Christian, J. T., Ladd, C. C., and Baecher, G. B. (1992), Reliability and probability on stability analysis, ASCE Geotechnical Special Publ. no. 31, pp. 1071-1111.

Christodoulatos, C., Korfiatis, G. P., Talimcioglu, N. M., and Mohiuddin, M. (1994), Adsorption of pentachlorophenol by natural soils, *J. Environ. Sci. Health*, Part A29(5), pp. 883-898.

Circeo, L. J. (1963), Engineering properties and applications of nuclear excavations, HRR no. 50, pp. 13-31.

Clarke, F. W. (1908), The Data of Geochemistry, USGS Bulletin 330.

Collins, K. and McGown, A. (1974), The form function of microfabric features in a variety of natural soils, *Geotechnique*, v. 14, no. 2, pp. 223-254.

Collins, A. G. and Johnson, A. I., eds. (1988), *Ground Water Contamination: Field Methods*, ASTM STP 963, 491p.

Collins, R. J. and Ciesielski, S. K. (1994), Recycling and Use of Waste Materials and By-Products in Highway Construction, NCHRP Synthesis of Highway Practice 199, TRB, 84p.

Conner, J. R. (1990), *Chemical Fixation and Solidification of Hazardous Wastes*, Van Nostrand Reinhold Co., NY.

Cowardin, L. M., Golet, F. C., and LaBoe, E. T. (1979), Classification of Wetlands and Deepwater Habitats of the United States, Fish and Wildlife Service, U.S. Department of the Interior.

Cothern, C. R. and Rebers, P. A., eds. (1990), *Radon, Radium and Uranium in Drinking Water*, Lewis Publ., Chelsea, MI.

Crandell, F. J. (1949), Ground vibrations due to blasting and its effect upon structures, *J. Boston Soc. Civil Eng.*, pp. 222-245.

Croney, D. and Coleman, J. D. (1961), Pore pressure and suction in soil, *Pore Pressure and Suction in Soils*, Butterworths, Inc., London, pp. 31-37.

CRRI (1976, 1979), *Collected Papers on Landslides*, v.1 and 2, Chinese Railway Research Institute, Lanzhou, China. (In Chinese).

Czeratzki, W. and Frese, H. (1958), Importance of water in formation of soil structure, HRB Special Report 40, pp. 200-211.

Dakshanamurthy, V. (1979), Stress-controlled study of swelling characteristics of compacted expansive clays, *ASTM Geotech. Test. J.*, v. 2, no. 1, pp. 57-60.

Daniel, D. E. and Koerner, R. M. (1991), Landfill liners from top to bottom, *Civil Eng.*, ASCE, NY, pp. 46-49.

Daniel, D. E. and Koerner, R. M. eds. (1995), *Waste Containment Facilities: Guidance for Construction, Quality Assurance, and Quality Control of Liner and Cover Systems*, ASCE, NY.

Darilek, G. T. and Parra, J. O. (1988), The Electrical Leak Location Method for Geomembrane Liners, U.S. EPA Report no. EPA/600/S2-88/035.

Davis, R. O. E. and Bennett, H. H. (1947), Grouping of Soil on the Basis of Mechanical Analysis, USDA Circular 419.

Davis, N. M., Hon, R. and Dillon, P. (1987), Determination of bulk radon emanation rates by high resolution gamma-ray spectroscopy, *Radon in Ground Water*, Lewis Publ., Chelsea, MI, pp. 111-128.

Davisson, M. T. (1970), Design pile capacity, *Design and Installation of Pile Foundations and Cellular Structures*, Envo Publ. Co., Bethlehem, PA, pp.75-86.

Dawson, P. R. and Tillerson, J. R. (1978), Salt motion following nuclear waste disposal, *Evaluation and Prediction of Subsidence*, ASCE, NY., pp. 459-478.

Day, M. C., Jr. and Selbin, J. (1969), *Theoretical Inorganic Chemistry*, 2nd. ed. Van Nostrand Reinhold Co., NY.

De Bano, L. F. and Letey, J., eds. (1969), Water Repellent Soils, *Proc. Symposium on Water-Repellent Soils*, Univ. of California at Riverside, 354p.

De Beer, E. E. (1969), Experimental data concerning clay slopes, *Proc. 7th ICSMFE*, v. 2.

De Boer, J. H. (1953), *The Dynamical Character of Adsorption*, Oxford Clarendon Press, London.

Deer, W. A., Howie, R. A. and Zussman, J. (1962), *Rock Forming Minerals*, Vols. 1-5, John Wiley & Sons, NY.

Deere, D. U. and Patton, F. D. (1971), Slope stability in residual soils, *Proc. 4th Panamerican Conf. SMFE*, v. 1, pp. 87-170.

Deetz, J. (1967), *Invitation to Archaelogy*, The Natural History Press, Garden City, NJ, p. 8.

Demars, K. R. and Long, R. P. (1991,) Geotechnical analysis of an ocean waste disposal mound, *Proc. 2nd. Int. Symp. Environmental Geotechnology*, v. 2, Envo Publ. Co., Bethlehem, PA., pp. 66-79.

DER, (1985a), Indoor Radon Gas Fact Sheet, Pennsylvania Department of Environmental Resources, Harrisburg, PA, Jan. 12p.

DER, (1985b), General Remedial Action Details for Radon Gas Mitigation, Pennsylvania Department of Environmental Resources, Harrisburg, PA, May, 38p.

DER, (1990), Acid Rain in Pennsylvania, Fact Sheet, Pennsylvania Department of Environmental Resources, Harrisburg, PA, PL2-0390, 2p.

Derjaguin, B. V. (1960), The force between molecules, *Sci. Am.*, v. 203, pp. 47-53.

Derjaguin, B. V. and Melnikova, N. K. (1958), Mechanism of moisture equilibrium and migration in soils, HRB Special Report 40, pp. 43-54.

De Simone, S. V. (1973), Distribution of wind loads to soil, *Proc. Intern. Conf. on Planning and Design of Tall Buildings*, ASCE, NY, v. 1a, pp. 35-59.

Devine, J. F. (1966), Avoiding damage to residences from blasting vibrations, HRR no. 135, pp. 35-42.

De Wiest, R. J. M. (1965), *Geohydrology*, John Wiley & Sons, NY.

Dickson, T. R. (1974), *Understanding Chemistry: From Atoms to Attitudes*, John Wiley & Sons, NY.

Dismuke, T. D. (1991), Durability and protection of foundations, ch. 25, *Foundation Engineering Handbook*, 2nd. ed. Van Nostrand Reinhold, NY, pp. 856-867.

Dobrovolny, E. and Scholl, H. R. (1976), Geology as applied to urban planning: an example from the greater Anchorage area borough, Alaska, ch. 2, *Focus on Environmental Geology*, Oxford Univ. Press, NY, pp. 11-27.

Doehring, D. O. (1981), *Geomorphology in Arid Regions*, Allen & Unwin Publ., England.

Dregne, H. E. (1983), *Desertification of Arid Lands*, Harwood Academic Publ., NY.

Druback, G. W. and Arlotta, S. V., Jr. (1985), Subsurface pollution containment using composite system vertical cutoff barrier, ASTM STP 874, pp. 24-33.

Drucker, D. C. and Prager, W. (1952), Soil mechanics and plastic analysis or limit design, *Q. Appl. Math.*, v. 10, pp. 157-165.

Du, B. L., Mikroudis, G. K., and Fang, H. Y. (1987), Effect of pore fluid pH on the dynamic shear modulus of clay, ASTM STP 933, pp. 226-239.

Dunn, R. J. (1995), Successful development of closed landfill sites, GREEN'93, U.K., *Waste Disposal by Landfill*, R. W. Sarsby, ed., A.A. Balkema, Rotterdam, pp. 527-533.

Dvirnoff, A. H. and Munion, D. W. (1986), Stability failure of sanitary landfill, *Proc. 1st. Int. Symp. Environ. Geotechnol.*, v. 1, Envo Publ. Co., Bethlehem, PA.pp. 25-35.

Eckholm, E. and Brown, L. R. (1977), *Spreading Deserts - The Hand of Man*, Worldwatch Paper no. 13, Worldwatch Institute, Washington, D.C. 40p.

Edil, T. B. and Dhowian, A. W. (1981), At-rest lateral pressure of peat soils, *J. Geotech. Eng. Div. Proc. ASCE*, v. 107, no. GT2, pp. 201-220.

Elton, G. A. H. (1948), Electroviscosity I: the flow of liquid between surfaces in close proximity and II: experimental demonstration of the electroviscous effect, *Proc. Royal Soc. London*, 194A: pp. 559-274 & pp. 275-287.

Elzeftawy, A. and Dempsey, B. J. (1976), Unsaturated transient and steady-state flow of moisture in subgrade soil, TRR 612, pp. 56-61.

Encyclopaedia Britannic (1974), *Subsidence*, v. 9, part 2, p. 635.

EPA (1982), 40 CFR Part 260, July 26.

EPA (1985), Federal Register (1985), 50(105), 23250-23258 (May 31, 1985).

EPA (1986), A Citizen's Guide to Radon: What it is and what to do about it, EPA-86-004, 13p.

EPA (1990), EPA Report on Definition, Potentially Incompatible Wastes and Hazardous Waste Determination.

EPA (1991), 1991 Innovative Radon Mitigation Design Competition Summary of Award Winners, U.S. EPA and Association of Energy Engineers, 12p.

Erickson, P. A. and Camougis, G. (1980), Interim Procedural Guidelines: Highways and Wetlands, FHWA Implementation Package, FHWA-IP-80-11.

Eshbach, O. W. (1952), *Handbook of Engineering Fundamentals*, John Wiley & Sons, NY.

Eucken, A. (1940), *Lehrbuch der Chemischen Physik*, Akad. Verlagsgesellschaft, Leipzig, p. 524.

Evans, J. C. (1991), Geotechnics of hazardous waste control systems, ch. 20, *Foundation Engineering Handbook*, 2nd ed. Van Nostrand Reinhold Co., NY, pp. 750-777.

Evans, J. C. and Fang, H. Y. (1982), Geotechnical aspects of the design and construction of waste containment systems, Proc. 3rd. National Conf. on Management of Uncontrolled Hazardous Waste Sites, Washington, D.C. pp. 175-182.

Evans, J. C., Kugelman, I. J. and Fang, H. Y. (1983), Influence of industrial wastes on the geotechnical oroperties of soils, *Proc. 5th Mid-Atlantic Industrial Waste Conf.*, pp. 557-568.

Evans, J. C., Fang, H. Y., and Kugelman, I. J. (1985), Containment of hazardous materials with soil-bentonite slurry walls, *Proc. 6th National Conf. on the Management of Uncontrolled Hazardous Waste Sites*, Washington, D. C. pp. 249-252.

Evans, J. C. and Fang, H. Y. (1986), Triaxial equipment for permeability testing with hazardous and toxic permeants, *ASTM Geotech. Test. J.*, v. 9, no. 3, pp. 126-132.

Evans, J. C. and Fang, H. Y. (1988), Triaxial permeability and strength testing of contaminated soils, ASTM STP 977, pp. 387-404.

Eyraus, J. C., Lareal, P. and Gielly, J. (1965), General report: Physico-chemical properties of water, *RILEM Bull.*, New Series no. 27, pp. 15-18.

Fang, H. Y. (1980), Geotechnical properties and foundation problems of Shanghai soft clays, ASCE Portland Convention and Exposition, April, Reprint 80-176, 23p.

Fang, H. Y. (1986), Introductory remarks on environmental geotechnology, *Proc. 1st. Int. Symp. Environ. Geotechnol.*, v. 1, Envo Publ. Co., Bethlehem, PA., pp. 1-14.

Fang, H. Y. (1987), Discussions, *1st. Int. Symposium on Environmental Geotechnology*, v. 2, D.1 and D.3, Envo Publ. Co., Bethlehem, PA., pp. 290, 297, 310 & 340.

Fang, H. Y. (1989a), Particle energy field theory: a unified approach for analyzing soil behavior, *Proc. 2nd Int. Symp. Environ. Geotechnol.*, v. 1, pp. 167-194.

Fang, H. Y. (1989b), Expert systems for assessment of radon gas, ASCE Environmental Engineering 1989 Specialty Conf. ASCE, NY. pp. 97-104.

Fang, H. Y. (1990), Environmental geotechnical approach for evaluation of radon gas, 83rd Annual Meeting and Exhibition, Air and Waste Management Association, Paper no. 90-153.10, Pittsburgh, PA, June, 19p.

Fang, H. Y. (1991a), Theory of particle energy field and its applications to environmental geotechnology, *Proc. 2nd Int. Symp. Environ. Geotechnol.*, v. 2, pp. 277-309.

Fang, H. Y. (1991b), Sensitivity of soil to environment, *Proc. 1st. Canadian Conf. on Environmental Geotechnics*, Montreal, Canada, May, 18p.

Fang, H. Y. (1991c), The use of bamboo inclusions for earth reinforcement, *Proc. Biotechnical Stabilization Workshop*, D. H. Gray, ed. Univ. of Michigan and U.S. NSF, pp. 33-62.

Fang, H. Y. (1992a), Environmental geotechnology: a perspective, *Proc. Mediterranean Conf. on Environmental Geotechnology*, M. A. Usmen and Y. B. Acar, eds., A.A. Balkema, Robberdam, pp. 11-19.

Fang, H. Y. (1992b), Topic 6, Geoenvironmental Engineering, Recent Accomplishments and Future Trends in Geomechanics in the 21th Century, U.S.-Canada Workshop, Univ. of Oklahoma and U.S. NSF, pp. 309-317.

Fang, H. Y. (1993), Identification and characterization of contaminated ground soil for engineering purposes, *Proc. Int. Conf. on Environment and Geotechnics,* Paris, April, 8p.

Fang, H. Y. (1994a), Cracking and fracture behavior of soil, *Fracture Mechanics Applied to Geotechnical Engineering,* ASCE, NY, pp. 102-117.

Fang, H. Y. (1994b), Pre-failure deformation characteristics of contaminated fine-grained soil, *Proc. Intern. Symposium on Pre-failure Deformation Characteristics of Geomaterials,* Japan, S. Shibuya, T. Mitachi and S. Miura, eds. pp. 299-304.

Fang. H. Y. (1994c), Mass transport phenomena of contaminated-unsaturated fine-grained soil, *Proc. 8th Intern. Conf. on Computer Methods and Advances in Geomechanics,* pp. 1049-1054.

Fang, H. Y. (1995a), Engineering behaviour of urban refuse, compaction control and slope stability analysis of landfill, *GREEN'93, U. K., Waste Disposal by Landfill,* R. W. Sarsby, ed., A.A. Balkema, Rotterdam, pp. 47-72.

Fang, H. Y. (1995b), Environmental geotechnology, Encyclopedia of Environmental Control Technology, v. 9: *Geotechnical,* Ch. 2, P. N. Cheremisinoff, ed., Gulf Publ. Co., Houston, TX, pp. 13-117.

Fang, H. Y. (1996), Leaking mechanism from landfill through impervious soil layer, *Proc. 2nd. Int. Symp. Environ. Geotechnol.,* IS-Osaka'96, Japan, 6p.

Fang, H. Y. (1997), Soil decontamination by electromagnetic wave excitation, *Proc. 3rd. Int. Symp. Environ. Geotechnol.,* v. 2, 10p.

Fang, H. Y. and Chen, W. F. (1971), New method for determining tensile strength of soils, HRR no. 354, pp. 62-68.

Fang. H. Y. and Hirst, T. J. (1973), A method for determination the strength parameters of soil, HRR no. 463, pp. 45-50.

Fang, H. Y. and Dismuke, T. D. (1975), Discussion of H-bearing piles in limestone and clay shales, JSMFE Div. Proc. ASCE, v. 101, no. SM6, pp. 594-598.

Fang, H. Y. and Cleary, T. (1976), Subsidence, ch. 15, *Analysis and Design of Building Foundations,* Envo Publ. Co., Bethlehem, PA., pp. 467-518.

Fang, H. Y., Slutter, R. C. and Steubben, G. A. (1976), Stress-strain characteristics of compacted waste disposal material, *New Horizons in Construction Materials,* v. 1, Envo Publ. Co., Bethlehem, PA. pp. 127-138.

Fang, H. Y. and Koerner, R. M. (1977), An instrument for measuring in situ soil-structure response during dynamic vibration, *Proc. 14th Annual Meeting of Society of Engineering Science,* pp. 1171-1180.

Fang, H. Y., Slutter, R. G., and Koerner, R. M. (1977), Load bearing capacity of compacted waste disposal materials, *Proc. Specialty Session of Geotechnical Engineering and Environmental Control,* Z. C. Moh, ed. 9th ICSMFE, v. 1, pp. 265-278.

Fang, H. Y. and Fernandez, J. (1981), Determination of tensile strength of soils by unconfined-penetration test, ASTM STP 740, pp. 130-144.

Fang, H. Y., Chaney, R. C., and Pandit, N. S. (1981), Dynamic shear modulus of soft silt, *Proc. Int. Conf. on Recent Advance in Geotechnical Earthquake Engineering, and Soil Dynamics,* Univ. of Missouri, v. 2, pp. 575-580.

Fang, H. Y. and Chaney, R. C. (1983), Mechanics of soil-heat interaction and its thermal storage capacity, *Proc. ASCE Engineering Mechanics Specialty Conf.,* Purdue Univ., v. 2, pp. 786-789.

Fang, H. Y., Lou, G. Y., and Chu, T. G. (1987), Radioactive toxic radon gas and its control methods, *Proc. 1st. Int. Symp. Environ. Geotechnol.,* v. 2, Envo Publ. Co., Bethlehem, PA., pp. 219-231.

Fang. H. Y. and Chaney, R. C. (1987), Desert-environment interaction and antidesertification measures, *Proc. 1st. Int. Symp. Environ. Geotechnol.,* Envo Publ. Co., Bethlehem, PA., v. 2, pp. 258-288.

Fang, H. Y. and Mikroudis, G. K. (1991), Stability of earth slopes, ch. 10, *Foundation Engineering Handbook,* 2nd ed. Van Nostrand Reinhold Co. NY, pp. 379-409.

Fang, H. Y. and Evans, J. C. (1994), Technique for controlling solid and liquid wastes, ch. 11, *Process Engineering for Pollution Control and Waste Minimization,* Wise and Trantolo, eds. Marcel Dekker, Inc., NY., pp. 247-270.

Fanning, D. S. and Fanning, M. C. B. (1989), *Soil Morphology, Genesis and Classification,* John Wiley & Sons, NY.

Fantechi, R. and Margaris, N. S., eds. (1986), Desertification in Europe, *Proc. Information System in the EEC Programs on Climatology* (1984), Mytilene, Greece, D. Reidel Publ. Co., Norwell, MA.

Farney, D. (1974), Ominous problems: what to do with radioactive waste, *Smithsonian,* April, pp. 20-27.

Farouki, O. T. and Winterkorn, H. F. (1964), Mechanical properties of granular systems, HRR no. 52, pp. 10-42.

Fernandez, F. and Quigley, R. B. (1985), Hydraulic conductivity of natural clays permeated with simple liquid hydrocarbons, *Can. Geotech. J.*, v. 22, pp. 205-214

Fernando, J., Smith, R. and Arulanandan, K. (1975), New approach to determination of expansion index, *J. Geotech. Eng. Div. Proc. ASCE*, v. 101, no. GT9, pp. 1003-1013.

FHWA (1978), Highways and Ecology: Impact assessment and mitigation, FHWA Report FHWA-RWE-OEP-78-2.

FHWA (1980), Highways and Wetlands: Impact assessment, migration and enhancement Measures, FHWA.

Findley, R. (1972), *Great American Desert*, National Geographic Society, Washington, D.C., 207p.

Finn, F. N. (1951), Effect of temperature on the consolidation characteristics of remolded clay, ASTM STP 126, p. 65.

Finn, W. D. L. and Lee, M. K. W. (1979), Seafloor stability under seismic and wave loading, *Proc. Soil Dynamics in the Marine Environment*, ASCE Reprint 3604.

Fleischer, R. L. (1983), Theory of alpha-recoil effects on radon release and isotopic disequilibrium: *Geochim. Cosmochim. Acta*, v. 47, pp. 779-784.

Fleischer, R. L. and Mogro-Camero, A. (1978), Mapping of integrated radon emanation for detection of long-distance migration of gases within the earth: techniques and principles, *J. Geophys. Res.*, v. 83, pp. 3539-3549.

Flexser, S., Wollenberg, H. A.,, and Smith, A. R. (1987), Radon in groundwater of the Long Valley Caldera, California, *Radon in Groundwater*, Lewis Publ., Chelsea, MI, pp. 131-152.

Foster, G. R. and Wischmeier, W. H. (1973), Evaluating irregular slopes for soil loss prediction, ASAE Annual Meeting, Paper no. 73-227,

Foster, J. B., Erickson, J. R. and Healy, R. W. (1984), Hydrogeology of a low-level radioactive waste site near Sheffield, Illionios, USGS, Water Resources Investigation Report 83-4125, 83p.

Fowkes, R. S. and Fritz, J. F. (1974), Theoretical and Experimental Studies on the Packing of Solid Particles: A Survey, Bureau of Mines Information Circular, IC no. 8623, 30p.

Frascoia, R. I. and Cauley, R. F. (1995), Tire chips in the base course of a local road, *Proc. 6th Intern. Conf. on Low-volume Roads,* TRB, v. 2, pp. 47-52.

Fredlund, D. G. and Rahardjo, H. (1993), *Soil Mechanics for Unsaturated Soils*, John Wiley & Sons, NY.

Freudenthal, (1950) (In German, see Winterkorn, 1974).

Friend, D. J. (1996), Remediation of Petroleum-Contaminated Soils, NCHRP Synthesis of Highway Practice 226, TRB, 88p.

Frye, S. C. (1970), The protection of piling, *Proc. Design and Installation of Pile Foundations and Cellular Structures*, Envo Publ. Inc. Bethlehem, PA, pp. 191-208.

Fukuoka, M. (1992), Mechanism of a landslide caused by rainfall, ASCE Geotechnical Special Publ. no. 31, pp. 342-357.

Fuller, W. H. (1978), Investigation of Landfill Leachate Pollutant Attenuation by Soils, U.S. EPA Report no. 600/2-78-158.

Furnas, C. C. (1931), Grading aggregates: I, mathematical relations for beds of broken solids of maximum density, *Ind. Eng. Chem.*, v. 23, pp. 1052-1058.

Fwa, T. F. and Tan, S. A. (1992), Laboratory density measurements of bituminous mixes by gamma-ray, TRR 1353, pp. 1-8.

Gabr, M. A., Hunter, T. J. and Collin, J. G. (1992), Stability of geogrid-reinforced landfill liners over sinkholes, *Proc. Int. Symp. Earth Reinforcement Practice*, Japan, pp. 595-600.

Gao, G. R. (1983), Microstructure of loess soil in China relate to geologic environment, ASCE Specialty Conf. NY, pp. 121-136.

Gao, D. Z., Wei, D. D. and Hu, Z. X. (1986), Geotechnical properties of Shanghai soils, and engineering applications, ASTM STP 923, pp. 161-177.

Gardner, W. R. (1958), Mathematics of isothermal water conduction in unsaturated soil HRB Special Report 40, pp. 78-87.

Garrels, R. M. and Christ, C. L. (1965), *Solutions, Minerals and Equilibria*, Harper & Row, NY.

Gary, M., McAfree, R. Jr., and Wolf, C. L., eds. (1972), *Subsidence, Glossary of Geology*, American Geological Institute, p. 708.

Gauffreau, P. E. (1987), A review of pile protection methods in a corrosive environment, *Proc. 1st. Int. Symp. Environ. Geotechnol.*, v. 2, Envo Publ. Co., Bethlehem, PA., pp. 372-378.

Gauffreau, P. E. (1989), Hydrophobic soil: a low-cost alternative to clay lining materials, *Proc. 2nd. Int. Symp. Environ. Geotechnol.*, v. 1, Envo Publ. Co., Bethlehem, PA., pp. 61-82.

Gautier, E. F. (1953), *Sahara, the Great Desert*, (Translated by D. F. Mayjew), Columbia Univ. Press, NY.

Geiger, R. (1965), *The Climate Near the Ground*, Harvard Univ. Press. Cambridge, MA.

Geuze, E.C.W.A. and Bruyn, B. (1948), The first law of thermodynamics and the consolidation process, Proc. 2nd ICSMFE, v. 3.

Gidigasu, M. D. (1975), *Laterite Soil Engineering, Pedogenesis and Engineering Principles*, Elsevier Scientific Publ., NY.

Gilbert, O. H., Jr. (1960), A device for measuring tensions in water, HRB Bulletin 245, pp. 1-9.

Giletti, B. J. and Kulp, J. L. (1955), Radon leakage from radioactive minerals, *Am. Mineralogist*, v. 40, pp. 481-496.

Gilkeson, R. H. and Cowart, J. B. (1987), Radium, radon and uranium isotopes in groundwater from cambrian-ordovician sandstone aquifers in Illinois, *Radon in Groundwater*, B. Graves ed., Lewis Publ., Chelsea, MI. pp. 403-422.

Golchert, N. W. and Sedlet, J. (1977), Radiological survey of site A and plot M: Argonne National Laboratory, Argonne, IL, prepared for DOE/EV/0005/7, 89p.

Gouy, G. (1910), Sur la constitution de la charge electrique a la surface d'un electrolyte, *Ann. Phys. (Paris)*, Series 4, v. 9, pp. 457-468.

Graham, M. J. (1981), The radionuclide groundwater monitoring program for the separations area, Hanford site, Washington State, *Groundwater Monitoring Rev.*, v. 1, no. 2, pp. 52-56.

Gray, H. (1936), Progress report on research on the consolidation of fine-grained soils, *Proc. 1st. ICSMFE*, v. 2, pp. 138-141.

Gray, R. E. and Meyers, J. (1970), Mine subsidence and support methods in Pittsburgh area, *JSMFE Div. Proc. ASCE*, v. 96, no. SM4, pp. 1267-1287.

Gray, D. H. and Mitchell, J. K. (1967), Fundamental aspects of electroosmosis in soils, *JSMFE Div. Proc. ASCE*, v. 93, no. SM6, pp. 209-236

Gray, D. H. and Leiser, A. T. (1982), *Biotechnical Slope Protection and Erosion Control*, Van Nostrand Reinhold Co., NY.

Gray, W. A. (1968), *The Packing of Solid Particles*, Chapman & Hall, London.

Graves, B. ed. (1987), Radon in Ground Water, *Proc. of the National Water Well Association*, Lewis Publ., Chelsea, MI, 546p.

Green, W. J., Lee, G. F. and Jones, R. A. (1980), The permeability of clay soils to water and organic solvents: Implications for the storage of Hazardous Wastes. Report submitted to U.S. EPA.

Griffin, (1976) (see Fuller, 1978).

Griffith, A. A. (1921), The phenomena of rupture and flow in solids, *Trans. Royal Soc., London*, Series A221, pp. 163-198.

Grim, R. E. (1940), The clay minerals in soils and their significance, *Proc. Purdue Conf. on Soil Mechanics and Its Applications*, Purdue Univ. pp. 216-223.

Grim, R. E. (1958), Organization of water on clay minerals surfaces and its implications for the properties of clay-water systems, HRB Special Report 40, pp. 17-23.

Grim, R. E. (1953, 1968), *Clay Mineralogy*, 2nd ed. (1st ed. 1953), McGraw-Hill Book Co., NY.

Gromko, G. J. (1974), Review of expansive soils, *J. Geotech. Eng., Proc. ASCE*, v. 100, no. GT6, pp. 667-687.

Gundersen, L. C. S. and Wanty, R. B. eds. (1992), *Field Studies of Radon in Rocks, Soils and Water*, Lewis Publ., Boca Raton, FL, 368p.

Gunnerson, C. G. and Willard, B. E., eds. (1979), *Acid Rain*, ASCE, NY, 166p.

Gupta, S. C. and Larson,, W. E. (1979), Estimating soil water retention characteristics from particle size distribution, organic matter percent and bulk density, *Water Res. Res.*, v. 15, pp. 1633-1635.

Habibagahi, K. (1976), Temperature effects on primary consolidation, *Geotech. Eng.*, v. 7, no. 2, pp. 95-108.

Haines, W. B. (1923), The volume-changes associated with variations of water content in soils, *J. Agric. Sci.*, v. 13, p. 296.

Haley, J. F. and Kaplar, C. W. (1952), Cold-room studies of frost action in soils, HRB Special Report 2, pp. 246-267.

Hallaire, M. (1958), Soil water movement in the film and vapor phase under the influence of evapotranspiration, HRB Special Report 40, pp. 88-105.

Hamilton, J. M., Daniel, D. E. and Olson, R. E. (1981), Measurement of hydraulic conductivity of partially saturated soils, ASTM STP 746.

Handy, R. L. and Fendon, T. E. (1977), Particle size and mineralogy in soil taxonomy, TRR 642, pp. 13-19.

Hanes, R.E., Zelazny, L.W., Verghese, K.G., Bosshart, R. P., Carson, E.W. Jr., Blaser, R. E. and Wolf, D.D. (1976), Effects of Deicing Salts on Plant Biota and Soil, NCHRP Report 170, TRB, 88p.

Hansen, S. M. and Toman, J. (1966), Aggregate production with nuclear explosives, HRR no. 107, pp. 37-53.

Hardin, G. (1968), The tragedy of the commons, *Science,* American Assoc. for the Advancement of Science, v. 162, pp. 1243-1248.

Hardin, B. O. and Black, W. L. (1968), Vibration modulus of normally consolidated clay, *JSMFE Div. Proc. ASCE,* v. 94, no. SM2, pp. 353-369.

Hardin, B. O. and Drnevich, V. P. (1972), Shear modulus and damping in soils, measurement and parameter effects, *JSMFE Div. Proc. ASCE,* v. 98, no. SM6, pp. 603-624.

Hashim, R. and Ang, E. K. (1996), Evaluation of slope stability using safety index, *Proc. 3rd. Int. Symp. Environ. Geotechnol.,* v. 1, Technomic Publ., Lancaster, PA, pp. 669-675.

Haxo, H. E., Jr. (1981), Durability of liner materials for hazardous waste disposal facilities, U.S. EPA Report no. 600/9-81-002b, pp. 140-156.

Hazen, A. (1911), Discussion of dams on sand foundations, *Trans. ASCE,* v. 73, p. 199.

Henkel, D. S. (1970), The role of waves in causing submarine slides, *Geotechnique,* v. 20, no. 1, pp. 75-80.

Hewett, D. F. (1917), The origin of bentonite, *J. Washington Acad. Sci.,* v. 7, pp. 196-198.

Hewlett-Packard Journal (1995), June, pp. 6-12.

Hilf, J. J. (1991), Compacted fills, ch. 8, *Foundation Engineering Handbook,* 2nd ed. Van Nostrand Reinhold Co., NY, pp. 249-316.

Hissink, D. J. (1924), Base exchange in soils, *Trans. Faraday Soc.,* v. 20, pp.551-566.

Hoek, E. and Bray, J. W. (1981), *Rock Slope Engineering,* 3rd. ed., The Institution of Mining and Metallurgy, London.

Hoekstra, P. (1969), The physics and chemistry of frozen soils, HRB Special Report 103, pp. 78-90.

Hofmann, U. and Klemen, (1950), cited by R. E. Grim, 1968.

Hogentogler, C. A., Jr. (1937a), Essentials of soil compaction, *Proc. HRB,* v. 16, pp. 309-316.

Hogentogler, C. A., Jr. (1937b), *Engineering Properties of Soil,* McGraw-Hill Book Co., NY.

Holl, A. (1969), Thermodynamics of granular systems, HRB Special Report 103, pp. 91-113.

Holtz, R. D. and Kovacs, W. D. (1981), *An Introduction to Geotechnical Engineering,* Prentice-Hall Co., NJ.

Hough, B. K. (1957), *Basic Soils Engineering,* The Ronald Press Co., NY.

Howard, A. K. (1984), The revised ASTM standard on the unified classification system, *ASTM Geotech. Testing J.,* v. 7, no. 4, pp. 216-222.

HRB (1943), The Use of Soil Cement and Soil Aggregate Cement Mixtures in the Construction of Base Courses, War Time Road Problems, HRB Bulletin 7.

HRB (1957), Glossary: pedologic (soils) and landform terminology, HRB Special Report 25, 32p.

HRB (1958), Water and Its Conduction in Soils, Highway Research Board Special Report 40, 338p.

HRB (1959), Subsurface Draingae of Highways and Airports, HRB Bulletin 209.

HRB (1962a), The AASHO Road Test Report 2: Materials and Construction, HRB Special Report 61B, 173p.

HRB (1962b), The AASHO Road Test Report 5: Pavement Research, HRB Special Report 61E, 352p.

HRB (1966), Nuclear Applications, HRR no. 107, 65p.

HRB (1969), Effect of Temperature and Heat on Engineering Behavior of Soils, HRB Special Report 103, 300p

HRB (1973), Soil Erosion: Causes and Mechanisms, Prevention and Control, HRB Special Report 135, 141p.

Huhta, R. S. (1991), Introduction to the aggregate industry, ch. 1, *The Aggregate Handbook,* R. D. Barksdale, ed. National Stone Assoc., pp. 1.1 to 1.14.

Hwang, J. M. and Wu, C. M. (1969), Land subsidence problems in Taipei basin, *Proc. Int. Symp. Land Subsidence, Tokyo,* v. 1, pp. 21-34.

ICE, (1961), *Pore Pressure and Suction in Soils, Institution of Civil Engineers,* Butterworths, Inc., London, 151p.

Ingles, C. G. (1968), Soil chemistry relevant to the engineering behavior of soils, ch. 1, *Soil Mechanics, Selected Topics,* I. K. Lee, ed., American Elsevier Publ. Co., NY., pp. 1-57.

Ingram, R. L. (1953), Fissility of mud rocks, *Bull., Geolog. Soc. Am.*, v. 64, August.

Inyang, H. I. (1992a), Application potential of dynamic compaction to contaminant migration control, Proc. HMC/Superfund'92, Hazardous Materials Control Resources Institute, pp. 626-630.

Inyang, H. I. (1992b), Aspects of landfill design stability in seismic zones, *J. Environ. Syst.*, v. 21, no. 3, pp. 223-235.

Inyang, H. I. and Bergeson, K. L. eds. (1992), *Utilization of Waste Materials in Civil Engineering Construction*, ASCE, NY. 347p.

Irwin, G. R. (1968), Linear fracture mechanics, *Engineering Fracture Mechanics*, v. 1, pp. 241-257.

Israelsen, C. E., Clyde, C. G., Fletcher, J. E., Israelsen, E. K., Haws, F. W., Packer, P. E. and Farmer, E. E. (1980a), Erosion Control During Highway Construction:Research Report, NCHRP Report 220, TRB, 30p.

Israelsen, C. E., Clyde, C. G., Fletcher, J. E., Israelsen, E. K., Haws, F. W., Packer, P. E. and Farmer, E. E. (1980b), Erosion Control During Highway Construction: Manual on Principles and Practices, NCHRP Report 221, TRB, 23p. plus appendix.

Jefferis. S. A. (1995), Old landfills: perception and remediation of problem sites, *GREEN'93, U.K., Waste Disposal by Landfill*, R. W. Sarsby, ed., A.A. Balkema, Rotterdam, pp. 93-106.

Jenny, H. and Overstreet, R. (1935), Surface migration of ion and contact exchange, *J. Phys. Chem.*, v. 43, pp. 1185-1196.

Joffe, J. S. (1949), *Pedology*, 2nd. ed., Rutgers Univ. Press, NJ.

Johnson, A. W. and Sallberg, J. R. (1962), Factors Influencing Compaction Test Results, HRB Bulletin 319, 148p.

Jones, P. C. T. (1955), Microbiological factors in soil stabilization, HRB Bulletin 108, pp. 81-95.

Jones, L. W. (1990), Interference Mechanisms in Waste Stabilization Solidification Processes, U.S. EPA Report no. EPA/600/S2-89/067.

Jordine, E.S.A. (1963), The effect of surface ions on the mutual interaction of montmorillonite particles, *Soil Sci.*, v. 96, pp. 149-151.

Jordine, E.S.A., Bodman, G. B. and Gold, A. H. (1962), Effect of surface ions on the mutual interaction of montmorillonite particles, *Soil Sci.*, v. 94, pp. 371-378.

Jumikis, A. R. (1955), *The Frost Penetration Problem in Highway Engineering*, Rutgers State Univ. Press, NJ.

Jumikis, A. R. (1958), Some concepts pertaining to the freezing soil systems, HRB Special Report 40, pp. 178-190.

Jumikis, A. R. (1966), *Thermal Soil Mechanics*, Rutgers State Univ. Press, NJ.

Kao, S. and Sun, G. F. (see Fang, 1989b and 1995b).

Kargbo, D. M., Fanning, D. S., Inyang, H. I. and Duell, R. W. (1993), Environmental significance of acid sulfate 'clay' as waste covers, *Environ. Geol.*, v. 22, pp. 218-226.

Karpoff, K. P. (1976), Stabilization of fine-grained soils by electro-osmotic and electrochemical methods, *New Horizons in Construction Materials*, Envo Publ. Co., Bethlehem, PA, pp. 265-273.

Katti, R. K. and Barve, A. G. (1962), Effect of inorganic chemical on the consistency properties of an expansive soil sample, HRB Bull. 349, pp. 1-8.

Kaya, A. and Fang, H. Y. (1996), Characterization of dielectric constant on fine-grained soil behavior, ASTM STP 1282, pp. 303-314.

Kaya, A. and Fang, H. Y. (1997), Identification of contaminated soils by dielectric constant and electrical conductivity, *J. Environ. Eng. Div. ASCE*, v. 123, no. 2, pp. 169-177.

Keller, G. V. (1974), Engineering applications of electrical geophysical methods, *Subsurface Exploration for Underground Excavation and Heavy Construction*, ASCE, NY., pp. 128-143.

Keller, E. A. (1984), *Environmental Geology*, 5th ed., Merrill Publ., Columbus, OH.

Kelley, W. P. (1948), *Cation Exchange in Soils*, Reinhold Publ. Co., NY.

Kenney, T. C. (1967), Influence of mineral composition on the residual strength of natural soils, *Proc. Geotechnical Conf. Oslo*, v. 1, pp. 123-129.

Kennish, M. J. ed. (1996), *Practical Handbook of Estuarine and Marine Pollution*, CRC Press, Inc., Boca Raton, FL.

Kent, D. M. ed. (1994), *Applied Wetlands Science and Technology*, CRC Press, Inc., Boca Raton, FL.

Kersten, M. S. (1949), Thermal Properties of Soils, Bulletin 28, Engineering Experiment Station, Univ. of Minnesota, 225p.

Kersten, M. S. and Cox, A. E. (1951), The effect of temperature on the bearing value of frozen soils, HRB Bulletin no. 40, pp. 32-38.

Kezdi, A. (1964), Discussion on mechanical properties of granular systems, HRR no. 52, pp. 42-58.

Khan, L. I., Pamukcu, S. and Kugelman, I. J. (1989), Electro-osmosis in fine grained soil, *Proc. 2nd Int. Symp. Environ. Geotechnol.*, v. 1, pp. 39-47.

Kiersch, G. A. ed. (1991), *The Heritage of Engineering Geology; The First Hundred Years*, The Geology Society of America, Centennial Volume 3.

Koerner, R. M. (1970), Effect of particle characteristics on soil strength, *JSMFE Div. Proc. ASCE*, v. 96, no. SM4, pp. 1221-1234.

Koerner, R. M. (1987), Keynote Address, *Proc. 1st Int. Symp. Environ. Geotechnol.*, v. 2, Envo Publ. Co., Bethlehem, PA., pp. 1-3.

Koerner, R. M. (1994), *Designing with Geosynthetics*, 3rd ed. Prentice Hall, Englewood Cliffs, NJ.

Koryta, J. (1991), *Ions, Electrodes and Membranes*, John Wiley & Sons, NY.

Kostecki, P. T. and Calabrese, E. J. (1990), *Petroleum Contaminated Soils*, v. 3, Lewis Publ., Boca Raton, FL.

Krebs, R. D. (1964), Discussion of study of reproducibility of Atterberg limits, HRR no. 63, pp. 29-30.

Krishnayya, A. V. G., Eisenstein, Z. and Morgenstern, N. R. (1974), Behavior of compacted soil in tension, *J. Geotech. Eng. Div. Proc. ASCE*, v. 100, no. GT9, Sept. pp. 1051-1061.

Krumbein, W. C. (1950), Geological aspects of beach engineering, *Application of Geology to Engineering Practice, Berkey Volume*, The Geological Soc. of America, pp. 195-223.

Ladd, C. C. (1960), Mechanisms of swelling by compacted clay, HRB Bulletin 245, pp. 10-26.

Ladd, C. C. (1971), Strength Parameters and Stress-strain Behavior of Saturated Clays, Research Report R71-23, Soils Publ. 278, MIT, 230p.

Lafavore, M. (1987), *Radon: The Invisible Threat*, Rodale Press, Emmaus, PA.

Laguros. J. G. (1969), Effect of temperature on some engineering properties of clay soils, HRB Special Report 103, pp. 186-203.

Laguros, J. G., Kumar, S. and Medhani, R. (1982), Failure of slopes in weathered overconsolidated clays, TRR 873, pp. 12-14.

Lambe, T. W. (1953), The effect of polymers on soil properties, *Proc. 3rd. ICSMFE*, v. 1, pp. 253-257.

Lambe, T. W. (1958), The structure of compacted clay, *JSMFE Div. Proc. ASCE*, v. 84, no. SM2, p. 1654.

Lambe, T. W. (1959), Physico-chemical properties of soils: role of soil technology, *JSMFE Div. Proc. ASCE*, v. 85, no. SM2, pp. 55.

Lambe, T. W. and Whitman, R. V. (1979), *Soil Mechanics, SI Version*, John Wiley & Sons, NY.

Lamberton, H. C., Jr., Sainz, A. J. Crawford, R. A., Ogletree, W. B., and Gunn, J. E. (1981), Underwater Inspection and Repair of Bridge Substructures, NCHRP Synthesis of Highway Practice 88, TRB, 75p.

Landva, A. O., Korpijaakko, E. O., and Pheeney, P. E. (1983), Geotechnical classification of peats and organic soils, ASTM STP 820, pp. 37-51.

Landin, M. C. ed. (1988), Beneficial Uses of Dredged Material, *Proc. North Atlantic Regional Conf.*, U.S. Army Corps of Engineers, Baltimore District, 231p.

Lane, K. S. and Washburn, R. V. (1946), Capillarity tests by capillarimeter and by soil filled tubes, *Proc. HRB*, v. 25, pp. 460-473.

Lassen, L., Lull, H. W. and Frank, B. (1952), Some plant-soil-water relations in watershed management, Circular 910, USDA, Washington, D.C. Oct.

Lee, J. Y. (1996), The evaluation of landfill cover systems in cold weather, *Proc. 3rd Int. Symp. Environ. Geotechnol.*, v. 1, Technomic Publ., Lancaster, PA, pp. 579-588.

Lee, K. L. and Shen, C. K. (1969), Horizontal movements related to subsidence, JSMFE Div. Proc. ASCE, v. 95, no. SM1, pp. 139-166.

Legget, R. F. (1973), *Cities and Geology*, McGraw-Hill Book Co., NY.

Leonards, G. A. (1979), Stability of slopes in soft clays, *Proc. 6th Panamerican Conf. SMFE*, v. 1, pp. 225-274.

Leonards, G. A. and Andersland, O. B. (1960), The clay-water system-the shear strength of clays, *Proc. ASCE Research Conf. on the Shear Strength of Cohesive Soils*, pp. 793-818.

Leonards, G. A. and Girault, P. (1961), A study of the one-dimensional consolidation test, *Proc. ICSMFE*, v.1, pp. 116-130.

Leonards, G. A., Cutter, W. A. and Holtz, R. D. (1980), Dynamic compaction of granular soils, *J. Geotech. Eng. Div. Proc. ASCE*, v. 106, no. GT1, pp. 17-34.

Leonards, G. A., Schmednecht, F., Chameau, J. L. and Diamond, S. (1985), Thin slurry cutoff walls installed by the vibrated beam method, ASTM STP 784, pp. 34-44.

Leopold, A. S. (1980), *The Desert*, Time-Life Books, Alexandria, VA, 192p.

Lewis, W. A. (1950), Effect of temperature on consolidation of soils, *Nature*, v. 166, pp. 614-615.

Li, C. Y. (1956), Basic concepts on the compaction of soil, *JSMFE Div. Proc. ASCE*, v. 82, no. SM1, paper 862, 20p.

Li, W. X. (1989), Fuzzy models for estimation of surface ground subsidence, *Proc. 2nd Int. Symp. Environ. Geotechnol.*, v. 1, pp. 317-329.

Liao, V. C. (1983) (see Fang, 1989b & 1995b).

Lilly, W. (1976), Probability study of sinkhole distribution in the Lehigh Valley, Pennsylvania, *Proc. ASCE Symposium on Engineering, Construction and Maintenance Problems in Limestone Regions*, Aug. pp. 161-180.

Linell, K. A. and Kaplar, C. W. (1959), The factor of soil and material type in frost action, HRB Bulletin 225, pp. 23-44.

Linell, K. A., Hennion, F. B. and Lobacz, E. F. (1963), Corps of engineers pavement design in areas of seasonal frost, HRR no. 33, pp. 76-136.

Ling, H. I. and Leshchinsky, D. (1996), Seismic stability of landfill cover system, *Proc. 3rd Int. Symp. Environ. Geotechnol.*, v. 1, Technomic Publ., Lancaster, PA., pp. 569-578.

Liou, Y. D. (1970), Hydraulic Erodibility of Two Pure Clay Systems, Ph.D. Dissertation, Dept. of Civil Engineering, Colorado State Univ. Fort Collins, CO.

Little, A. L. ((1967), Laterites, *Proc. 3rd. Asian Regional Conf. on SMFE*, Haifa, v. 2, pp. 61-71.

LMER (1992), Understanding changes in coastal environments: the LMER program, EOS, v. 73, no. 45, Nov. 10, AGU, p. 481.

Lo, K. Y. (1961), Secondary compression of clays, *JSMFE Div. Proc. ASCE,* v. 87, no. SM4, pp. 61-66.

Lo, K. Y. (1965), Stability of slopes in anisotropic soils, *JSMFE Div. Proc. ASCE*, v.91, no. SM4, pp. 85-106.

Locard, E. (1930), The analysis of dust traces, *Am. J. Police Sci.*, v. 1, p. 276.

Lockhart, N. C. (1983), Electroosmotic dewatering of fine tailings from mineral processing, *Int. J. Mineral Processing*, v. 10, pp. 131-140.

Lohman, S. W. (1961), Compression of elastic artesian aquifers USGE Prof. Paper 424-B art. 23, pp. B47-B49.

Longwell, C. R., Flint, R. F., and Sanders, J. E. (1969), *Physical Geology*, John Wiley & Sons, NY, pp. 288-313.

Lovell, C. W. (1957), Temperature effects on phase composition and strength of partially-frozen soil, HRB Bulletin 168, pp. 74-95.

Low, P. F. (1951), Force fields and chemical equilibrium in heterogeneous systems with special reference to soils, *Soil Sci.*, v. 71, no. 6, p. 409.

Low, P. F. (1958), Movement and equilibrium of water in soil systems as affected by soil-water forces, HRB Special Report 40, pp. 55-64.

Low, P. F. (1968), Mineralogical data requirements in soil physical investigations, SSSA Special Publ. Series no. 3, pp. 1-34.

Lukas, R. C. (1980), Densification of loose deposits by pounding, *J. Geotech. Eng. Div. Proc. ASCE*, v. 106, no. GT4, pp. 435-446.

Lumb, P. (1962), The properties of decomposed granite, *Geotechnique*, v. 12, no. 3, pp. 226-243.

Lumb, P. (1975), Slope failures in Hong Kong, *Q. Eng. Geol.*, v. 8, pp. 31-65.

Luo, G. Y., Wang, P. Q., Cai, Z. Y. and Wu, H. (1982), On the concept of two preferred planes and its analysis methods, *Chinese J. Geotech. Eng.*, v. 4, no. 2, pp. 57-66. (In Chinese with English summary).

Luo, G. Y., Wang, P. Q. and Wu, H. (1984), Main types and models of the deformation and failure of rock slopes in igneous areas, Proc. 27th Intern. Geological Congress, Moscow, USSR, pp. 4-14.

Luo, G. Y., Chu, T. Q., Wu, H. and Liu, G. C. (1991), Locating radon/water leaking source in rock formation based on the two-preferred planes principle, *Proc. 2nd Int. Symp. Environ. Geotechnol.*, v. 2, pp. 229-241.

Luongo, V. (1992), Dynamic compaction: predicting the depth of improvement, *Proc. Grouting, Soil Improvement and Geosynthetics*, ASCE, v. 2, pp. 927-939.

Luscher, (1967), (In Germam see Winterkorn, 1974).

Lyon, T. L. and Buckman, H. O. (1939), *The Nature and Properties of Soils*, The MacMillan Co., NY, pp. 104-111.

MAA Group (1980), Collection of Technical Papers Published in 1976-1980, Consulting Engineers, Taipei, Taiwan, 379p.

MacEwan, D.M.C. (1954), Short-range electrical forces between charged colloid particles, *Nature*, v. 174, pp. 39-40.

MacFarlane, I. C. and Rutka, A. (1962), An evaluation of pavement performance over muskeg in Northern Ontario, HRB Bullertin 316, pp. 32-43.

MacFarlane, I. C. (1969), *Muskeg, Engineering Handbook*, Univ. of Toronto Press, Toronto, Canada.

MacFarlane, I. C. and Williams, G. P. (1974), Some engineering aspects of peat soils, histosols: their characteristics, classificatioin and use, SSSA Special Publ. no. 6, pp. 79-93.

MacMahon, J. A. (1985), *Deserts*, A. A. Knopf, Inc., NY.

Marbut, C. F. (1920), The contribution of soil surveys to soil science, *Proc. Agric. Sci.*, v. 41, pp. 116-142.

Marbut, C. F. (1935), Soils of the U.S. Atlas of America Agriculture, USDA.

Marcuson, W. F., III, Hynes, M. E. and Franklin, A. G. (1992), Seismic stability and permanent deformation analysis, the last 25 years, ASCE Geotechnical Special Publ. no. 31, pp. 552-592.

Marion, J. B. (1976), *Physics in the Modern World*, Academic Press, NY.

Marshall, C. E. (1949), *The Colloidal Chemistry of the Silicate Minerals*, Academic Press, NY.

Marshall, T. J. (1959), Relations between water and soil, Technical Communication no. 50, Commonwealth Bureau of Soils, Harpenden, England.

Marshall, A. F. (1973), How much more will Houston sink? The disturbing problem of land surface subsidence, *The Slide Rule*, Houston Engineering & Scientific Society, v. 33, no. 2, 6p.

Martin, R. E. (1977), Estimating foundation settlement in residual soils, *J. Geotech. Eng. Div. Proc. ASCE*, v. 103, no. GT3, pp. 197-212.

Martin, C. C. and Hu, Z. X. (1987), The geomorphic processes of soil/rock with examples, *Proc. 1st Int. Symp. Environ. Geotechnol.*, v. 2, Envo Publ. Co., Bethlehem, PA., pp. 390-393.

Matsuo, S. (1957), A study of the effect of cation exchange on the stability of slopes, *Proc. 4th ICSMFE*, v. 2, pp. 330-333.

Mattson, S. (1929, 1932), The laws of soil colloidal behavior, *Soil Sci.*, part II, v. 28, pp. 373-409; part IX, v. 34, pp. 330-333.

McCarthy, G. J. (1975), Radioactive waste management: The nuclear waste form, *Earth and Mineral Sci.*, v. 45, no. 3, pp. 17-20.

McCarthy, G. J. (1978), Interactions between nuclear waste and surrounding rock, *Nature*, no. 273, pp. 216-217.

McCready, A. A. (1986), QC/QA for large geomembrane installations in the uranium milling industry, *Proc. 1st Int. Symp. Environ. Geotechnol.*, v. 1, Envo Publ. Co., Bethlehem, PA., pp. 257-261.

McGuffey, V. C. (1982), Design of cut slopes in overconsolidated clays, TRR 873, pp. 8-11.

McNabb, J. F., Dunlap, W. J., and Keeley, J. W. (1977), Nutrient, Bacterial and Virus Control as Related to Groundwater Contamination, U.S. EPA Report no. 600/8-77-010, 18p.

McNeal, B. L. and Coleman, N. T. (1966), Effect of solution composition on soil hydraulic conductivity, *J. Soil Sci.*, v. 6, no. 22, pp. 163-178.

McNeal, B. C., Norvell, W. A. and Coleman, N. T. (1966), Effect of solution composition on the swelling of extracted soil clays, *Proc. SSSA*, v. 30, pp. 313-317

Mead, W. J. (1936), Engineering geology of dam sites, *Trans. 2nd. Int. Congr. Large Dams*, v. 4, p. 183.

Means, J. L., Smith, L. A., Nehring, K. W. Brauning, S. E., Gavaskar, A. R., Sass, B. M., Wiles, C. C., and Mashni, C. I. (1995), *The Application of Solidification/Stabilization of Waste Materials*, CRC Press, Boca Raton, FL.

Meegoda, J. N. (1996), A state-of-the-art geoenvironmental engineering research facilities, paper presented at the *3rd. Int. Symp. Environmental Geotechnology*, Technomic Publ., Lancaster, PA, June.

Meegoda, J. N. and Rajapakse, R. A. (1993), Short-term and long-term permeabilities of contaminated clays, *J. Environ. Eng., Proc. ASCE*, v. 119, no. 4, pp. 725-743.

Mehlich, A. (1960) Charge characterization of soils, *Int. Congr. Soil Sci.*, v. II, pp. 292-302.

Menard, L. and Broise, Y. (1975), Theoretical and practical aspects of dynamic consolidation, *Geotechnique*, v. 15, no. 1, pp. 3-18.

Mesri, G. and Olson, R. E. (1971), Mechanisms controlling the permeability of clays, *Clays and Clay Minerals*, v. 10, pp. 151-158.

Michaels, A. S. and Lin, C. S. (1955), Effects of counter-electroosmosis andsodium ion exchange on permeability of kaolinite, *Ind. Chem.*, v. 47, pp. 1249-1253.

Mikroudis, G. K. and Fang, H. Y. (1987), GEOTOX-PC: A new hazardous waste management tool, *Microcomputer Knowledge-based Expert Systems on Civil Engineering*, H. Adeli, ed., ASCE, NY, pp. 101-117.

Millar, C. E. and Turk, L. M. (1943), *Fundamentals of Soil Science*, John Wiley & Sons, NY.

Mitchell, J. K. (1969), Temperature effects on the engineering properties and behavior of soils, HRB Special Report 103, pp. 9-28.

Mitchell, J. K. (1976, 1993), *Fundamentals of Soil Behavior*, 2nd ed. (1st. ed. 1976), John Wiley & Sons, NY.

Mitchell, J. K. and Woodward, R. J. (1973), Clay chemistry and slope stability, *JSMFE Div. Proc. ASCE*, v.99, no. SM10, pp. 905-912.

Mitchell, J. K. and Kao, T. C. (1978), Measurement of soil thermal resistivity, *J. Geotech. Eng. Div. Proc. ASCE*, v. 104, no. GT10, pp. 1307-1320.

Mitchell, J.K. and Jaber, M.(1990), Factors controlling the long-term properties of clay liners, ASCE Geotechnical Special Publ. no. 26, pp. 84-105.

Mitchell, R. A., Hatch, S. E. and Siegal, R. (1992), Stability and closure design for a landfill on soft clay and peat, ASCE Geotechnical Special Publ. no. 31, pp. 695-704.

Mogami, T. (1967), Mechanics of granular material composed of particles of various sizes, Japanese Society of Civil Engineering, Publ. no. 137.

Moh, Z. C. ed. (1977), Geotechnical Engineering and Environmental Control, Specialty Session, *Proc. 9th, ICSMFE*, Japan, v. 1, 480p and v. 2, 105p.

Moore, L. H. and McGrath, M. E. (1970), Highway construction on refuse landfills, *Highway Focus*, v. 2, no. 5, pp. 11-26.

Morgenstern, N. R. (1992), The evaluation of slope stability - a 25-years perspective, ASCE Geotechnical Special Publ. no. 31, pp. 1-26.

Morgenstern, N. R. and de Matos, M. M. (1975), Stability of slopes in residual soils, *Proc. 5th Panamerican Conf. SMFE*, v. 3, pp. 367-383.

Morrison, M., Senturk, C. and Fang, H. Y. (1993), Soil-cement scrap/waste tire mixture used as construction fill, *Proc. 25th Mid-Atlantic Industrial Waste Conf.*, pp. 381-388.

Moulton,, L. K. (1991), Aggregate for drainage, filtration and erosion control, ch. 12, *The Aggregate Handbook*, R. D. Barksdale, ed., National Stone Assoc., pp. 12.1 to 12.93.

Moum, J. and Rosequist, I. Th. (1961), The mechanical properties of montmorillonitic and illitic clays related to the electrolytes of the pore water, *Proc. 5th ICSMFE*, v. 1, p. 263.

Mowafy, Y. M. and Bauer, G. E. (1985), Prediction of swelling pressure and factors affecting the swell behavior of an expansive soil, TRR 1032, pp. 23-27.

Mulamoottil, G., Warner, B. G. and McBean, E. A., eds. (1996), *Wetlands: Environmental Gradients, Boundaries, and Buffers*, Lewis Publ., Boca Raton, FL.

Murayama, S. (1969), Effect of temperature on elasticity of clays, HRB Special Report 103, pp. 194-203.

Naik, D. (1986), Effect of temperature and pore fluid on shear characteristic of clay, *Proc. 1st Int. Symp. Environ. Geotechnol.*, v. 1, pp. 382-390.

Narain, J. and Rawat, P. C. (1970), Tensile strength of compacted soils, *JSMFE Div. Proc. ASCE*, v. 96, no. SM6, pp. 2185-2190.

NAS, (1974), *More Water for Arid Lands*, National Academy of Sciences, Washington, D.C., 153p.

Nelson, J. D. and Miller, D. J. (1992), *Expansive Soils: Problems and Practice in Foundation and Pavement Engineering*, John Wiley & Sons, NY.

Nero, A. (1989), Earth, air, radon and home, *Phys. Today*, v.42, no.4, pp.32-39.

Nester, E. W., Roberts, C. E., Pearsall, N. N., and McCarthy, B. J. (1978), *Microbiology*, 2nd ed. W.B. Saunders College Press, Philadelphia.

Niering, W. A. (1985), *Wetlands*, The Audubon Society Nature Guides, A. A. Knopf Publ., NY.

NISS (1978), Soils of China, Nanking Institute of Soil Science (NISS), *Academia Sinica*. (in Chinese).

Nordyke, M. D. and Circeo, L. J. (1960), Progress in nuclear excavation technology, HRR no. 107, pp. 54-65.

Norris, R. D., Hinchee, R. E., Brown, R., McCarty, P. L., Semprini, L., Wilson, J. T., Kampbell, D. H., Reinhard, M., Bouwer, E. J., Borden, R. C. Vogel, T. M., Thomas, J. M., and Ward, C. H. (1994), *Handbook of Bioremediation*, Lewis Publ., Boca Raton, FL.

NRC, (1957), Disposal of Radioactive Wastes on Land, Publ. 519, National Academy Press, Washington, D.C., 142p.

NRC, (1970), Disposal of Solid Radioactive Wastes in Bedded Salt Deposits, National Academy Press, Washington, D.C., 28p.

NRC, (1977), Methane Generation from Human, Animal and Agricultural Wastes, National Academy Press, Washington, D.C.

NRC, (1980), The Effects on Populations of Exposure to Low Levels of Ionizing Radiation:1980, National Academy Press, Washington, D.C., 524p.

NRC, (1984), Groundwater Contamination, National Academy Press, Washington, D.C., 179p.

NRC, (1988), Health Risks of Radon and Other Internally Deposited Alpha-Emitters, BEIR IV, National Academy Press, Washington, D.C., 602p.

Nunes, A. J de Coata, (1969), Landslides in soils of decomposed rock due to intense rainstorms, *Proc. 7th ICSMFE*, v. 2, pp. 547-554.

Ohira, Y. (1977), Laboratory and in-situ testing methods, *Engineering Problems of Organic Soils in Japan*, T. Yamanouchi, ed., Japanese Society of Soil Mechanics and Foundation Engineering, pp. 19-34.

Olmstead, L. B. and Smith, W. O. (1938), Water relation of soils, *Soils and Men*, Yearbook of Agriculture, USDA, pp. 897-910.

Ofer, Z. and Blight, G. E. (1985), Measurement of swelling pressure in the laboratory and in situ, TRR 1032, pp. 15-22.

Paaswell, R. E. (1967), Temperature effects on clay soil consolidation, *JSMFE Div. Proc. ASCE,* v. 93, no. SM3, pp. 9-22.

Paaswell, R. E. (1973), Causes and mechanisms of cohesive soil erosion: the state-of-the-art, HRB Special Report 135, pp. 52-74.

Palmer, C. M., Peterson, J. L., and Behnke, J. (1991), *Principles of Contaminant Hydrogeology*, Lewis Publ., Boca Raton, FL.

Pamukcu, S. and Topcu, I. B. (1991), Study of Causes of Random Cracking of Solidified Sludge Reused as Capping Material, Fritz Engineering Laboratory Report, Department of Civil Eng., Lehigh Univ., Bethlehem, PA, 33p.

Patin, T. R. ed. (1987), Management of Bottom Sediments Containing Toxic Substances, *Proc. 11th U.S.-Japan Experts Meeting*, U.S. Army Corps of Engineers, Seattle, Washington, 279p.

Paton,T.R. (1978), *The Formation of Soil Material*, George Allen & Unwin, U.K.

Pauling, L. (1945, 1960), *Nature of the Chemical Bond*, (1st. ed. 1945) 3rd. ed., Cornell Univ. Press, Ithaca, NY.

PCA (1962, 1992), *PCA Soil Primer*, Portland Cement Assoc. Skokie, IL, 40p.

Penner, E. (1963), The nature of frost heaving in soils, *Proc. Int. Conf. of Permeafrost*, NAS-NRC Publ. 1287.

PERF (1982), *Glossary of Hazardous Waste Terms*, Pennsylvania Environmental Research Foundation, Inc. 225 S. 15th st., Philadelphia, PA., 64p.

Perlow, M. Jr., Schadl, S. M., Kugelman, I. J. and Fang, H. Y. (1983), Waste disposal considerations in Lehigh Valley carbonate formations, *Proc. Conf. on the Disposal of Solid, Liquid and Hazardous Wastes*, pp. 17-1 to 17-35.

Perry, E. B. (1975), Piping in Earth Dams Constructed of Dispersive Clay; Literature Review and Design of Laboratory Tests, Waterway Experiment Station Technical Report S-75-15, 105p.

Petroleum Geology (1979), Beijing, p. 26, & p. 29. (In Chinese).

Phukan, A. (1985), Frozen Ground Engineering, McGraw-Hill Book Co. NY.

Phukan, A. (1991), Foundations in cold regions, ch. 19, *Foundation Engineering Handbook*, 2nd. ed., Van Nostrand Reinhold Co., NY, pp. 735-749.

Piwowar, S. B. and Work, D. (1997), Acid mine drainage in the United States, *Proc. 3rd Int. Symp. Environ. Geotechnol.*, v. 2, 10p.

Plum, R. L. and Esrig, M. I. (1969), Some temperature effects on soil compressibility and porewater pressure, HRB Special Report 103, pp. 231-242.

Poland, J. F. (1961), The coefficient of storage in a region of major subsidence caused by compaction of an aquifer systemm USGE Prof. Paper 424-B, art. 25, pp. B52-B54.

Poland, J. F. and Davis, G. H. (1969), Land subsidence due to withdrawal of fluids, *Rev. Eng. Geol.*, v. 2, pp. 187-269.

Porterfield, W. W. (1972), *Concepts of Chemistry*, W. W. Norton & Co., NY.

Post, J. L. and Paduana, J. A. (1969), Soil stabilization by incipient fusion, HRB Special Report 103, pp. 243-253.

Post, W. M., Peng, T. H., Emanuel, W. R., King, A. W., Dale, V. H. and DeAngelis, D. L. (1990), The global carbon cycle, *Am. Sci.*, v. 78, July-Aug. pp. 310-326.

Powers, T. C. (1966), Some observations on the interpretation of creep data, *RILEM Bulletin*, New Series no. 33, pp. 381-391.

Pritchard, D. W. (1952), Estuarine hydrography, *Adv. Geophys.*, pp. 243-280.

Proctor, R. R. (1933), Fundamental principles of soil compaction, *Engineering News-Record*, v. 111, no. 9.

Prokopovich, N. P. (1978), Genetic classification of land subsidence, *Evaluation and Prediction of Subsidence*, ASCE, NY., pp. 389-399.

Purvis, R. L. (1994), Underwater Bridge Maintenance and Repair, NCHRP Synthesis of Highway Practice 200, TRB, 54p.

Quirk, J. P. and Schofield, R. K. (1955), The effect of electrolyte concentration on soil permeability, *J. Soil Sci.*, v. 6, no. 22, pp. 163-178.

Rahn, P. H. (1986), Ground water contamination by radioactive waste, *Proc. 1st Int. Symp. Environ. Geotechnol.*, v. 1, Envo Publ. Co., Bethlehem, PA., pp. 233-246.

Ramana, V. (1996), Personal Communication, The Univ. of the West Indies, St. Augustine, Trinidad.

Rao, I. R. (1933), Constitution of water in different states, *Nature*, v. 132, no. 3334, Sept. 23, p. 480.

Rao, S. N. (1979), The influence of fabric on the shrinkage limit of clay, *Geotech. Eng.*, v. 19, no. 2, pp. 243-251.

Rau, G. and Chaney, R. C. (1988), Triaxial testing of marine sediments with high gas contents, ASTM, STP 977, pp. 338-352.

Remson, I., Fungaroli, A. A. and Lawrence, A. (1968), Water movement in an unsaturated sanitary landfill, *J. Sanitary Eng. Div. Proc. ASCE*, v. 94, no. SA2.

Reno, W. H. and Winterkorn, H. F. (1967), Thermal conductivity of kaolinite clay as a function of type of exchange ion, density and moisture content, HRR no. 209, pp. 79-85.

Reuss, J. O. (1978), Simulation of Nutrient Loss from Soils due to Rainfall Acidity, U.S. EPA Report 600/3-78-053.

Rhodes, H. T. F. (1931), *Some Persons Unknown*, John Murray, London, p. 11.

Rich, C. I. (1968), Applications of soil mineralogy in soil chemistry and fertility investigations, SSSA Special Publ. Series no. 3, pp. 61-90.

Richards, L. A. ed. (1954), Diagnosis and improvement of saline and alkali soils, *Agricultural Handbook,* no. 60, USDA, Washington, D.C.

Richards, A. F. and Chaney, (1982), Marine slope stability-a geological approach, Proc. NATO Conf. on Marine Slides and Other Mass Movements, pp. 163-172.

Ridgeway, H. H. (1976), Infiltration of water through the pavement surface, TRR no. 616, pp. 98-100.

Riker, R. E., Anderson, D. G. and Bacon, D. D. (1992), Slope displacement from pile driving, ASCE Geotechnical Special Publ. no. 31, pp. 292-309.

Robertson, J. B. (1984), Geological problems at low-level radioactive waste disposal sites, *Groundwater Contamination*, National Academy Press, Washington, D. C., pp. 104-108.

Robertson, R. K., Woeller, D. J. and Gillespie, D. (1990), Evaluation of excess pore pressures and damage conditions around driven piles using the cone penetration test with pore pressure measurements, *Can. Geotech. J.*, v. 27, pp. 249-254.

Rodebush, W. and Buswell, A. M. (1958), Properties of water substance, HRB Special Report 40, pp. 5-13.

Rose, A. W., Washington, J. W., and Greeman, D. J. (1988), Variability of radon with depth and season in a central Pennsylvania soil, *Northeastern Environ. Sci.*, v. 7, no. 1, pp. 35-39.

Rosenquist, I. Th. (1953), Considerations on the sensitivity of Norwegian quick-clays, *Geotechnique*, v. 3, pp. 195-200.

Ross, C. S. and Shannon, E. V. (1926), Minerals of bentonite and related clays and their physical properties, *J. Am. Ceram. Soc.*, v. 9, pp. 77-96.

Roy, B. (1980), Nuclear waste - the Penn State connection, *Earth and Mineral Sci.*, v. 50, no. 2, Nov.-Dec., p. 13.

Russell, E. W. (1934), The interaction of clay with water and organic liquids as measured by specific volume changes and its relation to the phenomena of crumb formation in soils, *Phil. Trans. Royal Soc. London*, Series A, v. 233, pp. 361-389.

Russell, E. R., Sr. (1994), Highway Maintenance Procedures Dealing with Hazardous Material Incidents, NCHRP Synthesis of Highway Practice 196, TRB, 60p.

Russell, O. R., Amato, R. V. and Leshendok, T. V. (1979), Remote sensing and mine subsidence in Pennsylvania, *Trans. Eng. J. ASCE, Proc. ASCE*, v. 105, no. TE2, pp. 185-198.

Salomone, L. A., Kovacs, W. D. and Kusuda, T. (1984), Thermal performance of fine-grained soils, *J. Geotech. Eng., Proc. ASCE*, v. 110, no. 3, pp. 359-374.

Samuels, S. G. and Chaney, J. E. (1975), Long-term heave of a building on clay due to tree removal, Building Research Establishment, U.K., Current Paper CP 73/75, 7p.

Sanger, F. J. (1963), Degree days and heat conduction in soils, Proc. Int. Conf. on Permanfrost, NAS-NRC Publ. 1287.

Sarsby, R. W. ed. (1995), *GREEN'93, U. K., Waste Disposal by Landfill*, A. A. Balkema Publ., Rotterdam, 692p.

Saxena, S. K. (1978), A review of the methods used in investigation of subsidence, *Evaluation and Prediction of Subsidence*, ASCE, NY., pp. 214-233.

Scheffer, F. and Schachtschabel, P. (1959), *Lehrbuch der Bodenkunde*, 5th ed. Stuttgart.

Schofield, R. K. (1935), The pF of the water in soil, *Trans. 3rd. Int. Congr. Soil Science*, v. 2, pp. 37-48.

Schofield, R. K. (1949), Effect of pH on electric charges carried by clay particles, *J. Soil Sci.*, v. 1, pp. 1-8.

Schram, M. (1981), Permeability of Soils to Four Organic Solvents and Water, MSCE Thesis, University of Arizona, Tuscon, AR.

Scott, R. F. (1978), Subsidence, a review, *Evaluation and Prediction of Subsidence*, ASCE, NY. pp. 1-25.

Sears, F. W., Zemansky, M. W., and Young, H. D. (1987), *College Physics*, 6th ed., Addison-Wesley Publ. Co., Reading, MA.

Seed, H. B. (1975, 1991), Earthquake effects on soil-foundation systems, part 1, ch. 16, *Foundation Engineeering Handbook*, 2nd. ed. (1st. ed. 1975), Van Nostrand Reinhold Co., NY, pp. 594-623.

Seed, H. B. and Goodman, R. E. (1964), Earthquake stability of slopes of cohesionless soils, *JSMFE Div. Proc. ASCE*, v. 90, no. SM6, pp. 43-74.

Seed, H. B. and Schnabel, P. B. (1972), Soil and geologic effects on site response during earthquake, Proc. Microzonation Conf., v. 1, pp. 61-85.

Seed, R. B. and Bonaparte, R. (1992), Seismic analysis and design of waste fills: current practice, *Stability and Performance of Slopes and Embankments-II*, ASCE, NY.

Segall, B. A., O'Bannon, C. E. and Mattias, J. A. (1980), Electro-osmosis chemistry and water quality, *J. Geotech. Eng. Div. Proc. ASCE*, v. 106, no. GT10, pp. 1143-1147.

Segall, B. A. and Bruell, C. J. (1992), Electroosmotic contaminant removal processes, *J. Environ. Eng. Proc. ASCE*, v. 118, no. EE2, pp. 84-100.

Seifert, R., Ehrenberg, J., Tiedemann, B., Endell, K.Hofmann, J. and Wilm, D. (1935), Relation between landslide slope and the chemistry of clay soils, *Mitt. preuss. versuchanstalt f. Wasserbau u. Schiffbau*, v. 20, p. 34.

Senior, E. ed. (1995), *Microbiology of Landfill Sites*, 2nd ed., CRC Press, Inc., Boca Raton, FL.

Shackley, M. (1981), *Environmental Archaeology*, George Allen & Unwin, London.

Shockley, W. G. (1978), Suggested practice for description of frozen soils (visual-manual procedure), *ASTM Geotech. Test. J.*, v. 1, no. 4, pp. 228-233.

Shang, J. Q. and Lo, K. Y. (1996), Electrokinetic dewatering of Florida phosphate clay, *Proc. 3rd. Int. Symp. Environ. Geotechnol.*, v. 1, Technomic Publ., Lancaster, PA, pp. 925-934.

Sharma, H. D. and Lewis, S. P. (1994), *Waste Containment Systems, Waste Stabilization, and Landfills, Design and Evaluation*, John Wiley & Sons, NY.

Sherard, J. L., Decker, R. S. and Ryker, N. L. (1972), Piping in earth dams of dispersive clay, *Proc. Specialty Conf. on Performance of Earth and Earth-Supported Structures*, ASCE, NY, v. 1, part 1, pp. 589-626.

Sherard, J. L., Dunnigan, L. P. and Decker, R. S. (1976), Identification and nature of dispersive soils, *J. Geotech. Eng. Div. Proc. ASCE*, v. 102, no. GT4, pp. 287-301.

Sheeler, J. B. (1968), Summarization and comparison of engineering properties of loess in the United States, HRR no. 212, pp. 1-9.

Shi, G. C. and Liebowitz, H. (1968), Mathematical theories of brittle fracture, *Mathematical Fundamental of Fracture*, Academic Press, NY.

Shi, G. C. and MacDonald, B. (1974), Fracture mechanics applied to engineering problems-strain energy density fracture criterion, *Eng. Fracture Mechanics*, v. 6, pp. 361-386.

Shi, M. X. (1985), *Ground Water*, Scientific Publ. Co., Beijing, 215p. (In Chinese).

Shibuya, T. (1973), Geological Study of Landslide Clay, KICT Report no. 10, Kajima Institute of Construction Technology, Tokyo, 182, Japan, 37p.

Shoemaker, H.D., Advani, S.H., Gmeindl, F.D., and Lin, Y.T. (1978), Studies of thermo-mechanical subsidence associated with underground coal gasification, *Evaluation and Prediction of Subsidence*, ASCE, NY. pp.140-153.

Shook, J. F. and Fang, H. Y. (1961), Cooperative materials testing program at the AASHO Road Test, HRB Special Report 66, pp. 59-102.

Shuldiner, P. W., Cope, D. F., and Newton, R. B. (1978), Ecological Effects of Highway Fills on Wetlands, NCHRP Report 218A, TRB, 34p.

Siddiqi, F. H., Seed, R. B., Chan, C. K., Seed, H. B., and Pyke, R. M. (1987), Strength Evaluation of Coarse-Grained Soils, Earthquake Engineering Research Center, Report no. UCB/EERC-87/22, Univ. of California, Berkeley, CA, 53p.

Sinacori, M. W. (1951), Compaction of unstable material with heavy pneumatic tired roller, HRB Bulletin 42.

Sinclair, W. A., Buller, F. H. and Benham, C. B. (1960), Soil thermal characteristics in relation to underground power cables, A Report of the AIEE Insulated Conductors Committee, Report no. 60-785, pp. 71-94.

Singh, A. and Mitchell, J. K. (1968), General stress-strain-time function for soils, *JSMFE Div. Proc. ASCE*, v. 94, no. SM1, pp. 21-46.

Sitar, N., Anderson, S. A. and Johnson, K. A. (1992), Conditions for initiation of rain-induced debris flows, ASCE Geotechnical Special Publ. no. 31, pp. 834-849.

Skempton, A. W. (1964), Long-term stability of clay slopes, *Geotechnique*, v. 14, no. 2, pp. 77-102.

Skempton, A. W. and Northey, R. D. (1952), The sensitivity of clays, *Geotechnique*, v. 3, no. 1, pp. 30-53.

Smith, W. O. (1943), Temperature transfer of moisture in soils, *Trans. AGU*, v. 24, pp. 511-524.

Smith, S. S. and Arulanandan, K. (1981), Relationship of electrical dispersion to soil properties, *J. Geotech. Eng., Proc. ASCE*, v. 107, no. GT5, pp. 591-604.

SMSS (1990), *Keys to Soil Taxonomy*, Soil Management Support Services Survey.

Staff, SMSS Technical Monograph no. 6, 4th ed. Blacksburg, VA (This reference is updated approximately every 2 years).

Sowers, G. F. (1968), Foundation problems in sanitary landfill, *J. Sanitary Eng. Div. Proc. ASCE*, v. 94, no. 1, pp. 102-116.

Sowers, G. F. (1973), Settlement of waste disposal fill, *Proc. 9th ICSMFE*, v. 4, pp. 297-310.

Sowers, G. F. (1976), Mechanisms of subsidence due to underground openings, TRR 612, pp. 2-8.

Sowers, G. B. and Sowers, G. F. (1970), *Introductory Soil Mechanics and Foundations*, 3rd. ed., The MacMillan Co., NY.

Sposito, G. (1989), *The Surface Chemistry of Soils*, Oxford Univ. Press, NY.

Spangler, M. G. (1951), *Soil Engineering*, Intern. Textbook Co., Scranton, PA.

Sridharan, A. and Rao, G. V. (1972), Surface area determination of clays, *Geotech. Eng.*, v. 3, no. 2, pp. 127-132.

Stefanko, R., Linden, K. V., and Tilton, J. G. (1965), Subsurface Disposal of Acid Mine Water by Injection Wells, Coal Research Board, Commonwealth of Pennsylvania, Special Research Report no. SR-52, 70p.

Stein, L. (1987), Chemical properties of radon: radon and its decay products, P. K. Hopke, ed., American Chemical Society Symposium Series 331, pp. 10-29.

Stern, O. (1924), Zur Theorie der Elektrolytischen Doppelschrift, *Zietschrift Electrochem.*, v. 30, pp. 508-516.

Stevenson, F. J. (1982), *Humus Chemistry*, John Wiley & Sons, NY.

Strauss, P. J. and Hugo, F. (1979), Innovations in design and construction of a low volume road on windblown sands, TRR 702, pp. 208-214.

Street, N. (1959), Electrokinetics II, electroviscosity and the flow of reservoir fluids, Illinois Geol. Survey Circular no. 263.

Su, H. Y. (1979), Investigation of the deformation characteristics of various soil strata under the influence of pumping-out-and-back of groundwater in Shanghai, *Chinese J. Geotech. Eng.*, v. 1, no. 1, pp. 24-35. (In Chinese with English summary).

Sudo, T., Shimoda, S., Yotsumoto, H., and Aita, S. (1981), *Electron Micrographs of Clay Minerals*, Elsevier Scientific Publ. Co., NY.

Sun, Z. C. (1989), The effects of acids on physico-mechanical properties of laterite soil, *Chinese J. Geotech. Eng.*, v. 11, no. 4, July, pp. 89-93. (In Chinese with English summary).

Takenaka, H. and Yasuhara, K. (1977), Physical and mechanical properties, *Engineering Problems of Organic Soils in Japan*, T. Yamanouchi, ed., Japanese Society of Soil Mechanics and Foundation Engineering, pp. 35-67.

Tamez, E. (1957), Some factors affecting the dynamic compaction test, ASTM STP 232, pp. 54-66.

Tammann, G. (1925), *The States of Aggregation*; the exchanges in the state of matter in their dependence upon pressure and temperature. (Translated by R. F. Mehl), D. Van Nostrand Co., NY.

Tan, K. H., Hajek, B. F., and Barshad, I. (1986), Thermal analysis techniques, methods of soil analysis, ch. 7, *Agronomy*, no. 9, part 1, 2nd ed. American Society of Agronomy, Madison, WI.

Taylor, D. W. (1948), *Fundamentals of Soil Mechanics*, John Wiley & Sons, NY.

Tepordei, V. V. (1990), Crushed stone and sand and gravel in the fourth quarter of 1989, Mineral Surveys, U.S. Department of the Interior, Bureau of Mines, Washington, D.C.

Terzaghi, K. (1936), Stability of slopes of natural clay, *Proc. 1st. ICSMFE*, v. 1, p. 161.

Terzaghi, K. (1942), Soil moisture and capillary phenomena in soils, *Hydrology*, O. E. Meinzer, ed., McGraw-Hill Book Co., NY, pp. 331-363.

Terzaghi, K. (1943), *Theoretical Soil Mechanics*, John Wiley & Sons, NY.

Terzaghi, K. and Peck, R. B. (1967), *Soil Mechanics in Engineering Practice*, John Wiley & Sons, NY.

Thom, C. and Smith, N. R. (1938), Fauna and flora of soil, *Soils and Man*, Yearbook of Agriculture, USDA, pp. 940-947.

Thomas, G. W. (1960), Effects of electrolyte imhibition upon cation exchange behavior of soils, *Proc. SSSA*, v. 24, pp. 329-332.

Thompson, D. A. W. (1942), *Growth and Form*, Cambridge Univ. Press, Cambridge, England.

Thornthwaite, C. W. (1931), Climate of North America, *Geograph. Rev.*, v. 21, pp. 633-655.

Thorp, J. and Smith, G. D. (1949), Higher categrories of soil classification: orders, suborder, and great soil groups, *Soil Sci.*, v. 67, pp. 117-126.

Thrasher, M. H. (1983), Highway impacts on wetlands: assessment, mitigation, and enhancement measures, TRR 948, pp. 17-20.

Tianjin Univ. (1978), *Soil Mechanics and Foundation Engineering*, Chinese Construction Publ., Beijing, (In Chinese).

Tilley, D. E. and Thumm, W. (1971), *College Physics: A Text with Applications to the Life Sciences*, Cummings Publ. Co., Menlo Park, CA.

Todd, D. K. and McNulty, D. E. O. (1976), *Polluted Groundwater*, Water Information Center, Port Washington, NY.

Tongji Univ. (1979), *Soil Properties and Mechanics*, Shanghai, China. (In Chinese)

Torrance, J. K. (1975), On the role of chemistry in the development and behavior of the sensitive marine clays of Canada and Scandinavia, *Can. Geotech. J.*, v. 12, no. 3.

Townsend, F. C. and Gilbert, P. A. (1973), Tests to measure residual strength of some clay shales, *Geotechnique*, v. 23, no. 2, pp. 267--271.

Towhasta, I., Kuntiwattanakul, P., Seko, I., and Ohishi, K. (1993), Volume change of clays induced by heating as observed in consolidation tests, *Soils and Foundations*, v. 33, no. 4, pp. 170-183.

Traughber, E. B., Snowden, J. O., and Simmons, W. B. (1978), Differential subsidence on reclaimed marshland peat in metropolitan New Orleans, Louisiana, *Evaluation and Prediction of Subsidence*, ASCE, NY, pp. 479-499.

Trautz, (1922) (In German see Winterkorn, 1974).

TRB (1974), Roadway Design in Seasonal Frost Areas, NCHRP, Synthesis of Highway Practice 26, TRB, 104p.

TRB (1976), Estimation of Consolidation Settlement, TRB Special Report 163, 26p.

TRB (1979), Snow Removal and Ice Control Research TRB Special Report 185, 355p.

TRB (1986), Recent Advances in Hazardous Materials Transportation Research, An International Exchange, State-of-the-Art Report 3, TRB, 212p.

Tscheborarioff, G. P. (1951), *Soil Mechanics, Foundations, and Earth Structures*, McGraw-Hill Book Co., NY.

Tscheborarioff, G. P. and Winterkorn, F. H. (1947), Sensitivity of clay to remolding and its possible causes, *Proc. HRB Annual Meeting*, pp. 435-442.

Tsuchida, T., Kobayashii, M. and Mizukami, J. (1991), Effect of aging of marine clay and its duplication by high temperature consolidation, *Soils and Foundations*, v. 31, no. 4, pp. 133-147.

Tuncan, M., Pamukcu, S., and Hu, Z. X. (1989), Development of multi-purpose triaxial apparatus for testing of soils under coupled influence of thermal-chemical-hydraulic and electrical potential, *Proc. 2nd. Int. Symp. Environ. Geotechnol.*, v. 1, Envo Publ. Co., Bethlehem, PA., pp. 111-123.

Tuncan, A. (1992), Mechanics of A Marine Clay Based on Physico-chemical Interactions with Organic Contaminants, Ph.D. Dissertation, Department of Civil Engineering, Lehigh Univ., Bethlehem, PA, 223p.

Tuncan, A. and Pamukcu, S. (1992), Predicted mechanism of crude oil and marine clay interactions in salt water environment, *Proc. Mediterranean Conf. on Environmental Geotechnology*, M. A. Usmen and Y. B. Acar, ed., A.A. Balkema, Rotterdam, pp. 109-124.

Turnbull, W. J. (1968), Construction problems experienced with loess soils, HRR no. 212, pp. 10-27.

Underwood, L. B. (1967), Classification and identification of shales, *JSMFE Div. Proc. ASCE*, v. 93, no. SM6, pp. 97-116.

US Army Corps of Engineers (1984), *3rd U.S.-The Netherlands Meeting on Dredging and Related Technology*, Charleston, SC, 305p.

USBR (1973), *Earth Manual*, U.S. Bureau of Reclamation, Denver, CO.

USDA (1938), *Soils and Men*, Yearbook of Agriculture, U.S. Department of Agriculture, Washington, D.C.

USDA (1969), *Wildlife Habital Improvement Handbook*, Forest Service, U.S. Department of Agriculture, Washington, D.C.

USDA (1981), *Soil Survey Manual*, 430-V, Issue 1, USDA, Soil Conservation Service, June.

USDA (1986), Designations for Master Horizons and Layers of Soils, USDA & AID. Vallerga, B. A. and Van Til, C. J. (1970), Classification and engineering properties of lateritic material, HRR no. 310, pp. 52-67.

Van Rooyen, M. and Winterkorn, H. F. (1957) Theoretical and practical aspects of the thermal conductivity of soils and similar granular systems, HRB Bulletin 168, pp. 143-205.

Van Rooyen, M. and Winterkorn, H. F. (1959), Structural and textural influences on thermal conductivity of soils, Proc. HRB, v. 38, pp. 576-621.

Van Olphen, H. (1954), Interlayer forces in bentonite, *Clays and Clay Minerals*, v. 2, pp. 418-438.

Van Olphen, H. (1977), *An Introduction to Clay Colloidal Chemistry for Clay Technologists, Geologists and Soil Scientists*, 2nd ed., John Wiley & Sons, NY.

Van Zyl, D. and Shackelford, C. (1988), Electro-kinetic treatment applications in environmental geotechnical engineering, *Geotech. News*, v. 6, no. 3, pp. 19-25.

Vargas, M. (1953), Some engineering properties of residual clay soils occuring in southern Brazil, *Proc. 3rd. ICSMFE*, v. 1, pp. 67-71.

Vees, E. and Winterkorn, H. F. (1967), Engineering properties of several pure clays as functions of mineral type, exchange ions and phase composition, HRR no. 209, pp. 55-65.

Velde, B. (1985), *Clay Minerals: A Physico-Chemical Explanation of Their Occurrence*, Elsevier Scientific Publ. NY.

Velde, B. (1992), *Introduction to Clay Minerals, Chemistry, Origins, Uses and Environmental Significance*, Chapman & Hall, London.

Voight, B. (1973), Correlation between Atterberg plasticity limits and residual shear strength of natural soils, *Geotechnique*, v. 23, no. 2, pp. 265-267.

Voight, B. and Pariseau, W. (1970), State of predictive art in subsidence engineering, *JSMFE Div. Proc. ASCE*, v. 96, no. SM2, pp. 721-750.

Wachinski, A. M. and Etzel, J. E. (1996), *Environmental Ion Exchange, Principles and Design*, Lewis Publ., Boca Raton, FL.

Waidelich, W. C. (1958), Influence of liquid and clay mineral type on consolidation of clay-liquid systems, HRB Special Report 40, pp. 24-42.

Walch, M. and Mitchell, R. (1984), Biological aspects of corrosion of offshore structures, *Naval Research Reviews*, no. 3, pp. 13-19.

Walker, A. S. (1982), Deserts of China, *Am. Sci.*, v. 70, pp. 366-376.

Wang, C. H. (1985), *Water Pollution*, Scientific Publ., Beijing, 267p. (In Chinese).

Wang, J., Cheng, G. J. and Zhang, C. Y. (1996), The research of salty soil, *Proc. 3rd. Int. Symp. Environ. Geotechnol.*, Technomic Publ., Lancaster, PA, v. 1, pp. 189-197.

Wang, J. Z. (1983), The effect of free iron oxides on the engineering properties of red clay, *Chinese J. Geotech. Eng.*, v. 5, no. 1, pp. 147-156. (In Chinese with English summary).

Warshaw, C. M. (1962), *McGraw-Hill Yearbook of Science and Technology, Annual Review of Scientific Developments*, McGraw-Hill Book Co., NY, p. 462.

Watanabe, S. (1977), Design and execution of works in depth, *Engineering Problems of Organic Soils in Japan*, T. Yamanouchi, ed., Japanese Society of SMFE, pp. 69-82.

Weaver, K. D., Evans, J. C. and Pancoski, S. E. (1990), Grout testing for a hazardous waste application, Concrete International, p. 45.

Wei, R. P. (1976), Environmental considerations in fatigue and fracture of construction steels, *New Horizons in Construction Materials*, Envo Publ. Co., Bethlehem, PA, v. 1, pp. 583-600.

Wenzel, L. K. (1942), Methods for determining permeability of water-bearing materials with special reference to discharging well methods, USGS Water-Supply Paper 887.

White, W. B. (1980), Stabilities of nuclear waste forms and their geotechnical interactions in repositories, *Earth and Mineral Sci.*, v. 50, no. 2, pp.13-19.

White, W. B., Culver, D. C., Herman, J. S., Kane, T. C. and Mylroie, J. E. (1995), Karst lands,*Am. Sci.*, v. 83, no. 5, pp. 450-459.

Wichens, G. E. ed. (1985), *Plants for Arid Lands*, Allen and Unwin Publ., England.

Williams, P. J. (1964), Unfrozen water content of frozen soils and soil moisture suction,*Geotechnique*, v. 14, p. 2.

Wilson, G. (1984), Pedotechnical aspects of organic soil classification and interpretation, presented at the TRB Annual Meeting. Session 42, 15p.

Winograd, I. J. (1974), Radioactive waste storage in the arid zone, GOS, *Trans. AGU*, v. 55, no. 10, pp. 884-894.

Winterkorn, H. F. (1937), The application of base exchange and soil physics to problems of highway construction, *Proc. SSSA*, v. 1, pp. 93-99.

Winterkorn, H. F. (1942), Mechanism of water attack on dry cohesive soil systems,*Soil Sci.*, v. 54, pp. 259-273.

Winterkorn, H. F. (1944), Climate and highways,*Trans. AGU*, June, pp. 405-411.

Winterkorn, H. F. (1947), Fundamental similarities between electro-osmosis and thermo-osmosis,*Proc. 27th Annual Meeting*, HRB, pp. 443-455.

Winterkorn, H. F. (1953), Macromeritic liquids, ASTM STP 156, pp. 77-89.

Winterkorn, H. F. (1955), The science of soil stabilization, HRB Bulletin 108, pp. 1-24.

Winterkorn, H. F. (1958), Mass transport phenomena in moist porous systems as viewed from the thermodynamic of irreversible processes, HRB Special Report 40, pp. 324-338.

Winterkorn, H. F. (1963), Soil water interaction and its bearing on water conduction in soils,*The Princeton Univ. Conf. on Engineering and World Water Resources*, Princeton Univ. Princeton, NJ, pp. 15-34.

Winterkorn, H. F. (1970), Analogies between macromeritic and molecular liquids and the mechanical properties of sand and gravel assemblies, *Chemical Dynamics*, Wiley-Intersciences, NY., pp. 751-766.

Winterkorn, H. F. (1974), Notes on Construction Materials, Department of Civil and Geological Engineering, Princeton Univ. Princeton, NJ.

Winterkorn, H. F. and Baver, L. D. (1934), Sorption of liquids by soil colloids, I: liquid intake and swelling by soil colloidal materials, *Soil Sci.*, v. 38, no. 4, pp. 291-298.

Winterkorn, H. F. and Moorman, R. B. B. (1941), A study of changes in physical properties of Putnam soil induced by ionic substitution, *Proc. 21st Annual Meeting*, HRB, pp. 415-434.

Winterkorn, H. F. and Fang, H. Y. (1975, 1991), Soil technology and engineering properties of soils, ch 3, *Foundation Engineering Handbook*, 2nd ed. (1st. ed. 1975) Van Nostrand Reinhold Co., NY, pp. 88-143.

Wischmeier, W. H. and Smith, D. D. (1978), Predicting Rainfall Erosion Losses-A Guide to Conservation Planning, USDA Agric. Handbook no. 537, Dec.

Wiss, J. F. and Nicholls, H. R. (1974), *A Study of Damage to A Residential Structure from Blasting Vibration*, ASCE, NY, 73p.

Witmer, K. (1987), Geomorphic process: from fresh garbage to organic soil,*Proc. 1st. Int. Symp. Environ. Geotechnol.*, v. 2, pp. 389.

Wohlbier, H. and Henning, D. (1969), Effect of preliminary heat treatment on the shear strength of kaolinite clay, HRB Special Report 103, pp. 287-300.

Woods, K. B. (1940), Design and construction of highway embankment, *Proc. Purdue Conf. on Soil Mechanics, and Its Applications*, pp. 355-366.

Wooltorton, F. L. D. (1954), The scientific basis of road design, *The Road Maker's Library*, v. 14, Ed. Arnold Ltd., London.

Wooltorton, F. L. D. (1955), Engineering pedology and soil stabilization, HRB Bulletin 108, pp. 29-57.

Wu, C. M. (1976), Groundwater depletion and land subsidence in Taipei basin, *Proc. 2nd Int. Symp. Land Subsidence*, Anaheim, CA, pp. 389-398.

Wu, T. H. (1984), Effect of vegetation on slope stability, TRR 965, pp. 37-46.

Xanthakos, P., Bruce, D. A., and Abramson, L. W. (1994), *Ground Control and Improvement*, John Wiley & Sons, NY.

Yamamura, K. (1983), Current status of waste management in Japan, *Waste Management and Res.*, v. 1, pp. 1-15.

Yamanouchi, T. ed. (1977), *Engineering Problems of Organic Soils in Japan*, Research Committee on Organic Soils, Japanese Society of Soil Mechanics and Foundation Engineering, 97p.

Yen, B. C. and Scanlon, B. (1975), Sanitary landfill settlement rates, *J. Geotech. Eng. Div. Proc. ASCE*, v. 101, no. GT5, pp. 475-487.

Yen, B. C. and Wang, W. L. (1977), Seismically induced shallow hillside slope failure, *Proc. 6th World Conf. on Earthquake Engineering*, v. 7, pp. 43-48.

Yeung, A. T. and Datla, S. (1995), Fundamental formulation of electrokinetic extration of contaminants from soil, *Can. Geotech. J.*, v. 32, pp. 569-583.

Yong, R. N. and Warkentin, B. P. (1966), *Introduction to Soil Behavior*, MacMillan Book Co. NY.

Yong, R. N., Chang, R. K. and Warkentin, B. P. (1969), Temperature effect on water retention and swelling pressure of clay soils, HRB Special Report 103, pp. 132-138.

Yong, R. N. and Sheeran, D. E. (1973), Fabric unit interaction and soil behavior, *Proc. Int. Symp. Soil Structure*, Sweden, pp. 176-183.

Yong, R. N. and Townsend, F. C. eds. (1981), *Symp. Laboratory Shear Strength of Soil*, ASTM STP 740, 717p.

Young, S. C. (1984), A Review of Leachate Induced Permeability Changes, Tennessee Valley Authority, Water Systems Development Branch, Research Project 2280-1. Prepared for Electrical Power Research Institute, Palo Alto, CA.

Youssef, M. S., Sabry, A., and El Ramli, A. H. (1961), Temperature changes and their effects on some physical properties of soils, *Proc. ICSMFE*, v. 1, pp. 419-421.

Yu, T. Y. (1964), *Electro-Chemical Properties of Soils*, Scientific Publ. Co., Beijing. (In Chinese).

Yu, T. Y. and Chiang, H. Y. (1984), *Electro-Chemical Properties of Soils*, Scientific Publ. Co., Beijing. (In Chinese).

Zeevaert, L. (1972, 1983), *Foundation Engineering for Difficult Subsoil Conditions*, 2nd ed. (1st. ed. 1972), Van Nostrand Reinhold Co., NY.

Zeevaert, L. (1996), The seismic geodynamics in the design of foundations in difficult subsoil conditions, Keynote Address, *Proc. 3rd. Int. Symp. Environ. Geotechnol.*, v. 1, Technomic Publ., Lancaster, PA, pp. 19-69.

Zhang, H. C. (1989), Research on seismic thixotropy in muck foundations, *Chinese J. Geotech. Eng.*, v. 11, no. 3, pp. 78-85. (In Chinese with English summary).

Zhu, Z. D. (1987), Desert research activities at the Chinese Academy of Sciences, *Proc.1st. Int. Symp. Environ. Geotechnol.*, v. 2, pp. 232-235.

Zhu, Z. D., Wu, C. and Liu, S. (1980), *Desertification in China*, Chinese Academy of Science, Desert Research Institute, 107p. (In Chinese).

Zimmie, T. F., Moo-Young, H., and LaPlante, K. (1995), The use of waste paper sludge for landfill cover material, *GREEN'93, U.K., Waste Disposal by Landfill*, R. W. Sarsby, ed., A.A. Balkema, Rotterdam, pp. 487-495.

Index

F